Information Theory for
Data Communications and Processing

Information Theory for Data Communications and Processing

Editors

Shlomo Shamai (Shitz)
Abdellatif Zaidi

MDPI • Basel • Beijing • Wuhan • Barcelona • Belgrade • Manchester • Tokyo • Cluj • Tianjin

Editors
Shlomo Shamai (Shitz)
Technion—Israel Institute of
Technology, EE Department
Israel

Abdellatif Zaidi
Institut Gaspard Monge,
Université Paris-Est
France

Editorial Office
MDPI
St. Alban-Anlage 66
4052 Basel, Switzerland

This is a reprint of articles from the Special Issue published online in the open access journal *Entropy* (ISSN 1099-4300) (available at: https://www.mdpi.com/journal/entropy/special_issues/ Information_Communications).

For citation purposes, cite each article independently as indicated on the article page online and as indicated below:

LastName, A.A.; LastName, B.B.; LastName, C.C. Article Title. *Journal Name* **Year**, *Volume Number*, Page Range.

ISBN 978-3-03943-817-4 (Hbk)
ISBN 978-3-03943-818-1 (PDF)

Contents

About the Editors

Shlomo Shamai (Shitz) is with the Viterbi Department of Electrical Engineering, Technion—Israel Institute of Technology, where he is a Technion Distinguished Professor, and holds the William Fondiller Chair of Telecommunications. He is an IEEE Life Fellow, a URSI Fellow, a member of the Israeli Academy of Sciences and Humanities and a foreign member of the US National Academy of Engineering. He was the recipient of the 2011 Claude E. Shannon Award, the 2014 Rothschild Prize in Mathematics/Computer Sciences and Engineering, and the 2017 IEEE Richard W. Hamming Medal. He was also a co-recipient of the 2018 Third Bell Labs Prize for Shaping the Future of Information and Communications Technology and other awards and recognitions.

Abdellatif Zaidi received his B.S. degree in Electrical Engineering from ENTSTA ParisTech, Paris, in 2002 and his M. Sc. and Ph.D. degrees in Electrical Engineering from TELECOM ParisTech, Paris, France, in 2003 and 2006, respectively. From December 2003 to March 2006, he was with the Communications and Electronics Dept., TELECOM ParisTech, Paris, France, and the Signals and Systems Lab, CNRS/Suplec, France, pursuing his PhD degree. From May 2006 to September 2010, he was at Ecole Polytechnique de Louvain, Universite Catholique de Louvain, Belgium, working as a senior researcher. Dr. Zaidi was a Research Visitor at the University of Notre Dame, Indiana, USA, during 2007 and 2008, the Technical University of Munich during Summer 2014, and the Ecole Polytechnique Federale de Lausanne, EFPL, Switzerland. He is currently an Associate Professor at Universite Paris-Est, France; and with the Mathematics and Algorithmic Sciences Lab, Paris Research Center, Huawei France. His research interests lie broadly in network information theory and its interactions with other fields, including communication and coding, statistics, security, and privacy, as well as learning, with applications for diverse problems of data transmission and compression in networks. Dr. Zaidi is an IEEE senior member. From 2013 to 2016, he served as an Associate Editor for the Eurasip Journal on Wireless Communication and Networking (EURASIP JWCN); and, since 2016, as Associate Editor for the IEEE Transactions on Wireless Communications. He was the recipient of the French Excellence in Research Award (in 2011); and co-recipient (with Shlomo Shamai (Shitz)) of the N# Best Paper Award (in 2014).

Editorial

Information Theory for Data Communications and Processing

Shlomo Shamai (Shitz) [1,*] and Abdellatif Zaidi [2,*]

[1]   The Viterbi Faculty of Electrical Engineering, Technion—Israel Institute of Technology, Haifa 3200003, Israel
[2]   Institut Gaspard Monge, Université Paris-Est, 05 Boulevard Descartes, Cité Descartes,
      77454 Champs sur Marne, France
*   Correspondence: sshlomo@ee.technion.ac.il (S.S.); abdellatif.zaidi@univ-mlv.fr (A.Z.)

Received: 20 October 2020; Accepted: 30 October 2020; Published: 3 November 2020

Keywords: information theory; data communications; data processing

---

This book, composed of the collection of papers that have appeared in the Special Issue of the *Entropy* journal dedicated to "Information Theory for Data Communications and Processing", reflects, in its eleven chapters, novel contributions based on the firm basic grounds of information theory. The book chapters [1–11] address timely theoretical and practical aspects that carry both interesting and relevant theoretical contributions, as well as direct implications for modern current and future communications systems.

Information theory has started with the monumental work of Shannon: Shannon, C.E. "A Mathematical Theory of Communications", *Bell System Technical Journal*, vol. 27, pp. 379–423, 623–656, 1948, and it provided from its very start the mathematical/theoretical framework which facilitated addressing information related problems, in all respects: starting with the basic notion of what is information, going through basic features of how to convey information in the best possible way and how to process it given actual and practical constraints. Shannon himself not only fully realized the power of the basic theory he has developed but further in his profound contributions addressed practical constraints of communications systems, such as bandwidth, possible signaling limits (as peak limited signals), motivating from the very start to address practical constraints via theoretical tools, see, for example: Jelonek, Z. A comparison of transmission systems, In Proc. Symp. Appl. Commun. Theory, E.E. Dep., Imperial College, Buttenvorths Scientific Press, London, September 1952, pp. 45–81. Shannon has contributed fundamentally also to most relevant aspects as source coding under a fidelity (distortion) measure, finite code lengths (error exponents) as well as network aspects of information theory (the multiple-access channel), see: Sloane, N.J.A. and Wyner, A.D., Eds., Collected Papers of Claude Elwood Shannon. IEEE Press: New York, 1993.

While at its beginning and through the first decades, information theory, as is reflected in the basic 1948 work of Shannon, was a mathematical tool that pointed out the best that can be achieved (as channel capacity for point-to-point communications), which with past technology could not even be imagined to be approached. Now, the power of information theory is way greater as it is able to theoretically address network problems and not only point out the limits of communications/signal processing, but with current technology, those limits can, in general, be decently approached. This is classically demonstrated by the capacity of the point-to-point Gaussian channel, which is actually achieved within fractions of dB in signal-to-noise (snr) ratio by advanced coding techniques (Low-Density-Parity-Check, Turbo and Polar codes). In our times, current advanced technology turns information theory into a practical important tool that is capable also to provide basic guidelines how to come close to ultimate optimal performance.

Modern, current and future communications/processing aspects motivate in fact basic information theoretic research for a wide variety of systems for which we yet do not have the ultimate theoretical solutions (for example a variety of problems in network information theory as the broadcast/interference and relay channels, which mostly are yet unsolved in terms of determining capacity regions and the like). Technologies as 5/6G cellular communications, Internet of Things (IoT), Mobile Edge Networks and others place in center not only features of reliable rates of information measured by the relevant capacity, and capacity regions, but also notions such as latency vs. reliability, availability of system state information, priority of information, secrecy demands, energy consumption per mobile equipment, sharing of communications resource (time/frequency/space) and the like.

This book focuses on timely and relevant features, and the contributions in the eleven book chapters [1–11], summarized below, address the information theoretical frameworks that have important practical implications.

The basic contributions of this book could be divided into three basic parts:

(1) The first part Chapters [1–5] considers central notions such as the Information Bottleneck, overviewed in the first chapter, pointing out basic connections to a variety of classical information theoretic problems, such as remote source coding. This subject covering timely novel information theoretic results demonstrates the role information theory plays in current top technology. These chapters, on one hand, provide application to 'deep learning', and, on the other, they present the basic theoretical framework of future communications systems such as Cloud and Fog Radio Access Networks (CRAN, FRAN). The contributions in this part directly address aspects such as ultra-reliable low-latency communications, impacts of congestion, and non-orthogonal access strategies.

(2) The second part of the contributions in this book Chapters [6–8] addresses classical communications systems, point-to-point Multiple-Input-Multiple-Output (MIMO) channels subjected to practical constraints, as well as network communications models. Specifically, relay and multiple access channels are discussed.

(3) The third part of the contributions of this book Chapters [9–11] focuses mainly on caching, which, for example, is the center component in FRAN. Information theory indeed provides the classical tool to address network features of caching, as demonstrated in the contributions summarized below (and references therein).

Chapter 1: "On the Information Bottleneck Problems: Models, Connections, Applications and Information Theoretic Views" provides a tutorial that addresses from an information theoretic viewpoint variants of the information bottleneck problem. It provides an overview emphasizing variational inference, representation learning and presents a broad spectrum of inherent connections to classical information theoretic notions such as: remote source-coding, information combining, common reconstruction, the Wyner–Ahlswede–Korner problem and others. The distributed information bottleneck overviewed in this tutorial sets the theoretical grounds for the uplink CRAN, with oblivious processing.

Chapter 2: "Variational Information Bottleneck for Unsupervised Clustering: Deep Gaussian Mixture Embedding" develops an unsupervised generative clustering framework that combines the variational information bottleneck and the Gaussian mixture model. Among other results, this approach that models the latent space as a mixture of Gaussians generates inference-type algorithms for exact computation, and generalizes the so-called evidence lower bound, which is useful in a variety of unsupervised learning problems.

Chapter 3: "Asymptotic Rate-Distortion Analysis of Symmetric Remote Gaussian Source Coding: Centralized Encoding vs. Distributed Encoding" addresses remote multivariate source coding, which is a CEO problem and, as indicated in Chapter 1, connects directly to the distributed bottleneck problem. The distortion measure considered here is minimum-mean-square-error, which can be connected to the logarithmic distortion via classical information–estimation relations. Both cases—the distributed and joint remote source coding (all terminals cooperate)—are studied.

Chapter 4: "Non-Orthogonal eMBB-URLLC Radio Access for Cloud Radio Access Networks with Analog Fronthauling" provides an information-theoretic perspective of the performance of Ultra-Reliable

Low-Latency Communications (URLLC) and enhanced Mobile BroadBand (eMBB) traffic under both Orthogonal and Non-Orthogonal multiple access procedures. The work here considers CRAN based on the relaying of radio signals over analog fronthaul links.

Chapter 5: "Robust Baseband Compression against Congestion in Packet-Based Fronthaul Networks Using Multiple Description Coding" also addresses CRAN and considers the practical scenario when the fronthaul transport network is packet based and it may have a multi-hop architecture. The timely information theoretic concepts of multiple description coding are employed, and demonstrated to provide advantageous performance over conventional packet-based multi-route reception or coding.

Chapter 6: "Amplitude Constrained MIMO Channels: Properties of Optimal Input Distributions and Bounds on the Capacity" studies the classical information theoretic setting where input signals are subjected to practical constraints, with focus on amplitude constraints. Followed by a survey of available results for Gaussian MIMO channels, which are of direct practical importance, it is shown that the support of a capacity-achieving input distribution is a small set in both a topological and a measure theoretical sense. Bounds on the respective capacities are developed and demonstrated to be tight in the high amplitude regime (high snr).

Chapter 7: "Quasi-Concavity for Gaussian Multicast Relay Channels" addresses the classical model of a relay channel, which is one of the classical information theoretic problems that are not yet fully solved. This work identifies useful features of quasi-concavity of relevant bounds (as the cut-set bound) that are useful in addressing communications schemes based on relaying.

Chapter 8: "Gaussian Multiple Access Channels with One-Bit Quantizer at the Receiver" investigates the practical setting when the received input is sampled and here it employs a zero-threshold one-bit analogue-to-digital converter. It is shown that the optimal capacity achieving signal distribution is discrete, and bounds on the respective capacity are reported.

Chapter 9: "Efficient Algorithms for Coded Multicasting in Heterogeneous Caching Networks" addresses crucial performance–complexity tradeoffs in a heterogeneous caching network setting, where edge caches with possibly different storage capacities collect multiple content requests that may follow distinct demand distributions. The basic known performance-efficient coded multicasting schemes suffer from inherent complexity issues, which makes them impractical. This chapter demonstrates that the proposed approach provides a compelling step towards the practical achievability of the promising multiplicative caching gain in future-generation access networks.

Chapter 10: "Cross-Entropy Method for Content Placement and User Association in Cache-Enabled Coordinated Ultra-Dense Networks" focuses on ultra-dense networks, which play a central role for future wireless technologies. In Coordinated Multi-Point-based Ultra-Dense Networks, a great challenge is to tradeoff between the gain of network throughput and the degraded backhaul latency, and caching popular files has been identified as a promising method to reduce the backhaul traffic load. This chapter investigated Cross-Entropy methodology for content placement strategies and user association algorithms for the proactive caching ultra-dense networks, and demonstrates advantageous performance.

Chapter 11: "Symmetry, Outer Bounds, and Code Constructions: A Computer-Aided Investigation on the Fundamental Limits of Caching" also focuses on caching, which, as mentioned, is a fundamental procedure for future efficient networks. Most known analyses and bounds developed are based on information theoretic arguments and techniques. This work illustrates how computer-aided methods can be applied to investigate the fundamental limits of the caching systems, which are significantly different from the conventional analytic approach usually seen in the information theory literature. The methodology discussed and suggested here allows, among other things, to compute performance bounds for multi-user/terminal schemes, which were believed to require unrealistic computation scales.

In closing, one can view all the above three categories of the eleven chapters, in a unified way, as all are relevant to future wireless networks. The massive growth of smart devices and the advent of many new applications dictates not only having better systems, such as coding and modulation on the point-to-point channel, classically characterized by channel capacity, but a change of the network/communications paradigms (as demonstrated for example, by the notions of CRAN and FRAN) and performance measures. New architectures and concepts are a must in current and future communications systems, and information theory provides the basic tools to address these, developing concepts and results, which actually are not only of essential theoretical value, but are also of direct practical importance. We trust that this book provides a sound glimpse to these aspects.

Funding: This research received no external funding.

Acknowledgments: We express our thanks to the authors of the above contributions, and to the journal *Entropy* and MDPI for their support during this work.

Conflicts of Interest: The authors declare no conflict of interest.

References

1. Zaidi, A.; Estella-Aguerri, I.; Shamai, S. On the Information Bottleneck Problems: Models, Connections, Applications and Information Theoretic Views. *Entropy* **2020**, *22*, 151. [CrossRef]
2. Uğur, Y.; Arvanitakis, G.; Zaidi, A. Variational Information Bottleneck for Unsupervised Clustering: Deep Gaussian Mixture Embedding. *Entropy* **2020**, *22*, 213. [CrossRef]
3. Wang, Y.; Xie, L.; Zhou, S.; Wang, M.; Chen, J. Asymptotic Rate-Distortion Analysis of Symmetric Remote Gaussian Source Coding: Centralized Encoding vs. Distributed Encoding. *Entropy* **2019**, *21*, 213. [CrossRef]
4. Matera, A.; Kassab, R.; Simeone, O.; Spagnolini, U. Non-Orthogonal eMBB-URLLC Radio Access for Cloud Radio Access Networks with Analog Fronthauling. *Entropy* **2018**, *20*, 661. [CrossRef]
5. Park, S.-H.; Simeone, O.; Shamai, S. Robust Baseband Compression Against Congestion in Packet-Based Fronthaul Networks Using Multiple Description Coding. *Entropy* **2019**, *21*, 433. [CrossRef]
6. Dytso, A.; Goldenbaum, M.; Poor, H.V.; Shamai, S. Amplitude Constrained MIMO Channels: Properties of Optimal Input Distributions and Bounds on the Capacity. *Entropy* **2019**, *21*, 200. [CrossRef]
7. Thakur, M.; Kramer, G. Quasi-Concavity for Gaussian Multicast Relay Channels. *Entropy* **2019**, *21*, 109. [CrossRef]
8. Rassouli, B.; Varasteh, M.; Gündüz, D. Gaussian Multiple Access Channels with One-Bit Quantizer at the Receiver. *Entropy* **2018**, *20*, 686. [CrossRef]
9. Vettigli, G.; Ji, M.; Shanmugam, K.; Llorca, J.; Tulino, A.M.; Caire, G. Efficient Algorithms for Coded Multicasting in Heterogeneous Caching Networks. *Entropy* **2019**, *21*, 324. [CrossRef]
10. Yu, J.; Wang, Y.; Gu, S.; Zhang, Q.; Chen, S.; Zhang, Y. Cross-Entropy Method for Content Placement and User Association in Cache-Enabled Coordinated Ultra-Dense Networks. *Entropy* **2019**, *21*, 576. [CrossRef]
11. Tian, C. Symmetry, Outer Bounds, and Code Constructions: A Computer-Aided Investigation on the Fundamental Limits of Caching. *Entropy* **2018**, *20*, 603. [CrossRef]

 *entropy*

Tutorial

On the Information Bottleneck Problems: Models, Connections, Applications and Information Theoretic Views

Abdellatif Zaidi [1,2,*], Iñaki Estella-Aguerri [2] and Shlomo Shamai (Shitz) [3]

[1] Institut d'Électronique et d'Informatique Gaspard-Monge, Université Paris-Est, 77454 Champs-sur-Marne, France
[2] Mathematics and Algorithmic Sciences Lab, Paris Research Center, Huawei Technologies France, 92100 Boulogne-Billancourt, France; inaki.estella@gmail.com
[3] Technion Institute of Technology, Technion City, Haifa 32000, Israel; sshlomo@ee.technion.ac.il
* Correspondence: abdellatif.zaidi@u-pem.fr

Received: 16 October 2019; Accepted: 21 January 2020; Published: 27 January 2020

Abstract: This tutorial paper focuses on the variants of the bottleneck problem taking an information theoretic perspective and discusses practical methods to solve it, as well as its connection to coding and learning aspects. The intimate connections of this setting to remote source-coding under logarithmic loss distortion measure, information combining, common reconstruction, the Wyner–Ahlswede–Korner problem, the efficiency of investment information, as well as, generalization, variational inference, representation learning, autoencoders, and others are highlighted. We discuss its extension to the distributed information bottleneck problem with emphasis on the Gaussian model and highlight the basic connections to the uplink Cloud Radio Access Networks (CRAN) with oblivious processing. For this model, the optimal trade-offs between relevance (i.e., information) and complexity (i.e., rates) in the discrete and vector Gaussian frameworks is determined. In the concluding outlook, some interesting problems are mentioned such as the characterization of the optimal inputs ("features") distributions under power limitations maximizing the "relevance" for the Gaussian information bottleneck, under "complexity" constraints.

Keywords: information bottleneck; rate distortion theory; logarithmic loss; representation learning

1. Introduction

A growing body of works focuses on developing learning rules and algorithms using information theoretic approaches (e.g., see [1–6] and references therein). Most relevant to this paper is the Information Bottleneck (IB) method of Tishby et al. [1], which seeks the right balance between data fit and generalization by using the mutual information as both a cost function and a regularizer. Specifically, IB formulates the problem of extracting the relevant information that some signal $X \in \mathcal{X}$ provides about another one $Y \in \mathcal{Y}$ that is of interest as that of finding a representation U that is maximally informative about Y (i.e., large mutual information $I(U;Y)$) while being minimally informative about X (i.e., small mutual information $I(U;X)$). In the IB framework, $I(U;Y)$ is referred to as the *relevance* of U and $I(U;X)$ is referred to as the *complexity* of U, where complexity here is measured by the minimum description length (or rate) at which the observation is compressed. Accordingly, the performance of learning with the IB method and the optimal mapping of the data are found by solving the Lagrangian formulation

$$\mathcal{L}_{\beta}^{\text{IB},*} := \max_{P_{U|X}} I(U;Y) - \beta I(U;X), \tag{1}$$

where $P_{U|X}$ is a stochastic map that assigns the observation X to a representation U from which Y is inferred and β is the Lagrange multiplier. Several methods, which we detail below, have been proposed to obtain solutions $P_{U|X}$ to the IB problem in Equation (4) in several scenarios, e.g., when the distribution of the sources (X, Y) is perfectly known or only samples from it are available.

The IB approach, as a method to both characterize performance limits as well as to design mapping, has found remarkable applications in supervised and unsupervised learning problems such as classification, clustering, and prediction. Perhaps key to the analysis and theoretical development of the IB method is its elegant connection with information-theoretic rate-distortion problems, as it is now well known that the IB problem is essentially a remote source coding problem [7–9] in which the distortion is measured under logarithmic loss. Recent works show that this connection turns out to be useful for a better understanding the fundamental limits of learning problems, including the performance of deep neural networks (DNN) [10], the emergence of invariance and disentanglement in DNN [11], the minimization of PAC-Bayesian bounds on the test error [11,12], prediction [13,14], or as a generalization of the evidence lower bound (ELBO) used to train variational auto-encoders [15,16], geometric clustering [17], or extracting the Gaussian "part" of a signal [18], among others. Other connections that are more intriguing exist also with seemingly unrelated problems such as privacy and hypothesis testing [19–21] or multiterminal networks with oblivious relays [22,23] and non-binary LDPC code design [24]. More connections with other coding problems such as the problems of information combining and common reconstruction, the Wyner–Ahlswede–Korner problem, and the efficiency of investment information are unveiled and discussed in this tutorial paper, together with extensions to the distributed setting.

The abstract viewpoint of IB also seems instrumental to a better understanding of the so-called *representation learning* [25], which is an active research area in machine learning that focuses on identifying and disentangling the underlying explanatory factors that are hidden in the observed data in an attempt to render learning algorithms less dependent on feature engineering. More specifically, one important question, which is often controversial in statistical learning theory, is the choice of a "good" loss function that measures discrepancies between the true values and their estimated fits. There is however numerical evidence that models that are trained to maximize mutual information, or equivalently minimize the error's entropy, often outperform ones that are trained using other criteria such as mean-square error (MSE) and higher-order statistics [26,27]. On this aspect, we also mention Fisher's dissertation [28], which contains investigation of the application of information theoretic metrics to blind source separation and subspace projection using Renyi's entropy as well as what appears to be the first usage of the now popular Parzen windowing estimator of information densities in the context of learning. Although a complete and rigorous justification of the usage of mutual information as cost function in learning is still awaited, recently, a partial explanation appeared in [29], where the authors showed that under some natural data processing property Shannon's mutual information uniquely quantifies the reduction of prediction risk due to side information. Along the same line of work, Painsky and Wornell [30] showed that, for binary classification problems, by minimizing the logarithmic-loss (log-loss), one actually minimizes an upper bound to any choice of loss function that is smooth, proper (i.e., unbiased and Fisher consistent), and convex. Perhaps, this justifies partially why mutual information (or, equivalently, the corresponding loss function, which is the log-loss fidelity measure) is widely used in learning theory and has already been adopted in many algorithms in practice such as the *infomax* criterion [31], the tree-based algorithm of Quinlan [32], or the well known Chow–Liu algorithm [33] for learning tree graphical models, with various applications in genetics [34], image processing [35], computer vision [36], etc. The logarithmic loss measure also plays a central role in the theory of prediction [37] (Ch. 09) where it is often referred to as the *self-information* loss function, as well as in Bayesian modeling [38] where priors are usually designed to maximize the mutual information between the parameter to be estimated and the observations. The goal of learning, however, is not merely to learn model parameters accurately for previously seen data. Rather, in essence, it is the ability to successfully apply rules that are extracted from previously seen data to characterize new unseen data. This is often captured through the notion of "generalization error". The generalization capability of a learning algorithm hinges on how sensitive the output of the algorithm is to modifications of the input dataset, i.e., its *stability* [39,40]. In the context of deep learning, it can be seen as a measure of how much the algorithm overfits the model parameters to the seen data. In fact, efficient algorithms should strike a good balance between their ability to fit training dataset and that to generalize well to unseen data. In statistical learning theory [37], such a dilemma is reflected through that the minimization of the "population risk" (or "test error" in the deep learning literature) amounts to the minimization of the sum of the two terms that are generally difficult to minimize simultaneously, the "empirical risk" on the training data and the generalization error. To prevent over-fitting, regularization methods can be employed, which include parameter penalization, noise injection, and averaging over multiple models trained with distinct sample sets. Although it is not yet very well understood how to optimally control model complexity, recent works [41,42] show that the generalization error can be upper-bounded using the mutual information between the input dataset and the output of the algorithm. This result actually formalizes the intuition that the less information a learning algorithm extracts from the input dataset the less it is likely to overfit, and justifies, partly, the use of mutual information also as a regularizer term. The interested reader may refer to [43] where it is shown that regularizing with mutual information alone does not always capture all desirable properties of a latent representation. We also point out that there exists an extensive literature on building optimal estimators of information quantities (e.g., entropy, mutual information), as well as their Matlab/Python implementations, including in the high-dimensional regime (see, e.g., [44–49] and references therein).

This paper provides a review of the information bottleneck method, its classical solutions, and recent advances. In addition, in the paper, we unveil some useful connections with coding problems such as remote source-coding, information combining, common reconstruction, the Wyner–Ahlswede–Korner problem, the efficiency of investment information, CEO source coding under logarithmic-loss distortion measure, and learning problems such as inference, generalization, and representation learning. Leveraging these connections, we discuss its extension to the distributed information bottleneck problem with emphasis on its solutions and the Gaussian model and highlight the basic connections to the uplink Cloud Radio Access Networks (CRAN) with oblivious processing. For this model, the optimal trade-offs between relevance and complexity in the discrete and vector Gaussian frameworks is determined. In the concluding outlook, some interesting problems are mentioned such as the characterization of the optimal inputs distributions under power limitations maximizing the "relevance" for the Gaussian information bottleneck under "complexity" constraints.

Notation

Throughout, uppercase letters denote random variables, e.g., X; lowercase letters denote realizations of random variables, e.g., x; and calligraphic letters denote sets, e.g., $\mathcal{X}$. The cardinality of a set is denoted by $|\mathcal{X}|$. For a random variable X with probability mass function (pmf) P_X, we use $P_X(x) = p(x)$, $x \in \mathcal{X}$ for short. Boldface uppercase letters denote vectors or matrices, e.g., $\mathbf{X}$, where context should make the distinction clear. For random variables $(X_1, X_2, \cdots)$ and a set of integers $\mathcal{K} \subseteq \mathbb{N}$, $X_\mathcal{K}$ denotes the set of random variables with indices in the set $\mathcal{K}$, i.e., $X_\mathcal{K} = \{X_k : k \in \mathcal{K}\}$. If $\mathcal{K} = \varnothing$, $X_\mathcal{K} = \varnothing$. For $k \in \mathcal{K}$, we let $X_{\mathcal{K}/k} = (X_1, \cdots, X_{k-1}, X_{k+1}, \cdots, X_K)$, and assume that $X_0 = X_{K+1} = \varnothing$. In addition, for zero-mean random vectors $\mathbf{X}$ and $\mathbf{Y}$, the quantities $\mathbf{\Sigma_x}$, $\mathbf{\Sigma_{x,y}}$ and $\mathbf{\Sigma_{x|y}}$ denote, respectively, the covariance matrix of the vector $\mathbf{X}$, the covariance matrix of vector $(\mathbf{X}, \mathbf{Y})$, and the conditional covariance matrix of $\mathbf{X}$, conditionally on $\mathbf{Y}$, i.e., $\mathbf{\Sigma_x} = \mathbb{E}[\mathbf{XX}^H]$ $\mathbf{\Sigma_{x,y}} := \mathbb{E}[\mathbf{XY}^H]$, and $\mathbf{\Sigma_{x|y}} = \mathbf{\Sigma_x} - \mathbf{\Sigma_{x,y}}\mathbf{\Sigma_y^{-1}}\mathbf{\Sigma_{y,x}}$. Finally, for two probability measures P_X and Q_X on the random variable $X \in \mathcal{X}$, the relative entropy or Kullback–Leibler divergence is denoted as $D_{\mathrm{KL}}(P_X \| Q_X)$. That is, if P_X is absolutely continuous with respect to Q_X, $P_X \ll Q_X$ (i.e., for every $x \in \mathcal{X}$, if $P_X(x) > 0$, then $Q_X(x) > 0$), $D_{\mathrm{KL}}(P_X \| Q_X) = \mathbb{E}_{P_X}[\log(P_X(X)/Q_X(X))]$, otherwise $D_{\mathrm{KL}}(P_X \| Q_X) = \infty$.

2. The Information Bottleneck Problem

The Information Bottleneck (IB) method was introduced by Tishby et al. [1] as a method for extracting the information that some variable $X \in \mathcal{X}$ provides about another one $Y \in \mathcal{Y}$ that is of interest, as shown in Figure 1.

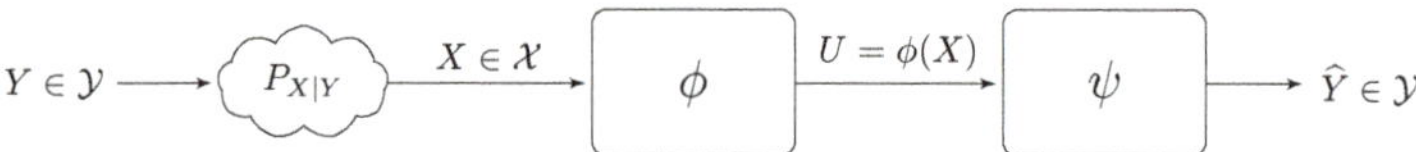

Figure 1. Information bottleneck problem.

Specifically, the IB method consists of finding the stochastic mapping $P_{U|X} : \mathcal{X} \to \mathcal{U}$ that from an observation X outputs a representation $U \in \mathcal{U}$ that is maximally informative about Y, i.e., large mutual information $I(U;Y)$, while being minimally informative about X, i.e., small mutual information $I(U;X)$ (As such, the usage of Shannon's mutual information seems to be motivated by the intuition that such a measure provides a natural quantitative approach to the questions of meaning, relevance, and common-information, rather than the solution of a well-posed information-theoretic problem—a connection with source coding under logarithmic loss measure appeared later on in [50].) The auxiliary random variable U satisfies that $U \multimap X \multimap Y$ is a Markov Chain in this order; that is, that the joint distribution of (X, U, Y) satisfies

$$p(x,u,y) = p(x)p(y|x)p(u|x), \qquad (2)$$

and the mapping $P_{U|X}$ is chosen such that U strikes a suitable balance between the degree of *relevance* of the representation as measured by the mutual information $I(U;Y)$ and its degree of *complexity* as measured by the mutual information $I(U;X)$. In particular, such U, or effectively the mapping $P_{U|X}$, can be determined to maximize the IB-Lagrangian defined as

$$\mathcal{L}_\beta^{\text{IB}}(P_{U|X}) := I(U;Y) - \beta I(U;X) \qquad (3)$$

over all mappings $P_{U|X}$ that satisfy $U \multimap X \multimap Y$ and the trade-off parameter β is a positive Lagrange multiplier associated with the constraint on $I(U;Y)$.

Accordingly, for a given β and source distribution $P_{X,Y}$, the optimal mapping of the data, denoted by $P_{U|X}^{*,\beta}$, is found by solving the IB problem, defined as

$$\mathcal{L}_\beta^{\text{IB},*} := \max_{P_{U|X}} I(U;Y) - \beta I(U;X)., \qquad (4)$$

over all mappings $P_{U|Y}$ that satisfy $U \multimap X \multimap Y$. It follows from the classical application of Carathéodory's theorem [51] that without loss of optimality, U can be restricted to satisfy $|\mathcal{U}| \leq |\mathcal{X}| + 1$.

In Section 3 we discuss several methods to obtain solutions $P_{U|X}^{*,\beta}$ to the IB problem in Equation (4) in several scenarios, e.g., when the distribution of (X, Y) is perfectly known or only samples from it are available.

2.1. The Ib Relevance–Complexity Region

The minimization of the IB-Lagrangian $\mathcal{L}_\beta$ in Equation (4) for a given $\beta \geq 0$ and $P_{X,Y}$ results in an optimal mapping $P_{U|X}^{*,\beta}$ and a relevance–complexity pair (Δ_β, R_β) where $\Delta_\beta = I(U_\beta, X)$ and $R_\beta = I(U_\beta, Y)$ are, respectively, the relevance and the complexity resulting from generating U_β with the solution $P_{U|X}^{*,\beta}$. By optimizing over all $\beta \geq 0$, the resulting relevance–complexity pairs (Δ_β, R_β) characterize the boundary of the region of simultaneously achievable relevance–complexity pairs for a distribution $P_{X,Y}$ (see Figure 2). In particular, for a fixed $P_{X,Y}$, we define this region as the union of relevance–complexity pairs (Δ, R) that satisfy

$$\Delta \leq I(U, Y), \quad R \geq I(X, U) \tag{5}$$

where the union is over all $P_{U|X}$ such that U satisfies $U \multimap X \multimap Y$ form a Markov Chain in this order. Any pair (Δ, R) outside of this region is not simultaneously achievable by any mapping $P_{U|X}$.

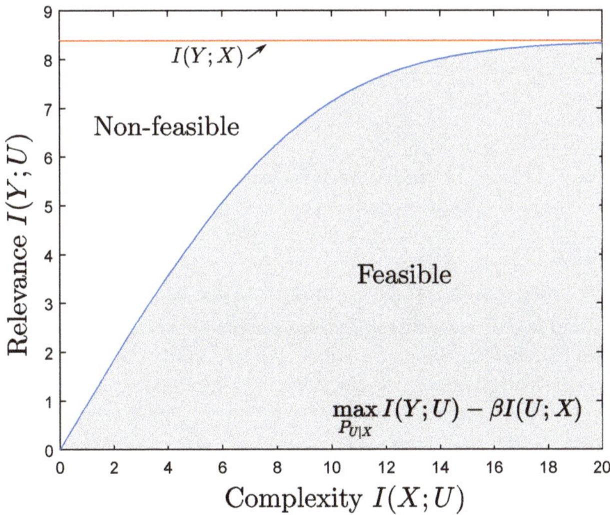

Figure 2. Information bottleneck relevance–complexity region. For a given β, the solution $P_{U|X}^{*,\beta}$ to the minimization of the IB-Lagrangian in Equation (3) results in a pair (Δ_β, R_β) on the boundary of the IB relevance–complexity region (colored in grey).

3. Solutions to the Information Bottleneck Problem

As shown in the previous region, the IB problem provides a methodology to design mappings $P_{U|X}$ performing at different relevance–complexity points within the region of feasible (Δ, R) pairs, characterized by the IB relevance–complexity region, by minimizing the IB-Lagrangian in Equation (3) for different values of β. However, in general, this optimization is challenging as it requires computation of mutual information terms.

In this section, we describe how, for a fixed parameter β, the optimal solution $P_{U|X}^{\beta,*}$, or an efficient approximation of it, can be obtained under: (i) particular distributions, e.g., Gaussian and binary symmetric sources; (ii) known general discrete memoryless distributions; and (iii) unknown memory distributions and only samples are available.

3.1. Solution for Particular Distributions: Gaussian and Binary Symmetric Sources

In certain cases, when the joint distribution $P_{X,Y}$ is know, e.g., it is binary symmetric or Gaussian, information theoretic inequalities can be used to minimize the IB-Lagrangian in (4) in closed form.

3.1.1. Binary IB

Let X and Y be a doubly symmetric binary sources (DSBS), i.e., $(X, Y) \sim \text{DSBS}(p)$ for some $0 \leq p \leq 1/2$. (A DSBS is a pair (X, Y) of binary random variables $X \sim \text{Bern}(1/2)$ and $Y \sim \text{Bern}(1/2)$ and $X \oplus Y \sim \text{Bern}(p)$, where $\oplus$ is the sum modulo 2. That is, Y is the output of a binary symmetric channel with crossover probability p corresponding to the input X, and X is the output of the same channel with input Y.) Then, it can be shown that the optimal U in (4) is such that $(X, U) \sim \text{DSBS}(q)$ for some $0 \leq q \leq 1$. Such a U can be obtained with the mapping $P_{U|X}$ such that

$$U = X \oplus Q, \quad \text{with } Q \sim \text{DSBS}(q). \tag{6}$$

In this case, straightforward algebra leads to that the complexity level is given by

$$I(U; X) = 1 - h_2(q), \tag{7}$$

where, for $0 \leq x \leq 1$, $h_2(x)$ is the entropy of a Bernoulli-(x) source, i.e., $h_2(x) = -x \log_2(x) - (1 - x) \log_2(1 - x)$, and the relevance level is given by

$$I(U; Y) = 1 - h_2(p \star q) \tag{8}$$

where $p \star q = p(1 - q) + q(1 - p)$. The result extends easily to discrete symmetric mappings $Y \longrightarrow X$ with binary X (one bit output quantization) and discrete non-binary Y.

3.1.2. Vector Gaussian IB

Let $(\mathbf{X}, \mathbf{Y}) \in \mathbb{C}^{N_x} \times \mathbb{C}^{N_y}$ be a pair of jointly Gaussian, zero-mean, complex-valued random vectors, of dimension $N_x > 0$ and $N_y > 0$, respectively. In this case, the optimal solution of the IB-Lagrangian in Equation (3) (i.e., test channel $P_{U|X}$) is a noisy linear projection to a subspace whose dimensionality is determined by the tradeoff parameter β. The subspaces are spanned by basis vectors in a manner similar to the well known canonical correlation analysis [52]. For small β, only the vector associated to the dimension with more energy, i.e., corresponding to the largest eigenvalue of a particular hermitian matrix, will be considered in U. As β increases, additional dimensions are added to U through a series of critical points that are similar to structural phase transitions. This process continues until U becomes rich enough to capture all the relevant information about Y that is contained in X. In particular, the boundary of the optimal relevance–complexity region was shown in [53] to be achievable using a test channel $P_{U|X}$, which is such that $(\mathbf{U}, \mathbf{X})$ is Gaussian. Without loss of generality, let

$$\mathbf{U} = \mathbf{AX} + \boldsymbol{\xi} \tag{9}$$

where $\mathbf{A} \in \mathcal{M}_{N_u, N_x}(\mathbb{C})$ is an $N_u \times N_x$ complex valued matrix and $\boldsymbol{\xi} \in \mathbb{C}^{N_u}$ is a Gaussian noise that is independent of $(\mathbf{X}, \mathbf{Y})$ with zero-mean and covariance matrix $\mathbf{I}_{N_u}$. For a given non-negative trade-off parameter β, the matrix $\mathbf{A}$ has a number of rows that depends on β and is given by [54] (Theorem 3.1)

$$\mathbf{A} = \begin{cases} \left[\mathbf{0}^T; \ldots; \mathbf{0}^T \right], & 0 \leq \beta < \beta_1^c \\ \left[\alpha_1 \mathbf{v}_1^T; \mathbf{0}^T; \ldots; \mathbf{0}^T \right], & \beta_1^c \leq \beta < \beta_2^c \\ \left[\alpha_1 \mathbf{v}_1^T; \alpha_2 \mathbf{v}_2^T; \mathbf{0}^T; \ldots; \mathbf{0}^T \right], & \beta_2^c \leq \beta < \beta_3^c \\ \vdots & \end{cases} \tag{10}$$

where $\{\mathbf{v}_1^T, \mathbf{v}_2^T, \cdots, \mathbf{v}_{N_x}^T\}$ are the left eigenvectors of $\boldsymbol{\Sigma}_{X|Y} \boldsymbol{\Sigma}_X^{-1}$ sorted by their corresponding ascending eigenvalues $\lambda_1, \lambda_2, \cdots, \lambda_{N_x}$. Furthermore, for $i = 1, \cdots, N_x$, $\beta_i^c = \frac{1}{1-\lambda_i}$ are critical β-values, $\alpha_i = \sqrt{\frac{\beta(1-\lambda_i)-1}{\lambda_i r_i}}$ with $r_i = \mathbf{v}_i^T \boldsymbol{\Sigma}_X \mathbf{v}_i$, $\mathbf{0}^T$ denotes the N_x-dimensional zero vector and semicolons separate the rows of the matrix. It is interesting to observe that the optimal projection consists of eigenvectors of $\boldsymbol{\Sigma}_{X|Y} \boldsymbol{\Sigma}_X^{-1}$, combined in a judicious manner: for values of β that are smaller than β_1^c, reducing complexity is of prime importance, yielding extreme compression $\mathbf{U} = \boldsymbol{\xi}$, i.e., independent noise and no information preservation at all about $\mathbf{Y}$. As β increases, it undergoes a series of critical points $\{\beta_i^c\}$, at each of which a new eigenvector is added to the matrix $\mathbf{A}$, yielding a more complex but richer representation—the rank of $\mathbf{A}$ increases accordingly.

For the specific case of scalar Gaussian sources, that is $N_x = N_y = 1$, e.g., $X = \sqrt{\mathrm{snr}}\,Y + N$ where N is standard Gaussian with zero-mean and unit variance, the above result simplifies considerably. In this case, let without loss of generality the mapping $P_{U|X}$ be given by

$$X = \sqrt{a}X + Q \tag{11}$$

where Q is standard Gaussian with zero-mean and variance σ_q^2. In this case, for $I(U;X) = R$, we get

$$I(U;Y) = \frac{1}{2}\log(1+\mathrm{snr}) - \frac{1}{2}\log\left(1+\mathrm{snr}\exp(-2R)\right). \tag{12}$$

3.2. Approximations for Generic Distributions

Next, we present an approach to obtain solutions to the the information bottleneck problem for generic distributions, both when this solution is known and when it is unknown. The method consists in defining a variational (lower) bound on the IB-Lagrangian, which can be optimized more easily than optimizing the IB-Lagrangian directly.

3.2.1. A Variational Bound

Recall the IB goal of finding a representation U of X that is maximally informative about Y while being concise enough (i.e., bounded $I(U;X)$). This corresponds to optimizing the IB-Lagrangian

$$\mathcal{L}_\beta^{\mathrm{IB}}(P_{U|X}) := I(U;Y) - \beta I(U;X) \tag{13}$$

where the maximization is over all stochastic mappings $P_{U|X}$ such that $U \multimap X \multimap Y$ and $|\mathcal{U}| \le |\mathcal{X}| + 1$. In this section, we show that minimizing Equation (13) is equivalent to optimizing the variational cost

$$\mathcal{L}_\beta^{\mathrm{VIB}}(P_{U|X}, Q_{Y|U}, S_U) := \mathrm{E}_{P_{U|X}}\left[\log Q_{Y|U}(Y|U)\right] - \beta D_{\mathrm{KL}}(P_{U|X}\|S_U), \tag{14}$$

where $Q_{Y|U}(y|u)$ is an given stochastic map $Q_{Y|U} : \mathcal{U} \to [0,1]$ (also referred to as the variational approximation of $P_{Y|U}$ or decoder) and $S_U(u) : \mathcal{U} \to [0,1]$ is a given stochastic map (also referred to as the variational approximation of P_U), and $D_{\mathrm{KL}}(P_{U|X}\|S_U)$ is the relative entropy between $P_{U|X}$ and S_U.

Then, we have the following bound for a any valid $P_{U|X}$, i.e., satisfying the Markov Chain in Equation (2),

$$\mathcal{L}_\beta^{\mathrm{IB}}(P_{U|X}) \ge \mathcal{L}_\beta^{\mathrm{VIB}}(P_{U|X}, Q_{Y|U}, S_U), \tag{15}$$

where the equality holds when $Q_{Y|U} = P_{Y|U}$ and $S_U = P_U$, i.e., the variational approximations correspond to the true value.

In the following, we derive the variational bound. Fix $P_{U|X}$ (an encoder) and the variational decoder approximation $Q_{Y|U}$. The relevance $I(U;Y)$ can be lower-bounded as

$$I(U;Y) = \oint_{u\in\mathcal{U},\,y\in\mathcal{Y}} P_{U,Y}(u,y) \log \frac{P_{Y|U}(y|u)}{P_Y(y)}\, dy\, du \tag{16}$$

$$\overset{(a)}{=} \oint_{u\in\mathcal{U},\,y\in\mathcal{Y}} P_{U,Y}(u,y) \log \frac{Q_{Y|U}(y|u)}{P_Y(y)}\, dy\, du + D\left(P_Y\|Q_Y|U\right) \tag{17}$$

$$\overset{(b)}{\ge} \oint_{u\in\mathcal{U},\,y\in\mathcal{Y}} P_{U,Y}(u,y) \log \frac{Q_{Y|U}(y|u)}{P_Y(y)}\, dy\, du \tag{18}$$

$$= H(Y) + \oint_{u\in\mathcal{U},\,y\in\mathcal{Y}} P_{U,Y}(u,y) \log Q_{Y|U}(y|u)\, dy\, du \tag{19}$$

$$\overset{(c)}{\ge} \oint_{u\in\mathcal{U},\,y\in\mathcal{Y}} P_{U,Y}(u,y) \log Q_{Y|U}(y|u)\, dy\, du \tag{20}$$

$$\overset{(d)}{=} \oint_{u \in \mathcal{U}, x \in \mathcal{X}, y \in \mathcal{Y}} P_X(x) P_{Y|X}(y|x) P_{U|X}(u|x) \log Q_{Y|U}(y|u) dx dy du, \tag{21}$$

where in (a) the term $D\left(P_Y \| Q_Y | U\right)$ is the conditional relative entropy between P_Y and Q_Y, given P_U; (b) holds by the non-negativity of relative entropy; (c) holds by the non-negativity of entropy; and (d) follows using the Markov Chain $U \multimap X \multimap Y$.

Similarly, let S_U be a given the variational approximation of P_U. Then, we get

$$I(U;X) = \oint_{u \in \mathcal{U}, x \in \mathcal{X}} P_{U,X}(u,x) \log \frac{P_{U|X}(u|x)}{P_U(u)} dx du \tag{22}$$

$$= \oint_{u \in \mathcal{U}, x \in \mathcal{X}} P_{U,X}(u,x) \log \frac{P_{U|X}(u|x)}{S_U(u)} dx du - D\left(P_U \| S_U\right) \tag{23}$$

$$\leq \oint_{u \in \mathcal{U}, x \in \mathcal{X}} P_{U,X}(u,x) \log \frac{P_{U|X}(u|x)}{S_U(u)} dx du \tag{24}$$

where the inequality follows since the relative entropy is non-negative.

Combining Equations (21) and (24), we get

$$I(U;Y) - \beta I(U;X) \geq \oint_{u \in \mathcal{U}, x \in \mathcal{X}, y \in \mathcal{Y}} P_X(x) P_{Y|X}(y|x) P_{U|X}(u|x) \log Q_{Y|U}(y|u) dx dy du$$

$$- \beta \oint_{u \in \mathcal{U}, x \in \mathcal{X}} P_{U,X}(u,x) \log \frac{P_{U|X}(u|x)}{S_U(u)} dx du. \tag{25}$$

The use of the variational bound in Equation (14) over the IB-Lagrangian in Equation (13) shows some advantages. First, it allows the derivation of alternating algorithms that allow to obtain a solution by optimizing over the encoders and decoders. Then, it is easier to obtain an empirical estimate of Equation (14) by sampling from: (i) the joint distribution $P_{X,Y}$; (ii) the encoder $P_{U|X}$; and (iii) the prior S_U. Additionally, as noted in Equation (15), when evaluated for the optimal decoder $Q_{Y|U}$ and prior S_U, the variational bound becomes tight. All this allows obtaining algorithms to obtain good approximate solutions to the IB problem, as shown next. Further theoretical implications of this variational bound are discussed in [55].

3.2.2. Known Distributions

Using the variational formulation in Equation (14), when the data model is discrete and the joint distribution $P_{X,Y}$ is known, the IB problem can be solved by using an iterative method that optimizes the variational IB cost function in Equation (14) alternating over the distributions $P_{U|X}, Q_{Y|U}$, and S_U. In this case, the maximizing distributions $P_{U|X}, Q_{Y|U}$, and S_U can be efficiently found by an alternating optimization procedure similar to the expectation-maximization (EM) algorithm [56] and the standard Blahut–Arimoto (BA) method [57]. In particular, a solution $P_{U|X}$ to the constrained optimization problem is determined by the following self-consistent equations, for all $(u, x, y) \in \mathcal{U} \times \mathcal{X} \times \mathcal{Y}$, [1]

$$P_{U|X}(u|x) = \frac{P_U(u)}{Z(\beta, x)} \exp\left(-\beta D_{KL}\left(P_{Y|X}(\cdot|x) \| P_{Y|U}(\cdot|u)\right)\right) \tag{26a}$$

$$P_U(u) = \sum_{x \in \mathcal{X}} P_X(x) P_{U|X}(u|x) \tag{26b}$$

$$P_{Y|U}(y|u) = \sum_{x \in \mathcal{X}} P_{Y|X}(y|x) P_{X|U}(x|u) \tag{26c}$$

where $P_{X|U}(x|u) = P_{U|X}(u|x) P_X(x) / P_U(u)$ and $Z(\beta, x)$ is a normalization term. It is shown in [1] that alternating iterations of these equations converges to a solution of the problem for any initial $P_{U|X}$. However, by opposition to the standard Blahut–Arimoto algorithm [57,58], which is classically used in the computation of rate-distortion functions of discrete memoryless sources for which convergence to the optimal solution is guaranteed, convergence here may be to a local optimum only. If $\beta = 0$, the optimization is non-constrained and one can set $U = \varnothing$, which yields minimal relevance and complexity levels. Increasing the value of β steers towards more accurate and more complex representations, until $U = X$ in the limit of very large (infinite) values of β for which the relevance reaches its maximal value $I(X;Y)$.

For discrete sources with (small) alphabets, the updating equations described by Equation (26) are relatively easy computationally. However, if the variables X and Y lie in a continuum, solving the equations described by Equation (26) is very challenging. In the case in which X and Y are joint multivariate Gaussian, the problem of finding the optimal representation U is analytically tractable in [53] (see also the related [54,59]), as discussed in Section 3.1.2. Leveraging the optimality of Gaussian mappings $P_{U|X}$ to restrict the optimization of $P_{U|X}$ to Gaussian distributions as in Equation (9), allows reducing the search of update rules to those of the associated parameters, namely covariance matrices. When Y is a deterministic function of X, the IB curve cannot be explored, and other Lagrangians have been proposed to tackle this problem [60].

3.3. Unknown Distributions

The main drawback of the solutions presented thus far for the IB principle is that, in the exception of small-sized discrete (X, Y) for which iterating Equation (26) converges to an (at least local) solution and jointly Gaussian (X, Y) for which an explicit analytic solution was found, solving Equation (3) is generally computationally costly, especially for high dimensionality. Another important barrier in solving Equation (3) directly is that IB necessitates knowledge of the joint distribution $P_{X,Y}$. In this section, we describe a method to provide an approximate solution to the IB problem in the case in which the joint distribution is unknown and only a give training set of N samples $\{(x_i, y_i)\}_{i=1}^{N}$ is available.

A major step ahead, which widened the range of applications of IB inference for various learning problems, appeared in [48], where the authors used neural networks to parameterize the variational inference lower bound in Equation (14) and show that its optimization can be done through the classic and widely used stochastic gradient descendent (SGD). This method, denoted by Variational IB in [48] and detailed below, has allowed handling handle high-dimensional, possibly continuous, data, even in the case in which the distributions are unknown.

3.3.1. Variational IB

The goal of the variational IB when only samples $\{(x_i, y_i)\}_{i=1}^{N}$ are available is to solve the IB problem optimizing an approximation of the cost function. For instance, for a given training set $\{(x_i, y_i)\}_{i=1}^{N}$, the right hand side of Equation (14) can be approximated as

$$\mathcal{L}_{\text{low}} \approx \frac{1}{N} \sum_{i=1}^{N} \left[\oint_{u \in \mathcal{U}} \left(P_{U|X}(u|x_i) \log Q_{Y|U}(y_i|u) - \beta P_{U|X}(u|x_i) \log \frac{P_{U|X}(u|x_i)}{S_U(u)} \right) du \right]. \tag{27}$$

However, in general, the direct optimization of this cost is challenging. In the variational IB method, this optimization is done by parameterizing the encoding and decoding distributions $P_{U|X}$, $Q_{Y|U}$, and S_U that are to optimize using a family of distributions whose parameters are determined by DNNs. This allows us to formulate Equation (14) in terms of the DNN parameters, i.e., its weights, and optimize it by using the reparameterization trick [15], Monte Carlo sampling, and stochastic gradient descent (SGD)-type algorithms.

Let $P_{\theta}(u|x)$ denote the family of encoding probability distributions $P_{U|X}$ over $\mathcal{U}$ for each element on $\mathcal{X}$, parameterized by the output of a DNN f_{θ} with parameters θ. A common example is the family of multivariate Gaussian distributions [15], which are parameterized by the mean μ^{θ} and covariance matrix Σ^{θ}, i.e., $\gamma := (\mu^{\theta}, \Sigma^{\theta})$. Given an observation X, the values of $(\mu^{\theta}(x), \Sigma^{\theta}(x))$ are determined by the output of the DNN f_{θ}, whose input is X, and the corresponding family member is given by

$P_\theta(u|x) = \mathcal{N}(u; \boldsymbol{\mu}^\theta(x), \boldsymbol{\Sigma}^\theta(x))$. For discrete distributions, a common example are concrete variables [61] (or Gumbel-Softmax [62]). Some details are given below.

Similarly, for decoder $Q_{Y|U}$ over $\mathcal{Y}$ for each element on $\mathcal{U}$, let $Q_\psi(y|u)$ denote the family of distributions parameterized by the output of the DNNs f_{ψ_k}. Finally, for the prior distributions $S_U(u)$ over $\mathcal{U}$ we define the family of distributions $S_\varphi(u)$, which do not depend on a DNN.

By restricting the optimization of the variational IB cost in Equation (14) to the encoder, decoder, and prior within the families of distributions $P_\theta(u|x)$, $Q_\psi(y|u)$, and $S_\varphi(u)$, we get

$$\max_{P_{U|X}} \max_{Q_{Y|U}, S_U} \mathcal{L}_\beta^{\mathrm{VIB}}(P_{U|X}, Q_{Y|U}, S_U) \geq \max_{\theta, \phi, \varphi} \mathcal{L}_\beta^{\mathrm{NN}}(\theta, \phi, \varphi), \tag{28}$$

where θ, ϕ, and φ denote the DNN parameters, e.g., its weights, and the cost in Equation (29) is given by

$$\mathcal{L}_\beta^{\mathrm{NN}}(\theta, \phi, \varphi) := \mathbb{E}_{P_{Y,X}} \mathbb{E}_{\{P_\theta(U|X)\}} \Big[\log Q_{\phi(Y|U)}(Y|U) \Big] - \beta D_{\mathrm{KL}}(P_\theta(U|X) \| S_\varphi(U)). \tag{29}$$

Next, using the training samples $\{(x_i, y_i)\}_{i=1}^N$, the DNNs are trained to maximize a Monte Carlo approximation of Equation (29) over $\boldsymbol{\theta}, \boldsymbol{\phi}, \boldsymbol{\varphi}$ using optimization methods such as SGD or ADAM [63] with backpropagation. However, in general, the direct computation of the gradients of Equation (29) is challenging due to the dependency of the averaging with respect to the encoding P_θ, which makes it hard to approximate the cost by sampling. To circumvent this problem, the reparameterization trick [15] is used to sample from $P_\theta(U|X)$. In particular, consider $P_\theta(U|X)$ to belong to a parametric family of distributions that can be sampled by first sampling a random variable Z with distribution $P_Z(z)$, $z \in \mathcal{Z}$ and then transforming the samples using some function $g_\theta : \mathcal{X} \times \mathcal{Z} \to \mathcal{U}$ parameterized by θ, such that $U = g_\theta(x, Z) \sim P_\theta(U|x)$. Various parametric families of distributions fall within this class for both discrete and continuous latent spaces, e.g., the Gumbel-Softmax distributions and the Gaussian distributions. Next, we detail how to sample from both examples:

1. Sampling from Gaussian Latent Spaces: When the latent space is a continuous vector space of dimension D, e.g., $\mathcal{U} = \mathbb{R}^D$, we can consider multivariate Gaussian parametric encoders of mean ($\boldsymbol{\mu}^\theta$, and covariance $\boldsymbol{\Sigma}^\theta$), i.e., $P_\theta(u|x) = \mathcal{N}(u; \boldsymbol{\mu}^\theta, \boldsymbol{\Sigma}^\theta)$. To sample $\mathbf{U} \sim \mathcal{N}(u; \boldsymbol{\mu}^\theta, \boldsymbol{\Sigma}^\theta)$, where $\boldsymbol{\mu}^\theta(x) = f_{e,\theta}^\mu(x)$ and $\boldsymbol{\Sigma}^\theta(x) = f_{e,\theta}^\Sigma(x)$ are determined as the output of a NN, sample a random variable $Z \sim \mathcal{N}(z; \mathbf{0}, \mathbf{I})$ i.i.d. and, given data sample $x \in \mathcal{X}$, and generate the jth sample as

    $$u_j = f_{e,\theta}^\mu(x) + f_{e,\theta}^\Sigma(x) z_j \tag{30}$$

    where z_j is a sample of $Z \sim \mathcal{N}(\mathbf{0}, \mathbf{I})$, which is an independent Gaussian noise, and $f_e^\mu(x)$ and $f_e^\Sigma(x)$ are the output values of the NN with weights θ for the given input sample x.

    An example of the resulting DIB architecture to optimize with an encoder, a latent space, and a decoder parameterized by Gaussian distributions is shown in Figure 3.

2. Sampling from a discrete latent space with the Gumbel-Softmax:

    If U is categorical random variable on the finite set $\mathcal{U}$ of size D with probabilities $\pi := (\pi_1, \dots, \pi_D)$), we can encode it as D-dimensional one-hot vectors lying on the corners of the $(D-1)$-dimensional simplex, Δ_{D-1}. In general, costs functions involving sampling from categorical distributions are non-differentiable. Instead, we consider Concrete variables [62] (or Gumbel-Softmax [61]), which are continuous differentiable relaxations of categorical variables on the interior of the simplex, and are easy to sample. To sample from a Concrete random variable $U \in \Delta_{D-1}$ at temperature $\lambda \in (0, \infty)$, with probabilities $\pi \in (0, 1)^D$, sample $G_d \sim \mathrm{Gumbel}(0, 1)$ i.i.d.

(The Gumbel$(0,1)$ distribution can be sampled by drawing $u \sim \text{Uniform}(0,1)$ and calculating $g = -\log(-\log(u))$.), and set for each of the components of $U = (U_1, \ldots, U_D)$

$$U_d = \frac{\exp((\log(\pi_d + G_d)/\lambda))}{\sum_{j=1}^{D} \exp((\log(\pi_j + G_j)/\lambda))}, \quad d = 1, \ldots, D. \tag{31}$$

We denote by $Q_{\pi,\lambda(u,x)}$ the Concrete distribution with parameters $(\pi(x),\lambda)$. When the temperature λ approaches 0, the samples from the concrete distribution become one-hot and $\Pr\{\lim_{\lambda \to 0} U_d\} = \pi_d$ [61]. Note that, for discrete data models, standard application of Caratheodory's theorem [64] shows that the latent variables U that appear in Equation (3) can be restricted to be with bounded alphabets size.

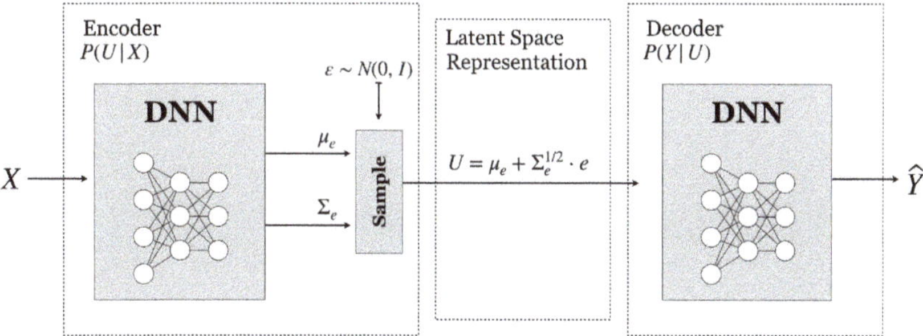

Figure 3. Example parametrization of Variational Information Bottleneck using neural networks.

The reparametrization trick transforms the cost function in Equation (29) into one which can be to approximated by sampling M independent samples $\{u_m\}_{m=1}^{M} \sim P_\theta(u|x_i)$ for each training sample (x_i, y_i), $i = 1, \ldots, N$ and allows computing estimates of the gradient using backpropagation [15]. Sampling is performed by using $u_{i,m} = g_\phi(x_i, z_m)$ with $\{z_m\}_{m=1}^{M}$ i.i.d. sampled from P_Z. Altogether, we have the empirical-DIB cost for the ith sample in the training dataset:

$$\mathcal{L}_{\beta,i}^{\text{emp}}(\theta, \phi, \varphi) := \frac{1}{M} \sum_{m=1}^{M} \left[\log Q_\phi(y_i|u_{i,m}) - \beta D_{\text{KL}}(P_\theta(U_i|x_i) \| Q_\varphi(U_i)) \right]. \tag{32}$$

Note that, for many distributions, e.g., multivariate Gaussian, the divergence $D_{\text{KL}}(P_\theta(U_i|x_i) \| Q_\varphi(U_i))$ can be evaluated in closed form. Alternatively, an empirical approximation can be considered.

Finally, we maximize the empirical-IB cost over the DNN parameters θ, ϕ, φ as,

$$\max_{\theta, \phi, \varphi} \frac{1}{N} \sum_{i=1}^{N} \mathcal{L}_{\beta,i}^{\text{emp}}(\theta, \phi, \varphi). \tag{33}$$

By the law of large numbers, for large N, M, we have $1/N \sum_{i=1}^{M} \mathcal{L}_{\beta,i}^{\text{emp}}(\theta, \phi, \varphi) \to \mathcal{L}_{\beta}^{\text{NN}}(\theta, \phi, \varphi)$ almost surely. After convergence of the DNN parameters to $\theta^*, \phi^*, \varphi^*$, for a new observation X, the representation U can be obtained by sampling from the encoders $P_{\theta_k^*}(U_k|X_k)$. In addition, note that a soft estimate of the remote source Y can be inferred by sampling from the decoder $Q_{\phi^*}(Y|U)$. The notion of encoder and decoder in the IB-problem will come clear from its relationship with lossy source coding in Section 4.1.

4. Connections to Coding Problems

The IB problem is a one-shot coding problem, in the sense that the operations are performed letter-wise. In this section, we consider now the relationship between the IB problem and (asymptotic) coding problem in which the coding operations are performed over blocks of size n, with n assumed to be large and the joint distribution of the data $P_{X,Y}$ is in general assumed to be known a priori. The connections between these problems allow extending results from one setup to another, and to consider generalizations of the classical IB problem to other setups, e.g., as shown in Section 6.

4.1. Indirect Source Coding under Logarithmic Loss

Let us consider the (asymptotic) indirect source coding problem shown in Figure 4, in which Y designates a memoryless remote source and X a noisy version of it that is observed at the encoder.

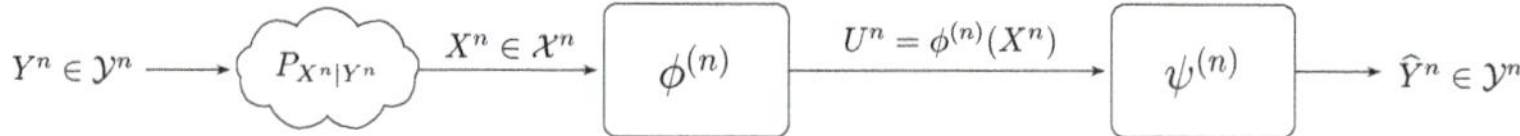

Figure 4. A remote source coding problem.

A sequence of n samples $X^n = (X_1, \ldots, X_n)$ is mapped by an encoder $\phi^{(n)} : \mathcal{X}^n \to \{1, \ldots, 2^{nR}\}$ which outputs a message from a set $\{1, \ldots, 2^{nR}\}$, that is, the encoder uses at most R bits per sample to describe its observation and the range of the encoder map is allowed to grow with the size of the input sequence as

$$\|\phi^{(n)}\| \leq nR. \tag{34}$$

This message is mapped with a decoder $\phi^{(n)} : \{1, \ldots, 2^{nR}\} \to \hat{\mathcal{Y}}$ to generate a reconstruction of the source sequence Y^n as $\mathcal{Y}^n \in \hat{\mathcal{Y}}^n$. As already observed in [50], the IB problem in Equation (3) is essentially equivalent to a remote point-to-point source coding problem in which distortion between Y^n as $\mathcal{Y}^n \in \hat{\mathcal{Y}}^n$ is measured under the logarithm loss (log-loss) fidelity criterion [65]. That is, rather than just assigning a deterministic value to each sample of the source, the decoder gives an assessment of the degree of confidence or reliability on each estimate. Specifically, given the output description $m = \phi^{(n)}(x^n)$ of the encoder, the decoder generates a soft-estimate $\hat{y}^n$ of y^n in the form of a probability distribution over $\mathcal{Y}^n$, i.e., $\hat{y}^n = \hat{P}_{Y^n|M}(\cdot)$. The incurred discrepancy between y^n and the estimation $\hat{y}^n$ under log-loss for the observation x^n is then given by the per-letter logarithmic loss distortion, which is defined as

$$\ell_{\log}(y, \hat{y}) := \log \frac{1}{\hat{y}(y)}. \tag{35}$$

for $y \in \mathcal{Y}$ and $\hat{y} \in \mathcal{P}(\mathcal{Y})$ designates here a probability distribution on $\mathcal{Y}$ and $\hat{y}(y)$ is the value of that distribution evaluated at the outcome $y \in \mathcal{Y}$.

That is, the encoder uses at most R bits per sample to describe its observation to a decoder which is interested in reconstructing the remote source Y^n to within an average distortion level D, using a per-letter distortion metric, i.e.,

$$\mathbb{E}[\ell_{\log}^{(n)}(Y^n, \hat{Y}^n)] \leq D \tag{36}$$

where the incurred distortion between two sequences Y^n and $\hat{Y}^n$ is measured as

$$\ell_{\log}^{(n)}(Y^n, \hat{Y}^n) = \frac{1}{n} \sum_{i=1}^{n} \ell_{\log}(y_i, \hat{y}_i) \tag{37}$$

and the per-letter distortion is measured in terms of that given by the logarithmic loss in Equation (53). The rate distortion region of this model is given by the union of all pairs (R, D) that satisfy [7,9]

$$R \geq I(U; X) \tag{38a}$$

$$D \geq H(Y|U) \tag{38b}$$

where the union is over all auxiliary random variables U that satisfy that $U \multimap X \multimap Y$ forms a Markov Chain in this order. Invoking the support lemma [66] (p. 310), it is easy to see that this region is not altered if one restricts U to satisfy $|\mathcal{U}| \leq |\mathcal{X}| + 1$. In addition, using the substitution $\Delta := H(Y) - D$, the region can be written equivalently as the union of all pairs $(R, H(Y) - \Delta)$ that satisfy

$$R \geq I(U; X) \tag{39a}$$
$$\Delta \leq I(U; Y) \tag{39b}$$

where the union is over all Us with pmf $P_{U|X}$ that satisfy $U \multimap X \multimap Y$, with $|\mathcal{U}| \leq |\mathcal{X}| + 1$.

The boundary of this region is equivalent to the one described by the IB principle in Equation (3) if solved for all β, and therefore the IB problem is essentially a remote source coding problem in which the distortion is measured under the logarithmic loss measure. Note that, operationally, the IB problem is equivalent to that of finding an encoder $P_{U|X}$ which maps the observation X to a representation U that satisfies the bit rate constraint R and such that U captures enough relevance of Y so that the posterior probability of Y given U satisfies an average distortion constraint.

4.2. Common Reconstruction

Consider the problem of source coding with side information at the decoder, i.e., the well known Wyner–Ziv setting [67], with the distortion measured under logarithmic-loss. Specifically, a memoryless source X is to be conveyed lossily to a decoder that observes a statistically correlated side information Y. The encoder uses R bits per sample to describe its observation to the decoder which wants to reconstruct an estimate of X to within an average distortion level D, where the distortion is evaluated under the log-loss distortion measure. The rate distortion region of this problem is given by the set of all pairs (R, D) that satisfy

$$R + D \geq H(X|Y). \tag{40}$$

The optimal coding scheme utilizes standard Wyner–Ziv compression [67] at the encoder and the decoder map $\psi : \mathcal{U} \times \mathcal{Y} \to \hat{\mathcal{X}}$ is given by

$$\psi(U, Y) = \Pr[X = x|U, Y] \tag{41}$$

for which it is easy to see that

$$\mathbb{E}[\ell_{\log}(X, \psi(U, Y))] = H(X|U, Y). \tag{42}$$

Now, assume that we constrain the coding in a manner that the encoder is be able to produce an exact copy of the compressed source constructed by the decoder. This requirement, termed *common reconstruction* constraint (CR), was introduced and studied by Steinberg [68] for various source coding models, including the Wyner–Ziv setup, in the context of a "general distortion measure". For the Wyner–Ziv problem under log-loss measure that is considered in this section, such a CR constraint causes some rate loss because the reproduction rule in Equation (41) is no longer possible. In fact, it is not difficult to see that under the CR constraint the above region reduces to the set of pairs (R, D) that satisfy

$$R \leq I(U; X|Y) \tag{43a}$$
$$D \geq H(X|U) \tag{43b}$$

for some auxiliary random variable for which $U \multimap X \multimap Y$ holds. Observe that Equation (43b) is equivalent to $I(U;X) \geq H(X) - D$ and that, for a given prescribed fidelity level D, the minimum rate is obtained for a description U that achieves the inequality in Equation (43b) with equality, i.e.,

$$R(D) = \min_{P_{U|X}\, :\, I(U;X)=H(X)-D} I(U;X|Y). \tag{44}$$

Because $U \multimap X \multimap Y$, we have

$$I(U;Y) = I(U;X) - I(U;X|Y). \tag{45}$$

Under the constraint $I(U;X) = H(X) - D$, it is easy to see that minimizing $I(U;X|Y)$ amounts to maximizing $I(U;Y)$, an aspect which bridges the problem at hand with the IB problem.

In the above, the side information Y is used for binning but not for the estimation at the decoder. If the encoder ignores whether Y is present or not at the decoder side, the benefit of binning is reduced—see the Heegard–Berger model with common reconstruction studied in [69,70].

4.3. Information Combining

Consider again the IB problem. Assume one wishes to find the representation U that maximizes the relevance $I(U;Y)$ for a given prescribed complexity level, e.g., $I(U;X) = R$. For this setup, we have

$$I(X;U,Y) = I(U;X) + I(Y;X) - I(U;Y) \tag{46}$$
$$= R + I(Y;X) - I(U;Y) \tag{47}$$

where the first equality holds since $U \multimap X \multimap Y$ is a Markov Chain. Maximizing $I(U;Y)$ is then equivalent to minimizing $I(X;U,Y)$. This is reminiscent of the problem of *information combining* [71,72], where X can be interpreted as a source information that is conveyed through two channels: the channel $P_{Y|X}$ and the channel $P_{U|X}$. The outputs of these two channels are conditionally independent given X, and they should be processed in a manner such that, when combined, they preserve as much information as possible about X.

4.4. Wyner–Ahlswede–Korner Problem

Here, the two memoryless sources X and Y are encoded separately at rates R_X and R_Y, respectively. A decoder gets the two compressed streams and aims at recovering Y losslessly. This problem was studied and solved separately by Wyner [73] and Ahlswede and Körner [74]. For given $R_X = R$, the minimum rate R_Y that is needed to recover Y losslessly is

$$R_Y^\star(R) = \min_{P_{U|X}\, :\, I(U;X) \leq R} H(Y|U). \tag{48}$$

Thus, we get

$$\max_{P_{U|X}\, :\, I(U;X) \leq R} I(U;Y) = H(Y) - R_Y^\star(R),$$

and therefore, solving the IB problem is equivalent to solving the Wyner–Ahlswede–Korner Problem.

4.5. The Privacy Funnel

Consider again the setting of Figure 4, and let us assume that the pair (Y,X) models data that a user possesses and which have the following properties: the data Y are some sensitive (private) data that are not meant to be revealed at all, or else not beyond some level Δ; and the data X are non-private and are meant to be shared with another user (analyst). Because X and Y are correlated, sharing the non-private data X with the analyst possibly reveals information about Y. For this reason, there is a trade off between the amount of information that the user shares about X and the information that he keeps private about Y. The data X are passed through a randomized mapping ϕ whose purpose is to make $U = \phi(X)$ maximally informative about X while being minimally informative about Y.

The analyst performs an inference attack on the private data Y based on the disclosed information U. Let $\ell : \mathcal{Y} \times \hat{\mathcal{Y}} \longrightarrow \mathbb{R}$ be an arbitrary loss function with reconstruction alphabet $\hat{\mathcal{Y}}$ that measures the cost of inferring Y after observing U. Given $(X, Y) \sim P_{X,Y}$ and under the given loss function ℓ, it is natural to quantify the difference between the prediction losses in predicting $Y \in \mathcal{Y}$ prior and after observing $U = \phi(X)$. Let

$$C(\ell, P) = \inf_{\hat{y} \in \hat{\mathcal{Y}}} \mathbb{E}_P[\ell(Y, \hat{y})] - \inf_{\hat{Y}(\phi(X))} \mathbb{E}_P[\ell(Y, \hat{Y})] \tag{49}$$

where $\hat{y} \in \hat{\mathcal{Y}}$ is deterministic and $\hat{Y}(\phi(X))$ is any measurable function of $U = \phi(X)$. The quantity $C(\ell, P)$ quantifies the reduction in the prediction loss under the loss function ℓ that is due to observing $U = \phi(X)$, i.e., the inference cost gain. In [75] (see also [76]), it is shown that that under some mild conditions the inference cost gain $C(\ell, P)$ as defined by Equation (49) is upper-bounded as

$$C(\ell, P) \le 2\sqrt{2}L\sqrt{I(U; Y)} \tag{50}$$

where L is a constant. The inequality in Equation (50) holds irrespective to the choice of the loss function ℓ, and this justifies the usage of the logarithmic loss function as given by Equation (53) in the context of finding a suitable trade off between utility and privacy, since

$$I(U; Y) = H(Y) - \inf_{\hat{Y}(U)} \mathbb{E}_P[\ell_{\log}(Y, \hat{Y})]. \tag{51}$$

Under the logarithmic loss function, the design of the mapping $U = \phi(X)$ should strike a right balance between the utility for inferring the non-private data X as measured by the mutual information $I(U; X)$ and the privacy metric about the private date Y as measured by the mutual information $I(U; Y)$.

4.6. Efficiency of Investment Information

Let Y model a stock market data and X some correlated information. In [77], Erkip and Cover investigated how the description of the correlated information X improves the investment in the stock market Y. Specifically, let $\Delta(C)$ denote the maximum increase in growth rate when X is described to the investor at rate C. Erkip and Cover found a single-letter characterization of the incremental growth rate $\Delta(C)$. When specialized to the horse race market, this problem is related to the aforementioned source coding with side information of Wyner [73] and Ahlswede-Körner [74], and, thus, also to the IB problem. The work in [77] provides explicit analytic solutions for two horse race examples, jointly binary and jointly Gaussian horse races.

5. Connections to Inference and Representation Learning

In this section, we consider the connections of the IB problem with learning, inference and generalization, for which, typically, the joint distribution $P_{X,Y}$ of the data is not known and only a set of samples is available.

5.1. Inference Model

Let a measurable variable $X \in \mathcal{X}$ and a target variable $Y \in \mathcal{Y}$ with unknown joint distribution $P_{X,Y}$ be given. In the classic problem of statistical learning, one wishes to infer an accurate predictor of the target variable $Y \in \mathcal{Y}$ based on observed realizations of $X \in \mathcal{X}$. That is, for a given class $\mathcal{F}$ of admissible predictors $\psi : \mathcal{X} \to \hat{\mathcal{Y}}$ and a loss function $\ell : \mathcal{Y} \to \hat{\mathcal{Y}}$ that measures discrepancies between true values and their estimated fits, one aims at finding the mapping $\psi \in \mathcal{F}$ that minimizes the expected (population) risk

$$C_{P_{X,Y}}(\psi, \ell) = \mathbb{E}_{P_{X,Y}}[\ell(Y, \psi(X))]. \tag{52}$$

An abstract inference model is shown in Figure 5.

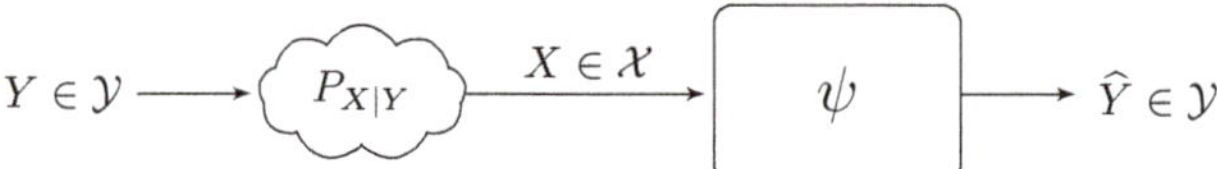

Figure 5. An abstract inference model for learning.

The choice of a "good" loss function $\ell(\cdot)$ is often controversial in statistical learning theory. There is however numerical evidence that models that are trained to minimize the error's entropy often outperform ones that are trained using other criteria such as mean-square error (MSE) and higher-order statistics [26,27]. This corresponds to choosing the loss function given by the logarithmic loss, which is defined as

$$\ell_{\log}(y, \hat{y}) := \log \frac{1}{\hat{y}(y)} \tag{53}$$

for $y \in \mathcal{Y}$, where $\hat{y} \in \mathcal{P}(\mathcal{Y})$ designates here a probability distribution on $\mathcal{Y}$ and $\hat{y}(y)$ is the value of that distribution evaluated at the outcome $y \in \mathcal{Y}$. Although a complete and rigorous justification of the usage of the logarithmic loss as distortion measure in learning is still awaited, recently a partial explanation appeared in [30] where Painsky and Wornell showed that, for binary classification problems, by minimizing the logarithmic-loss one actually minimizes an upper bound to any choice of loss function that is smooth, proper (i.e., unbiased and Fisher consistent), and convex. Along the same line of work, the authors of [29] showed that under some natural data processing property Shannon's mutual information uniquely quantifies the reduction of prediction risk due to side information. Perhaps, this justifies partially why the logarithmic-loss fidelity measure is widely used in learning theory and has already been adopted in many algorithms in practice such as the *infomax* criterion [31], the tree-based algorithm of Quinlan [32], or the well known Chow–Liu algorithm [33] for learning tree graphical models, with various applications in genetics [34], image processing [35], computer vision [36], and others. The logarithmic loss measure also plays a central role in the theory of prediction [37] (Ch. 09), where it is often referred to as the *self-information* loss function, as well as in Bayesian modeling [38] where priors are usually designed to maximize the mutual information between the parameter to be estimated and the observations.

When the join distribution $P_{X,Y}$ is known, the optimal predictor and the minimum expected (population) risk can be characterized. Let, for every $x \in \mathcal{X}$, $\psi(x) = Q(\cdot|x) \in \mathcal{P}(\mathcal{Y})$. It is easy to see that

$$\mathbb{E}_{P_{X,Y}}[\ell_{\log}(Y, Q)] = \sum_{x \in \mathcal{X}, y \in \mathcal{Y}} P_{X,Y}(x, y) \log\left(\frac{1}{Q(y|x)}\right) \tag{54a}$$

$$= \sum_{x \in \mathcal{X}, y \in \mathcal{Y}} P_{X,Y}(x, y) \log\left(\frac{1}{P_{Y|X}(y|x)}\right) + \sum_{x \in \mathcal{X}, y \in \mathcal{Y}} P_{X,Y}(x, y) \log\left(\frac{P_{Y|X}(y|x)}{Q(y|x)}\right) \tag{54b}$$

$$= H(Y|X) + D\left(P_{Y|X} \| Q\right) \tag{54c}$$

$$\geq H(Y|X) \tag{54d}$$

with equality iff the predictor is given by the conditional posterior $\psi(x) = P_Y(Y|X = x)$. That is, the minimum expected (population) risk is given by

$$\min_{\psi} C_{P_{X,Y}}(\psi, \ell_{\log}) = H(Y|X). \tag{55}$$

If the joint distribution $P_{X,Y}$ is unknown, which is most often the case in practice, the population risk as given by Equation (56) cannot be computed directly, and, in the standard approach, one usually resorts to choosing the predictor with minimal risk on a training dataset consisting of n labeled samples $\{(x_i, y_i)\}_{i=1}^n$ that are drawn independently from the unknown joint distribution $P_{X,Y}$. In this case, one is interested in optimizing the empirical population risk, which for a set of n i.i.d. samples from $P_{X,Y}$, $\mathcal{D}_n := \{(x_i, y_i)\}_{i=1}^n$, is defined as

$$\hat{\mathcal{C}}_{P_{X,Y}}(\psi, \ell, \mathcal{D}_n) = \frac{1}{n} \sum_{i=1}^n \ell(y_i, \psi(x_i)). \tag{56}$$

The difference between the empirical and population risks is normally measured in terms of the generalization gap, defined as

$$\text{gen}_{P_{X,Y}}(\psi, \ell, \mathcal{D}_n) := \mathcal{C}_{P_{X,Y}}(\psi, \ell_{\log}) - \hat{\mathcal{C}}_{P_{X,Y}}(\psi, \ell, \mathcal{D}_n). \tag{57}$$

5.2. Minimum Description Length

One popular approach to reducing the generalization gap is by restricting the set $\mathcal{F}$ of admissible predictors to a low-complexity class (or constrained complexity) to prevent over-fitting. One way to limit the model's complexity is by restricting the range of the prediction function, as shown in Figure 6. This is the so-called minimum description length complexity measure, often used in the learning literature to limit the description length of the weights of neural networks [78]. A connection between the use of the minimum description complexity for limiting the description length of the input encoding and accuracy studied in [79] and with respect to the weight complexity and accuracy is given in [11]. Here, the stochastic mapping $\phi : \mathcal{X} \longrightarrow \mathcal{U}$ is a compressor with

$$\|\phi\| \leq R \tag{58}$$

for some prescribed "input-complexity" value R, or equivalently prescribed average description-length.

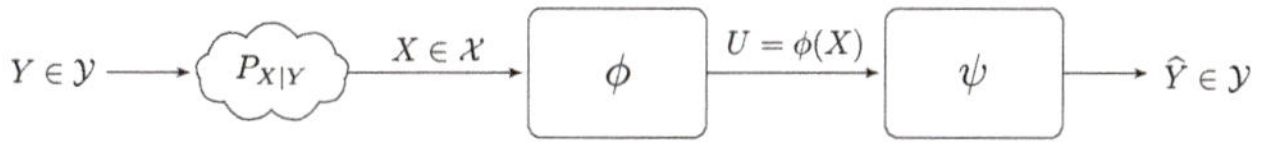

Figure 6. Inference problem with constrained model's complexity.

Minimizing the constrained description length population risk is now equivalent to solving

$$\mathcal{C}_{P_{X,Y},\text{DLC}}(R) = \min_{\phi} \mathbb{E}_{P_{X,Y}}[\ell_{\log}(Y^n, \psi(U^n))] \tag{59}$$

$$\text{s.t. } \|\phi(X^n)\| \leq nR. \tag{60}$$

It can be shown that this problem takes its minimum value with the choice of $\psi(U) = P_{Y|U}$ and

$$\mathcal{C}_{P_{X,Y},\text{DLC}}(R) = \min_{P_{U|X}} H(Y|U) \quad \text{s.t. } R \geq I(U; X), \tag{61}$$

The solution to Equation (61) for different values of R is effectively equivalent to the IB-problem in Equation (4). Observe that the right-hand side of Equation (61) is larger for small values of R; it is clear that a good predictor ϕ should strike a right balance between reducing the model's complexity and reducing the error's entropy, or, equivalently, maximizing the mutual information $I(U; Y)$ about the target variable Y.

5.3. Generalization and Performance Bounds

The IB-problem appears as a relevant problem in fundamental performance limits of learning. In particular, when $P_{X,Y}$ is unknown, and instead n samples i.i.d from $P_{X,Y}$ are available, the optimization of the empirical risk in Equation (56) leads to a mismatch between the true loss given by the population risk and the empirical risk. This gap is measured by the generalization gap in Equation (57). Interestingly, the relationship between the true loss and the empirical loss can be bounded (in high probability) in terms of the IB-problem as [80]

$$C_{P_{X,Y}}(\psi, \ell_{\log}) \leq \hat{C}_{P_{X,Y}}(\psi, \ell, \mathcal{D}_n) + \text{gen}_{P_{X,Y}}(\psi, \ell, \mathcal{D}_n)$$

$$= \underbrace{H_{\hat{P}_{X,Y}^{(n)}}(Y|U)}_{\hat{C}_{P_{X,Y}}(\psi,\ell,\mathcal{D}_n)} + \underbrace{A\sqrt{I(\hat{P}_X^{(n)};P_{U|X})} \cdot \frac{\log n}{n} + \frac{B\sqrt{\Lambda(P_{U|X}, \hat{P}_{Y|U}, P_{\hat{Y}|U})}}{\sqrt{n}} + \mathcal{O}\left(\frac{\log n}{n}\right)}_{\text{Bound on gen}_{P_{X,Y}}(\psi,\ell,\mathcal{D}_n)}$$

where $\hat{P}_{U|X}$ and $\hat{P}_{Y|U}$ are the empirical encoder and decoder and $P_{\hat{Y}|U}$ is the optimal decoder. $H_{\hat{P}_{X,Y}^{(n)}}(Y|U)$ and $I(\hat{P}_X^{(n)};P_{U|X})$ are the empirical loss and the mutual information resulting from the dataset $\mathcal{D}_n$ and $\Lambda(P_{U|X}, \hat{P}_{Y|U}, P_{\hat{Y}|U})$ is a function that measures the mismatch between the optimal decoder and the empirical one.

This bound shows explicitly the trade-off between the empirical relevance and the empirical complexity. The pairs of relevance and complexity simultaneously achievable is precisely characterized by the IB-problem. Therefore, by designing estimators based on the IB problem, as described in Section 3, one can perform at different regimes of performance, complexity and generalization.

Another interesting connection between learning and the IB-method is the connection of the logarithmic-loss as metric to common performance metrics in learning:

- The logarithmic-loss gives an upper bound on the probability of miss-classification (accuracy):

$$\epsilon_{Y|X}(Q_{\hat{Y}|X}) := 1 - \mathbb{E}_{P_{XY}}[Q_{\hat{Y}|X}] \leq 1 - \exp\left(-\mathbb{E}_{P_{X,Y}}[\ell_{\log}(Y, Q_{\hat{Y}|X})]\right)$$

- The logarithmic-loss is equivalent to maximum likelihood for large n:

$$-\frac{1}{n}\log P_{Y^n|X^n}(y^n|x^n) = -\frac{1}{n}\sum_{i=1}^{n}\log P_{Y|X}(y_i|x_i) \xrightarrow{n\to\infty} \mathbb{E}_{X,Y}[-\log P_{Y|X}(Y|X)]$$

- The true distribution P minimizes the expected logarithmic-loss:

$$P_{Y|X} = \arg\min_{Q_{\hat{Y}|X}} \mathbb{E}_P \log \frac{1}{Q_{\hat{Y}|X}} \quad \text{and} \quad \min_{Q_{\hat{Y}|X}} \mathbb{E}[\ell_{\log}(Y, Q_{\hat{Y}|X})] = H(Y|X)$$

Since for $n \to \infty$ the joint distribution P_{XY} can be perfectly learned, the link between these common criteria allows the use of the IB-problem to derive asymptotic performance bounds, as well as design criteria, in most of the learning scenarios of classification, regression, and inference.

5.4. Representation Learning, Elbo and Autoencoders

The performance of machine learning algorithms depends strongly on the choice of data representation (or features) on which they are applied. For that reason, feature engineering, i.e., the set of all pre-processing operations and transformations applied to data in the aim of making them support effective machine learning, is important. However, because it is both data- and task-dependent, such feature-engineering is labor intensive and highlights one of the major weaknesses of current learning algorithms: their inability to extract discriminative information from the data themselves instead of hand-crafted transformations of them. In fact, although it may sometimes appear useful to deploy feature engineering in order to take advantage of human know-how and prior domain knowledge, it is highly desirable to make learning algorithms less dependent on feature engineering to make progress towards true artificial intelligence.

Representation learning is a sub-field of learning theory that aims at learning representations of the data that make it easier to extract useful information, possibly without recourse to any feature engineering. That is, the goal is to identify and disentangle the underlying explanatory factors that are hidden in the observed data. In the case of probabilistic models, a good representation is one that captures the posterior distribution of the underlying explanatory factors for the observed input. For related works, the reader may refer, e.g., to the proceedings of the International Conference on Learning Representations (ICLR), see https://iclr.cc/.

The use of the Shannon's mutual information as a measure of similarity is particularly suitable for the purpose of learning a good representation of data [81]. In particular, a popular approach to representation learning are autoencoders, in which neural networks are designed for the task of representation learning. Specifically, we design a neural network architecture such that we impose a bottleneck in the network that forces a compressed knowledge representation of the original input, by optimizing the Evidence Lower Bound (ELBO), given as

$$\mathcal{L}^{\text{ELBO}}(\theta,\phi,\varphi) := \frac{1}{N}\sum_{i=1}^{N}\Big[\log Q_\phi(x_i|u_i) - D_{\text{KL}}(P_\theta(U_i|x_i)\|Q_\varphi(U_i))\Big]. \tag{62}$$

over the neural network parameters θ,ϕ,φ. Note that this is precisely the variational-IB cost in Equation (32) for $\beta = 1$ and $Y = X$, i.e., the IB variational bound when particularized to distributions whose parameters are determined by neural networks. In addition, note that the architecture shown in Figure 3 is the classical neural network architecture for autoencoders, and that is coincides with the variational IB solution resulting from the optimization of the IB-problem in Section 3.3.1. In addition, note that Equation (32) provides an operational meaning to the β-VAE cost [82], as a criterion to design estimators on the relevance–complexity plane for different β values, since the β-VAE cost is given as

$$\mathcal{L}^{\beta\text{-VAE}}(\theta,\phi,\varphi) := \frac{1}{N}\sum_{i=1}^{N}\Big[\log Q_\phi(x_i|u_i) - \beta D_{\text{KL}}(P_\theta(U_i|x_i)\|Q_\varphi(U_i))\Big], \tag{63}$$

which coincides with the empirical version of the variational bound found in Equation (32).

5.5. Robustness to Adversarial Attacks

Recent advances in deep learning has allowed the design of high accuracy neural networks. However, it has been observed that the high accuracy of trained neural networks may be compromised under nearly imperceptible changes in the inputs [83–85]. The information bottleneck has also found applications in providing methods to improve robustness to adversarial attacks when training models. In particular, the use of the variational IB method of Alemi et al. [48] showed the advantages of the resulting neural network for classification in terms of robustness to adversarial attacks. Recently, alternatives strategies for extracting features in supervised learning are proposed in [86] to construct classifiers robust to small perturbations in the input space. Robustness is measured in terms of the (statistical)-Fisher information, given for two random variables (Y, Z) as

$$\Phi(Z|Y) = \mathbb{E}_{Y,Z}\left|\frac{\partial}{\partial y}\log p(Z|Y)\right|^2. \tag{64}$$

The method in [86] builds upon the idea of the information bottleneck by introducing an additional penalty term that encourages the Fisher information in Equation (64) of the extracted features to be small, when parametrized by the inputs. For this problem, under jointly Gaussian vector sources

(X, Y), the optimal representation is also shown to be Gaussian, in line with the results in Section 6.2.1 for the IB without robustness penalty. For general source distributions, a variational method is proposed similar to the variational IB method in Section 3.3.1. The problem shows connections with the I-MMSE [87], de Brujin identity [88,89], Cramér–Rao inequality [90], and Fano's inequality [90].

6. Extensions: Distributed Information Bottleneck

Consider now a generalization of the IB problem in which the prediction is to be performed in a distributed manner. The model is shown in Figure 7. Here, the prediction of the target variable $Y \in \mathcal{Y}$ is to be performed on the basis of samples of statistically correlated random variables $(X_1, \cdots, X_K)$ that are observed each at a distinct predictor. Throughout, we assume that the following Markov Chain holds for all $k \in \mathcal{K} := \{1, \cdots, K\}$,

$$X_k \hspace{0.5em}\multimap\hspace{0.5em} Y \hspace{0.5em}\multimap\hspace{0.5em} X_{\mathcal{K}/k}. \tag{65}$$

The variable Y is a target variable and we seek to characterize how accurately it can be predicted from a measurable random vector $(X_1, \cdots, X_K)$ when the components of this vector are processed separately, each by a distinct encoder.

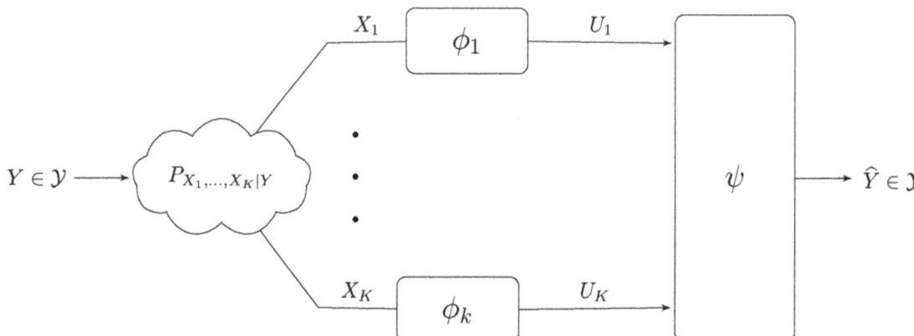

Figure 7. A model for distributed, e.g., multi-view, learning.

6.1. The Relevance–Complexity Region

The distributed IB problem of Figure 7 is studied in [91,92] from information-theoretic grounds. For both discrete memoryless (DM) and memoryless vector Gaussian models, the authors established fundamental limits of learning in terms of optimal trade-offs between relevance and complexity, leveraging on the connection between the IB-problem and source coding. The following theorem states the result for the case of discrete memoryless sources.

Theorem 1 ([91,92]). *The relevance–complexity region $\mathcal{IR}_{\mathrm{DIB}}$ of the distributed learning problem is given by the union of all non-negative tuples $(\Delta, R_1, \ldots, R_K) \in \mathbb{R}_+^{K+1}$ that satisfy*

$$\Delta \leq \sum_{k \in \mathcal{S}} [R_k - I(X_k; U_k | Y, T)] + I(Y; U_{\mathcal{S}^c} | T), \quad \forall \mathcal{S} \subseteq \mathcal{K} \tag{66}$$

for some joint distribution of the form $P_T P_Y \prod_{k=1}^K P_{X_k|Y} \prod_{k=1}^K P_{U_k|X_k,T}$.

Proof. The proof of Theorem 1 can be found in Section 7.1 of [92] and is reproduced in Section 8.1 for completeness. $\square$

For a given joint data distribution $P_{X_{\mathcal{K}},Y}$, Theorem 1 extends the single encoder IB principle of Tishby in Equation (3) to the distributed learning model with K encoders, which we denote by Distributed Information Bottleneck (DIB) problem. The result characterizes the optimal relevance–complexity trade-off as a region of achievable tuples $(\Delta, R_1, \ldots, R_K)$ in terms of a distributed representation learning problem involving the optimization over K conditional pmfs $P_{U_k|X_k,T}$ and a pmf

P_T. The pmfs $P_{U_k|X_k,T}$ correspond to stochastic encodings of the observation X_k to a latent variable, or representation, U_k which captures the relevant information of Y in observation X_k. Variable T corresponds to a time-sharing among different encoding mappings (see, e.g., [51]). For such encoders, the optimal decoder is implicitly given by the conditional pmf of Y from $U_1, \ldots, U_K$, i.e., $P_{Y|U_K,T}$.

The characterization of the relevance–complexity region can be used to derive a cost function for the D-IB similarly to the IB-Lagrangian in Equation (3). For simplicity, let us consider the problem of maximizing the relevance under a sum-complexity constraint. Let $R_{\text{sum}} = \sum_{k=1}^{K} R_k$ and

$$\mathcal{RI}_{\text{DIB}}^{\text{sum}} := \left\{ (\Delta, R_{\text{sum}}) \in \mathbb{R}_+^2 : \exists (R_1, \ldots, R_K) \in \mathbb{R}_+^K \text{ s.t. } \sum_{k=1}^{K} R_k = R_{\text{sum}} \text{ and } (\Delta, R_1, \ldots, R_K) \in \mathcal{RI}_{\text{DIB}} \right\}.$$

We define the DIB-Lagrangian (under sum-rate) as

$$\mathcal{L}_s(\mathbf{P}) := -H(Y|U_K) - s \sum_{k=1}^{K} [H(Y|U_k) + I(X_k; U_k)]. \tag{67}$$

The optimization of Equation (67) over the encoders $P_{U_k|X_k,T}$ allows obtaining mappings that perform on the boundary of the relevance–sum complexity region $\mathcal{RI}_{\text{DIB}}^{\text{sum}}$. To see that, note that it is easy to see that the relevance–sum complexity region $\mathcal{RI}_{\text{DIB}}^{\text{sum}}$ is composed of all the pairs $(\Delta, R_{\text{sum}}) \in \mathbb{R}_+^2$ for which $\Delta \leq \Delta(R_{\text{sum}}, P_{X_K,Y})$, with

$$\Delta(R_{\text{sum}}, P_{X_K,Y}) = \max_{\mathbf{P}} \min \left\{ I(Y; U_K), R_{\text{sum}} - \sum_{k=1}^{K} I(X_k; U_k|Y) \right\}, \tag{68}$$

where the maximization is over joint distributions that factorize as $P_Y \prod_{k=1}^{K} P_{X_k|Y} \prod_{k=1}^{K} P_{U_k|X_k}$. The pairs (Δ, R_{sum}) that lie on the boundary of $\mathcal{RI}_{\text{DIB}}^{\text{sum}}$ can be characterized as given in the following proposition.

Proposition 1. *For every pair $(\Delta, R_{\text{sum}}) \in \mathbb{R}_+^2$ that lies on the boundary of the region $\mathcal{RI}_{\text{DIB}}^{\text{sum}}$, there exists a parameter $s \geq 0$ such that $(\Delta, R_{\text{sum}}) = (\Delta_s, R_s)$, with*

$$\Delta_s = \frac{1}{(1+s)} \left[(1+sK)H(Y) + sR_s + \max_{\mathbf{P}} \mathcal{L}_s(\mathbf{P}) \right], \tag{69}$$

$$R_s = I(Y; U_K^*) + \sum_{k=1}^{K} [I(X_k; U_k^*) - I(Y; U_k^*)], \tag{70}$$

where $\mathbf{P}^*$ is the set of conditional pmfs $\mathbf{P} = \{P_{U_1|X_1}, \cdots, P_{U_K|X_K}\}$ that maximize the cost function in Equation (67).

Proof. The proof of Proposition 1 can be found in Section 7.3 of [92] and is reproduced here in Section 8.2 for completeness. $\square$

The optimization of the distributed IB cost function in Equation (67) generalizes the centralized Tishby's information bottleneck formulation in Equation (3) to the distributed learning setting. Note that for $K = 1$ the optimization in Equation (69) reduces to the single encoder cost in Equation (3) with a multiplier $s/(1+s)$.

6.2. Solutions to the Distributed Information Bottleneck

The methods described in Section 3 can be extended to the distributed information bottleneck case in order to find the mappings $P_{U_1|X_1,T}, \cdots, P_{U_K|X_K,T}$ in different scenarios.

6.2.1. Vector Gaussian Model

In this section, we show that for the jointly vector Gaussian data model it is enough to restrict to Gaussian auxiliaries $(\mathbf{U}_1, \cdots, \mathbf{U}_K)$ in order to exhaust the entire relevance–complexity region. In addition, we provide an explicit analytical expression of this region. Let $(\mathbf{X}_1, \ldots, \mathbf{X}_K, \mathbf{Y})$ be a jointly vector Gaussian vector that satisfies the Markov Chain in Equation (83). Without loss of generality, let the target variable be a complex-valued, zero-mean multivariate Gaussian $\mathbf{Y} \in \mathbb{C}^{n_y}$ with covariance matrix $\boldsymbol{\Sigma}_{\mathbf{y}}$, i.e., $\mathbf{Y} \sim \mathcal{CN}(\mathbf{y}; \mathbf{0}, \boldsymbol{\Sigma}_{\mathbf{y}})$, and $\mathbf{X}_k \in \mathbb{C}^{n_k}$ given by

$$\mathbf{X}_k = \mathbf{H}_k \mathbf{Y} + \mathbf{N}_k, \tag{71}$$

where $\mathbf{H}_k \in \mathbb{C}^{n_k \times n_y}$ models the linear model connecting $\mathbf{Y}$ to the observation at encoder k and $\mathbf{N}_k \in \mathbb{C}^{n_k}$ is the noise vector at encoder k, assumed to be Gaussian with zero-mean, covariance matrix $\boldsymbol{\Sigma}_k$, and independent from all other noises and $\mathbf{Y}$.

For the vector Gaussian model Equation (71), the result of Theorem 1, which can be extended to continuous sources using standard techniques, characterizes the relevance–complexity region of this model. The following theorem characterizes the relevance–complexity region, which we denote hereafter as $\mathcal{RI}_{\mathrm{DIB}}^{\mathrm{G}}$. The theorem also shows that in order to exhaust this region it is enough to restrict to no time sharing, i.e., $T = \varnothing$ and multivariate Gaussian test channels

$$\mathbf{U}_k = \mathbf{A}_k \mathbf{X}_k + \mathbf{Z}_k \sim \mathcal{CN}(\mathbf{u}_k; \mathbf{A}_k \mathbf{X}_k, \boldsymbol{\Sigma}_{z,k}), \tag{72}$$

where $\mathbf{A}_k \in \mathbb{C}^{n_k \times n_k}$ projects $\mathbf{X}_k$ and $\mathbf{Z}_k$ is a zero-mean Gaussian noise with covariance $\boldsymbol{\Sigma}_{z,k}$.

Theorem 2. *For the vector Gaussian data model, the relevance–complexity region $\mathcal{RI}_{\mathrm{DIB}}^{\mathrm{G}}$ is given by the union of all tuples $(\Delta, R_1, \ldots, R_L)$ that satisfy*

$$\Delta \leq \sum_{k \in \mathcal{S}} \left(R_k + \log \left| \mathbf{I} - \boldsymbol{\Sigma}_k^{1/2} \boldsymbol{\Omega}_k \boldsymbol{\Sigma}_k^{1/2} \right| \right) + \log \left| \mathbf{I} + \sum_{k \in \mathcal{S}^c} \boldsymbol{\Sigma}_{\mathbf{y}}^{1/2} \mathbf{H}_k^\dagger \boldsymbol{\Omega}_k \mathbf{H}_k \boldsymbol{\Sigma}_{\mathbf{y}}^{1/2} \right|, \quad \forall \mathcal{S} \subseteq \mathcal{K},$$

for some matrices $\mathbf{0} \preceq \boldsymbol{\Omega}_k \preceq \boldsymbol{\Sigma}_k^{-1}$.

Proof. The proof of Theorem 2 can be found in Section 7.5 of [92] and is reproduced here in Section 8.4 for completeness. $\quad \square$

Theorem 2 extends the result of [54,93] on the relevance–complexity trade-off characterization of the single-encoder IB problem for jointly Gaussian sources to K encoders. The theorem also shows that the optimal test channels $P_{U_k | X_k}$ are multivariate Gaussian, as given by Equation (72).

Consider the following symmetric distributed scalar Gaussian setting, in which $Y \sim \mathcal{N}(0, 1)$ and

$$X_1 = \sqrt{\mathrm{snr}}\, Y + N_1 \tag{73a}$$
$$X_2 = \sqrt{\mathrm{snr}}\, Y + N_2 \tag{73b}$$

where N_1 and N_2 are standard Gaussian with zero-mean and unit variance, both independent of Y. In this case, for $I(U_1; X_1) = R$ and $I(U; X_2) = R$, the optimal relevance is

$$\Delta^\star(R, \mathrm{snr}) = \frac{1}{2} \log \left(1 + 2\,\mathrm{snr} \exp(-4R) \left(\exp(4R) + \mathrm{snr} - \sqrt{\mathrm{snr}^2 + (1 + 2\,\mathrm{snr}) \exp(4R)} \right) \right). \tag{74}$$

An easy upper bound on the relevance can be obtained by assuming that X_1 and X_2 are encoded jointly at rate $2R$, to get

$$\Delta_{\mathrm{ub}}(R, \mathrm{snr}) = \frac{1}{2} \log(1 + 2\,\mathrm{snr}) - \frac{1}{2} \log \left(1 + 2\,\mathrm{snr} \exp(-4R) \right). \tag{75}$$

The reader may notice that, if X_1 and X_2 are encoded independently, an achievable relevance level is given by

$$\Delta_{\text{lb}}(R, \text{snr}) = \frac{1}{2}\log(1 + 2\text{snr} - \text{snr}\exp(-2R)) - \frac{1}{2}\log\left(1 + \text{snr}\exp(-2R)\right). \tag{76}$$

6.3. Solutions for Generic Distributions

Next, we present how the distributed information bottleneck can be solved for generic distributions. Similar to the case of single encoder IB-problem, the solutions are based on a variational bound on the DIB-Lagrangian. For simplicity, we look at the D-IB under sum-rate constraint [92].

6.4. A Variational Bound

The optimization of Equation (67) generally requires computing marginal distributions that involve the descriptions $U_1, \cdots, U_K$, which might not be possible in practice. In what follows, we derive a variational lower bound on $\mathcal{L}_s(\mathbf{P})$ on the DIB cost function in terms of families of stochastic mappings $Q_{Y|U_1,\ldots,U_K}$ (a decoder), $\{Q_{Y|U_k}\}_{k=1}^K$ and priors $\{Q_{U_k}\}_{k=1}^K$. For the simplicity of the notation, we let

$$\mathbf{Q} := \{Q_{Y|U_1,\ldots,U_K}, Q_{Y|U_1}, \ldots, Q_{Y|U_K}, Q_{U_1}, \ldots, Q_{U_K}\}. \tag{77}$$

The variational D-IB cost for the DIB-problem is given by

$$\mathcal{L}_s^{\text{VB}}(\mathbf{P}, \mathbf{Q}) := \underbrace{\mathbb{E}[\log Q_{Y|U_K}(Y|U_K)]}_{\text{av. logarithmic-loss}} + s\underbrace{\sum_{k=1}^K \left(\mathbb{E}[\log Q_{Y|U_k}(Y|U_k)] - D_{\text{KL}}(P_{U_k|X_k}\|Q_{U_k})\right)}_{\text{regularizer}}. \tag{78}$$

Lemma 1. *For fixed* $\mathbf{P}$*, we have*

$$\mathcal{L}_s(\mathbf{P}) \geq \mathcal{L}_s^{\text{VB}}(\mathbf{P}, \mathbf{Q}), \qquad \text{for all pmfs } \mathbf{Q}. \tag{79}$$

In addition, there exists a unique $\mathbf{Q}$ *that achieves the maximum* $\max_{\mathbf{Q}} \mathcal{L}_s^{\text{VB}}(\mathbf{P}, \mathbf{Q}) = \mathcal{L}_s(\mathbf{P})$*, and is given by,* $\forall k \in \mathcal{K}$,

$$Q_{U_k}^* = P_{U_k} \tag{80a}$$

$$Q_{Y|U_k}^* = P_{Y|U_k} \tag{80b}$$

$$Q_{Y|U_1,\ldots,U_k}^* = P_{Y|U_1,\ldots,U_K}, \tag{80c}$$

where the marginals P_{U_k} *and the conditional marginals* $P_{Y|U_k}$ *and* $P_{Y|U_1,\ldots,U_K}$ *are computed from* $\mathbf{P}$*.*

Proof. The proof of Lemma 1 can be found in Section 7.4 of [92] and is reproduced here in Section 8.3 for completeness. $\square$

Then, the optimization in Equation (69) can be written in terms of the variational DIB cost function as follows,

$$\max_{\mathbf{P}} \mathcal{L}_s(\mathbf{P}) = \max_{\mathbf{P}} \max_{\mathbf{Q}} \mathcal{L}_s^{\text{VB}}(\mathbf{P}, \mathbf{Q}). \tag{81}$$

The variational DIB cost in Equation (78) is a generalization to distributed learning with K-encoders of the evidence lower bound (ELBO) of the target variable Y given the representations $U_1, \cdots, U_K$ [15]. If $Y = (X_1, \ldots, X_K)$, the bound generalizes the ELBO used for VAEs to the setting of $K \geq 2$ encoders. In addition, note that Equation (78) also generalizes and provides an operational meaning to the β-VAE cost [82] with $\beta = s/(1+s)$, as a criteria to design estimators on the relevance–complexity plane for different β values.

6.5. Known Memoryless Distributions

When the data model is discrete and the joint distribution $P_{X,Y}$ is known, the DIB problem can be solved by using an iterative method that optimizes the variational IB cost function in Equation (81) alternating over the distributions $\mathbf{P}, \mathbf{Q}$. The optimal encoders and decoders of the D-IB under sum-rate constraint satisfy the following self consistent equations,

$$p(u_k|y_k) = \frac{p(u_k)}{Z(\beta, u_k)} \exp\left(-\psi_s(u_k, y_k)\right),$$

$$p(x|u_k) = \sum_{y_k \in \mathcal{Y}_k} p(y_k|u_k)p(x|y_k)$$

$$p(x|u_1, \ldots, u_K) = \sum_{y_K \in \mathcal{Y}_K} p(y_K)p(u_K|y_K)p(x|y_K)/p(u_K)$$

where $\psi_s(u_k, y_k) := D_{\mathrm{KL}}(P_{X|y_k}\|Q_{X|u_k}) + \frac{1}{s}E_{U_{K\setminus k}|y_k}[D_{\mathrm{KL}}(P_{X|U_{K\setminus k},y_k}\|Q_{X|U_{K\setminus k},u_k})]$.

Alternating iterations of these equations converge to a solution for any initial $p(u_k|x_k)$, similarly to a Blahut–Arimoto algorithm and the EM.

6.5.1. Distributed Variational IB

When the data distribution is unknown and only data samples are available, the variational DIB cost in Equation (81) can be optimized following similar steps as for the variational IB in Section 3.3.1 by parameterizing the encoding and decoding distributions $\mathbf{P}, \mathbf{Q}$ using a family of distributions whose parameters are determined by DNNs. This allows us to formulate Equation (81) in terms of the DNN parameters, i.e., its weights, and optimize it by using the reparameterization trick [15], Monte Carlo sampling, and stochastic gradient descent (SGD)-type algorithms.

Considering encoders and decoders $\mathbf{P}, \mathbf{Q}$ parameterized by DNN parameters θ, ϕ, φ, the DIB cost in Equation (81) can be optimized by considering the following empirical Monte Carlo approximation:

$$\max_{\theta, \phi, \varphi} \frac{1}{n} \sum_{i=1}^{n} \left[\log Q_{\phi_K}(y_i|u_{1,i,j}, \ldots, u_{K,i,j}) + s \sum_{k=1}^{K}\left(\log Q_{\phi_k}(y_i|u_{k,i,j}) - D_{\mathrm{KL}}(P_{\theta_k}(U_{k,i}|x_{k,i})\|Q_{\varphi_k}(U_{k,i}))\right)\right], \quad (82)$$

where $u_{k,i,j} = g_{\phi_k}(x_{k,i}, z_{k,j})$ are samples obtained from the reparametrization trick by sampling from K random variables P_{Z_k}. The details of the method can be found in [92]. The resulting architecture is shown in Figure 8. This architecture generalizes that from autoencoders to the distributed setup with K encoders.

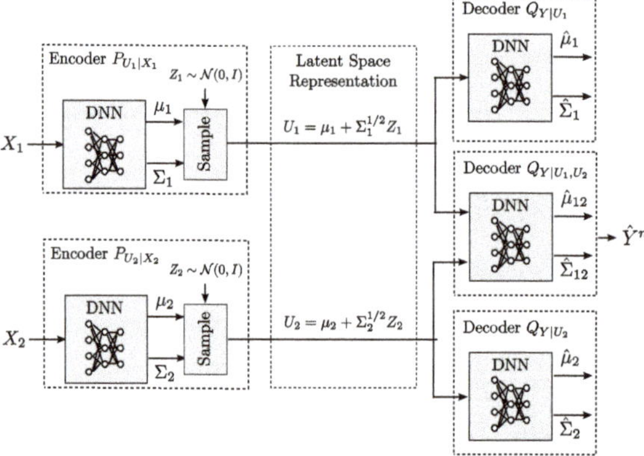

Figure 8. Example parameterization of the Distributed Variational Information Bottleneck method using neural networks.

6.6. Connections to Coding Problems and Learning

Similar to the point-to-point IB-problem, the distributed IB problem also has abundant connections with (asymptotic) coding and learning problems.

6.6.1. Distributed Source Coding under Logarithmic Loss

Key element to the proof of the converse part of Theorem 3 is the connection with the Chief Executive Officer (CEO) source coding problem. For the case of $K \geq 2$ encoders, while the characterization of the optimal rate-distortion region of this problem for general distortion measures has eluded the information theory for now more than four decades, a characterization of the optimal region in the case of logarithmic loss distortion measure has been provided recently in [65]. A key step in [65] is that the log-loss distortion measure admits a lower bound in the form of the entropy of the source conditioned on the decoders' input. Leveraging this result, in our converse proof of Theorem 3, we derive a single letter upper bound on the entropy of the channel inputs conditioned on the indices J_K that are sent by the relays, in the absence of knowledge of the codebooks indices F_L. In addition, the rate region of the vector Gaussian CEO problem under logarithmic loss distortion measure has been found recently in [94,95].

6.6.2. Cloud RAN

Consider the discrete memoryless (DM) CRAN model shown in Figure 9. In this model, L users communicate with a common destination or central processor (CP) through K relay nodes, where $L \geq 1$ and $K \geq 1$. Relay node k, $1 \leq k \leq K$, is connected to the CP via an error-free finite-rate fronthaul link of capacity C_k. In what follows, we let $\mathcal{L} := [1:L]$ and $\mathcal{K} := [1:K]$ indicate the set of users and relays, respectively. Similar to Simeone et al. [96], the relay nodes are constrained to operate without knowledge of the users' codebooks and only know a time-sharing sequence Q^n, i.e., a set of time instants at which users switch among different codebooks. The obliviousness of the relay nodes to the actual codebooks of the users is modeled via the notion of *randomized encoding* [97,98]. That is, users or transmitters select their codebooks at random and the relay nodes are *not* informed about the currently selected codebooks, while the CP is given such information.

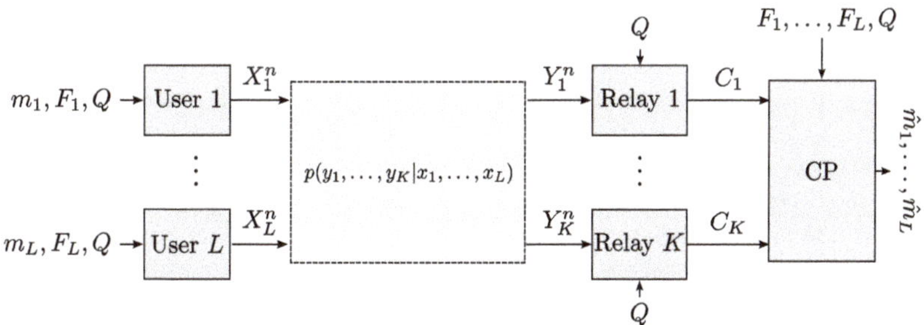

Figure 9. CRAN model with oblivious relaying and time-sharing.

Consider the following class of DM CRANs in which the channel outputs at the relay nodes are independent conditionally on the users' inputs. That is, for all $k \in \mathcal{K}$ and all $i \in [1:n]$,

$$Y_{k,i} \; \multimap \; X_{\mathcal{L},i} \; \multimap \; Y_{\mathcal{K}/k,i} \tag{83}$$

forms a Markov Chain in this order.

The following theorem provides a characterization of the capacity region of this class of DM CRAN problem under oblivious relaying.

Theorem 3 ([22,23]). *For the class of DM CRANs with oblivious relay processing and enabled time-sharing for which Equation (83) holds, the capacity region $\mathcal{C}(C_{\mathcal{K}})$ is given by the union of all rate tuples $(R_1, \ldots, R_L)$ which satisfy*

$$\sum_{t \in \mathcal{T}} R_t \leq \sum_{s \in \mathcal{S}} [C_s - I(Y_s; U_s | X_{\mathcal{L}}, Q)] + I(X_{\mathcal{T}}; U_{\mathcal{S}^c} | X_{\mathcal{T}^c}, Q),$$

for all non-empty subsets $\mathcal{T} \subseteq \mathcal{L}$ and all $\mathcal{S} \subseteq \mathcal{K}$, for some joint measure of the form

$$p(q) \prod_{l=1}^{L} p(x_l | q) \prod_{k=1}^{K} p(y_k | x_{\mathcal{L}}) \prod_{k=1}^{K} p(u_k | y_k, q). \tag{84}$$

The direct part of Theorem 3 can be obtained by a coding scheme in which each relay node compresses its channel output by using Wyner–Ziv binning to exploit the correlation with the channel outputs at the other relays, and forwards the bin index to the CP over its rate-limited link. The CP jointly decodes the compression indices (within the corresponding bins) and the transmitted messages, i.e., Cover-El Gamal compress-and-forward [99] (Theorem 3) with joint decompression and decoding (CF-JD). Alternatively, the rate region of Theorem 3 can also be obtained by a direct application of the noisy network coding (NNC) scheme of [64] (Theorem 1).

The connection between this problem, source coding and the distributed information bottleneck is discussed in [22,23], particularly in the derivation of the converse part of the theorem. Note also the similarity between the resulting capacity region in Theorem 3 and the relevance complexity region of the distributed information bottleneck in Theorem 1, despite the significant differences of the setups.

6.6.3. Distributed Inference, ELBO and Multi-View Learning

In many data analytics problems, data are collected from various sources of information or feature extractors and are intrinsically *heterogeneous*. For example, an image can be identified by its color or texture features and a document may contain text and images. Conventional machine learning approaches concatenate all available data into one big row vector (or matrix) on which a suitable algorithm is then applied. Treating different observations as a single source might cause overfitting and is not physically meaningful because each group of data may have different statistical properties. Alternatively, one may partition the data into groups according to samples homogeneity, and each group of data be regarded as a separate *view*. This paradigm, termed *multi-view learning* [100], has received growing interest, and various algorithms exist, sometimes under references such as *co-training* [101–104], *multiple kernel learning* [104], and *subspace learning* [105]. By using distinct encoder mappings to represent distinct groups of data, and jointly optimizing over all mappings to remove redundancy, multi-view learning offers a degree of flexibility that is not only desirable in practice but is also likely to result in better learning capability. Actually, as shown in [106], local learning algorithms produce fewer errors than global ones. Viewing the problem as that of function approximation, the intuition is that it is usually not easy to find a unique function that holds good predictability properties in the entire data space.

Besides, the distributed learning of Figure 7 clearly finds application in all those scenarios in which learning is performed collaboratively but distinct learners either only access subsets of the entire dataset (e.g., due to physical constraints) or access independent noisy versions of the entire dataset.

In addition, similar to the single encoder case, the distributed IB also finds applications in fundamental performance limits and formulation of cost functions from an operational point of view. One of such examples is the generalization of the commonly used ELBO and given in Equation (62) to the setup with K views or observations, as formulated in Equation (78). Similarly, from the formulation of the DIB problem, a natural generalization of the classical autoencoders emerge, as given in Figure 8.

7. Outlook

A variant of the bottleneck problem in which the encoder's output is constrained in terms of its entropy, rather than its mutual information with the encoder's input as done originally in [1], was considered in [107]. The solution of this problem turns out to be a deterministic encoder map as opposed to the stochastic encoder map that is optimal under the IB framework of Tishby et al. [1], which results in a reduction of the algorithm's complexity. This idea was then used and extended to the case of available resource (or time) sharing in [108].

In the context of privacy against inference attacks [109], the authors of [75,76] considered a dual of the information bottleneck problem in which $X \in \mathcal{X}$ represents some private data that are correlated with the non-private data $Y \in \mathcal{Y}$. A legitimate receiver (analyst) wishes to infer as much information as possible about the non-private data Y but does not need to infer any information about the private data X. Because X and Y are correlated, sharing the non-private data X with the analyst possibly reveals information about Y. For this reason, there is a trade-off between the amount of information that the user shares about X as measured by the mutual information $I(U; X)$ and the information that he keeps private about Y as measured by the mutual information $I(U; Y)$, where $U = \phi(X)$.

Among interesting problems that are left unaddressed in this paper is that of characterizing optimal input distributions under rate-constrained compression at the relays where, e.g., discrete signaling is already known to sometimes outperform Gaussian signaling for single-user Gaussian CRAN [97]. It is conjectured that the optimal input distribution is discrete. Other issues might relate to extensions to continuous time filtered Gaussian channels, in parallel to the regular bottleneck problem [108], or extensions to settings in which fronthauls may be not available at some radio-units, and that is unknown to the systems. That is, the more radio units are connected to the central unit, the higher is the rate that could be conveyed over the CRAN uplink [110]. Alternatively, one may consider finding the worst-case noise under given input distributions, e.g., Gaussian, and rate-constrained compression at the relays. Furthermore, there are interesting aspects that address processing constraints of continuous waveforms, e.g., sampling at a given rate [111,112] with focus on remote logarithmic distortion [65], which in turn boils down to the distributed bottleneck problem [91,92]. We also mention finite-sample size analysis (i.e., finite block length n, which relates to the literature on finite block length coding in information theory). Finally, it is interesting to observe that the bottleneck problem relates to interesting problem when R is not necessarily scaled with the block length n.

8. Proofs

8.1. Proof of Theorem 1

The proof relies on the equivalence of the studied distributed learning problem with the Chief-Executive Officer (CEO) problem under logarithmic-loss distortion measure, which was studied in [65] (Theorem 10). For the K-encoder CEO problem, let us consider K encoding functions $\phi_k : \mathcal{X}_k \to \mathcal{M}_k^{(n)}$ satisfying $nR_k \geq \log|\phi_k(X_k^n)|$ and a decoding function $\tilde{\psi} : \mathcal{M}_1^{(n)} \times \cdots \times \mathcal{M}_K^{(n)} \to \hat{\mathcal{Y}}^n$, which produces a probabilistic estimate of Y from the outputs of the encoders, i.e., $\hat{\mathcal{Y}}^n$ is the set of distributions on $\mathcal{Y}$. The quality of the estimation is measured in terms of the average log-loss.

Definition 1. *A tuple $(D, R_1, \ldots, R_K)$ is said to be achievable in the K-encoder CEO problem for $P_{X_{\mathcal{K}}, Y}$ for which the Markov Chain in Equation (83) holds, if there exists a length n, encoders ϕ_k for $k \in \mathcal{K}$, and a decoder $\tilde{\psi}$, such that*

$$D \geq \mathbb{E}\left[\frac{1}{n} \log \frac{1}{\hat{P}_{Y^n|J_{\mathcal{K}}}(Y^n|\phi_1(X_1^n), \cdots, \phi_K(X_K^n))}\right], \tag{85}$$

$$R_k \geq \frac{1}{n} \log|\phi_k(X_k^n)| \quad for\ all\ k \in \mathcal{K}. \tag{86}$$

The rate-distortion region $\mathcal{RD}_{\mathrm{CEO}}$ is given by the closure of all achievable tuples $(D, R_1, \ldots, R_K)$.

The following lemma shows that the minimum average logarithmic loss is the conditional entropy of Y given the descriptions. The result is essentially equivalent to [65] (Lemma 1) and it is provided for completeness.

Lemma 2. *Let us consider $P_{X_{\mathcal{K}}, Y}$ and the encoders $J_k = \phi_k(X_k^n)$, $k \in \mathcal{K}$ and the decoder $\hat{Y}^n = \tilde{\psi}(J_{\mathcal{K}})$. Then,*

$$\mathrm{E}[\ell_{\log}(Y^n, \hat{Y}^n)] \geq H(Y^n | J_{\mathcal{K}}), \tag{87}$$

with equality if and only if $\tilde{\psi}(J_{\mathcal{K}}) = \{P_{Y^n | J_{\mathcal{K}}}(y^n | J_{\mathcal{K}})\}_{y^n \in \mathcal{Y}^n}$.

Proof. Let $Z := (J_1, \ldots, J_K)$ be the argument of $\tilde{\psi}$ and $\hat{P}(y^n | z)$ be a distribution on $\mathcal{Y}^n$. We have for $Z = z$:

$$\mathrm{E}[\ell_{\log}(Y^n, \hat{Y}^n) | Z = z] = \sum_{y^n \in \mathcal{Y}^n} P(y^n | z) \log\left(\frac{1}{\hat{P}(y^n | z)}\right) \tag{88}$$

$$= \sum_{y^n \in \mathcal{Y}^n} P(y^n | z) \log\left(\frac{P(y^n | z)}{\hat{P}(y^n | z)}\right) + H(Y^n | Z = z) \tag{89}$$

$$= D_{\mathrm{KL}}(P(y^n | z) \| \hat{P}(y^n | z)) + H(Y^n | Z = z) \tag{90}$$

$$\geq H(Y^n | Z = z), \tag{91}$$

where Equation (91) is due to the non-negativity of the KL divergence and the equality holds if and only if for $\hat{P}(y^n | z) = P(y^n | z)$ where $P(y^n | z) = \Pr\{Y^n = y^n | Z = z\}$ for all z and $y^n \in \mathcal{Y}^n$. Averaging over Z completes the proof. $\square$

Essentially, Lemma 2 states that minimizing the average log-loss is equivalent to maximizing relevance as given by the mutual information $I\left(Y^n; \psi\left(\phi_1(X_1^n), \cdots, \phi_K(X_K^n)\right)\right)$. Formally, the connection between the distributed learning problem under study and the K-encoder CEO problem studied in [65] can be formulated as stated next.

Proposition 2. *A tuple $(\Delta, R_1, \ldots, R_K) \in \mathcal{RI}_{\mathrm{DIB}}$ if and only if $(H(Y) - \Delta, R_1, \ldots, R_K) \in \mathcal{RD}_{\mathrm{CEO}}$.*

Proof. Let the tuple $(\Delta, R_1, \ldots, R_K) \in \mathcal{RI}_{\mathrm{DIB}}$ be achievable for some encoders ϕ_k. It follows by Lemma 2 that, by letting the decoding function $\tilde{\psi}(J_{\mathcal{K}}) = \{P_{Y^n | J_{\mathcal{K}}}(y^n | J_{\mathcal{K}})\}$, we have $\mathrm{E}[\ell_{\log}(Y^n, \hat{Y}^n) | J_{\mathcal{K}}] = H(Y^n | J_{\mathcal{K}})$, and hence $(H(Y) - \Delta, R_1, \ldots, R_K) \in \mathcal{RD}_{\mathrm{CEO}}$.
Conversely, assume the tuple $(D, R_1, \ldots, R_K) \in \mathcal{RD}_{\mathrm{CEO}}$ is achievable. It follows by Lemma 2 that $H(Y) - D \leq H(Y^n) - H(Y^n | J_{\mathcal{K}}) = I(Y^n; J_{\mathcal{K}})$, which implies $(\Delta, R_1, \ldots, R_K) \in \mathcal{RI}_{\mathrm{DIB}}$ with $\Delta = H(Y) - D$. $\square$

The characterization of rate-distortion region $\mathcal{R}_{\mathrm{CEO}}$ has been established recently in [65] (Theorem 10). The proof of the theorem is completed by noting that Proposition 2 implies that the result in [65] (Theorem 10) can be applied to characterize the region $\mathcal{RI}_{\mathrm{DIB}}$, as given in Theorem 1.

8.2. Proof of Proposition 1

Let $\mathbf{P}^*$ be the maximizing in Equation (69). Then,

$$(1 + s)\Delta_s = (1 + sK)H(Y) + sR_s + \mathcal{L}_s(\mathbf{P}^*) \tag{92}$$

$$= (1 + sK)H(Y) + sR_s + \left(-H(Y | U_{\mathcal{K}}^*) - s \sum_{k=1}^{K} [H(Y | U_k^*) + I(X_k; U_k^*)]\right) \tag{93}$$

$$= (1 + sK)H(Y) + sR_s + (-H(Y | U_{\mathcal{K}}^*) - s(R_s - I(Y; U_{\mathcal{K}}^*) + KH(Y))) \tag{94}$$

$$= (1+s)I(Y;U_{\mathcal{K}}^*) \tag{95}$$

$$\leq (1+s)\Delta(R_s, P_{X_{\mathcal{K}},Y}), \tag{96}$$

where Equation (94) is due to the definition of $\mathcal{L}_s(\mathbf{P})$ in Equation (67); Equation (95) holds since $\sum_{k=1}^{K}[I(X_k;U_k^*) + H(Y|U_k^*)] = R_s - I(Y;U_{\mathcal{K}}^*) + KH(Y)$ using Equation (70); and Equation (96) follows by using Equation (68).

Conversely, if $\mathbf{P}^*$ is the solution to the maximization in the function $\Delta(R_{\mathrm{sum}}, P_{X_{\mathcal{K}},Y})$ in Equation (68) such that $\Delta(R_{\mathrm{sum}}, P_{X_{\mathcal{K}},Y}) = \Delta_s$, then $\Delta_s \leq I(Y;U_{\mathcal{K}}^*)$ and $\Delta_s \leq R - \sum_{k=1}^{K} I(X_k;U_k^*|Y)$ and we have, for any $s \geq 0$, that

$$\Delta(R_{\mathrm{sum}}, P_{X_{\mathcal{K}},Y}) = \Delta_s \tag{97}$$

$$\leq \Delta_s - (\Delta_s - I(Y;U_{\mathcal{K}}^*)) - s\left(\Delta_s - R_{\mathrm{sum}} + \sum_{k=1}^{K} I(X_k;U_k^*|Y)\right) \tag{98}$$

$$= I(Y;U_{\mathcal{K}}^*) - s\Delta_s + sR_{\mathrm{sum}} - s\sum_{k=1}^{K} I(X_k;U_k^*|Y) \tag{99}$$

$$= H(Y) - s\Delta_s + sR_{\mathrm{sum}} - H(Y|U_{\mathcal{K}}^*) - s\sum_{k=1}^{K}[I(X_k;U_k^*) + H(Y|U_k^*)] + sKH(Y) \tag{100}$$

$$\leq H(Y) - s\Delta_s + sR_{\mathrm{sum}} + \mathcal{L}_s^* + sKH(Y) \tag{101}$$

$$= H(Y) - s\Delta_s + sR_{\mathrm{sum}} + sKH(Y) - ((1+sK)H(Y) + sR_s - (1+s)\Delta_s) \tag{102}$$

$$= \Delta_s + s(R_{\mathrm{sum}} - R_s), \tag{103}$$

where in Equation (100) we use that $\sum_{k=1}^{K} I(X_k;U_k|Y) = -KH(Y) + \sum_{k=1}^{K} I(X_k;U_k) + H(Y|U_k)$. which follows by using the Markov Chain $U_k \, \text{o-o} \, X_k \, \text{o-o} \, Y \, \text{o-o} \, (X_{\mathcal{K}\setminus k}, U_{\mathcal{K}\setminus k})$; Equation (101) follows since $\mathcal{L}_s^*$ is the maximum over all possible distributions $\mathbf{P}$ (possibly distinct from the $\mathbf{P}^*$ that maximizes $\Delta(R_{\mathrm{sum}}, P_{X_{\mathcal{K}},Y})$); and Equation (102) is due to Equation (69). Finally, Equation (103) is valid for any $R_{\mathrm{sum}} \geq 0$ and $s \geq 0$. Given s, and hence (Δ_s, R_s), letting $R = R_s$ yields $\Delta(R_s, P_{X_{\mathcal{K}},Y}) \leq \Delta_s$. Together with Equation (96), this completes the proof of Proposition 1.

8.3. Proof of Lemma 1

Let, for a given random variable Z and $z \in \mathcal{Z}$, a stochastic mapping $Q_{Y|Z}(\cdot|z)$ be given. It is easy to see that

$$H(Y|Z) = \mathbb{E}[-\log Q_{Y|Z}(Y|Z)] - D_{\mathrm{KL}}(P_{Y|Z}\|Q_{Y|Z}). \tag{104}$$

In addition, we have

$$I(X_k;U_k) = H(U_k) - H(U_k|X_k) \tag{105}$$

$$= D_{\mathrm{KL}}(P_{U_k|X_k}\|Q_{U_k}) - D_{\mathrm{KL}}(P_{U_k}\|Q_{U_k}). \tag{106}$$

Substituting it into Equation (67), we get

$$\mathcal{L}_s(\mathbf{P}) = \mathcal{L}_s^{\mathrm{VB}}(\mathbf{P},\mathbf{Q}) + D_{\mathrm{KL}}(P_{Y|U_{\mathcal{K}}}\|Q_{Y|U_{\mathcal{K}}}) + s\sum_{k=1}^{K}(D_{\mathrm{KL}}(P_{Y|U_k}\|Q_{Y|U_k}) + D_{\mathrm{KL}}(P_{U_k}\|Q_{U_k})) \tag{107}$$

$$\geq \mathcal{L}_s^{\mathrm{VB}}(\mathbf{P},\mathbf{Q}), \tag{108}$$

where Equation (108) follows by the non-negativity of relative entropy. In addition, note that the inequality in Equation (108) holds with equality iff $\mathbf{Q}^*$ is given by Equation (80).

8.4. Proof of Theorem 2

The proof of Theorem 2 relies on deriving an outer bound on the relevance–complexity region, as given by Equation (66), and showing that it is achievable with Gaussian pmfs and without time-sharing. In doing so, we use the technique of [89] (Theorem 8), which relies on the de Bruijn identity and the properties of Fisher information and MMSE.

Lemma 3 ([88,89]). *Let* $(\mathbf{X}, \mathbf{Y})$ *be a pair of random vectors with pmf* $p(\mathbf{x}, \mathbf{y})$. *We have*

$$\log|(\pi e)\mathbf{J}^{-1}(\mathbf{X}|\mathbf{Y})| \le h(\mathbf{X}|\mathbf{Y}) \le \log|(\pi e)\mathrm{mmse}(\mathbf{X}|\mathbf{Y})|, \tag{109}$$

where the conditional Fischer information matrix is defined as

$$\mathbf{J}(\mathbf{X}|\mathbf{Y}) := \mathrm{E}[\nabla \log p(\mathbf{X}|\mathbf{Y})\nabla \log p(\mathbf{X}|\mathbf{Y})^\dagger] \tag{110}$$

and the minimum mean square error (MMSE) matrix is

$$\mathrm{mmse}(\mathbf{X}|\mathbf{Y}) := \mathrm{E}[(\mathbf{X} - \mathrm{E}[\mathbf{X}|\mathbf{Y}])(\mathbf{X} - \mathrm{E}[\mathbf{X}|\mathbf{Y}])^\dagger]. \tag{111}$$

For $t \in \mathcal{T}$ and fixed $\prod_{k=1}^{K} p(\mathbf{u}_k|\mathbf{x}_k, t)$, choose $\mathbf{\Omega}_{k,t}$, $k = 1, \ldots, K$ satisfying $0 \le \mathbf{\Omega}_{k,t} \le \mathbf{\Sigma}_k^{-1}$ such that

$$\mathrm{mmse}(\mathbf{Y}_k|\mathbf{X}, \mathbf{U}_{k,t}, t) = \mathbf{\Sigma}_k - \mathbf{\Sigma}_k\mathbf{\Omega}_{k,t}\mathbf{\Sigma}_k. \tag{112}$$

Note that such $\mathbf{\Omega}_{k,t}$ exists since $0 \le \mathrm{mmse}(\mathbf{X}_k|\mathbf{Y}, \mathbf{U}_{k,t}, t) \le \mathbf{\Sigma}_k^{-1}$, for all $t \in \mathcal{T}$, and $k \in \mathcal{K}$. Using Equation (66), we get

$$I(\mathbf{X}_k; \mathbf{U}_k|\mathbf{Y}, t) \ge \log|\mathbf{\Sigma}_k| - \log|\mathrm{mmse}(\mathbf{X}_k|\mathbf{Y}, \mathbf{U}_{k,t}, t)|$$
$$= -\log|\mathbf{I} - \mathbf{\Sigma}_k^{1/2}\mathbf{\Omega}_{k,t}\mathbf{\Sigma}_k^{1/2}|, \tag{113}$$

where the inequality is due to Lemma 3, and Equation (113) is due to Equation (112). In addition, we have

$$I(\mathbf{Y}; \mathbf{U}_{\mathcal{S}^c, t}|t) \le \log|\mathbf{\Sigma_y}| - \log|\mathbf{J}^{-1}(\mathbf{Y}|\mathbf{U}_{\mathcal{S}^c, t}, t)| \tag{114}$$

$$= \log\left|\sum_{k \in \mathcal{S}^c} \mathbf{\Sigma_y}^{1/2}\mathbf{H}_k^\dagger\mathbf{\Omega}_{k,t}\mathbf{H}_k\mathbf{\Sigma_y}^{1/2} + \mathbf{I}\right|, \tag{115}$$

where Equation (114) is due to Lemma 3 and Equation (115) is due to to the following equality, which relates the MMSE matrix in Equation (112) and the Fisher information, the proof of which follows,

$$\mathbf{J}(\mathbf{Y}|\mathbf{U}_{\mathcal{S}^c, t}, t) = \sum_{k \in \mathcal{S}^c} \mathbf{H}_k^\dagger\mathbf{\Omega}_{k,t}\mathbf{H}_k + \mathbf{\Sigma_y}^{-1}. \tag{116}$$

To show Equation (116), we use de Brujin identity to relate the Fisher information with the MMSE as given in the following lemma, the proof of which can be found in [89].

Lemma 4. *Let* $(\mathbf{V}_1, \mathbf{V}_2)$ *be a random vector with finite second moments and* $\mathbf{N} \sim \mathcal{CN}(0, \mathbf{\Sigma}_N)$ *independent of* $(\mathbf{V}_1, \mathbf{V}_2)$. *Then,*

$$\mathrm{mmse}(\mathbf{V}_2|\mathbf{V}_1, \mathbf{V}_2 + \mathbf{N}) = \mathbf{\Sigma}_N - \mathbf{\Sigma}_N\mathbf{J}(\mathbf{V}_2 + \mathbf{N}|\mathbf{V}_1)\mathbf{\Sigma}_N. \tag{117}$$

From the MMSE of Gaussian random vectors [51],

$$\mathbf{Y} = \mathrm{E}[\mathbf{Y}|\mathbf{X}_{\mathcal{S}^c}] + \mathbf{Z}_{\mathcal{S}^c} = \sum_{k \in \mathcal{S}^c} \mathbf{G}_k\mathbf{X}_k + \mathbf{Z}_{\mathcal{S}^c}, \tag{118}$$

where $\mathbf{G}_k = \mathbf{\Sigma}_{\mathbf{y}|\mathbf{x}_{S^c}} \mathbf{H}_k^\dagger \mathbf{\Sigma}_k^{-1}$ and $\mathbf{Z}_{S^c} \sim \mathcal{CN}(\mathbf{0}, \mathbf{\Sigma}_{\mathbf{y}|\mathbf{x}_{S^c}})$, and

$$\mathbf{\Sigma}_{\mathbf{y}|\mathbf{x}_{S^c}}^{-1} = \mathbf{\Sigma}_{\mathbf{y}}^{-1} + \sum_{k \in S^c} \mathbf{H}_k^\dagger \mathbf{\Sigma}_k^{-1} \mathbf{H}_k. \tag{119}$$

Note that $\mathbf{Z}_{S^c}$ is independent of $\mathbf{Y}_{S^c}$ due to the orthogonality principle of the MMSE and its Gaussian distribution. Hence, it is also independent of $\mathbf{U}_{S^c, q}$.
Thus, we have

$$\mathrm{mmse}\left(\sum_{k \in S^c} \mathbf{G}_k \mathbf{X}_k \Big| \mathbf{Y}, \mathbf{U}_{S^c, t}, t\right) = \sum_{k \in S^c} \mathbf{G}_k \mathrm{mmse}\left(\mathbf{X}_k | \mathbf{Y}, \mathbf{U}_{S^c, t}, t\right) \mathbf{G}_k^\dagger \tag{120}$$

$$= \mathbf{\Sigma}_{\mathbf{y}|\mathbf{x}_{S^c}} \sum_{k \in S^c} \mathbf{H}_k^\dagger \left(\mathbf{\Sigma}_k^{-1} - \mathbf{\Omega}_k\right) \mathbf{H}_k \mathbf{\Sigma}_{\mathbf{y}|\mathbf{x}_{S^c}}, \tag{121}$$

where Equation (120) follows since the cross terms are zero due to the Markov Chain $(\mathbf{U}_{k,t}, \mathbf{X}_k) \multimap \mathbf{Y} \multimap (\mathbf{U}_{\mathcal{K}/k,t}, \mathbf{X}_{\mathcal{K}/k})$ (see Appendix V of [89]); and Equation (121) follows due to Equation (112) and $\mathbf{G}_k$.
Finally, we have

$$\mathbf{J}(\mathbf{Y}|\mathbf{U}_{S^c, t}, t) = \mathbf{\Sigma}_{\mathbf{y}|\mathbf{x}_{S^c}}^{-1} - \mathbf{\Sigma}_{\mathbf{y}|\mathbf{x}_{S^c}}^{-1} \mathrm{mmse}\left(\sum_{k \in S^c} \mathbf{G}_k \mathbf{X}_k \Big| \mathbf{Y}, \mathbf{U}_{S^c, t}, t\right) \mathbf{\Sigma}_{\mathbf{y}|\mathbf{x}_{S^c}}^{-1} \tag{122}$$

$$= \mathbf{\Sigma}_{\mathbf{y}|\mathbf{x}_{S^c}}^{-1} - \sum_{k \in S^c} \mathbf{H}_k^\dagger \left(\mathbf{\Sigma}_k^{-1} - \mathbf{\Omega}_{k,t}\right) \mathbf{H}_k \tag{123}$$

$$= \mathbf{\Sigma}_{\mathbf{y}}^{-1} + \sum_{k \in S^c} \mathbf{H}_k^\dagger \mathbf{\Omega}_{k,t} \mathbf{H}_k, \tag{124}$$

where Equation (122) is due to Lemma 4; Equation (123) is due to Equation (121); and Equation (124) follows due to Equation (119).

Then, averaging over the time sharing random variable T and letting $\bar{\mathbf{\Omega}}_k := \sum_{t \in \mathcal{T}} p(t) \mathbf{\Omega}_{k,t}$, we get, using Equation (113),

$$I(\mathbf{X}_k; \mathbf{U}_k | \mathbf{Y}, T) \geq - \sum_{t \in \mathcal{T}} p(t) \log \left|\mathbf{I} - \mathbf{\Sigma}_k^{1/2} \mathbf{\Omega}_{k,t} \mathbf{\Sigma}_k^{1/2}\right|$$

$$\geq - \log \left|\mathbf{I} - \mathbf{\Sigma}_k^{1/2} \bar{\mathbf{\Omega}}_k \mathbf{\Sigma}_k^{1/2}\right|, \tag{125}$$

where Equation (125) follows from the concavity of the log-det function and Jensen's inequality.
Similarly, using Equation (115) and Jensen's Inequality, we have

$$I(\mathbf{Y}; \mathbf{U}_{S^c} | T) \leq \log \left|\sum_{k \in S^c} \mathbf{\Sigma}_{\mathbf{y}}^{1/2} \mathbf{H}_k^\dagger \bar{\mathbf{\Omega}}_k \mathbf{H}_k \mathbf{\Sigma}_{\mathbf{y}}^{1/2} + \mathbf{I}\right|. \tag{126}$$

The outer bound on $\mathcal{RI}_{\mathrm{DIB}}$ is obtained by substituting into Equation (66), using Equations (125) and (126), noting that $\mathbf{\Omega}_k = \sum_{t \in \mathcal{T}} p(t) \mathbf{\Omega}_{k,t} \preceq \mathbf{\Sigma}_k^{-1}$ since $\mathbf{0} \preceq \mathbf{\Omega}_{k,t} \preceq \mathbf{\Sigma}_k^{-1}$, and taking the union over $\mathbf{\Omega}_k$ satisfying $\mathbf{0} \preceq \mathbf{\Omega}_k \preceq \mathbf{\Sigma}_k^{-1}$.

Finally, the proof is completed by noting that the outer bound is achieved with $T = \varnothing$ and multivariate Gaussian distributions $p^*(\mathbf{u}_k | \mathbf{x}_k, t) = \mathcal{CN}(\mathbf{x}_k, \mathbf{\Sigma}_k^{1/2}(\mathbf{\Omega}_k - \mathbf{I})\mathbf{\Sigma}_k^{1/2})$.

Author Contributions: A.Z., I.E.-A. and S.S.(S.) equally contributed to the published work. All authors have read and agreed to the published version of the manuscript.

Funding: The work of S. Shamai was supported by the European Union's Horizon 2020 Research And Innovation Programme, grant agreement No. 694630, and by the WIN consortium via the Israel minister of economy and science.

Acknowledgments: The authors would like to thank the anonymous reviewers for the constructive comments and suggestions, which helped us improve this manuscript.

Conflicts of Interest: The authors declare no conflict of interest.

References

1. Tishby, N.; Pereira, F.; Bialek, W. The information bottleneck method. In Proceedings of the Thirty-Seventh Annual Allerton Conference on Communication, Control, and Computing, Allerton House, Monticello, IL, USA, 22–24 September 1999; pp. 368–377.
2. Pratt, W.K. *Digital Image Processing*; John Willey & Sons Inc.: New York, NY, USA, 1991.
3. Yu, S.; Principe, J.C. Understanding Autoencoders with Information Theoretic Concepts. *arXiv* **2018**, arXiv:1804.00057.
4. Yu, S.; Jenssen, R.; Principe, J.C. Understanding Convolutional Neural Network Training with Information Theory. *arXiv* **2018**, arXiv:1804.06537.
5. Kong, Y.; Schoenebeck, G. Water from Two Rocks: Maximizing the Mutual Information. *arXiv* **2018**, arXiv:1802.08887.
6. Ugur, Y.; Aguerri, I.E.; Zaidi, A. A generalization of Blahut-Arimoto algorithm to computing rate-distortion regions of multiterminal source coding under logarithmic loss. In Proceedings of the IEEE Information Theory Workshop, ITW, Kaohsiung, Taiwan, 6–10 November 2017.
7. Dobrushin, R.L.; Tsybakov, B.S. Information transmission with additional noise. *IRE Trans. Inf. Theory* **1962**, *85*, 293–304. [CrossRef]
8. Witsenhausen, H.S.; Wyner, A.D. A conditional Entropy Bound for a Pair of Discrete Random Variables. *IEEE Trans. Inf. Theory* **1975**, *IT-21*, 493–501. [CrossRef]
9. Witsenhausen, H.S. Indirect Rate Distortion Problems. *IEEE Trans. Inf. Theory* **1980**, *IT-26*, 518–521. [CrossRef]
10. Shwartz-Ziv, R.; Tishby, N. Opening the Black Box of Deep Neural Networks via Information. *arXiv* **2017**, arXiv:1703.00810.
11. Achille, A.; Soatto, S. Emergence of Invariance and Disentangling in Deep Representations. *arXiv* **2017**, arXiv:1706.01350.
12. McAllester, D.A. A PAC-Bayesian Tutorial with a Dropout Bound. *arXiv* **2013**, arXiv:1307.2118.
13. Alemi, A.A. Variational Predictive Information Bottleneck. *arXiv* **2019**, arXiv:1910.10831.
14. Mukherjee, S. Machine Learning using the Variational Predictive Information Bottleneck with a Validation Set. *arXiv* **2019**, arXiv:1911.02210.
15. Kingma, D.P.; Welling, M. Auto-Encoding Variational Bayes. *arXiv* **2013**, arXiv:1312.6114.
16. Mukherjee, S. General Information Bottleneck Objectives and their Applications to Machine Learning. *arXiv* **2019**, arXiv:1912.06248.
17. Strouse, D.; Schwab, D.J. The information bottleneck and geometric clustering. *Neural Comput.* **2019**, *31*, 596–612. [CrossRef] [PubMed]
18. Painsky, A.; Tishby, N. Gaussian Lower Bound for the Information Bottleneck Limit. *J. Mach. Learn. Res. (JMLR)* **2018** *18*, 7908–7936.
19. Kittichokechai, K.; Caire, G. Privacy-constrained remote source coding. In Proceedings of the 2016 IEEE International Symposium on Information Theory (ISIT), Barcelona, Spain, 10–15 July 2016; pp. 1078–1082.
20. Tian, C.; Chen, J. Successive Refinement for Hypothesis Testing and Lossless One-Helper Problem. *IEEE Trans. Inf. Theory* **2008**, *54*, 4666–4681. [CrossRef]
21. Sreekumar, S.; Gündüz, D.; Cohen, A. Distributed Hypothesis Testing Under Privacy Constraints. *arXiv* **2018**, arXiv:1807.02764.
22. Aguerri, I.E.; Zaidi, A.; Caire, G.; Shamai (Shitz), S. On the Capacity of Cloud Radio Access Networks with Oblivious Relaying. In Proceedings of the 2017 IEEE International Symposium on Information Theory (ISIT), Aachen, Germany, 25–30 June 2017; pp. 2068–2072.
23. Aguerri, I.E.; Zaidi, A.; Caire, G.; Shamai (Shitz), S. On the capacity of uplink cloud radio access networks with oblivious relaying. *IEEE Trans. Inf. Theory* **2019**, *65*, 4575–4596. [CrossRef]

24. Stark, M.; Bauch, G.; Lewandowsky, J.; Saha, S. Decoding of Non-Binary LDPC Codes Using the Information Bottleneck Method. In Proceedings of the ICC 2019-2019 IEEE International Conference on Communications (ICC), Shanghai, China, 20–24 May 2019; pp. 1–6.
25. Bengio, Y.; Courville, A.; Vincent, P. Representation learning: A review and new perspectives. *IEEE Trans. Pattern Anal. Mach. Intell.* **2013**, *35*, 1798–1828. [CrossRef]
26. Erdogmus, D. Information Theoretic Learning: Renyi's Entropy and Its Applications to Adaptive System Training. Ph.D. Thesis, University of Florida Gainesville, Florida, FL, USA, 2002.
27. Principe, J.C.; Euliano, N.R.; Lefebvre, W.C. *Neural and Adaptive Systems: Fundamentals Through Simulations*; Wiley: New York, NY, USA, 2000; Volume 672.
28. Fisher, J.W. *Nonlinear Extensions to the Minumum Average Correlation Energy Filter*; University of Florida: Gainesville, FL, USA, 1997.
29. Jiao, J.; Courtade, T.A.; Venkat, K.; Weissman, T. Justification of logarithmic loss via the benefit of side information. *IEEE Trans. Inf. Theory* **2015**, *61*, 5357–5365. [CrossRef]
30. Painsky, A.; Wornell, G.W. On the Universality of the Logistic Loss Function. *arXiv* **2018**, arXiv:1805.03804.
31. Linsker, R. Self-organization in a perceptual network. *Computer* **1988**, *21*, 105–117. [CrossRef]
32. Quinlan, J.R. *C4. 5: Programs for Machine Learning*; Elsevier: Amsterdam, The Netherlands, 2014.
33. Chow, C.; Liu, C. Approximating discrete probability distributions with dependence trees. *IEEE Trans. Inf. Theory* **1968**, *14*, 462–467. [CrossRef]
34. Olsen, C.; Meyer, P.E.; Bontempi, G. On the impact of entropy estimation on transcriptional regulatory network inference based on mutual information. *EURASIP J. Bioinf. Syst. Biol.* **2008**, *2009*, 308959. [CrossRef]
35. Pluim, J.P.; Maintz, J.A.; Viergever, M.A. Mutual-information-based registration of medical images: A survey. *IEEE Trans. Med. Imaging* **2003**, *22*, 986–1004. [CrossRef]
36. Viola, P.; Wells, W.M., III. Alignment by maximization of mutual information. *Int. J. Comput. Vis.* **1997**, *24*, 137–154. [CrossRef]
37. Cesa-Bianchi, N.; Lugosi, G. *Prediction, Learning and Games*; Cambridge University Press: New York, NY, USA, 2006.
38. Lehmann, E.L.; Casella, G. *Theory of Point Estimation*; Springer Science & Business Media: Berlin, Germany, 2006.
39. Bousquet, O.; Elisseeff, A. Stability and generalization. *J. Mach. Learn. Res.* **2002**, *2*, 499–526.
40. Shalev-Shwartz, S.; Shamir, O.; Srebro, N.; Sridharan, K. Learnability, stability and uniform convergence. *J. Mach. Learn. Res.* **2010**, *11*, 2635–2670.
41. Xu, A.; Raginsky, M. Information-theoretic analysis of generalization capability of learning algorithms. In Proceedings of the 31st International Conference on Neural Information Processing Systems, Long Beach, CA, USA, 4–9 December 2017; pp. 2521–2530.
42. Russo, D.; Zou, J. How much does your data exploration overfit? Controlling bias via information usage. *arXiv* **2015**, arXiv:1511.05219.
43. Amjad, R.A.; Geiger, B.C. Learning Representations for Neural Network-Based Classification Using the Information Bottleneck Principle. *IEEE Trans. Pattern Anal. Mach. Intell.* **2019**. [CrossRef]
44. Paninski, L. Estimation of entropy and mutual information. *Neural Comput.* **2003**, *15*, 1191–1253. [CrossRef]
45. Jiao, J.; Venkat, K.; Han, Y.; Weissman, T. Minimax estimation of functionals of discrete distributions. *IEEE Trans. Inf. Theory* **2015**, *61*, 2835–2885. [CrossRef] [PubMed]
46. Valiant, P.; Valiant, G. Estimating the unseen: improved estimators for entropy and other properties. In Proceedings of the Advances in Neural Information Processing Systems 26 (NIPS 2013), Lake Tahoe, NV, USA, 5–10 December 2013; pp. 2157–2165.
47. Chalk, M.; Marre, O.; Tkacik, G. Relevant sparse codes with variational information bottleneck. *arXiv* **2016**, arXiv:1605.07332.
48. Alemi, A.; Fischer, I.; Dillon, J.; Murphy, K. Deep Variational Information Bottleneck. In Proceedings of the International Conference on Learning Representations, ICLR 2017, Toulon, France, 24–26 April 2017.
49. Achille, A.; Soatto, S. Information Dropout: Learning Optimal Representations Through Noisy Computation. *IEEE Trans. Pattern Anal. Mach. Intell.* **2018**, *40*, 2897–2905. [CrossRef]
50. Harremoes, P.; Tishby, N. The information bottleneck revisited or how to choose a good distortion measure. In Proceedings of the 2007 IEEE International Symposium on Information Theory, Nice, France, 24–29 June 2007; pp. 566–570.
51. Gamal, A.E.; Kim, Y.H. *Network Information Theory*; Cambridge University Press: Cambridge, UK, 2011.

52. Hotelling, H. The most predictable criterion. *J. Educ. Psycol.* **1935**, *26*, 139–142. [CrossRef]
53. Globerson, A.; Tishby, N. *On the Optimality of the Gaussian Information Bottleneck Curve*; Technical Report; Hebrew University: Jerusalem, Israel, 2004.
54. Chechik, G.; Globerson, A.; Tishby, N.; Weiss, Y. Information Bottleneck for Gaussian Variables. *J. Mach. Learn. Res.* **2005**, *6*, 165–188.
55. Wieczorek, A.; Roth, V. On the Difference Between the Information Bottleneck and the Deep Information Bottleneck. *arXiv* **2019**, arXiv:1912.13480.
56. Dempster, A.P.; Laird, N.M.; Rubin, D.B. Maximum likelihood from incomplete data via the EM algorithm. *J. R. Stat. Soc. Ser. B* **1977**, *39*, 1–38.
57. Blahut, R. Computation of channel capacity and rate-distortion functions. *IEEE Trans. Inf. Theory* **1972**, *18*, 460–473. [CrossRef]
58. Arimoto, S. An algorithm for computing the capacity of arbitrary discrete memoryless channels. *IEEE Trans. Inf. Theory* **1972**, *IT-18*, 12–20. [CrossRef]
59. Winkelbauer, A.; Matz, G. Rate-information-optimal gaussian channel output compression. In Proceedings of the 48th Annual Conference on Information Sciences and Systems (CISS), Princeton, NJ, USA, 19–21 March 2014; pp. 1–5.
60. Gálvez, B.R.; Thobaben, R.; Skoglund, M. The Convex Information Bottleneck Lagrangian. *Entropy* **2020**, *20*, 98. [CrossRef]
61. Jang, E.; Gu, S.; Poole, B. Categorical Reparameterization with Gumbel-Softmax. *arXiv* **2017**, arXiv:1611.01144.
62. Maddison, C.J.; Mnih, A.; Teh, Y.W. The Concrete Distribution: A Continuous Relaxation of Discrete Random Variables. *arXiv* **2016**, arXiv:1611.00712.
63. Kingma, D.P.; Ba, J. Adam: A Method for Stochastic Optimization. *arXiv* **2014**, arXiv:1412.6980.
64. Lim, S.H.; Kim, Y.H.; Gamal, A.E.; Chung, S.Y. Noisy Network Coding. *IEEE Trans. Inf. Theory* **2011**, *57*, 3132–3152. [CrossRef]
65. Courtade, T.A.; Weissman, T. Multiterminal source coding under logarithmic loss. *IEEE Trans. Inf. Theory* **2014**, *60*, 740–761. [CrossRef]
66. Csiszár, I.; Körner, J. *Information Theory: Coding Theorems for Discrete Memoryless Systems*; Academic Press: London, UK, 1981.
67. Wyner, A.D.; Ziv, J. The rate-distortion function for source coding with side information at the decoder. *IEEE Trans. Inf. Theory* **1976**, *22*, 1–10. [CrossRef]
68. Steinberg, Y. Coding and Common Reconstruction. *IEEE Trans. Inf. Theory* **2009**, *IT-11*, 4995–5010. [CrossRef]
69. Benammar, M.; Zaidi, A. Rate-Distortion of a Heegard-Berger Problem with Common Reconstruction Constraint. In Proceedings of the International Zurich Seminar on Information and Communication, Cambridge, MA, USA, 1–6 July 2016.
70. Benammar, M.; Zaidi, A. Rate-distortion function for a heegard-berger problem with two sources and degraded reconstruction sets. *IEEE Trans. Inf. Theory* **2016**, *62*, 5080–5092. [CrossRef]
71. Sutskover, I.; Shamai, S.; Ziv, J. Extremes of Information Combining. *IEEE Trans. Inf. Theory* **2005**, *51*, 1313–1325. [CrossRef]
72. Land, I.; Huber, J. Information Combining. *Found. Trends Commun. Inf. Theory* **2006**, *3*, 227–230. [CrossRef]
73. Wyner, A.D. On source coding with side information at the decoder. *IEEE Trans. Inf. Theory* **1975**, *21*, 294–300. [CrossRef]
74. Ahlswede, R.; Korner, J. Source coding with side information and a converse for degraded broadcast channels. *IEEE Trans. Inf. Theory* **1975**, *21*, 629–637. [CrossRef]
75. Makhdoumi, A.; Salamatian, S.; Fawaz, N.; Médard, M. From the information bottleneck to the privacy funnel. In Proceedings of the IEEE Information Theory Workshop, ITW, Hobart, Tasamania, Australia, 2–5 November 2014; pp. 501–505.
76. Asoodeh, S.; Diaz, M.; Alajaji, F.; Linder, T. Information Extraction Under Privacy Constraints. *IEEE Trans. Inf. Theory* **2019**, *65*, 1512–1534. [CrossRef]
77. Erkip, E.; Cover, T.M. The efficiency of investment information. *IEEE Trans. Inf. Theory* **1998**, *44*, 1026–1040. [CrossRef]
78. Hinton, G.E.; van Camp, D. Keeping the Neural Networks Simple by Minimizing the Description Length of the Weights. In *Proceedings of the Sixth Annual Conference on Computational Learning Theory*; ACM: New York, NY, USA, 1993; pp. 5–13.

79. Gilad-Bachrach, R.; Navot, A.; Tishby, N. An Information Theoretic Tradeoff between Complexity and Accuracy. In *Learning Theory and Kernel Machines*; Lecture Notes in Computer Science; Springer: Berlin/Heidelberg, Germany, 2003; pp. 595–609.

80. Vera, M.; Piantanida, P.; Vega, L.R. The Role of Information Complexity and Randomization in Representation Learning. *arXiv* **2018**, arXiv:1802.05355.

81. Huang, S.L.; Makur, A.; Wornell, G.W.; Zheng, L. On Universal Features for High-Dimensional Learning and Inference. *arXiv* **2019**, arXiv:1911.09105.

82. Higgins, I.; Matthey, L.; Pal, A.; Burgess, C.; Glorot, X.; Botvinick, M.; Mohamed, S.; Lerchner, A. β-VAE: Learning basic visual concepts with a constrained variational framework. In Proceedings of the International Conference on Learning Representations, Toulon, France, 24–26 April 2017.

83. Kurakin, A.; Goodfellow, I.; Bengio, S. Adversarial Machine Learning at Scale. *arXiv* **2016**, arXiv:1611.01236.

84. Madry, A.; Makelov, A.; Schmidt, L.; Tsipras, D.; Vladu, A. Towards Deep Learning Models Resistant to Adversarial Attacks. *arXiv* **2017**, arXiv:1706.06083.

85. Engstrom, L.; Tran, B.; Tsipras, D.; Schmidt, L.; Madry, A. A rotation and a translation suffice: Fooling cnns with simple transformations. *arXiv* **2017**, arXiv:1712.02779.

86. Pensia, A.; Jog, V.; Loh, P.L. Extracting robust and accurate features via a robust information bottleneck. *arXiv* **2019**, arXiv:1910.06893.

87. Guo, D.; Shamai, S.; Verdu, S. Mutual Information and Minimum Mean-Square Error in Gaussian Channels. *IEEE Trans. Inf. Theory* **2005**, *51*, 1261–1282. [CrossRef]

88. Dembo, A.; Cover, T.M.; Thomas, J.A. Information theoretic inequalities. *IEEE Trans. Inf. Theory* **1991**, *37*, 1501–1518. [CrossRef]

89. Ekrem, E.; Ulukus, S. An Outer Bound for the Vector Gaussian CEO Problem. *IEEE Trans. Inf. Theory* **2014**, *60*, 6870–6887. [CrossRef]

90. Cover, T.M.; Thomas, J.A. *Elements of Information Theory*; John Willey & Sons Inc.: New York, NY, USA, 1991.

91. Aguerri, I.E.; Zaidi, A. Distributed information bottleneck method for discrete and Gaussian sources. In Proceedings of the International Zurich Seminar on Information and Communication, IZS, Zurich, Switzerland, 21–23 February 2018.

92. Aguerri, I.E.; Zaidi, A. Distributed Variational Representation Learning. *arXiv* **2018**, arxiv:1807.04193.

93. Winkelbauer, A.; Farthofer, S.; Matz, G. The rate-information trade-off for Gaussian vector channels. In Proceedings of the The 2014 IEEE International Symposium on Information Theory, Honolulu, HI, USA, 29 June–4 July 2014; pp. 2849–2853.

94. Ugur, Y.; Aguerri, I.E.; Zaidi, A. Rate region of the vector Gaussian CEO problem under logarithmic loss. In Proceedings of the 2018 IEEE Information Theory Workshop (ITW), Guangzhou, China, 25–29 November 2018.

95. Ugur, Y.; Aguerri, I.E.; Zaidi, A. Vector Gaussian CEO Problem under Logarithmic Loss. *arXiv* **2018**, arXiv:1811.03933.

96. Simeone, O.; Erkip, E.; Shamai, S. On Codebook Information for Interference Relay Channels With Out-of-Band Relaying. *IEEE Trans. Inf. Theory* **2011**, *57*, 2880–2888. [CrossRef]

97. Sanderovich, A.; Shamai, S.; Steinberg, Y.; Kramer, G. Communication Via Decentralized Processing. *IEEE Tran. Inf. Theory* **2008**, *54*, 3008–3023. [CrossRef]

98. Lapidoth, A.; Narayan, P. Reliable communication under channel uncertainty. *IEEE Trans. Inf. Theory* **1998**, *44*, 2148–2177. [CrossRef]

99. Cover, T.M.; El Gamal, A. Capacity Theorems for the Relay Channel. *IEEE Trans. Inf. Theory* **1979**, *25*, 572–584. [CrossRef]

100. Xu, C.; Tao, D.; Xu, C. A survey on multi-view learning. *arXiv* **2013**, arXiv:1304.5634.

101. Blum, A.; Mitchell, T. Combining labeled and unlabeled data with co-training. In Proceedings of the Eleventh Annual Conference on Computational Learning Theory, Madison, WI, USA, 24–26 July 1998; pp. 92–100.

102. Dhillon, P.; Foster, D.P.; Ungar, L.H. Multi-view learning of word embeddings via CCA. In Proceedings of the 2011 Advances in Neural Information Processing Systems, Granada, Spain, 12–17 December 2011; pp. 199–207.

103. Kumar, A.; Daumé, H. A co-training approach for multi-view spectral clustering. In Proceedings of the 28th International Conference on Machine Learning (ICML-11), Bellevue, WA, USA, 28 June–2 July 2011; pp. 393–400.

104. Gönen, M.; Alpaydın, E. Multiple kernel learning algorithms. *J. Mach. Learn. Res.* **2011**, *12*, 2211–2268.
105. Jia, Y.; Salzmann, M.; Darrell, T. Factorized latent spaces with structured sparsity. In Proceedings of the Advances in Neural Information Processing Systems, Brno, Czech, 24–25 June 2010; pp. 982–990.
106. Vapnik, V. *The Nature of Statistical Learning Theory*; Springer Science & Business Media: Berlin, Germany, 2013.
107. Strouse, D.J.; Schwab, D.J. The deterministic information bottleneck. *Mass. Inst. Tech. Neural Comput.* **2017**, *26*, 1611–1630. [CrossRef] [PubMed]
108. Homri, A.; Peleg, M.; Shitz, S.S. Oblivious Fronthaul-Constrained Relay for a Gaussian Channel. *IEEE Trans. Commun.* **2018**, *66*, 5112–5123. [CrossRef]
109. du Pin Calmon, F.; Fawaz, N. Privacy against statistical inference. In Proceedings of the 2012 50th Annual Allerton Conference on Communication, Control, and Computing (Allerton), Monticello, IL, USA, 1–5 October 2012.
110. Karasik, R.; Simeone, O.; Shamai, S. Robust Uplink Communications over Fading Channels with Variable Backhaul Connectivity. *IEEE Trans. Commun.* **2013**, *12*, 5788–5799.
111. Chen, Y.; Goldsmith, A.J.; Eldar, Y.C. Channel capacity under sub-Nyquist nonuniform sampling. *IEEE Trans. Inf. Theory* **2014**, *60*, 4739–4756. [CrossRef]
112. Kipnis, A.; Eldar, Y.C.; Goldsmith, A.J. Analog-to-Digital Compression: A New Paradigm for Converting Signals to Bits. *IEEE Signal Process. Mag.* **2018**, *35*, 16–39. [CrossRef]

Article

Variational Information Bottleneck for Unsupervised Clustering: Deep Gaussian Mixture Embedding

Yiğit Uğur [1,2,*], George Arvanitakis [2] and Abdellatif Zaidi [1,*]

[1]   Laboratoire d'informatique Gaspard-Monge, Université Paris-Est, 77454 Champs-sur-Marne, France
[2]   Mathematical and Algorithmic Sciences Lab, Paris Research Center, Huawei Technologies,
     92100 Boulogne-Billancourt, France; george.arvanitakis@huawei.com
*   Correspondence: ygtugur@gmail.com (Y.U.); abdellatif.zaidi@u-pem.fr (A.Z.)

Received: 3 December 2019; Accepted: 9 February 2020; Published: 13 February 2020

Abstract: In this paper, we develop an unsupervised generative clustering framework that combines the variational information bottleneck and the Gaussian mixture model. Specifically, in our approach, we use the variational information bottleneck method and model the latent space as a mixture of Gaussians. We derive a bound on the cost function of our model that generalizes the Evidence Lower Bound (ELBO) and provide a variational inference type algorithm that allows computing it. In the algorithm, the coders' mappings are parametrized using neural networks, and the bound is approximated by Markov sampling and optimized with stochastic gradient descent. Numerical results on real datasets are provided to support the efficiency of our method.

Keywords: clustering; unsupervised learning; Gaussian mixture model; information bottleneck

---

1. Introduction

Clustering consists of partitioning a given dataset into various groups (clusters) based on some similarity metric, such as the Euclidean distance, L_1 norm, L_2 norm, L_∞ norm, the popular logarithmic loss measure, or others. The principle is that each cluster should contain elements of the data that are closer to each other than to any other element outside that cluster, in the sense of the defined similarity measure. If the joint distribution of the clusters and data is not known, one should operate blindly in doing so, i.e., using only the data elements at hand; and the approach is called unsupervised clustering [1,2]. Unsupervised clustering is perhaps one of the most important tasks of unsupervised machine learning algorithms currently, due to a variety of application needs and connections with other problems.

Clustering can be formulated as follows. Consider a dataset that is composed of N samples $\{x_i\}_{i=1}^N$, which we wish to partition into $|\mathcal{C}| \geq 1$ clusters. Let $\mathcal{C} = \{1, \ldots, |\mathcal{C}|\}$ be the set of all possible clusters and C designate a categorical random variable that lies in $\mathcal{C}$ and stands for the index of the actual cluster. If $\mathbf{X}$ is a random variable that models elements of the dataset, given that $\mathbf{X} = x_i$ induces a probability distribution on C, which the learner should learn, thus mathematically, the problem is that of estimating the values of the unknown conditional probability $P_{C|\mathbf{X}}(\cdot|x_i)$ for all elements x_i of the dataset. The estimates are sometimes referred to as the assignment probabilities.

Examples of unsupervised clustering algorithms include the very popular K-means [3] and Expectation Maximization (EM) [4]. The K-means algorithm partitions the data in a manner that the Euclidean distance among the members of each cluster is minimized. With the EM algorithm, the underlying assumption is that the data comprise a mixture of Gaussian samples, namely a Gaussian Mixture Model (GMM); and one estimates the parameters of each component of the GMM while simultaneously associating each data sample with one of those components. Although they offer some advantages in the context of clustering, these algorithms suffer from some strong limitations. For example, it is well known that the K-means is highly sensitive to both the order of the data and scaling; and the obtained accuracy depends strongly on the initial seeds (in addition to that, it does not predict the number of clusters or K-value). The EM algorithm suffers mainly from slow convergence, especially for high-dimensional data.

Recently, a new approach has emerged that seeks to perform inference on a transformed domain (generally referred to as latent space), not the data itself. The rationale is that because the latent space often has fewer dimensions, it is more convenient computationally to perform inference (clustering) on it rather than on the high-dimensional data directly. A key aspect then is how to design a latent space that is amenable to accurate low-complexity unsupervised clustering, i.e., one that preserves only those features of the observed high-dimensional data that are useful for clustering while removing all redundant or non-relevant information. Along this line of work, we can mention [5], which utilized Principal Component Analysis (PCA) [6,7] for dimensionality reduction followed by K-means for clustering the obtained reduced dimension data; or [8], which used a combination of PCA and the EM algorithm. Other works that used alternatives for the linear PCA include kernel PCA [9], which employs PCA in a non-linear fashion to maximize variance in the data.

Thisby's Information Bottleneck (IB) method [10] formulates the problem of finding a good representation $\mathbf{U}$ that strikes the right balance between capturing all information about the categorical variable C that is contained in the observation $\mathbf{X}$ and using the most concise representation for it. The IB problem can be written as the following Lagrangian optimization:

$$\min_{P_{\mathbf{U}|\mathbf{X}}} \; I(\mathbf{X};\mathbf{U}) - sI(C;\mathbf{U}) \, , \tag{1}$$

where $I(\cdot\,;\,\cdot)$ denotes Shannon's mutual information and s is a Lagrange-type parameter, which controls the trade-off between accuracy and regularization. In [11,12], a text clustering algorithm is introduced for the case in which the joint probability distribution of the input data is known. This text clustering algorithm uses the IB method with an annealing procedure, where the parameter s is increased gradually. When $s \to 0$, the representation $\mathbf{U}$ is designed with the most compact form, i.e., $|\mathcal{U}| = 1$, which corresponds to the maximum compression. By gradually increasing the parameter s, the emphasis on the relevance term $I(C;\mathbf{U})$ increases, and at a critical value of s, the optimization focuses on not only the compression, but also the relevance term. To fulfill the demand on the relevance term, this results in the cardinality of $\mathbf{U}$ bifurcating. This is referred as a phase transition of the system. The further increases in the value of s will cause other phase transitions, hence additional splits of U until it reaches the desired level, e.g., $|\mathcal{U}| = |\mathcal{C}|$.

However, in the real-world applications of clustering with large-scale datasets, the joint probability distributions of the datasets are unknown. In practice, the usage of Deep Neural Networks (DNN) for unsupervised clustering of high-dimensional data on a lower dimensional latent space has attracted considerable attention, especially with the advent of Autoencoder (AE) learning and the development of powerful tools to train them using standard backpropagation techniques [13,14]. Advanced forms include Variational Autoencoders (VAE) [13,14], which are generative variants of AE that regularize the structure of the latent space, and the more general Variational Information Bottleneck (VIB) of [15], which is a technique that is based on the information bottleneck method and seeks a better trade-off between accuracy and regularization than VAE via the introduction of a Lagrange-type parameter s, which controls that trade-off and whose optimization is similar to deterministic annealing [12] or stochastic relaxation.

In this paper, we develop an unsupervised generative clustering framework that combines VIB and the Gaussian mixture model. Specifically, in our approach, we use the variational information bottleneck method and model the latent space as a mixture of Gaussians. We derive a bound on the cost function of our model that generalizes the Evidence Lower Bound (ELBO) and provide a variational inference type algorithm that allows computing it. In the algorithm, the coders' mappings are parametrized using Neural Networks (NN), and the bound is approximated by Markov sampling and optimized with stochastic gradient descent. Furthermore, we show how tuning the hyper-parameter s

appropriately by gradually increasing its value with iterations (number of epochs) results in a better accuracy. Furthermore, the application of our algorithm to the unsupervised clustering of various datasets, including the MNIST [16], REUTERS [17], and STL-10 [18], allows a better clustering accuracy than previous state-of-the-art algorithms. For instance, we show that our algorithm performs better than the Variational Deep Embedding (VaDE) algorithm of [19], which is based on VAE and performs clustering by maximizing the ELBO. Our algorithm can be seen as a generalization of the VaDE, whose ELBO can be recovered by setting $s = 1$ in our cost function. In addition, our algorithm also generalizes the VIB of [15], which models the latent space as an isotropic Gaussian, which is generally not expressive enough for the purpose of unsupervised clustering. Other related works, which are of lesser relevance to the contribution of this paper, are the Deep Embedded Clustering (DEC) of [20] and the Improved Deep Embedded Clustering (IDEC) of [21,22]. For a detailed survey of clustering with deep learning, the readers may refer to [23].

To the best of our knowledge, our algorithm performs the best in terms of clustering accuracy by using deep neural networks without any prior knowledge regarding the labels (except the usual assumption that the number of classes is known) compared to the state-of-the-art algorithms of the unsupervised learning category. In order to achieve the outperforming accuracy: (i) we derive a cost function that contains the IB hyperparameter s that controls optimal trade-offs between the accuracy and regularization of the model; (ii) we use a lower bound approximation for the KL term in the cost function, that does not depend on the clustering assignment probability (note that the clustering assignment is usually not accurate in the beginning of the training process); and (iii) we tune the hyperparameter s by following an annealing approach that improves both the convergence and the accuracy of the proposed algorithm.

Throughout this paper, we use the following notation. Uppercase letters are used to denote random variables, e.g., X; lowercase letters are used to denote realizations of random variables, e.g., x; and calligraphic letters denote sets, e.g., $\mathcal{X}$. The cardinality of a set $\mathcal{X}$ is denoted by $|\mathcal{X}|$. Probability mass functions (pmfs) are denoted by $P_X(x) = \Pr\{X = x\}$ and, sometimes, for short, as $p(x)$. Boldface uppercase letters denote vectors or matrices, e.g., $\mathbf{X}$, where context should make the distinction clear. We denote the covariance of a zero mean, complex-valued, vector $\mathbf{X}$ by $\mathbf{\Sigma_x} = \mathbb{E}[\mathbf{XX}^\dagger]$, where $(\cdot)^\dagger$ indicates the conjugate transpose. For random variables X and Y, the entropy is denoted as $H(X)$, i.e., $H(X) = \mathbb{E}_{P_X}[-\log P_X]$, and the mutual information is denoted as $I(X;Y)$, i.e., $I(X;Y) = H(X) - H(X|Y) = H(Y) - H(Y|X) = \mathbb{E}_{P_{X,Y}}[\log \frac{P_{X,Y}}{P_X P_Y}]$. Finally, for two probability measures P_X and Q_X on a random variable $X \in \mathcal{X}$, the relative entropy or Kullback–Leibler divergence is denoted as $D_{\mathrm{KL}}(P_X \| Q_X)$, i.e., $D_{\mathrm{KL}}(P_X \| Q_X) = \mathbb{E}_{P_X}[\log \frac{P_X}{Q_X}]$.

2. Proposed Model

In this section, we explain the proposed model, the so-called Variational Information Bottleneck with Gaussian Mixture Model (VIB-GMM), in which we use the VIB framework and model the latent space as a GMM. The proposed model is depicted in Figure 1, where the parameters $\pi_c, \boldsymbol{\mu}_c, \mathbf{\Sigma}_c$, for all values of $c \in \mathcal{C}$, are to be optimized jointly with those of the employed NNs as instantiation of the coders. Furthermore, the assignment probabilities are estimated based on the values of latent space vectors instead of the observations themselves, i.e., $P_{C|X} = Q_{C|U}$. In the rest of this section, we elaborate on the inference and generative network models for our method.

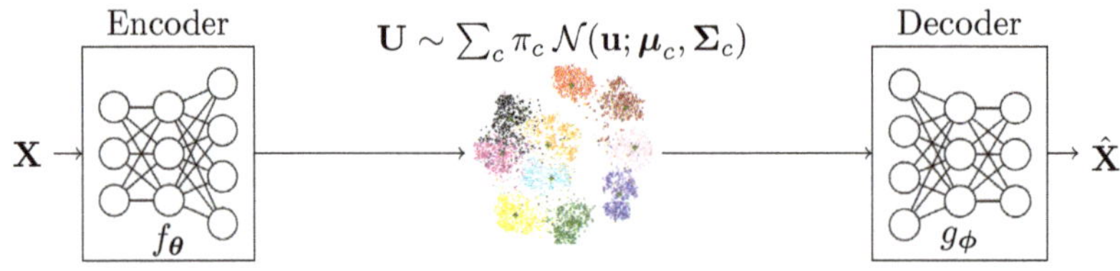

Figure 1. Variational Information Bottleneck with Gaussian mixtures.

2.1. Inference Network Model

We assume that observed data $\mathbf{x}$ are generated from a GMM with $|\mathcal{C}|$ components. Then, the latent representation $\mathbf{u}$ is inferred according to the following procedure:

1. One of the components of the GMM is chosen according to a categorical variable C.
2. The data $\mathbf{x}$ are generated from the cth component of the GMM, i.e., $P_{\mathbf{X}|C} \sim \mathcal{N}(\mathbf{x}; \tilde{\boldsymbol{\mu}}_c, \tilde{\boldsymbol{\Sigma}}_c)$.
3. Encoder maps $\mathbf{x}$ to a latent representation $\mathbf{u}$ according to $P_{\mathbf{U}|\mathbf{X}} \sim \mathcal{N}(\boldsymbol{\mu}_\theta, \boldsymbol{\Sigma}_\theta)$.

    3.1. The encoder is modeled with a DNN f_θ, which maps $\mathbf{x}$ to the parameters of a Gaussian distribution, i.e., $[\boldsymbol{\mu}_\theta, \boldsymbol{\Sigma}_\theta] = f_\theta(\mathbf{x})$.
    3.2. The representation $\mathbf{u}$ is sampled from $\mathcal{N}(\boldsymbol{\mu}_\theta, \boldsymbol{\Sigma}_\theta)$.

For the inference network, shown in Figure 2, the following Markov chain holds:

$$C \multimap \mathbf{X} \multimap \mathbf{U} . \tag{2}$$

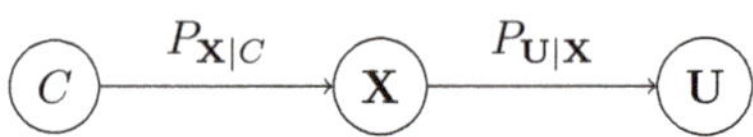

Figure 2. Inference network.

2.2. Generative Network Model

Since the encoder extracts useful representations of the dataset and we assume that the dataset is generated from a GMM, we model our latent space also with a mixture of Gaussians. To do so, the categorical variable C is embedded with the latent variable $\mathbf{U}$. The reconstruction of the dataset is generated according to the following procedure:

1. One of the components of the GMM is chosen according to a categorical variable C, with a prior distribution Q_C.
2. The representation $\mathbf{u}$ is generated from the cth component, i.e., $Q_{\mathbf{U}|C} \sim \mathcal{N}(\mathbf{u}; \boldsymbol{\mu}_c, \boldsymbol{\Sigma}_c)$.
3. The decoder maps the latent representation $\mathbf{u}$ to $\hat{\mathbf{x}}$, which is the reconstruction of the source $\mathbf{x}$ by using the mapping $Q_{\mathbf{X}|\mathbf{U}}$.

    3.1 The decoder is modeled with a DNN g_ϕ that maps $\mathbf{u}$ to the estimate $\hat{\mathbf{x}}$, i.e., $[\hat{\mathbf{x}}] = g_\phi(\mathbf{u})$.

For the generative network, shown in Figure 3, the following Markov chain holds:

$$C \multimap \mathbf{U} \multimap \mathbf{X} . \tag{3}$$

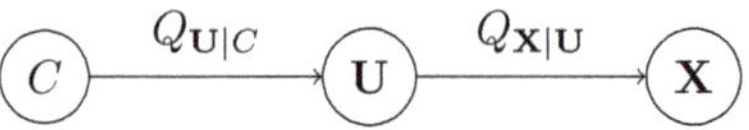

Figure 3. Generative network.

3. Proposed Method

In this section, we present our clustering method. First, we provide a general cost function for the problem of the unsupervised clustering that we study here based on the variational IB framework; and we show that it generalizes the ELBO bound developed in [19]. We then parametrize our model using NNs whose parameters are optimized jointly with those of the GMM. Furthermore, we discuss the influence of the hyperparameter s that controls optimal trade-offs between accuracy and regularization.

3.1. Brief Review of Variational Information Bottleneck for Unsupervised Learning

As described in Section 2, the stochastic encoder $P_{U|X}$ maps the observed data $\mathbf{x}$ to a representation $\mathbf{u}$. Similarly, the stochastic decoder $Q_{X|U}$ assigns an estimate $\hat{\mathbf{x}}$ of $\mathbf{x}$ based on the vector $\mathbf{u}$. As per the IB method [10], a suitable representation $\mathbf{U}$ should strike the right balance between capturing all information about the categorical variable C that is contained in the observation $\mathbf{X}$ and using the most concise representation for it. This leads to maximizing the following Lagrange problem:

$$\mathcal{L}_s(\mathbf{P}) = I(C; \mathbf{U}) - sI(\mathbf{X}; \mathbf{U}) , \tag{4}$$

where $s \geq 0$ designates the Lagrange multiplier and, for convenience, $\mathbf{P}$ denotes the conditional distribution $P_{U|X}$.

Instead of (4), which is not always computable in our unsupervised clustering setting, we find it convenient to maximize an upper bound of $\mathcal{L}_s(\mathbf{P})$ given by:

$$\tilde{\mathcal{L}}_s(\mathbf{P}) := I(\mathbf{X}; \mathbf{U}) - sI(\mathbf{X}; \mathbf{U}) \overset{(a)}{=} H(\mathbf{X}) - H(\mathbf{X}|\mathbf{U}) - s[H(\mathbf{U}) - H(\mathbf{U}|\mathbf{X})] , \tag{5}$$

where (a) is due to the definition of mutual information (using the Markov chain $C \;\text{–}\!\!\circ\!\!\text{–}\; \mathbf{X} \;\text{–}\!\!\circ\!\!\text{–}\; \mathbf{U}$, it is easy to see that $\tilde{\mathcal{L}}_s(\mathbf{P}) \geq \mathcal{L}_s(\mathbf{P})$ for all values of $\mathbf{P}$). Noting that $H(\mathbf{X})$ is constant with respect to $P_{U|X}$, maximizing $\tilde{\mathcal{L}}_s(\mathbf{P})$ over $\mathbf{P}$ is equivalent to maximizing:

$$\mathcal{L}'_s(\mathbf{P}) := -H(\mathbf{X}|\mathbf{U}) - s[H(\mathbf{U}) - H(\mathbf{U}|\mathbf{X})] \tag{6}$$

$$= \mathbb{E}_{P_X}\left[\mathbb{E}_{P_{U|X}}[\log P_{X|U} + s \log P_{U} - s \log P_{U|X}]\right] . \tag{7}$$

For a variational distribution Q_U on $\mathcal{U}$ (instead of the unknown P_U) and a variational stochastic decoder $Q_{X|U}$ (instead of the unknown optimal decoder $P_{X|U}$), let $\mathbf{Q} := \{Q_{X|U}, Q_U\}$. Furthermore, let:

$$\mathcal{L}_s^{\mathrm{VB}}(\mathbf{P}, \mathbf{Q}) := \mathbb{E}_{P_X}\left[\mathbb{E}_{P_{U|X}}[\log Q_{X|U}] - sD_{\mathrm{KL}}(P_{U|X}\|Q_U)\right] . \tag{8}$$

Lemma 1. *For given* $\mathbf{P}$, *we have:*

$$\mathcal{L}_s^{\mathrm{VB}}(\mathbf{P}, \mathbf{Q}) \leq \mathcal{L}'_s(\mathbf{P}) , \quad \textit{for all } \mathbf{Q} .$$

In addition, there exists a unique $\mathbf{Q}$ *that achieves the maximum* $\max_{\mathbf{Q}} \mathcal{L}_s^{\mathrm{VB}}(\mathbf{P}, \mathbf{Q}) = \mathcal{L}'_s(\mathbf{P})$ *and is given by:*

$$Q^*_{X|U} = P_{X|U} , \quad Q^*_U = P_U .$$

Proof. The proof of Lemma 1 is given in Appendix A. □

Using Lemma 1, maximization of (6) can be written in term of the variational IB cost as follows:

$$\max_{\mathbf{P}} \mathcal{L}'_s(\mathbf{P}) = \max_{\mathbf{P}} \max_{\mathbf{Q}} \mathcal{L}_s^{\mathrm{VB}}(\mathbf{P}, \mathbf{Q}) . \tag{9}$$

Next, we develop an algorithm that solves the maximization problem (9), where the encoding mapping $P_{\mathbf{U}|\mathbf{X}}$, the decoding mapping $Q_{\mathbf{X}|\mathbf{U}}$, as well as the prior distribution of the latent space $Q_{\mathbf{U}}$ are optimized jointly.

Remark 1. *As we already mentioned in the beginning of this section, the related work [19] performed unsupervised clustering by combining VAE with GMM. Specifically, it maximizes the following ELBO bound:*

$$\mathcal{L}_1^{\text{VaDE}} := \mathbb{E}_{P_\mathbf{X}}\left[\mathbb{E}_{P_{\mathbf{U}|\mathbf{X}}}[\log Q_{\mathbf{X}|\mathbf{U}}] - D_{\text{KL}}(P_{C|\mathbf{X}}\|Q_C) - \mathbb{E}_{P_{C|\mathbf{X}}}[D_{\text{KL}}(P_{\mathbf{U}|\mathbf{X}}\|Q_{\mathbf{U}|C})]\right]. \tag{10}$$

Let, for an arbitrary non-negative parameter s, $\mathcal{L}_s^{\text{VaDE}}$ be a generalization of the ELBO bound in (10) of [19] given by:

$$\mathcal{L}_s^{\text{VaDE}} := \mathbb{E}_{P_\mathbf{X}}\left[\mathbb{E}_{P_{\mathbf{U}|\mathbf{X}}}[\log Q_{\mathbf{X}|\mathbf{U}}] - sD_{\text{KL}}(P_{C|\mathbf{X}}\|Q_C) - s\mathbb{E}_{P_{C|\mathbf{X}}}[D_{\text{KL}}(P_{\mathbf{U}|\mathbf{X}}\|Q_{\mathbf{U}|C})]\right]. \tag{11}$$

Investigating the right-hand side (RHS) of (11), we get:

$$\mathcal{L}_s^{\text{VB}}(\mathbf{P},\mathbf{Q}) = \mathcal{L}_s^{\text{VaDE}} + s\mathbb{E}_{P_\mathbf{X}}\left[\mathbb{E}_{P_{\mathbf{U}|\mathbf{X}}}[D_{\text{KL}}(P_{C|\mathbf{X}}\|Q_{C|\mathbf{U}})]\right], \tag{12}$$

where the equality holds since:

$$\mathcal{L}_s^{\text{VaDE}} = \mathbb{E}_{P_\mathbf{X}}\left[\mathbb{E}_{P_{\mathbf{U}|\mathbf{X}}}[\log Q_{\mathbf{X}|\mathbf{U}}] - sD_{\text{KL}}(P_{C|\mathbf{X}}\|Q_C) - s\mathbb{E}_{P_{C|\mathbf{X}}}[D_{\text{KL}}(P_{\mathbf{U}|\mathbf{X}}\|Q_{\mathbf{U}|C})]\right] \tag{13}$$

$$\overset{(a)}{=} \mathbb{E}_{P_\mathbf{X}}\left[\mathbb{E}_{P_{\mathbf{U}|\mathbf{X}}}[\log Q_{\mathbf{X}|\mathbf{U}}] - sD_{\text{KL}}(P_{\mathbf{U}|\mathbf{X}}\|Q_\mathbf{U}) - s\mathbb{E}_{P_{\mathbf{U}|\mathbf{X}}}[D_{\text{KL}}(P_{C|\mathbf{X}}\|Q_{C|\mathbf{U}})]\right] \tag{14}$$

$$\overset{(b)}{=} \mathcal{L}_s^{\text{VB}}(\mathbf{P},\mathbf{Q}) - s\mathbb{E}_{P_\mathbf{X}}\left[\mathbb{E}_{P_{\mathbf{U}|\mathbf{X}}}[D_{\text{KL}}(P_{C|\mathbf{X}}\|Q_{C|\mathbf{U}})]\right], \tag{15}$$

where (a) can be obtained by expanding and re-arranging terms under the Markov chain $C \ {-\!\!\!\circ\!\!\!-}\ \mathbf{X}\ {-\!\!\!\circ\!\!\!-}\ \mathbf{U}$ (for a detailed treatment, please look at Appendix B); and (b) follows from the definition of $\mathcal{L}_s^{\text{VB}}(\mathbf{P},\mathbf{Q})$ in (8).

Thus, by the non-negativity of relative entropy, it is clear that $\mathcal{L}_s^{\text{VaDE}}$ is a lower bound on $\mathcal{L}_s^{\text{VB}}(\mathbf{P},\mathbf{Q})$. Furthermore, if the variational distribution $\mathbf{Q}$ is such that the conditional marginal $Q_{C|\mathbf{U}}$ is equal to $P_{C|\mathbf{X}}$, the bound is tight since the relative entropy term is zero in this case.

3.2. Proposed Algorithm: VIB-GMM

In order to compute (9), we parametrize the distributions $P_{\mathbf{U}|\mathbf{X}}$ and $Q_{\mathbf{X}|\mathbf{U}}$ using DNNs. For instance, let the stochastic encoder $P_{\mathbf{U}|\mathbf{X}}$ be a DNN f_θ and the stochastic decoder $Q_{\mathbf{X}|\mathbf{U}}$ be a DNN g_ϕ. That is:

$$
\begin{aligned}
P_\theta(\mathbf{u}|\mathbf{x}) &= \mathcal{N}(\mathbf{u};\boldsymbol{\mu}_\theta,\boldsymbol{\Sigma}_\theta), \quad \text{where } [\boldsymbol{\mu}_\theta,\boldsymbol{\Sigma}_\theta] = f_\theta(\mathbf{x}), \\
Q_\phi(\mathbf{x}|\mathbf{u}) &= g_\phi(\mathbf{u}) = [\hat{\mathbf{x}}],
\end{aligned}
\tag{16}
$$

where θ and ϕ are the weight and bias parameters of the DNNs. Furthermore, the latent space is modeled as a GMM with $|\mathcal{C}|$ components with parameters $\psi := \{\pi_c,\boldsymbol{\mu}_c,\boldsymbol{\Sigma}_c\}_{c=1}^{|\mathcal{C}|}$, i.e.,

$$Q_\psi(\mathbf{u}) = \sum_c \pi_c\,\mathcal{N}(\mathbf{u};\boldsymbol{\mu}_c,\boldsymbol{\Sigma}_c). \tag{17}$$

Using the parametrizations above, the optimization of (9) can be rewritten as:

$$\max_{\theta,\phi,\psi}\ \mathcal{L}_s^{\text{NN}}(\theta,\phi,\psi) \tag{18}$$

where the cost function $\mathcal{L}_s^{\mathrm{NN}}(\theta, \phi, \psi)$ is given by:

$$\mathcal{L}_s^{\mathrm{NN}}(\theta, \phi, \psi) := \mathbb{E}_{P_{\mathbf{X}}}\left[\mathbb{E}_{P_\theta(\mathbf{U}|\mathbf{X})}[\log Q_\phi(\mathbf{X}|\mathbf{U})] - sD_{\mathrm{KL}}(P_\theta(\mathbf{U}|\mathbf{X})\|Q_\psi(\mathbf{U}))\right] . \tag{19}$$

Then, for given observations of N samples, i.e., $\{\mathbf{x}_i\}_{i=1}^N$, (18) can be approximated in terms of an empirical cost as follows:

$$\max_{\theta,\phi,\psi} \frac{1}{N} \sum_{i=1}^N \mathcal{L}_{s,i}^{\mathrm{emp}}(\theta, \phi, \psi) , \tag{20}$$

where $\mathcal{L}_{s,i}^{\mathrm{emp}}(\theta, \phi, \psi)$ is the empirical cost for the ith observation $\mathbf{x}_i$ and given by:

$$\mathcal{L}_{s,i}^{\mathrm{emp}}(\theta, \phi, \psi) = \mathbb{E}_{P_\theta(\mathbf{U}_i|\mathbf{X}_i)}[\log Q_\phi(\mathbf{X}_i|\mathbf{U}_i)] - sD_{\mathrm{KL}}(P_\theta(\mathbf{U}_i|\mathbf{X}_i)\|Q_\psi(\mathbf{U}_i)) . \tag{21}$$

Furthermore, the first term of the RHS of (21) can be computed using Monte Carlo sampling and the reparametrization trick [13]. In particular, $P_\theta(\mathbf{u}|\mathbf{x})$ can be sampled by first sampling a random variable $\mathbf{Z}$ with distribution $P_{\mathbf{Z}}$, i.e., $P_{\mathbf{Z}} = \mathcal{N}(\mathbf{0}, \mathbf{I})$, then transforming the samples using some function $\tilde{f}_\theta : \mathcal{X} \times \mathcal{Z} \to \mathcal{U}$, i.e., $\mathbf{u} = \tilde{f}_\theta(\mathbf{x}, \mathbf{z})$. Thus,

$$\mathbb{E}_{P_\theta(\mathbf{U}_i|\mathbf{X}_i)}[\log Q_\phi(\mathbf{X}_i|\mathbf{U}_i)] = \frac{1}{M} \sum_{m=1}^M \log q(\mathbf{x}_i|\mathbf{u}_{i,m}), \quad \mathbf{u}_{i,m} = \mu_{\theta,i} + \Sigma_{\theta,i}^{\frac{1}{2}} \cdot \epsilon_m, \quad \epsilon_m \sim \mathcal{N}(\mathbf{0}, \mathbf{I}) ,$$

where M is the number of samples for the Monte Carlo sampling step.

The second term of the RHS of (21) is the KL divergence between a single component multivariate Gaussian and a Gaussian mixture model with $|\mathcal{C}|$ components. An exact closed-form solution for the calculation of this term does not exist. However, a variational lower bound approximation [24] of it can be obtained as:

$$D_{\mathrm{KL}}(P_\theta(\mathbf{U}_i|\mathbf{X}_i)\|Q_\psi(\mathbf{U}_i)) = -\log \sum_{c=1}^{|\mathcal{C}|} \pi_c \exp\left(-D_{\mathrm{KL}}(\mathcal{N}(\mu_{\theta,i}, \Sigma_{\theta,i})\|\mathcal{N}(\mu_c, \Sigma_c))\right) . \tag{22}$$

In particular, in the specific case in which the covariance matrices are diagonal, i.e., $\Sigma_{\theta,i} := \mathrm{diag}(\{\sigma_{\theta,i,j}^2\}_{j=1}^{n_u})$ and $\Sigma_c := \mathrm{diag}(\{\sigma_{c,j}^2\}_{j=1}^{n_u})$, with n_u denoting the latent space dimension, (22) can be computed as follows:

$$D_{\mathrm{KL}}(P_\theta(\mathbf{U}_i|\mathbf{X}_i)\|Q_\psi(\mathbf{U}_i))$$
$$= -\log \sum_{c=1}^{|\mathcal{C}|} \pi_c \exp\left(-\frac{1}{2} \sum_{j=1}^{n_u} \left[\frac{(\mu_{\theta,i,j} - \mu_{c,j})^2}{\sigma_{c,j}^2} + \log \frac{\sigma_{c,j}^2}{\sigma_{\theta,i,j}^2} - 1 + \frac{\sigma_{\theta,i,j}^2}{\sigma_{c,j}^2}\right]\right) , \tag{23}$$

where $\mu_{\theta,i,j}$ and $\sigma_{\theta,i,j}^2$ are the mean and variance of the ith representation in the jth dimension of the latent space. Furthermore, $\mu_{c,j}$ and $\sigma_{c,j}^2$ represent the mean and variance of the cth component of the GMM in the jth dimension of the latent space.

Finally, we train NNs to maximize the cost function (19) over the parameters θ, ϕ, as well as those ψ of the GMM. For the training step, we use the ADAM optimization tool [25]. The training procedure is detailed in Algorithm 1.

Algorithm 1 VIB-GMM algorithm for unsupervised learning.

1: **input:** Dataset $\mathcal{D} := \{\mathbf{x}_i\}_{i=1}^{N}$, parameter $s \geq 0$.

2: **output:** Optimal DNN weights $\theta^\star, \phi^\star$ and GMM parameters $\psi^\star = \{\pi_c^\star, \boldsymbol{\mu}_c^\star, \boldsymbol{\Sigma}_c^\star\}_{c=1}^{|\mathcal{C}|}$.

3: **initialization** Initialize θ, ϕ, ψ.

4: **repeat**

5:      Randomly select b mini-batch samples $\{\mathbf{x}_i\}_{i=1}^{b}$ from $\mathcal{D}$.

6:      Draw m random i.i.d samples $\{\mathbf{z}_j\}_{j=1}^{m}$ from $P_\mathbf{Z}$.

7:      Compute m samples $\mathbf{u}_{i,j} = \tilde{f}_\theta(\mathbf{x}_i, \mathbf{z}_j)$

8:      For the selected mini-batch, compute gradients of the empirical cost (20).

9:      Update θ, ϕ, ψ using the estimated gradient (e.g., with SGD or ADAM).

10: **until** convergence of θ, ϕ, ψ.

Once our model is trained, we assign the given dataset into the clusters. As mentioned in Section 2, we do the assignment from the latent representations, i.e., $Q_{C|\mathbf{U}} = P_{C|\mathbf{X}}$. Hence, the probability that the observed data $\mathbf{x}_i$ belongs to the cth cluster is computed as follows:

$$p(c|\mathbf{x}_i) = q(c|\mathbf{u}_i) = \frac{q_{\psi^\star}(c) q_{\psi^\star}(\mathbf{u}_i|c)}{q_{\psi^\star}(\mathbf{u}_i)} = \frac{\pi_c^\star \mathcal{N}(\mathbf{u}_i; \boldsymbol{\mu}_c^\star, \boldsymbol{\Sigma}_c^\star)}{\sum_c \pi_c^\star \mathcal{N}(\mathbf{u}_i; \boldsymbol{\mu}_c^\star, \boldsymbol{\Sigma}_c^\star)}, \tag{24}$$

where * indicates the optimal values of the parameters as found at the end of the training phase. Finally, the right cluster is picked based on the largest assignment probability value.

Remark 2. *It is worth mentioning that with the use of the KL approximation as given by (22), our algorithm does not require the assumption $P_{C|\mathbf{U}} = Q_{C|\mathbf{U}}$ to hold (which is different from [19]). Furthermore, the algorithm is guaranteed to converge. However, the convergence may be to (only) local minima; and this is due to the problem (18) being generally non-convex. Related to this aspect, we mention that while without a proper pre-training, the accuracy of the VaDE algorithm may not be satisfactory, in our case, the above assumption is only used in the final assignment after the training phase is completed.*

Remark 3. *In [26], it is stated that optimizing the original IB problem with the assumption of independent latent representations amounts to disentangled representations. It is noteworthy that with such an assumption, the computational complexity can be reduced from $\mathcal{O}(n_u^2)$ to $\mathcal{O}(n_u)$. Furthermore, as argued in [26], the assumption often results only in some marginal performance loss; and for this reason, it is adopted in many machine learning applications.*

3.3. Effect of the Hyperparameter

As we already mentioned, the hyperparameter s controls the trade-off between the relevance of the representation $\mathbf{U}$ and its complexity. As can be seen from (19) for small values of s, it is the cross-entropy term that dominates, i.e., the algorithm trains the parameters so as to reproduce $\mathbf{X}$ as accurately as possible. For large values of s, however, it is most important for the NN to produce an encoded version of $\mathbf{X}$ whose distribution matches the prior distribution of the latent space, i.e., the term $D_{\mathrm{KL}}(P_\theta(\mathbf{U}|\mathbf{X}) \| Q_\phi(\mathbf{U}))$ is nearly zero.

In the beginning of the training process, the GMM components are randomly selected; and so, starting with a large value of the hyperparameter s is likely to steer the solution towards an irrelevant prior. Hence, for the tuning of the hyper-parameter s in practice, it is more efficient to start with a small value of s and gradually increase it with the number of epochs. This has the advantage of avoiding possible local minima, an aspect that is reminiscent of deterministic annealing [12], where s plays the role of the temperature parameter. The experiments that will be reported in the next section show that proceeding in the above-described manner for the selection of the parameter s helps in obtaining higher clustering accuracy and better robustness to the initialization (i.e., no need for a strong pretraining). The pseudocode for annealing is given in Algorithm 2.

Algorithm 2 Annealing algorithm pseudocode.

1: **input:** Dataset $\mathcal{D} := \{\mathbf{x}_i\}_{i=1}^{n}$, hyperparameter interval $[s_{\min}, s_{\max}]$.
2: **output:** Optimal DNN weights $\theta^\star, \phi^\star$, GMM parameters $\psi^\star = \{\pi_c^\star, \boldsymbol{\mu}_c^\star, \boldsymbol{\Sigma}_c^\star\}_{c=1}^{|\mathcal{C}|}$,
       assignment probability $P_{C|\mathbf{X}}$.
3: **initialization** Initialize θ, ϕ, ψ.
4: **repeat**
5:      Apply VIB-GMM algorithm.
6:      Update ψ, θ, ϕ.
7:      Update s, e.g., $s = (1 + \epsilon_s)s_{\text{old}}$.
8: **until** s does not exceed $s_{\max}$.

Remark 4. *As we mentioned before, a text clustering algorithm was introduced by Slonim et al. [11,12], which uses the IB method with an annealing procedure, where the parameter s is increased gradually. In [12], the critical values of s (so-called phase transitions) were observed such that if these values were missed during increasing s, the algorithm ended up with the wrong clusters. Therefore, how to choose the step size in the update of s is very important. We note that tuning s is also very critical in our algorithm, such that the step size ϵ_s in the update of s should be chosen carefully, otherwise phase transitions might be skipped that would cause a non-satisfactory clustering accuracy score. However, the choice of the appropriate step size (typically very small) is rather heuristic; and there exists no concrete method for choosing the right value. The choice of step size can be seen as a trade-off between the amount of computational resource spared for running the algorithm and the degree of confidence about scanning s values not to miss the phase transitions.*

4. Experiments

4.1. Description of the Datasets Used

In our empirical experiments, we applied our algorithm to the clustering of the following datasets.

MNIST: A dataset of gray-scale images of 70,000 handwritten digits of dimensions 28×28 pixel.

STL-10: A dataset of color images collected from 10 categories. Each category consisted of 1300 images of size of 96×96 (pixels) $\times 3$ (RGB code). Hence, the original input dimension n_x was 27,648. For this dataset, we used a pretrained convolutional NN model, i.e., ResNet-50 [27], to reduce the dimensionality of the input. This preprocessing reduced the input dimension to 2048. Then, our algorithm and other baselines were used for clustering.

REUTERS10K: A dataset that was composed of 810,000 English stories labeled with a category tree. As in [20], 4 root categories (corporate/industrial, government/social, markets, economics) were selected as labels, and all documents with multiple labels were discarded. Then, tf-idf features were computed on the 2000 most frequently occurring words. Finally, 10,000 samples were taken randomly, which were referred to as the REUTERS10K dataset.

4.2. Network Settings and Other Parameters

We used the following network architecture: the encoder was modeled with NNs with 3 hidden layers with dimensions $n_x - 500 - 500 - 2000 - n_u$, where n_x is the input dimension and n_u is the dimension of the latent space. The decoder consisted of NNs with dimensions $n_u - 2000 - 500 - 500 - n_x$. All layers were fully connected. For comparison purposes, we chose the architecture of the hidden layers, as well as the dimension of the latent space $n_u = 10$ to coincide with those made for the DEC algorithm of [20] and the VaDE algorithm of [19]. All except the last layers of the encoder and decoder were activated with the ReLU function. For the last (i.e., latent) layer of the encoder, we used a linear activation; and for the last (i.e., output) layer of the decoder, we used the sigmoid function for MNIST and linear activation for the remaining datasets. The batch size was 100, and the variational bound (20) was maximized by the ADAM optimizer of [25]. The learning rate was initialized with 0.002 and decreased gradually every 20 epochs with a decay rate of 0.9 until it reached a small value (0.0005 in our experiments). The reconstruction loss was calculated by using the cross-entropy criterion for MNIST and the mean squared error function for the other datasets.

4.3. Clustering Accuracy

We evaluated the performance of our algorithm in terms of the so-called unsupervised clustering Accuracy (ACC), which is a widely used metric in the context of unsupervised learning [23]. For comparison purposes, we also present those of algorithms from the previous state-of-the-art.

For each of the aforementioned datasets, we ran our VIB-GMM algorithm for various values of the hyper-parameter s inside an interval $[s_{\min}, s_{\max}]$, starting from the smaller valuer $s_{\min}$ and gradually increasing the value of s every n_{epoch} epochs. For the MNIST dataset, we set $(s_{\min}, s_{\max}, n_{\text{epoch}}) = (1, 5, 500)$; and for the STL-10 dataset and the REUTERS10K dataset, we chose these parameters to be $(1, 20, 500)$ and $(1, 5, 100)$, respectively. The obtained ACC accuracy results are reported in Tables 1 and 2. It is important to note that the reported ACC results were obtained by running each algorithm ten times. For the case in which there was no pretraining (in [19,20], the DEC and VaDE algorithms were proposed to be used with pretraining; more specifically, the DNNs were initialized with a stacked autoencoder [28]), Table 1 states the accuracies of the best case run and average case run for the MNIST and STL-10 datasets. It was seen that our algorithm outperformed significantly the DEC algorithm of [20], as well as the VaDE algorithm of [19] and GMM for both the best case run and average case run. Besides, in Table 1, the values in parentheses correspond to the standard deviations of clustering accuracies. As seen, the standard deviation of our algorithm VIB-GMM was lower than the VaDE; which could be expounded by the robustness of VIB-GMM to non-pretraining. For the case in which there was pretraining, Table 2 states the accuracies of the best case run and average case run for the MNIST and REUTERS10K datasets. A stacked autoencoder was used to pretrain the DNNs of the encoder and decoder before running algorithms (DNNs were initialized with the same weights and biases of [19]). It was seen that our algorithm outperformed significantly the DEC algorithm of [20], as well as the VaDE algorithm of [19] and GMM for both the best case run and average case run. The effect of pretraining can be observed comparing Tables 1 and 2 for MNIST. Using a stacked autoencoder prior to running the VaDE and VIB-GMM algorithms resulted in a higher accuracy, as well as a lower standard deviation of accuracies; therefore, supporting the algorithms with a stacked autoencoder was beneficial for a more robust system. Finally, for the STL-10 dataset, Figure 4 depicts the evolution of the best case ACC with iterations (number of epochs) for the four compared algorithms.

Table 1. Comparison of the clustering accuracy of various algorithms. The algorithms are run without pretraining. Each algorithm is run ten times. The values in $(\cdot)$ correspond to the standard deviations of clustering accuracies. DEC, Deep Embedded Clustering; VaDE, Variational Deep Embedding; VIB, Variational Information Bottleneck.

	MNIST		STL-10	
	Best Run	Average Run	Best Run	Average Run
GMM	44.1	40.5 (1.5)	78.9	73.3 (5.1)
DEC			80.6 [†]	
VaDE	91.8	78.8 (9.1)	85.3	74.1 (6.4)
VIB-GMM	**95.1**	**83.5** (5.9)	**93.2**	**82.1** (5.6)

[†] Values are taken from VaDE [19].

Table 2. Comparison of the clustering accuracy of various algorithms. A stacked autoencoder is used to pretrain the DNNs of the encoder and decoder before running algorithms (DNNs are initialized with the same weights and biases of [19]). Each algorithm is run ten times. The values in $(\cdot)$ correspond to the standard deviations of clustering accuracies.

	MNIST		REURTERS10K	
	Best Run	Average Run	Best Run	Average Run
DEC	84.3[‡]		72.2 [‡]	
VaDE	94.2	93.2 (1.5)	79.8	79.1 (0.6)
VIB-GMM	**96.1**	**95.8** (0.1)	**81.6**	**81.2** (0.4)

[‡] Values are taken from DEC [20].

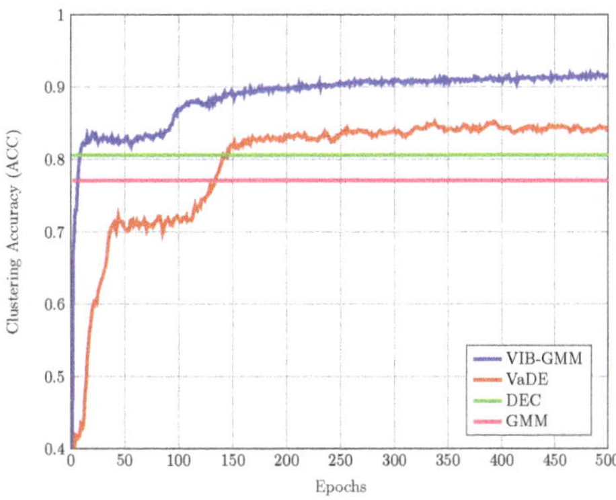

Figure 4. Accuracy vs. the number of epochs for the STL-10 dataset.

Figure 5 shows the evolution of the reconstruction loss of our VIB-GMM algorithm for the STL-10 dataset, as a function of simultaneously varying the values of the hyperparameter s and the number of epochs (recall that, as per the described methodology, we started with $s = s_{\min}$, and we increased its value gradually every $n_{\text{epoch}} = 500$ epochs). As can be seen from the figure, the few first epochs were spent almost entirely on reducing the reconstruction loss (i.e., a fitting phase), and most of the remaining epochs were spent making the found representation more concise (i.e., smaller KL divergence). This was reminiscent of the two-phase (fitting vs. compression) that was observed for supervised learning using VIB in [29].

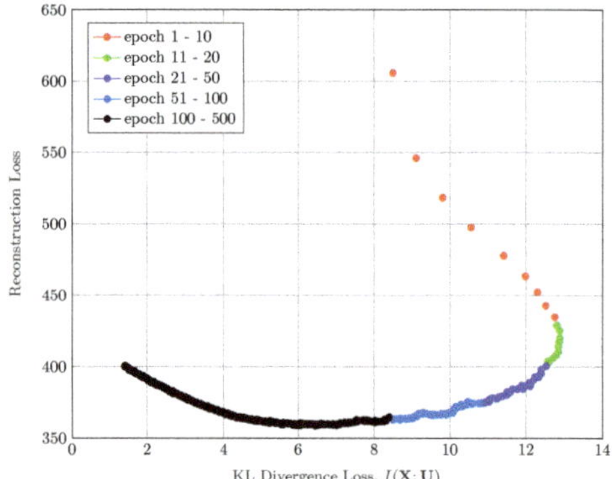

Figure 5. Information plane for the STL-10 dataset.

Remark 5. *For a fair comparison, our algorithm VIB-GMM and the VaDE of [19] were run for the same number of epochs, e.g., n_{epoch}. In the VaDE algorithm, the cost function (11) was optimized for a particular value of hyperparameter s. Instead of running n_{epoch} epochs for $s = 1$ as in VaDE, we ran n_{epoch} epochs by gradually increasing s to optimize the cost (21). In other words, the computational resources were distributed over a range of s values. Therefore, the computational complexity of our algorithm and the VaDE was equivalent.*

4.4. Visualization on the Latent Space

In this section, we investigate the evolution of the unsupervised clustering of the STL-10 dataset on the latent space using our VIB-GMM algorithm. For this purpose, we found it convenient to visualize the latent space through application of the t-SNEalgorithm of [30] in order to generate meaningful representations in a two-dimensional space. Figure 6 shows 4000 randomly chosen latent representations before the start of the training process and respectively after 1, 5, and 500 epochs. The shown points (with a · marker in the figure) represent latent representations of data samples whose labels are identical. Colors are used to distinguish between clusters. Crosses (with an **x** marker in the figure) correspond to the centroids of the clusters. More specifically, Figure 6a shows the initial latent space before the training process. If the clustering is performed on the initial representations, it allows ACC as small as 10%, i.e., as bad as a random assignment. Figure 6b shows the latent space after one epoch, from which a partition of some of the points starts to be already visible. With five epochs, that partitioning is significantly sharper, and the associated clusters can be recognized easily. Observe, however, that the cluster centers seem not to have converged yet. With 500 epochs, the ACC accuracy of our algorithm reached %91.6, and the clusters and their centroids were neater, as is visible from Figure 6d.

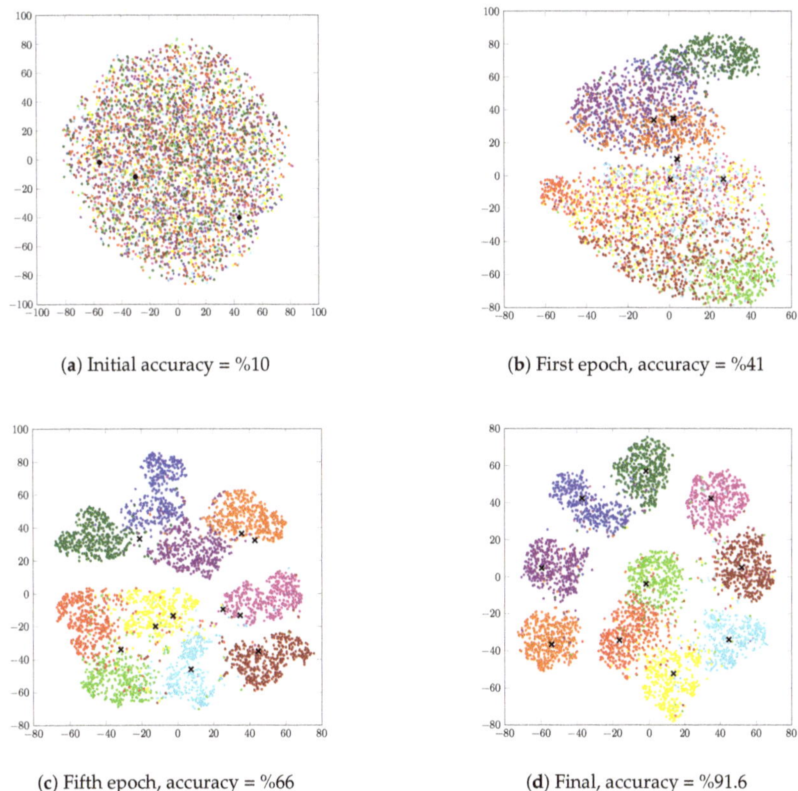

(**a**) Initial accuracy = %10

(**b**) First epoch, accuracy = %41

(**c**) Fifth epoch, accuracy = %66

(**d**) Final, accuracy = %91.6

Figure 6. Visualization of the latent space before training; and after 1, 5, and 500 epochs.

5. Conclusions and Future Work

In this paper, we proposed and analyzed the performance of an unsupervised algorithm for data clustering. The algorithm used the variational information bottleneck approach and modeled the latent space as a mixture of Gaussians. It was shown to outperform state-of-the-art algorithms such as the VaDE of [19] and the DEC of [20]. We note that although it was assumed that the number of classes was known beforehand (as was the case for almost all competing algorithms in its category), that number could be found (or estimated to within a certain accuracy) through inspection of the resulting bifurcations on the associated information-plane, as was observed for the standard information bottleneck method. Finally, we mention that among the interesting research directions in this line of work, one important question pertains to the distributed learning setting, i.e., along the counterpart, to the unsupervised setting, of the recent work [31–33], which contained distributed IB algorithms for both discrete and vector Gaussian data models.

Author Contributions: Software, Y.U.; Methodology, A.Z., Y.U. and G.A.; Supervision, A.Z.; Writing—original draft, A.Z., Y.U. and G.A. Writing—Review & Editing, Y.U. and A.Z. All authors have read and agreed to the published version of the manuscript.

Funding: The authors received no specific funding for this work.

Acknowledgments: The authors thank the reviewers for their comments that helped to improve the paper.

Conflicts of Interest: The authors declare no conflict of interest.

Appendix A. The Proof of Lemma 1

First, we expand $\mathcal{L}'_s(\mathbf{P})$ as follows:

$$\mathcal{L}'_s(\mathbf{P}) = -H(\mathbf{X}|\mathbf{U}) - sI(\mathbf{X};\mathbf{U}) = -H(\mathbf{X}|\mathbf{U}) - s[H(\mathbf{U}) - H(\mathbf{U}|\mathbf{X})]$$
$$= \iint_{\mathbf{ux}} p(\mathbf{u},\mathbf{x}) \log p(\mathbf{x}|\mathbf{u})\, d\mathbf{u}\, d\mathbf{x} + s\int_{\mathbf{u}} p(\mathbf{u}) \log p(\mathbf{u})\, d\mathbf{u} - s\iint_{\mathbf{ux}} p(\mathbf{u},\mathbf{x}) \log p(\mathbf{u}|\mathbf{x})\, d\mathbf{u}\, d\mathbf{x}\,.$$

Then, $\mathcal{L}^{\mathrm{VB}}_s(\mathbf{P},\mathbf{Q})$ is defined as follows:

$$\mathcal{L}^{\mathrm{VB}}_s(\mathbf{P},\mathbf{Q}) := \iint_{\mathbf{ux}} p(\mathbf{u},\mathbf{x}) \log q(\mathbf{x}|\mathbf{u})\, d\mathbf{u}\, d\mathbf{x} + s\int_{\mathbf{u}} p(\mathbf{u}) \log q(\mathbf{u})\, d\mathbf{u} - s\iint_{\mathbf{ux}} p(\mathbf{u},\mathbf{x}) \log p(\mathbf{u}|\mathbf{x})\, d\mathbf{u}\, d\mathbf{x}\,. \quad (\text{A1})$$

Hence, we have the following relation:

$$\mathcal{L}'_s(\mathbf{P}) - \mathcal{L}^{\mathrm{VB}}_s(\mathbf{P},\mathbf{Q}) = \mathbb{E}_{P_{\mathbf{U}}}[D_{\mathrm{KL}}(P_{\mathbf{X}|\mathbf{U}}\|Q_{\mathbf{X}|\mathbf{U}})] + sD_{\mathrm{KL}}(P_{\mathbf{U}}\|Q_{\mathbf{U}}) \geq 0\,,$$

where equality holds under equalities $Q_{\mathbf{X}|\mathbf{U}} = P_{\mathbf{X}|\mathbf{U}}$ and $Q_{\mathbf{U}} = P_{\mathbf{U}}$. We note that $s \geq 0$.

Now, we complete the proof by showing that (A1) is equal to (8). To do so, we proceed (A1) as follows:

$$\mathcal{L}^{\mathrm{VB}}_s(\mathbf{P},\mathbf{Q}) = \int_{\mathbf{x}} p(\mathbf{x}) \int_{\mathbf{u}} p(\mathbf{u}|\mathbf{x}) \log q(\mathbf{x}|\mathbf{u})\, d\mathbf{u}\, d\mathbf{x}$$
$$+ s\int_{\mathbf{x}} p(\mathbf{x}) \int_{\mathbf{u}} p(\mathbf{u}|\mathbf{x}) \log q(\mathbf{u})\, d\mathbf{u} - s\int_{\mathbf{x}} p(\mathbf{x}) \int_{\mathbf{u}} p(\mathbf{u}|\mathbf{x}) \log p(\mathbf{u}|\mathbf{x})\, d\mathbf{u}\, d\mathbf{x}$$
$$= \mathbb{E}_{P_{\mathbf{X}}}\Big[\mathbb{E}_{P_{\mathbf{U}|\mathbf{x}}}[\log Q_{\mathbf{X}|\mathbf{U}}] - sD_{\mathrm{KL}}(P_{\mathbf{U}|\mathbf{x}}\|Q_{\mathbf{U}})\Big]\,.$$

Appendix B. Alternative Expression $\mathcal{L}^{\mathrm{VaDE}}_s$

Here, we show that (13) is equal to (14).

To do so, we start with (14) and proceed as follows:

$$\mathcal{L}^{\mathrm{VaDE}}_s = \mathbb{E}_{P_{\mathbf{X}}}\Big[\mathbb{E}_{P_{\mathbf{U}|\mathbf{x}}}[\log Q_{\mathbf{X}|\mathbf{U}}] - sD_{\mathrm{KL}}(P_{\mathbf{U}|\mathbf{x}}\|Q_{\mathbf{U}}) - s\mathbb{E}_{P_{\mathbf{U}|\mathbf{x}}}\big[D_{\mathrm{KL}}(P_{\mathbf{C}|\mathbf{x}}\|Q_{\mathbf{C}|\mathbf{U}})\big]\Big]$$

$$= \mathbb{E}_{P_{\mathbf{X}}}[\mathbb{E}_{P_{\mathbf{U}|\mathbf{x}}}[\log Q_{\mathbf{X}|\mathbf{U}}]] - s\int_{\mathbf{x}} p(\mathbf{x}) \int_{\mathbf{u}} p(\mathbf{u}|\mathbf{x}) \log \frac{p(\mathbf{u}|\mathbf{x})}{q(\mathbf{u})}\, d\mathbf{u}\, d\mathbf{x}$$
$$- s\int_{\mathbf{x}} p(\mathbf{x}) \int_{\mathbf{u}} p(\mathbf{u}|\mathbf{x}) \sum_c p(c|\mathbf{x}) \log \frac{p(c|\mathbf{x})}{q(c|\mathbf{u})}\, d\mathbf{u}\, d\mathbf{x}$$

$$\stackrel{(a)}{=} \mathbb{E}_{P_{\mathbf{X}}}[\mathbb{E}_{P_{\mathbf{U}|\mathbf{x}}}[\log Q_{\mathbf{X}|\mathbf{U}}]] - s\iint_{\mathbf{ux}} p(\mathbf{x})p(\mathbf{u}|\mathbf{x}) \log \frac{p(\mathbf{u}|\mathbf{x})}{q(\mathbf{u})}\, d\mathbf{u}\, d\mathbf{x}$$
$$- s\iint_{\mathbf{ux}} \sum_c p(\mathbf{x})p(\mathbf{u}|c,\mathbf{x})p(c|\mathbf{x}) \log \frac{p(c|\mathbf{x})}{q(c|\mathbf{u})}\, d\mathbf{u}\, d\mathbf{x}$$

$$= \mathbb{E}_{P_{\mathbf{X}}}[\mathbb{E}_{P_{\mathbf{U}|\mathbf{x}}}[\log Q_{\mathbf{X}|\mathbf{U}}]] - s\iint_{\mathbf{ux}} \sum_c p(\mathbf{u},c,\mathbf{x}) \log \frac{p(\mathbf{u}|\mathbf{x})p(c|\mathbf{x})}{q(\mathbf{u})q(c|\mathbf{u})}\, d\mathbf{u}\, d\mathbf{x}$$

$$= \mathbb{E}_{P_{\mathbf{X}}}[\mathbb{E}_{P_{\mathbf{U}|\mathbf{x}}}[\log Q_{\mathbf{X}|\mathbf{U}}]] - s\iint_{\mathbf{ux}} \sum_c p(\mathbf{u},c,\mathbf{x}) \log \frac{p(c|\mathbf{x})}{q(c)} \frac{p(\mathbf{u}|\mathbf{x})}{q(\mathbf{u}|c)}\, d\mathbf{u}\, d\mathbf{x}$$

$$= \mathbb{E}_{P_{\mathbf{X}}}[\mathbb{E}_{P_{\mathbf{U}|\mathbf{x}}}[\log Q_{\mathbf{X}|\mathbf{U}}]] - s\int_{\mathbf{x}} \sum_c p(c,\mathbf{x}) \log \frac{p(c|\mathbf{x})}{q(c)}\, d\mathbf{x}$$
$$- s\iint_{\mathbf{ux}} \sum_c p(\mathbf{x})p(c|\mathbf{x})p(\mathbf{u}|c,\mathbf{x}) \log \frac{p(\mathbf{u}|\mathbf{x})}{q(\mathbf{u}|c)}\, d\mathbf{u}\, d\mathbf{x}$$

$$\overset{(b)}{=} \mathbb{E}_{P_{\mathbf{X}}}\left[\mathbb{E}_{P_{\mathbf{U}|\mathbf{X}}}[\log Q_{\mathbf{X}|\mathbf{U}}] - s D_{\mathrm{KL}}(P_{C|\mathbf{X}}\|Q_C) - s\mathbb{E}_{P_{C|\mathbf{X}}}[D_{\mathrm{KL}}(P_{\mathbf{U}|\mathbf{X}}\|Q_{\mathbf{U}|C})]\right],$$

where (a) and (b) follow due to the Markov chain $C \,\text{--o--}\, \mathbf{X} \,\text{--o--}\, \mathbf{U}$.

References

1. Sculley, D. Web-scale *K*-means clustering. In Proceedings of the 19th International Conference on World Wide Web, Raleigh, NC, USA, 26–30 April 2010; pp. 1177–1178.
2. Huang, Z. Extensions to the *k*-means algorithm for clustering large datasets with categorical values. *Data Min. Knowl. Disc.* **1998**, *2*, 283–304. [CrossRef]
3. Hartigan, J.A.; Wong, M.A. Algorithm AS 136: A *K*-means clustering algorithm. *J. R. Stat. Soc.* **1979**, *28*, 100–108. [CrossRef]
4. Dempster, A.P.; Laird, N.M.; Rubin, D.B. Maximum likelihood from incomplete data via the EM algorithm. *J. R. Stat. Soc.* **1977**, *39*, 1–38.
5. Ding, C.; He, X. *K*-means clustering via principal component analysis. In Proceedings of the 21st International Conference on Machine Learning, Banff, AB, Canada, 4–8 July 2004.
6. Pearson, K. On lines and planes of closest fit to systems of points in space. *Philos. Mag.* **1901**, *2*, 559–572. [CrossRef]
7. Wold, S.; Esbensen, K.; Geladi, P. Principal component analysis. *Chemom. Intell. Lab. Syst.* **1987**, *2*, 37–52. [CrossRef]
8. Roweis, S. EM algorithms for PCA and SPCA. In Proceedings of the Advances in Neural Information Processing Systems 10, Denver, CO, USA, 1–6 December 1997; pp. 626–632.
9. Hofmann, T.; Scholkopf, B.; Smola, A.J. Kernel methods in machine learning. *Ann. Stat.* **2008**, *36*, 1171–1220. [CrossRef]
10. Tishby, N.; Pereira, F.C.; Bialek, W. The information bottleneck method. In Proceedings of the 37th Annual Allerton Conference on Communication, Control and Computing, Monticello, IL, USA, 22–24 September 1999; pp. 368–377.
11. Slonim, N.; Tishby, N. Document clustering using word clusters via the information bottleneck method. In Proceedings of the 23rd Annual International ACM SIGIR Conference on Research and Development in Information Retrieval, Athens, Greece, 24–28 July 2000; pp. 208–215.
12. Slonim, N. The Information Bottleneck: Theory and Applications. Ph.D. Thesis, Hebrew University, Jerusalem, Israel, 2002.
13. Kingma, D.P.; Welling, M. Auto-encoding variational bayes. In Proceedings of the 2nd International Conference on Learning Representations, Banff, AB, Canada, 14–16 April 2014.
14. Rezende, D.J.; Mohamed, S.; Wierstra, D. Stochastic backpropagation and approximate inference in deep generative models. In Proceedings of the 31st International Conference on Machine Learning, Beijing, China, 21–26 June 2014; pp. 1278–1286.
15. Alemi, A.A.; Fischer, I.; Dillon, J.V.; Murphy, K. Deep variational information bottleneck. In Proceedings of the 5th International Conference on Learning Representations, Toulon, France, 24–26 April 2017.
16. Lecun, Y.; Bottou, L.; Bengio, Y.; Haffner, P. Gradient-based learning applied to document recognition. *Proc. IEEE* **1998**, *86*, 2278–2324. [CrossRef]
17. Lewis, D.D.; Yang, Y.; Rose, T.G.; Li, F. A new benchmark collection for text categorization research. *J. Mach. Learn. Res.* **2004**, *5*, 361–397.
18. Coates, A.; Ng, A.; Lee, H. An analysis of single-layer networks in unsupervised feature learning. In Proceedings of the 14th International Conference on Artificial Intelligence and Statistics, Fort Lauderdale, FL, USA, 11–13 April 2011; pp. 215–223.
19. Jiang, Z.; Zheng, Y.; Tan, H.; Tang, B.; Zhou, H. Variational deep embedding: An unsupervised and generative approach to clustering. In Proceedings of the 26th International Joint Conference on Artificial Intelligence, Melbourne, Australia, 19–25 August 2017; pp. 1965–1972.
20. Xie, J.; Girshick, R.; Farhadi, A. Unsupervised deep embedding for clustering analysis. In Proceedings of the 33rd International Conference on Machine Learning, New York, NY, USA, 19–24 June 2016; pp. 478–487.

21. Guo, X.; Gao, L.; Liu, X.; Yin, J. Improved deep embedded clustering with local structure preservation. In Proceedings of the 26th International Joint Conference on Artificial Intelligence, Melbourne, Australia, 19–25 August 2017; pp. 1753–1759.

22. Dilokthanakul, N.; Mediano, P.A.M.; Garnelo, M.; Lee, M.C.H.; Salimbeni, H.; Arulkumaran, K.; Shanahani, M. Deep unsupervised clustering with Gaussian mixture variational autoencoders. *arXiv* **2017**, arXiv:1611.02648.

23. Min, E.; Guo, X.; Liu, Q.; Zhang, G.; Cui, J.; Long, J. A survey of clustering with deep learning: From the perspective of network architecture. *IEEE Access* **2018**, *6*, 39501–39514. [CrossRef]

24. Hershey, J.R.; Olsen, P.A. Approximating the Kullback Leibler divergence between Gaussian mixture models. In Proceedings of the IEEE International Conference on Acoustics, Speech and Signal Processing, Honolulu, HI, USA, 15–20 April 2007; pp. 317–320.

25. Kingma, D.P.; Ba, J. Adam: A method for stochastic optimization. In Proceedings of the 3rd International Conference on Learning Representations, San Diego, CA, USA, 7–9 May 2015.

26. Achille, A.; Soatto, S. Information dropout: Learning optimal representations through noisy computation. *IEEE Trans. Pattern Anal. Mach. Intell.* **2018**, *40*, 2897–2905. [CrossRef] [PubMed]

27. He, K.; Zhang, X.; Ren, S.; Sun, J. Deep residual learning for image recognition. In Proceedings of the IEEE Conference on Computer Vision and Pattern Recognition, Las Vegas, NV, USA, 27–30 June 2016; pp. 770–778.

28. Vincent, P.; Larochelle, H.; Lajoie, I.; Bengio, Y.; Manzagol, P.-A. Stacked denoising autoencoders: Learning useful representations in a deep network with a local denoising criterion. *J. Mach. Learn. Res.* **2010**, *11*, 3371–3408.

29. Schwartz-Ziv, R.; Tishby, N. Opening the black box of deep neural networks via information. *arXiv* **2017**, arXiv:1703.00810.

30. van der Maaten, L.; Hinton, G. Visualizing Data using t-SNE. *J. Mach. Learn. Res* **2008**, *9*, 2579–2605.

31. Estella-Aguerri, I.; Zaidi, A. Distributed variational representation learning. *IEEE Trans. Pattern Anal. Mach. Intell.* **2020**, in press

32. Estella-Aguerri, I.; Zaidi, A. Distributed information bottleneck method for discrete and Gaussian sources. In Proceedings of the International Zurich Seminar on Information and Communication, Zürich, Switzerland, 21–23 February 2018.

33. Zaidi, A.; Estella-Aguerri, I.; Shamai (Shitz), S. On the information bottleneck problems: Models, connections, applications and information theoretic views. *Entropy* **2020**, *22*, 151. [CrossRef]

Article

Asymptotic Rate-Distortion Analysis of Symmetric Remote Gaussian Source Coding: Centralized Encoding vs. Distributed Encoding

Yizhong Wang [1], Li Xie [2], Siyao Zhou [3], Mengzhen Wang [3] and Jun Chen [1,3,*]

[1] College of Electronic Information and Automation, Tianjin University of Science and Technology, Tianjin 300222, China; yzwang@tust.edu.cn
[2] Department of Electrical System of Launch Vehicle, Institute of Aerospace System Engineering Shanghai, Shanghai Academy of Spaceflight Technology, Shanghai 201109, China; lixie.lx@gmail.com
[3] Department of Electrical and Computer Engineering, McMaster University, Hamilton, ON L8S 4K1, Canada; zhous58@mcmaster.ca (S.Z.); wangm43@mcmaster.ca (M.W.)
* Correspondence: junchen@ece.mcmaster.ca or chenjun@tust.edu.cn; Tel.: +1-905-525-9140 (ext. 20163)

Received: 10 January 2019; Accepted: 20 February 2019; Published: 23 February 2019

Abstract: Consider a symmetric multivariate Gaussian source with ℓ components, which are corrupted by independent and identically distributed Gaussian noises; these noisy components are compressed at a certain rate, and the compressed version is leveraged to reconstruct the source subject to a mean squared error distortion constraint. The rate-distortion analysis is performed for two scenarios: centralized encoding (where the noisy source components are jointly compressed) and distributed encoding (where the noisy source components are separately compressed). It is shown, among other things, that the gap between the rate-distortion functions associated with these two scenarios admits a simple characterization in the large ℓ limit.

Keywords: CEO problem; mean squared error; multiterminal source coding; rate-distortion; remote source coding

---

1. Introduction

Many applications involve collection and transmission of potentially noise-corrupted data. It is often necessary to compress the collected data to reduce the transmission cost. The remote source coding problem aims to characterize the optimal scheme for such compression and the relevant information-theoretic limit. In this work we study a quadratic Gaussian version of the remote source coding problem, where compression is performed on the noise-corrupted components of a symmetric multivariate Gaussian source. A prescribed mean squared error distortion constraint is imposed on the reconstruction of the noise-free source components; moreover, it is assumed that the noises across different source components are independent and obey the same Gaussian distribution. Two scenarios are considered: centralized encoding (see Figure 1) and distributed encoding (see Figure 2). It is worth noting that the distributed encoding scenario is closely related to the CEO problem, which has been studied extensively [1–18].

The present paper is primarily devoted to the comparison of the rate-distortion functions associated with the aforementioned two scenarios. We are particularly interested in understanding how the rate penalty for distributed encoding (relative to centralized encoding) depends on the target distortion as well as the parameters of source and noise models. Although the information-theoretic results needed for this comparison are available in the literature or can be derived in a relatively straightforward manner, the relevant expressions are too unwieldy to analyze. For this reason, we focus on the asymptotic regime where the number of source components, denoted by ℓ, is sufficiently large. Indeed, it will be seen that the gap between the two rate-distortion functions admits a simple characterization in the large ℓ limit, yielding useful insights into the fundamental difference between centralized encoding and distributed coding, which are hard to obtain otherwise.

The rest of this paper is organized as follows. We state the problem definitions and the main results in Section 2. The proofs are provided in Section 3. We conclude the paper in Section 4.

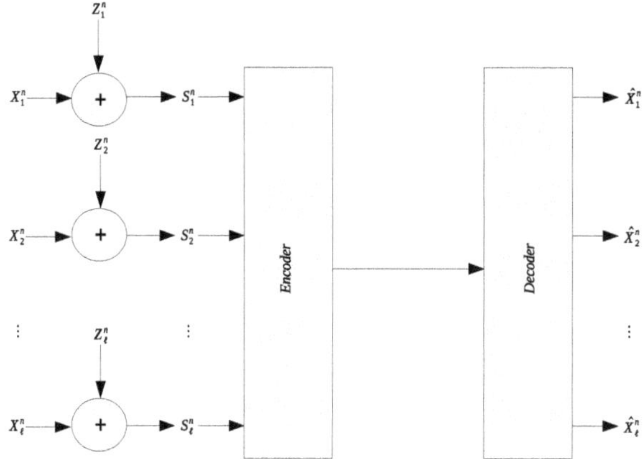

Figure 1. Symmetric remote Gaussian source coding with centralized encoding.

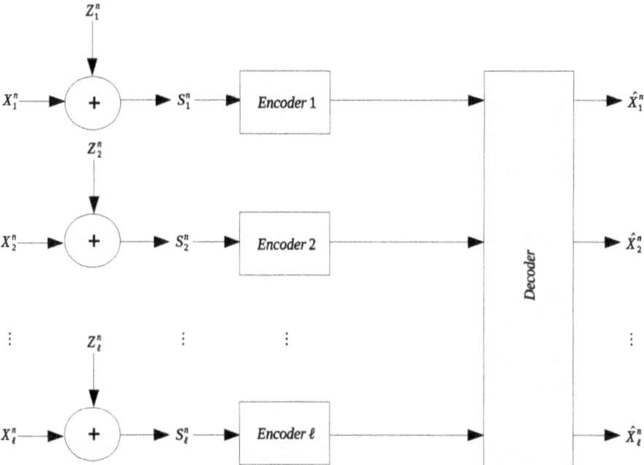

Figure 2. Symmetric remote Gaussian source coding with distributed encoding.

Notation: The expectation operator and the transpose operator are denoted by $\mathbb{E}[\cdot]$ and $(\cdot)^T$, respectively. An ℓ-dimensional all-one row vector is written as 1_ℓ. We use W^n as an abbreviation of $(W(1), \cdots, W(n))$. The cardinality of a set $\mathcal{C}$ is denoted by $|\mathcal{C}|$. We write $g(\ell) = O(f(\ell))$ if the absolute value of $\frac{g(\ell)}{f(\ell)}$ is bounded for all sufficiently large ℓ. Throughout this paper, the base of the logarithm function is e, and $\log^+ x \triangleq \max\{\log x, 0\}$.

2. Problem Definitions and Main Results

Let $S \triangleq (S_1, \cdots, S_\ell)^T$ be the sum of two mutually independent ℓ-dimensional ($\ell \geq 2$) zero-mean Gaussian random vectors, source $X \triangleq (X_1, \cdots, X_\ell)^T$ and noise $Z \triangleq (Z_1, \cdots, Z_\ell)^T$, with

$$
\mathbb{E}[X_i X_j] = \begin{cases} \gamma_X, & i = j, \\ \rho_X \gamma_X, & i \neq j, \end{cases}
$$

$$
\mathbb{E}[Z_i Z_j] = \begin{cases} \gamma_Z, & i = j, \\ 0, & i \neq j, \end{cases}
$$

where $\gamma_X > 0$, $\rho_X \in [\frac{1}{\ell-1}, 1]$, and $\gamma_Z \geq 0$. Moreover, let $\{(X(t), Z(t), S(t))\}_{t=1}^\infty$ be i.i.d. copies of (X, Z, S).

Definition 1 (Centralized encoding). *A rate-distortion pair (r, d) is said to be achievable with centralized encoding if, for any $\epsilon > 0$, there exists an encoding function $\phi^{(n)} : \mathbb{R}^{\ell \times n} \to \mathcal{C}^{(n)}$ such that*

$$
\frac{1}{n} \log |\mathcal{C}^{(n)}| \leq r + \epsilon,
$$

$$
\frac{1}{\ell n} \sum_{i=1}^{\ell} \sum_{t=1}^{n} \mathbb{E}[(X_i(t) - \hat{X}_i(t))^2] \leq d + \epsilon,
$$

where $\hat{X}_i(t) \triangleq \mathbb{E}[X_i(t)|(\phi^{(n)}(S^n))]$. For a given d, we denote by $\underline{r}(d)$ the minimum r such that (r, d) is achievable with centralized encoding.

Definition 2 (Distributed encoding). *A rate-distortion pair (r, d) is said to be achievable with distributed encoding if, for any $\epsilon > 0$, there exist encoding functions $\phi_i^{(n)} : \mathbb{R}^n \to \mathcal{C}_i^{(n)}$, $i = 1, \cdots, \ell$, such that*

$$
\frac{1}{n} \sum_{i=1}^{\ell} \log |\mathcal{C}_i^{(n)}| \leq r + \epsilon,
$$

$$
\frac{1}{\ell n} \sum_{i=1}^{\ell} \sum_{t=1}^{n} \mathbb{E}[(X_i(t) - \hat{X}_i(t))^2] \leq d + \epsilon,
$$

where $\hat{X}_i(t) \triangleq \mathbb{E}[X_i(t)|(\phi_1^{(n)}(S_1^n), \cdots, \phi_\ell^{(n)}(S_\ell^n))]$. For a given d, we denote by $\bar{r}(d)$ the minimum r such that (r, d) is achievable with distributed encoding.

We will refer to $\underline{r}(d)$ as the rate-distortion function of symmetric remote Gaussian source coding with centralized encoding, and $\bar{r}(d)$ as the rate-distortion function of symmetric remote Gaussian source coding with distributed encoding. It is clear that $\underline{r}(d) \leq \bar{r}(d)$ for any d since distributed encoding can be simulated by centralized encoding. Moreover, it is easy to show that $\underline{r}(d) = \bar{r}(d) = 0$ for $d \geq \gamma_X$ (since the distortion constraint is trivially satisfied with the reconstruction set to be zero) and $\underline{r}(d) = \bar{r}(d) = \infty$ for $d \leq d_{\min}$ (since $d_{\min}$ is the minimum achievable distortion when $\{S(t)\}_{t=1}^\infty$ is directly available at the decoder), where (see Section 3.1 for a detailed derivation)

$$
d_{\min} \triangleq \frac{1}{\ell} \mathbb{E}[(X - \mathbb{E}[X|S])^T (X - \mathbb{E}[X|S])] = \begin{cases} \frac{(\ell-1)\gamma_X \gamma_Z}{\ell \gamma_X + (\ell-1)\gamma_Z}, & \rho_X = -\frac{1}{\ell-1}, \\ \frac{(\ell \rho_X \gamma_X + \lambda_X)\gamma_Z}{\ell(\ell \rho_X \gamma_X + \lambda_X + \gamma_Z)} + \frac{(\ell-1)\lambda_X \gamma_Z}{\ell(\lambda_X + \gamma_Z)}, & \rho_X \in (-\frac{1}{\ell-1}, 1), \\ \frac{\gamma_X \gamma_Z}{\ell \gamma_X + \gamma_Z}, & \rho_X = 1, \end{cases}
$$

with $\lambda_X \triangleq (1 - \rho_X)\gamma_X$. Henceforth we shall focus on the case $d \in (d_{\min}, \gamma_X)$.

Lemma 1. *For $d \in (d_{\min}, \gamma_X)$,*

$$
\underline{r}(d) = \begin{cases}
\frac{\ell-1}{2} \log \frac{\ell(\ell-1)\gamma_X^2}{(\ell\gamma_X + (\ell-1)\gamma_Z)((\ell-1)d - \gamma_X)}, & \rho_X = -\frac{1}{\ell-1}, \\[2mm]
\frac{1}{2}\log^+ \frac{(\ell\rho_X\gamma_X + \lambda_X)^2}{(\ell\rho_X\gamma_X + \lambda_X + \gamma_Z)\xi} + \frac{\ell-1}{2}\log^+ \frac{\lambda_X^2}{(\lambda_X + \gamma_Z)\xi}, & \rho_X \in (-\frac{1}{\ell-1}, 1), \\[2mm]
\frac{1}{2}\log \frac{\ell\gamma_X^2}{(\ell\gamma_X + \gamma_Z)d - \gamma_X\gamma_Z}, & \rho_X = 1,
\end{cases}
$$

where

$$
\xi \triangleq \begin{cases}
d - d_{\min}, & d \leq \min\{\frac{(\ell\rho_X\gamma_X + \lambda_X)^2}{\ell\rho_X\gamma_X + \lambda_X + \gamma_Z}, \frac{\lambda_X^2}{\lambda_X + \gamma_Z}\} + d_{\min}, \\[2mm]
\frac{\ell(d - d_{\min})}{\ell-1} - \frac{(\ell\rho_X\gamma_X + \lambda_X)^2}{(\ell-1)(\ell\rho_X\gamma_X + \lambda_X + \gamma_Z)}, & d > \frac{(\ell\rho_X\gamma_X + \lambda_X)^2}{\ell\rho_X\gamma_X + \lambda_X + \gamma_Z} + d_{\min}, \\[2mm]
\ell(d - d_{\min}) - \frac{(\ell-1)\lambda_X^2}{\lambda_X + \gamma_Z}, & d > \frac{\lambda_X^2}{\lambda_X + \gamma_Z} + d_{\min}.
\end{cases}
$$

Proof. See Section 3.1. $\square$

The following result can be deduced from ([19] Theorem 1) (see also [11,15]).

Lemma 2. *For $d \in (d_{\min}, \gamma_X)$,*

$$
\bar{r}(d) = \frac{1}{2}\log \frac{\ell\rho_X\gamma_X + \lambda_X + \gamma_Z + \lambda_Q}{\lambda_Q} + \frac{\ell-1}{2}\log \frac{\lambda_X + \gamma_Z + \lambda_Q}{\lambda_Q},
$$

where

$$
\lambda_Q \triangleq \frac{-b + \sqrt{b^2 - 4ac}}{2a}
$$

with

$$
a \triangleq \ell(\gamma_X - d),
$$
$$
b \triangleq (\ell\rho_X\gamma_X + \lambda_X)(\lambda_X + 2\gamma_Z) + (\ell-1)\lambda_X(\ell\rho_X\gamma_X + \lambda_X + 2\gamma_Z) - \ell(\ell\rho_X\gamma_X + 2\lambda_X + 2\gamma_Z)d,
$$
$$
c \triangleq \ell(\ell\rho_X\gamma_X + \lambda_X + \gamma_Z)(\lambda_X + \gamma_Z)(d_{\min} - d).
$$

The expressions of $\underline{r}(d)$ and $\bar{r}(d)$ as shown in Lemmas 1 and 2 are quite complicated, rendering it difficult to make analytical comparisons. Fortunately, they become significantly simplified in the asymptotic regime where $\ell \to \infty$ (with d fixed). To perform this asymptotic analysis, it is necessary to restrict attention to the case $\rho_X \in [0,1]$; moreover, without loss of generality, we assume $d \in (d_{\min}^{(\infty)}, \gamma_X)$, where

$$
d_{\min}^{(\infty)} \triangleq \lim_{\ell \to \infty} d_{\min} = \begin{cases}
\frac{\lambda_X\gamma_Z}{\lambda_X + \gamma_Z}, & \rho_X \in [0,1), \\[2mm]
0, & \rho_X = 1.
\end{cases}
$$

Theorem 1 (Centralized encoding).

1. $\rho_X = 0$: *For $d \in (d_{\min}^{(\infty)}, \gamma_X)$,*

$$
\underline{r}(d) = \frac{\ell}{2}\log \frac{\gamma_X^2}{(\gamma_X + \gamma_Z)d - \gamma_X\gamma_Z}.
$$

2. $\rho_X \in (0, 1]$: For $d \in (d_{min}^{(\infty)}, \gamma_X)$,

$$\underline{r}(d) = \begin{cases} \frac{\ell}{2} \log \frac{\lambda_X^2}{(\lambda_X + \gamma_Z)d - \lambda_X \gamma_Z} + \frac{1}{2} \log \ell + \underline{\alpha} + O(\frac{1}{\ell}), & d < \lambda_X, \\ \frac{1}{2} \log \ell + \frac{1}{2} \log \frac{\rho_X \gamma_X (\lambda_X + \gamma_Z)}{\lambda_X^2} + \frac{\gamma_Z^2}{2\lambda_X^2} + O(\frac{1}{\ell}), & d = \lambda_X, \\ \frac{1}{2} \log \frac{\rho_X \gamma_X}{d - \lambda_X} + O(\frac{1}{\ell}), & d > \lambda_X, \end{cases}$$

where

$$\underline{\alpha} \triangleq \frac{1}{2} \log \frac{\rho_X \gamma_X (\lambda_X + \gamma_Z)}{\lambda_X^2} + \frac{\gamma_Z^2}{2((\lambda_X + \gamma_Z)d - \lambda_X \gamma_Z)}.$$

Proof. See Section 3.2. $\square$

Theorem 2 (Distributed encoding).

1. $\rho_X = 0$: For $d \in (d_{min}^{(\infty)}, \gamma_X)$,

$$\bar{r}(d) = \frac{\ell}{2} \log \frac{\gamma_X^2}{(\gamma_X + \gamma_Z)d - \gamma_X \gamma_Z}.$$

2. $\rho_X \in (0, 1]$: For $d \in (d_{min}^{(\infty)}, \gamma_X)$,

$$\bar{r}(d) = \begin{cases} \frac{\ell}{2} \log \frac{\lambda_X^2}{(\lambda_X + \gamma_Z)d - \lambda_X \gamma_Z} + \frac{1}{2} \log \ell + \bar{\alpha} + O(\frac{1}{\ell}), & d < \lambda_X, \\ \frac{(\lambda_X + \gamma_Z)\sqrt{\ell}}{2\lambda_X} + \frac{1}{4} \log \ell + \frac{1}{2} \log \frac{\rho_X}{1 - \rho_X} - \frac{(\lambda_X + \gamma_Z)(\lambda_X - \rho_X \gamma_Z)}{4\rho_X \lambda_X^2} + O(\frac{1}{\sqrt{\ell}}), & d = \lambda_X, \\ \frac{1}{2} \log \frac{\rho_X \gamma_X}{d - \lambda_X} + \frac{(\lambda_X + \gamma_Z)(\gamma_X - d)}{2\rho_X \gamma_X (d - \lambda_X)} + O(\frac{1}{\ell}), & d > \lambda_X, \end{cases}$$

where

$$\bar{\alpha} \triangleq \frac{1}{2} \log \frac{\rho_X \gamma_X (\lambda_X - d)}{\lambda_X^2} + \frac{(\lambda_X + \gamma_Z)d^2}{2(\lambda_X - d)((\lambda_X + \gamma_Z)d - \lambda_X \gamma_Z)}.$$

Proof. See Section 3.3. $\square$

Remark 1. *One can readily recover ([20] Theorem 3) for the case $m = 1$ (see [20] for the definition of parameter m) and Oohama's celebrated result for the quadratic Gaussian CEO problem ([3] Corollary 1) by setting $\gamma_Z = 0$ and $\rho_X = 1$, respectively, in Theorem 2.*

The following result is a simple corollary of Theorems 1 and 2.

Corollary 1 (Asymptotic gap).

1. $\rho_X = 0$: For $d \in (d_{min}^{(\infty)}, \gamma_X)$,

$$\bar{r}(d) - \underline{r}(d) = 0.$$

2. $\rho_X \in (0, 1]$: For $d \in (d_{min}^{(\infty)}, \gamma_X)$,

$$\lim_{\ell \to \infty} \bar{r}(d) - \underline{r}(d) = \psi(d) \triangleq \begin{cases} \frac{1}{2} \log \frac{\lambda_X - d}{\lambda_X + \gamma_Z} + \frac{\gamma_Z + d}{2(\lambda_X - d)}, & d < \lambda_X, \\ \infty, & d = \lambda_X, \\ \frac{(\lambda_X + \gamma_Z)(\gamma_X - d)}{2\rho_X \gamma_X (d - \lambda_X)}, & d > \lambda_X. \end{cases}$$

Remark 2. *When $\rho_X = 0$, we have $\psi(d) = \frac{\gamma_Z(\gamma_X - d)}{2\gamma_X d}$, which is a monotonically decreasing function over $(0, \gamma_X)$, converging to ∞ (here we assume $\gamma_Z > 0$) and 0 as $d \to 0$ and γ_X, respectively. When $\rho_X \in (0, 1)$, it is clear that the function $\psi(d)$ is monotonically decreasing over (λ_X, γ_X), converging to ∞ and 0 as $d \to \lambda_X$ and γ_X, respectively; moreover, since $\psi'(d) = \frac{\gamma_Z + d}{2(\lambda_X - d)^2} > 0$ for $d \in (d_{\min}^{(\infty)}, \lambda_X)$, the function $\psi(d)$ is monotonically increasing over $(d_{\min}^{(\infty)}, \lambda_X)$, converging to $\tau(\gamma_Z) \triangleq \frac{1}{2} \log \frac{\lambda_X^2}{(\lambda_X + \gamma_Z)^2} + \frac{2\lambda_X \gamma_Z + \gamma_Z^2}{2\lambda_X^2}$ and ∞ as $d \to d_{\min}^{(\infty)}$ and λ_X, respectively. Note that $\tau'(\gamma_Z) = \frac{2\lambda_X \gamma_Z + \gamma_Z^2}{\lambda_X^2(\lambda_X + \gamma_Z)} \geq 0$ for $\gamma_Z \in [0, \infty)$; therefore, the minimum value of $\tau(\gamma_Z)$ over $[0, \infty)$ is 0, which is attained at $\gamma_Z = 0$. See Figures 3 and 4 for some graphical illustrations of $\psi(d)$.*

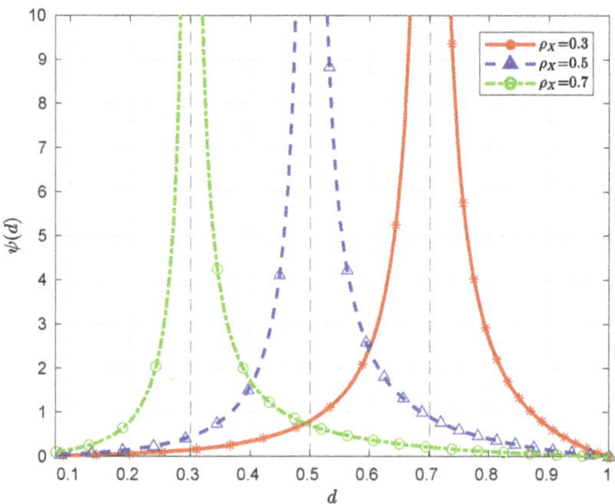

Figure 3. Illustration of $\psi(d)$ with $\gamma_X = 1$ and $\gamma_Z = 0.1$ for different ρ_X.

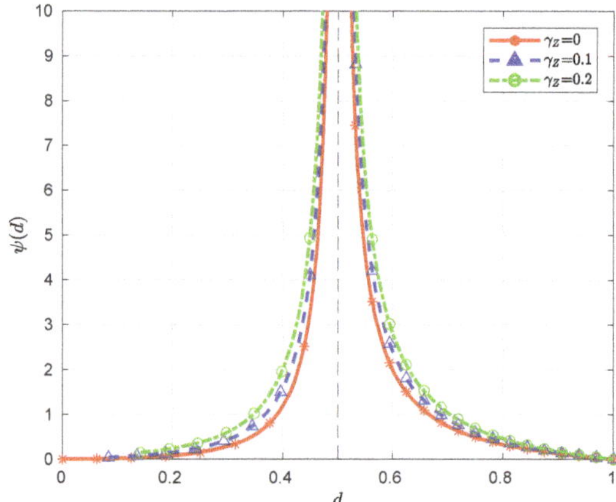

Figure 4. Illustration of $\psi(d)$ with $\gamma_X = 1$ and $\rho_X = 0.5$ for different γ_Z.

3. Proofs

3.1. Proof of Lemma 1

It is known [21] that $\underline{r}(d)$ is given by the solution to the following optimization problem:

$$(\mathbf{P}_1) \quad \min_{p_{\hat{X}|S}} I(S; \hat{X})$$

$$\text{subject to} \quad \mathbb{E}[(X - \hat{X})^T(X - \hat{X})] \leq \ell d,$$
$$X \leftrightarrow S \leftrightarrow \hat{X} \text{ form a Markov chain.}$$

Let $\tilde{X} \triangleq \Theta X$, $\tilde{Z} \triangleq \Theta Z$, and $\tilde{S} \triangleq \Theta S$, where Θ is an arbitrary (real) unitary matrix with the first row being $\frac{1}{\sqrt{\ell}}\mathbf{1}_\ell$. Since unitary transformations are invertible and preserve the Euclidean norm, we can write $(\mathbf{P}_1)$ equivalently as

$$(\mathbf{P}_2) \quad \min_{p_{\hat{X}|S}} I(\tilde{S}; \hat{X})$$

$$\text{subject to} \quad \mathbb{E}[(\tilde{X} - \hat{X})^T(\tilde{X} - \hat{X})] \leq \ell d,$$
$$\tilde{X} \leftrightarrow \tilde{S} \leftrightarrow \hat{X} \text{ form a Markov chain.}$$

For the same reason, we have

$$\ell d_{\min} = \mathbb{E}[(\tilde{X} - \mathbb{E}[\tilde{X}|\tilde{S}])^T(\tilde{X} - \mathbb{E}[\tilde{X}|\tilde{S}])]. \tag{1}$$

Denote the i-th components of $\tilde{X}$, $\tilde{Z}$, and $\tilde{S}$ by $\tilde{X}_i$, $\tilde{Z}_i$, and $\tilde{S}_i$, respectively, $i = 1, \cdots, \ell$. Clearly, $\tilde{S}_i = \tilde{X}_i + \tilde{Z}_i$, $i = 1, \cdots, \ell$. Moreover, it can be verified that $\tilde{X}_1, \cdots, \tilde{X}_\ell, \tilde{Z}_1, \cdots, \tilde{Z}_\ell$ are independent zero-mean Gaussian random variables with

$$\mathbb{E}[(\tilde{X}_1)^2] = \ell \rho_X \gamma_X + \lambda_X, \tag{2}$$
$$\mathbb{E}[(\tilde{X}_i)^2] = \lambda_X, \quad i = 2, \cdots, \ell, \tag{3}$$
$$\mathbb{E}[(\tilde{Z}_1)^2] = \gamma_Z, \quad i = 1, \cdots, \ell.$$

Now denote the i-th component of $\hat{S} \triangleq \mathbb{E}[\tilde{X}|\tilde{S}]$ by $\hat{S}_i$, $i = 1, \cdots, \ell$. We have

$$\hat{S}_i = \mathbb{E}[\tilde{X}_i|\tilde{S}_i], \quad i = 1, \cdots, \ell,$$

and

$$\mathbb{E}[(\hat{S}_1)^2] = \begin{cases} 0, & \rho_X = -\frac{1}{\ell-1}, \\ \frac{(\ell \rho_X \gamma_X + \lambda_X)^2}{\ell \rho_X \gamma_X + \lambda_X + \gamma_Z}, & \rho_X \in (-\frac{1}{\ell-1}, 1], \end{cases} \tag{4}$$

$$\mathbb{E}[(\hat{S}_i)^2] = \begin{cases} \frac{\lambda_X^2}{\lambda_X + \gamma_Z}, & \rho \in [-\frac{1}{\ell-1}, 1), \\ 0, & \rho_X = 1, \end{cases} \quad i = 2, \cdots, \ell. \tag{5}$$

Note that

$$\mathbb{E}[(\tilde{X} - \hat{S})^T(\tilde{X} - \hat{S})] = \sum_{i=1}^\ell \mathbb{E}[(\tilde{X}_i)^2] - \sum_{i=1}^\ell \mathbb{E}[(\hat{S}_i)^2],$$

which, together with (1)–(5), proves

$$d_{\min} = \frac{1}{\ell}\mathbb{E}[(\tilde{X}-\hat{S})^T(\tilde{X}-\hat{S})] = \begin{cases} \frac{(\ell-1)\gamma_X\gamma_Z}{\ell\gamma_X+(\ell-1)\gamma_Z}, & \rho_X = -\frac{1}{\ell-1}, \\ \frac{(\ell\rho_X\gamma_X+\lambda_X)\gamma_Z}{\ell(\ell\rho_X\gamma_X+\lambda_X+\gamma_Z)} + \frac{(\ell-1)\lambda_X\gamma_Z}{\ell(\lambda_X+\gamma_Z)}, & \rho_X \in (-\frac{1}{\ell-1},1), \\ \frac{\gamma_X\gamma_Z}{\ell\gamma_X+\gamma_Z}, & \rho_X = 1. \end{cases}$$

Clearly, $\hat{S}$ is determined by $\tilde{S}$; moreover, for any ℓ-dimensional random vector $\hat{X}$ jointly distributed with $(\tilde{X},\tilde{S})$ such that $\tilde{X} \leftrightarrow \tilde{S} \leftrightarrow \hat{X}$ form a Markov chain, we have

$$\mathbb{E}[(\tilde{X}-\hat{X})^T(\tilde{X}-\hat{X})] = \mathbb{E}[(\hat{S}-\hat{X})^T(\hat{S}-\hat{X})^2] + \mathbb{E}[(\tilde{X}-\hat{S})^T(\tilde{X}-\hat{S})^2]$$
$$= \mathbb{E}[(\hat{S}-\hat{X})^T(\hat{S}-\hat{X})^2] + \ell d_{\min}.$$

Therefore, $(\mathbf{P_2})$ is equivalent to

$$(\mathbf{P_3}) \quad \min_{p_{\hat{X}|\hat{S}}} I(\hat{S};\hat{X})$$
$$\text{subject to} \quad \mathbb{E}[(\hat{S}-\hat{X})^T(\hat{S}-\hat{X})] \leq \ell(d-d_{\min}).$$

One can readily complete the proof of Lemma 1 by recognizing that the solution to $(\mathbf{P_3})$ is given by the well-known reverse water-filling formula ([22] Theorem 13.3.3).

3.2. Proof of Theorem 1

Setting $\rho_X = 0$ in Lemma 1 gives

$$\underline{r}(d) = \frac{\ell}{2}\log\frac{\gamma_X^2}{(\gamma_X+\gamma_Z)d-\gamma_X\gamma_Z}$$

for $d \in (\frac{\gamma_X\gamma_Z}{\gamma_X+\gamma_Z},\gamma_X)$. Setting $\rho_X = 1$ in Lemma 1 gives

$$\underline{r}(d) = \frac{1}{2}\log\frac{\ell^2\gamma_X^2}{\ell(\ell\gamma_X+\gamma_Z)d-\gamma_X\gamma_Z}$$

for $d \in (\frac{\gamma_X\gamma_Z}{\ell\gamma_X+\gamma_Z},\gamma_X)$; moreover, we have

$$\frac{1}{2}\log\frac{\ell^2\gamma_X^2}{\ell(\ell\gamma_X+\gamma_Z)d-\gamma_X\gamma_Z} = \frac{1}{2}\log\frac{\gamma_X}{d}+O(\frac{1}{\ell}),$$

and $\frac{\gamma_X\gamma_Z}{\ell\gamma_X+\gamma_Z} \to 0$ as $\ell \to \infty$.

It remains to treat the case $\rho_X \in (0,1)$. In this case, it can be deduced from Lemma 1 that

$$\underline{r}(d) = \begin{cases} \frac{1}{2}\log\frac{(\ell\rho_X\gamma_X+\lambda_X)^2(\lambda_X+\gamma_Z)}{\lambda_X^2(\ell\rho_X\gamma_X+\lambda_X+\gamma_Z)} + \frac{\ell}{2}\log\frac{\lambda_X^2}{(\lambda_X+\gamma_Z)(d-d_{\min})}, & d \in (d_{\min},\frac{\lambda_X^2}{\lambda_X+\gamma_Z}+d_{\min}], \\ \frac{1}{2}\log\frac{(\ell\rho_X\gamma_X+\lambda_X)^2(\lambda_X+\gamma_Z)}{(\ell\rho_X\gamma_X+\lambda_X+\gamma_Z)(\ell(\lambda_X+\gamma_Z)(d-d_{\min})-(\ell-1)\lambda_X^2)}, & d \in (\frac{\lambda_X^2}{\lambda_X+\gamma_Z}+d_{\min},\gamma_X), \end{cases}$$

and we have

$$d_{\min} = \frac{(\ell\rho_X\gamma_X+\lambda_X)\gamma_Z}{\ell(\ell\rho_X\gamma_X+\lambda_X+\gamma_Z)} + \frac{(\ell-1)\lambda_X\gamma_Z}{\ell(\lambda_X+\gamma_Z)}$$
$$= \frac{\lambda_X\gamma_Z}{\lambda_X+\gamma_Z} + \frac{\rho_X\gamma_X\gamma_Z^2}{(\ell\rho_X\gamma_X+\lambda_X+\gamma_Z)(\lambda_X+\gamma_Z)} \tag{6}$$
$$= \frac{\lambda_X\gamma_Z}{\lambda_X+\gamma_Z} + \frac{\gamma_Z^2}{(\lambda_X+\gamma_Z)\ell} + O(\frac{1}{\ell^2}). \tag{7}$$

Consider the following two subcases separately.

- $d \in (\frac{\lambda_X \gamma_Z}{\lambda_X + \gamma_Z}, \lambda_X]$

  It can be seen from (6) that $d_{\min}$ is a monotonically decreasing function of ℓ and converges to $\frac{\lambda_X \gamma_Z}{\lambda_X + \gamma_Z}$ as $\ell \to \infty$. Therefore, we have $d \in (d_{\min}, \frac{\lambda_X^2}{\lambda_X + \gamma_Z} + d_{\min}]$ and consequently

$$\underline{r}(d) = \frac{1}{2} \log \frac{(\ell \rho_X \gamma_X + \lambda_X)^2 (\lambda_X + \gamma_Z)}{\lambda_X^2 (\ell \rho_X \gamma_X + \lambda_X + \gamma_Z)} + \frac{\ell}{2} \log \frac{\lambda_X^2}{(\lambda_X + \gamma_Z)(d - d_{\min})}, \tag{8}$$

  when ℓ is sufficiently large. Note that

$$\frac{1}{2} \log \frac{(\ell \rho_X \gamma_X + \lambda_X)^2}{\ell \rho_X \gamma_X + \lambda_X + \gamma_Z} = \frac{1}{2} \log \ell + \frac{1}{2} \log(\rho_X \gamma_X) + O(\frac{1}{\ell}) \tag{9}$$

  and

$$\frac{1}{2} \log(d - d_{\min}) = \frac{1}{2} \log \left(d - \frac{\lambda_X \gamma_Z}{\lambda_X + \gamma_Z} - \frac{\gamma_Z^2}{(\lambda_X + \gamma_Z)\ell} - O(\frac{1}{\ell^2}) \right) \tag{10}$$

$$= \frac{1}{2} \log \frac{(\lambda_X + \gamma_Z)d - \lambda_X \gamma_Z}{\lambda_X + \gamma_Z} - \frac{\gamma_Z^2}{2((\lambda_X + \gamma_Z)d - \lambda_X \gamma_Z)\ell} + O(\frac{1}{\ell^2}), \tag{11}$$

  where (10) is due to (7). Substituting (9) and (11) into (8) gives

$$\underline{r}(d) = \frac{\ell}{2} \log \frac{\lambda_X^2}{(\lambda_X + \gamma_Z)d - \lambda_X \gamma_Z} + \frac{1}{2} \log \ell + \frac{1}{2} \log \frac{\rho_X \gamma_X (\lambda_X + \gamma_Z)}{\lambda_X^2}$$

$$+ \frac{\gamma_Z^2}{2((\lambda_X + \gamma_Z)d - \lambda_X \gamma_Z)} + O(\frac{1}{\ell}).$$

  In particular, we have

$$\underline{r}(\lambda_X) = \frac{1}{2} \log \ell + \frac{1}{2} \log \frac{\rho_X \gamma_X (\lambda_X + \gamma_Z)}{\lambda_X^2} + \frac{\gamma_Z^2}{2\lambda_X^2} + O(\frac{1}{\ell}).$$

- $d \in (\lambda_X, \gamma_X)$

  Since $d_{\min}$ converges to $\frac{\lambda_X \gamma_Z}{\lambda_X + \gamma_Z}$ as $\ell \to \infty$, it follows that $d \in (\frac{\lambda_X^2}{\lambda_X + \gamma_Z} + d_{\min}, \gamma_X)$ and consequently

$$\underline{r}(d) = \frac{1}{2} \log \frac{(\ell \rho_X \gamma_X + \lambda_X)^2 (\lambda_X + \gamma_Z)}{(\ell \rho_X \gamma_X + \lambda_X + \gamma_Z)(\ell(\lambda_X + \gamma_Z)(d - d_{\min}) - (\ell - 1)\lambda_X^2)} \tag{12}$$

  when ℓ is sufficiently large. One can readily verify that

$$\frac{1}{2} \log \frac{(\ell \rho_X \gamma_X + \lambda_X)^2}{(\ell \rho_X \gamma_X + \lambda_X + \gamma_Z)(\ell(\lambda_X + \gamma_Z)(d - d_{\min}) - (\ell - 1)\lambda_X^2)}$$

$$= \frac{1}{2} \log \frac{\rho_X \gamma_X}{(\lambda_X + \gamma_Z)(d - \lambda_X)} + O(\frac{1}{\ell}). \tag{13}$$

  Substituting (13) into (12) gives

$$\underline{r}(d) = \frac{1}{2} \log \frac{\rho_X \gamma_X}{d - \lambda_X} + O(\frac{1}{\ell}).$$

This completes the proof of Theorem 1.

3.3. Proof of Theorem 2

One can readily prove part one of Theorem 2 by setting $\rho_X = 0$ in Lemma 2. So only part two of Theorem 2 remains to be proved. Note that

$$b = g_1 \ell^2 + g_2 \ell,$$
$$c = h_1 \ell^2 + h_2 \ell,$$

where

$$g_1 \triangleq \rho_X \gamma_X (\lambda_X - d),$$
$$g_2 \triangleq \lambda_X^2 + 2\gamma_X \gamma_Z - 2(\lambda_X + \gamma_Z)d,$$
$$h_1 \triangleq \rho_X \gamma_X (\lambda_X + \gamma_Z)(d_{\min}^{(\infty)} - d),$$
$$h_2 \triangleq \rho_X \gamma_X \gamma_Z^2 + \lambda_X \gamma_Z (\lambda_X + \gamma_Z) - (\lambda_X + \gamma_Z)^2 d.$$

We shall consider the following three cases separately.

- $d < \lambda_X$

  In this case $g_1 > 0$ and consequently

$$\lambda_Q = \frac{-b + b\sqrt{1 - \frac{4ac}{b^2}}}{2a} \tag{14}$$

  when ℓ is sufficiently large. Note that

$$\sqrt{1 - \frac{4ac}{b^2}} = 1 - \frac{2ac}{b^2} - \frac{2a^2 c^2}{b^4} + O(\frac{1}{\ell^3}). \tag{15}$$

  Substituting (15) into (14) gives

$$\lambda_Q = -\frac{c}{b} - \frac{ac^2}{b^3} + O(\frac{1}{\ell^2}). \tag{16}$$

  It is easy to show that

$$-\frac{c}{b} = -\frac{h_1}{g_1} - \frac{g_1 h_2 - g_2 h_1}{g_1^2 \ell} + O(\frac{1}{\ell^2}), \tag{17}$$

$$-\frac{ac^2}{b^3} = -\frac{(\gamma_X - d)h_1^2}{g_1^3 \ell} + O(\frac{1}{\ell^2}). \tag{18}$$

  Combining (16), (17) and (18) yields

$$\lambda_Q = \eta_1 + \frac{\eta_2}{\ell} + O(\frac{1}{\ell^2}),$$

where

$$\eta_1 \triangleq -\frac{h_1}{g_1},$$

$$\eta_2 \triangleq -\frac{g_1^2 h_2 - g_1 g_2 h_1 + (\gamma_X - d)h_1^2}{g_1^3}.$$

Moreover, it can be verified via algebraic manipulations that

$$\eta_1 = \frac{(\lambda_X + \gamma_Z)d - \lambda_X \gamma_Z}{\lambda_X - d},$$

$$\eta_2 = -\frac{\lambda_X^2 d^2}{(\lambda_X - d)^3}.$$

Now we write $\bar{r}(d)$ equivalently as

$$\bar{r}(d) = \frac{1}{2} \log \frac{\ell \rho_X \gamma_X + \lambda_X + \gamma_Z + \lambda_Q}{\lambda_X + \gamma_Z + \lambda_Q} + \frac{\ell}{2} \log \frac{\lambda_X + \gamma_Z + \lambda_Q}{\lambda_Q}. \tag{19}$$

Note that

$$\frac{1}{2} \log \frac{\ell \rho_X \gamma_X + \lambda_X + \gamma_Z + \lambda_Q}{\lambda_X + \gamma_Z + \lambda_Q} = \frac{1}{2} \log \ell + \frac{1}{2} \log \frac{\rho_X \gamma_X}{\lambda_X + \gamma_Z + \eta_1} + O(\tfrac{1}{\ell})$$

$$= \frac{1}{2} \log \ell + \frac{1}{2} \log \frac{\rho_X \gamma_X (\lambda_X - d)}{\lambda_X^2} + O(\tfrac{1}{\ell}) \tag{20}$$

and

$$\frac{1}{2} \log \frac{\lambda_X + \gamma_Z + \lambda_Q}{\lambda_Q}$$

$$= \frac{1}{2} \log \frac{\lambda_X + \gamma_Z + \eta_1}{\eta_1} - \frac{(\lambda_X + \gamma_Z)\eta_2}{2(\lambda_X + \gamma_Z + \eta_1)\eta_1 \ell} + O(\tfrac{1}{\ell^2})$$

$$= \frac{1}{2} \log \frac{\lambda_X^2}{(\lambda_X + \gamma_Z)d - \lambda_X \gamma_Z} + \frac{(\lambda_X + \gamma_Z)d^2}{2(\lambda_X - d)((\lambda_X + \gamma_X)d - \lambda_X \gamma_Z)\ell} + O(\tfrac{1}{\ell^2}). \tag{21}$$

Substituting (20) and (21) into (19) gives

$$\bar{r}(d) = \frac{\ell}{2} \log \frac{\lambda_X^2}{(\lambda_X + \gamma_Z)d - \lambda_X \gamma_Z} + \frac{1}{2} \log \ell + \frac{1}{2} \log \frac{\rho_X \gamma_X (\lambda_X - d)}{\lambda_X^2}$$

$$+ \frac{(\lambda_X + \gamma_Z)d^2}{2(\lambda_X - d)((\lambda_X + \gamma_X)d - \lambda_X \gamma_Z)} + O(\tfrac{1}{\ell}).$$

- $d = \lambda_X$

In this case $g_1 = 0$ and consequently

$$\lambda_Q = \frac{-g_2 + \sqrt{g_2^2 - 4(\gamma_X - \lambda_X)(h_1 \ell + h_2)}}{2(\gamma_X - \lambda_X)}. \tag{22}$$

Note that

$$\sqrt{g_2^2 - 4(\gamma_X - \lambda_X)(h_1\ell + h_2)} = \sqrt{-4(\gamma_X - \lambda_X)h_1\ell} + O(\frac{1}{\sqrt{\ell}}). \tag{23}$$

Substituting (23) into (22) gives

$$\lambda_Q = \mu_1\sqrt{\ell} + \mu_2 + O(\frac{1}{\sqrt{\ell}}),$$

where

$$\mu_1 \triangleq \sqrt{-\frac{h_1}{\gamma_X - \lambda_X}},$$

$$\mu_2 \triangleq -\frac{g_2}{2(\gamma_X - \lambda_X)}.$$

Moreover, it can be verified via algebraic manipulations that

$$\mu_1 = \lambda_X,$$

$$\mu_2 = \frac{(1-\rho_X)^2\gamma_X - 2\rho_X\gamma_Z}{2\rho_X}.$$

Now we proceed to derive an asymptotic expression of $\bar{r}(d)$. Note that

$$\frac{1}{2}\log\frac{\ell\rho_X\gamma_X + \lambda_X + \gamma_Z + \lambda_Q}{\lambda_X + \gamma_Z + \lambda_Q} = \frac{1}{4}\log\ell + \frac{1}{2}\log\frac{\rho_X\gamma_X}{\mu_1} + O(\frac{1}{\sqrt{\ell}})$$

$$= \frac{1}{4}\log\ell + \frac{1}{2}\log\frac{\rho_X}{1-\rho_X} + O(\frac{1}{\sqrt{\ell}}) \tag{24}$$

and

$$\frac{1}{2}\log\frac{\lambda_X + \gamma_Z + \lambda_Q}{\lambda_Q} = \frac{\lambda_X + \gamma_Z}{2\lambda_Q} - \frac{(\lambda_X + \gamma_Z)^2}{4\lambda_Q^2} + O(\frac{1}{\ell^{\frac{3}{2}}})$$

$$= \frac{\lambda_X + \gamma_Z}{2\mu_1\sqrt{\ell}} - \frac{(\lambda_X + \gamma_Z)(\lambda_X + \gamma_Z + 2\mu_2)}{4\mu_1^2\ell} + O(\frac{1}{\ell^{\frac{3}{2}}})$$

$$= \frac{\lambda_X + \gamma_Z}{2\lambda_X\sqrt{\ell}} - \frac{(\lambda_X + \gamma_Z)(\lambda_X - \rho_X\gamma_Z)}{4\rho_X\lambda_X^2\ell} + O(\frac{1}{\ell^{\frac{3}{2}}}). \tag{25}$$

Substituting (24) and (25) into (19) gives

$$\bar{r}(\lambda_X) = \frac{(\lambda_X + \gamma_Z)\sqrt{\ell}}{2\lambda_X} + \frac{1}{4}\log\ell + \frac{1}{2}\log\frac{\rho_X}{1-\rho_X} - \frac{(\lambda_X + \gamma_Z)(\lambda_X - \rho_X\gamma_Z)}{4\rho_X\lambda_X^2} + O(\frac{1}{\sqrt{\ell}}).$$

- $d > \lambda_X$

In this case $g_1 < 0$ and consequently

$$\lambda_Q = \frac{-b - b\sqrt{1 - \frac{4ac}{b^2}}}{2a} \tag{26}$$

when ℓ is sufficiently large. Note that

$$\sqrt{1 - \frac{4ac}{b^2}} = 1 + O(\frac{1}{\ell}).$$ (27)

Substituting (27) into (26) gives

$$\lambda_Q = -\frac{b}{a} + O(1).$$ (28)

It is easy to show that

$$-\frac{b}{a} = \frac{\rho_X \gamma_X (d - \lambda_X)\ell}{\gamma_X - d} + O(1).$$ (29)

Combining (28) and (29) yields

$$\lambda_Q = \frac{\rho_X \gamma_X (d - \lambda_X)\ell}{\gamma_X - d} + O(1).$$

Now we proceed to derive an asymptotic expression of $\bar{r}(d)$. Note that

$$\frac{1}{2} \log \frac{\ell \rho_X \gamma_X + \lambda_X + \gamma_Z + \lambda_Q}{\lambda_X + \gamma_Z + \lambda_Q} = \frac{1}{2} \log \frac{\rho_X \gamma_X}{d - \lambda_X} + O(\frac{1}{\ell})$$ (30)

and

$$\begin{aligned}
\frac{1}{2} \log \frac{\lambda_X + \gamma_Z + \lambda_Q}{\lambda_Q} &= \frac{\lambda_X + \gamma_Z}{2\lambda_Q} + O(\frac{1}{\ell^2}) \\
&= \frac{(\lambda_X + \gamma_Z)(\gamma_X - d)}{2\rho_X \gamma_X (d - \lambda_X)\ell} + O(\frac{1}{\ell^2}).
\end{aligned}$$ (31)

Substituting (30) and (31) into (19) gives

$$\bar{r}(d) = \frac{1}{2} \log \frac{\rho_X \gamma_X}{d - \lambda_X} + \frac{(\lambda_X + \gamma_Z)(\gamma_X - d)}{2\rho_X \gamma_X (d - \lambda_X)} + O(\frac{1}{\ell}).$$

This completes the proof of Theorem 2.

4. Conclusions

We have studied the problem of symmetric remote Gaussian source coding and made a systematic comparison of centralized encoding and distributed encoding in terms of the asymptotic rate-distortion performance. It is of great interest to extend our work by considering more general source and noise models.

Author Contributions: Conceptualization, Y.W. and J.C.; methodology, Y.W.; validation, L.X., S.Z. and M.W.; formal analysis, L.X., S.Z. and M.W.; investigation, L.X., S.Z. and M.W.; writing—original draft preparation, Y.W.; writing—review and editing, J.C.; supervision, J.C.

Funding: S.Z. was supported in part by the China Scholarship Council.

Acknowledgments: The authors wish to thank the anonymous reviewer for their valuable comments and suggestions.

References

1. Berger, T.; Zhang, Z.; Viswanathan, H. The CEO problem. *IEEE Trans. Inf. Theory* **1996**, *42*, 887–902. [CrossRef]
2. Viswanathan, H.; Berger, T. The quadratic Gaussian CEO problem. *IEEE Trans. Inf. Theory* **1997**, *43*, 1549–1559. [CrossRef]
3. Oohama, Y. The rate-distortion function for the quadratic Gaussian CEO problem. *IEEE Trans. Inf. Theory* **1998**, *44*, 1057–1070. [CrossRef]
4. Prabhakaran, V.; Tse, D.; Ramchandran, K. Rate region of the quadratic Gaussian CEO problem. In Proceedings of the IEEE International Symposium onInformation Theory, Chicago, IL, USA, 27 June–2 July 2004; p. 117.
5. Chen, J.; Zhang, X.; Berger, T.; Wicker, S.B. An upper bound on the sum-rate distortion function and its corresponding rate allocation schemes for the CEO problem. *IEEE J. Sel. Areas Commun.* **2004**, *22*, 977–987. [CrossRef]
6. Oohama, Y. Rate-distortion theory for Gaussian multiterminal source coding systems with several side informations at the decoder. *IEEE Trans. Inf. Theory* **2005**, *51*, 2577–2593. [CrossRef]
7. Chen, J.; Berger, T. Successive Wyner-Ziv coding scheme and its application to the quadratic Gaussian CEO problem. *IEEE Trans. Inf. Theory* **2008**, *54*, 1586–1603. [CrossRef]
8. Wagner, A.B.; Tavildar, S.; Viswanath, P. Rate region of the quadratic Gaussian two-encoder source-coding problem. *IEEE Trans. Inf. Theory* **2008**, *54*, 1938–1961. [CrossRef]
9. Tavildar, S.; Viswanath, P.; Wagner, A.B. The Gaussian many-help-one distributed source coding problem. *IEEE Trans. Inf. Theory* **2010**, *56*, 564–581. [CrossRef]
10. Wang, J.; Chen, J.; Wu, X. On the sum rate of Gaussian multiterminal source coding: New proofs and results. *IEEE Trans. Inf. Theory* **2010**, *56*, 3946–3960. [CrossRef]
11. Yang, Y.; Xiong, Z. On the generalized Gaussian CEO problem. *IEEE Trans. Inf. Theory* **2012**, *58*, 3350–3372. [CrossRef]
12. Wang, J.; Chen, J. Vector Gaussian two-terminal source coding. *IEEE Trans. Inf. Theory* **2013**, *59*, 3693–3708. [CrossRef]
13. Courtade, T.A.; Weissman, T. Multiterminal source coding under logarithmic loss. *IEEE Trans. Inf. Theory* **2014**, *60*, 740–761. [CrossRef]
14. Wang, J.; Chen, J. Vector Gaussian multiterminal source coding. *IEEE Trans. Inf. Theory* **2014**, *60*, 5533–5552. [CrossRef]
15. Oohama, Y. Indirect and direct Gaussian distributed source coding problems. *IEEE Trans. Inf. Theory* **2014**, *60*, 7506–7539. [CrossRef]
16. Nangir, M.; Asvadi, R.; Ahmadian-Attari, M.; Chen, J. Analysis and code design for the binary CEO problem under logarithmic loss. *IEEE Trans. Commun.* **2018**, *66*, 6003–6014. [CrossRef]
17. Ugur, Y.; Aguerri, I.-E.; Zaidi, A. Vector Gaussian CEO problem under logarithmic loss and applications. *arXiv* **2018**, arXiv:1811.03933.
18. Nangir, M.; Asvadi, R.; Chen, J.; Ahmadian-Attari, M.; Matsumoto, T. Successive Wyner-Ziv coding for the binary CEO problem under logarithmic loss. *arXiv* **2018**, arXiv:1812.11584.
19. Wang, Y.; Xie, L.; Zhang, X.; Chen, J. Robust distributed compression of symmetrically correlated Gaussian sources. *arXiv* **2018**, arXiv:1807.06799.
20. Chen, J.; Xie, L.; Chang, Y.; Wang, J.; Wang, Y. Generalized Gaussian multiterminal source coding: The symmetric case. *arXiv* **2017**, arXiv:1710.04750.
21. Dobrushin, R.; Tsybakov, B. Information transmission with additional noise. *IRE Trans. Inf. Theory* **1962**, *8*, 293–304. [CrossRef]
22. Cover, T.; Thomas, J.A. *Elements of Information Theory*; Wiley: New York, NY, USA, 1991.

Article

Non-Orthogonal eMBB-URLLC Radio Access for Cloud Radio Access Networks with Analog Fronthauling

Andrea Matera [1,*], **Rahif Kassab** [2], **Osvaldo Simeone** [2] **and Umberto Spagnolini** [1]

[1]   Dipartimento di Elettronica, Informazione e Bioingegneria (DEIB), Politecnico di Milano, 20133 Milano, Italy; umberto.spagnolini@polimi.it

[2]   Centre for Telecommunications Research (CTR), Department of Informatics, King's College London, London WC2B 4BG, UK; rahif.kassab@kcl.ac.uk (R.K.); osvaldo.simeone@kcl.ac.uk (O.S.)

*   Correspondence: andrea.matera@polimi.it; Tel.: +39-22-399-3462

Received: 17 July 2018; Accepted: 31 August 2018; Published: 2 September 2018

Abstract: This paper considers the coexistence of Ultra Reliable Low Latency Communications (URLLC) and enhanced Mobile BroadBand (eMBB) services in the uplink of Cloud Radio Access Network (C-RAN) architecture based on the relaying of radio signals over analog fronthaul links. While Orthogonal Multiple Access (OMA) to the radio resources enables the isolation and the separate design of different 5G services, Non-Orthogonal Multiple Access (NOMA) can enhance the system performance by sharing wireless and fronthaul resources. This paper provides an information-theoretic perspective in the performance of URLLC and eMBB traffic under both OMA and NOMA. The analysis focuses on standard cellular models with additive Gaussian noise links and a finite inter-cell interference span, and it accounts for different decoding strategies such as puncturing, Treating Interference as Noise (TIN) and Successive Interference Cancellation (SIC). Numerical results demonstrate that, for the considered analog fronthauling C-RAN architecture, NOMA achieves higher eMBB rates with respect to OMA, while guaranteeing reliable low-rate URLLC communication with minimal access latency. Moreover, NOMA under SIC is seen to achieve the best performance, while, unlike the case with digital capacity-constrained fronthaul links, TIN always outperforms puncturing.

Keywords: network slicing; RoC; URLLC; eMBB; C-RAN

---

1. Introduction

Accommodating the heterogeneity of users' requirements is one of the main challenges that both industry and academia are facing in order to make 5G a reality [1]. In fact, next-generation wireless communication systems must be designed to provision different services, each of which with distinct constraints in terms of latency, reliability, and information rate. In particular, 5G is expected to support three different macro-categories of services, namely enhanced Mobile BroadBand (eMBB), massive Machine-Type Communications (mMTC), and Ultra-Reliable and Low-Latency Communications (URLLC) [2–4].

eMBB service is meant to provide very-high data-rate communications as compared with current (4G) networks. This can be generally achieved by using codewords that spread over a large number of time-frequency resources, given that latency is not an issue. mMTC supports low-rate bursty communication between a massive number of uncoordinated devices and the network. Finally, URLLC is designed to ensure low-rate ultra-reliable radio access for a few nodes, while guaranteeing very low-latency. As a result, URLLC transmissions need to be localized in time, and hence URLLC packets should be short [5].

The coexistence among eMBB, mMTC and URLLC traffic types can be ensured by slicing the Radio Access Network (RAN) resources into non-overlapping, or orthogonal, blocks, and by assigning distinct resources to different services. With the resulting Orthogonal Multiple Access (OMA), the target quality-of-service guarantees can be achieved by designing each service separately [6,7]. However, when URLLC or mMTC traffic types are characterized by short and bursty transmissions at random time instants, resources allocated statically to these services are likely to be unused for most of time, and thus wasted. A more efficient use of radio resources can be accomplished by Non-Orthogonal Multiple Access (or NOMA), which allows multiple services to share the same physical resources.

By enabling an opportunistic shared use of the radio resources, NOMA can provide significant benefits in terms of spectrum efficiency, but it also poses the challenge of designing the system so that the heterogeneous requirements of the services are satisfied despite the mutual interference. The objective of this paper is to address this issue by considering a Cloud-RAN (C-RAN) architecture characterized by analog fronthaul links, referred to as Analog Radio-over-X, which is introduced in the next section.

1.1. C-RAN Based on Analog Radio-over-X Fronthauling

The advent of 5G is introducing advanced physical layer technologies and network deployment strategies such as massive MIMO, mmWave, small-cell densification, mobile edge computing, etc. (see [8,9] for an overview). C-RAN is an enabling technology that is based on the colocation of the Base Band Unit (BBU) of Edge Nodes (ENs) that are densely distributed in a given indoor or outdoor area. This solution has the advantages of allowing for centralized BBU signal processing, providing network scalability, increasing spectral efficiency, and reducing costs.

The most typical C-RAN architecture relies on digital optical fronthaul links to connect ENs to BBUs. This solution, known as Digital Radio-over-Fiber (D-RoF), is adopted in current 4G mobile networks, and is based on the transmission of in-phase and quadrature baseband signals, upon digitization and packetization according to the CPRI protocol [10].

Over the last years, several alternative C-RAN architectures have been proposed that redistribute the RAN functionalities between BBU and ENs, obtaining different trade-offs in terms of bandwidth and latency requirements, advanced Cooperative Multi-Point processing capabilities, and EN cost and complexity [11]. For scenarios with stringent cost and latency constraints, a promising solution is to use analog fronthauling.

With analog fronthauling, focusing on the uplink, the ENs directly relay the radio signals to the BBUs after frequency translation and, possibly, signal amplification. This has the advantages of avoiding any bandwidth expansion due to digitization; guaranteeing ENs synchronization; minimizing latency; reducing hardware cost; and improving energy efficiency [12–14]. A C-RAN architecture based on analog fronthauling is also known in the literature as Analog Radio-over-X (A-RoX), where X depends on the technology employed for the fronthaul, which can be either Fiber (A-RoF [12]), Radio (A-RoR [15]), or Copper (A-RoC [13]), as depicted in Figure 1.

In particular, A-RoF provides an effective example of analog fronthauling, due to its capability to support the transport of large bandwidths [12,16]. However, A-RoF requires the deployment of a fiber optic infrastructure whose installation is not always feasible, e.g., in dense urban areas. In such scenarios, a possible solution is to rely on the A-RoR concept, thus employing point-to-point wireless links, mainly based on mmWave or THz bands, with several advantages in terms of flexibility, resiliency, hardware complexity and cost [15,17]. Another application scenario where the installation of fiber links may be too expensive to provide satisfactory business cases is indoor coverage. For indoor deployments, A-RoC [18–20] has been recently proved to be an attractive solution, especially from the deployment costs perspective [21], since it leverages the pre-existing Local Area Network (LAN) cabling infrastructure of building and enterprises. Moreover, LAN cables are equipped with four twisted-pairs with a transport capability up to 500 MHz each, or 2 GHz overall, for radio signals, thus providing enough bandwidth for analog fronthaul applications [22]. Over the last years, A-RoC based on LAN cables has become a standard solution for in-building commercial C-RAN deployments, allowing to extend the indoor coverage over distances longer than 100 m [23].

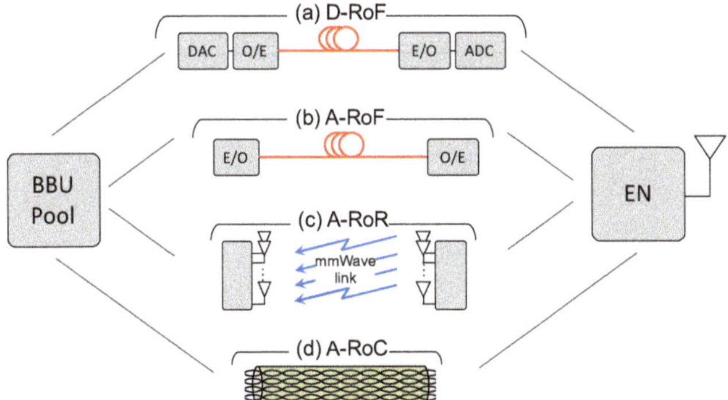

Figure 1. C-RAN architecture overview for uplink direction: (**a**) Digital Radio-over-Fiber, (**b**) Analog Radio-over-Fiber, (**c**) Analog Radio-over-Radio, (**d**) Analog Radio-over-Copper.

In this paper we study the coexistence of URLLC and eMBB services under both OMA and NOMA assuming a C-RAN multi-cell architecture based on analog fronthauling (see Figure 1) by using information theoretical tools.

1.2. Related Works

The concept of NOMA is well-known from the information theoretic literature [24], and its application to 5G dates back to [25], where authors demonstrated for a single-cell scenario that superimposing multiple users in the same resources achieves superior performance with respect to conventional LTE networks, provided that the resulting interference is properly taken care of. The extension of NOMA to multi-cell networks is presented in [26], which addresses several multi-cell NOMA challenges, including coordinated scheduling, beamforming, and practical implementation issues related to successive interference cancellation.

In contrast with most of the works on NOMA, which deal with homogeneous traffic conditions (see [27] for a recent review), here the focus is on NOMA techniques in the context of heterogeneous networks, such as the forthcoming 5G wireless systems, as discussed in [28–30]. In fact, NOMA represents an attractive solution to meet the distinct requirements of 5G services, as it improves spectral efficiency (eMBB), enables massive device connectivity (mMTC), and allows for low-transmission latency (URLLC) [31].

In [29] a communication-theoretic model was introduced to investigate the performance trade-offs for eMBB, mMTC and URLLC services in a single-cell scenario under both OMA and NOMA. This single-cell model has been later extended in [32] to the uplink of a multi-cell C-RAN architecture, in which the BBU communicates with multiple URLLC and eMBB users belonging to different cells through geographically distributed ENs. In the C-RAN system studied in [32], while the URLLC signals are locally decoded at the ENs due to latency constraints, the eMBB signals are quantized and forwarded over limited-capacity digital fronthaul links to the BBU, where centralized joint decoding is performed.

None of the aforementioned works considers the coexistence of different 5G services in a C-RAN architecture based on analog fronthaul links which is the focus of this paper.

1.3. Contributions

In this paper we study for the first time the coexistence between URLLC and eMBB services in the uplink of a C-RAN system with analog fronthauling in which the URLLC signals are still decoded locally at the EN, while the eMBB signals are forwarded to the BBU over analog fronthaul links.

In particular, the main contributions of this paper are three-fold:

- We extend the uplink C-RAN theoretic model proposed in [32] to the case of analog fronthaul, assuming that the fronthaul links are characterized by multiple, generally interfering, channels that carry the received radio signals;
- By leveraging information theoretical tools, we investigate the performance trade-offs between URLLC and eMBB services under both OMA and NOMA, by considering different interference management strategies such as puncturing, considered for the standardization of 5G New Radio [33,34], Treating Interference as Noise (TIN), and Successive Interference Cancellation (SIC);
- The analysis demonstrates that NOMA allows for higher eMBB information rates with respect to OMA, while guaranteeing a reliable low-rate URLLC communication with minimal access latency. Moreover, differently from the case of conventional digital C-RAN architecture based on limited-capacity fronthaul links [32], in analog C-RAN, TIN always outperforms puncturing, while the best performance is still achieved by NOMA with SIC.

1.4. Organization

The remainder of the paper is organized as follows. The considered system model is introduced in Section 2. Section 3 details the fronthaul signal processing techniques employed to cope with the impairments of the A-RoC fronthaul links. The eMBB rand URLLC information rates are discussed in Sections 4 and 5 for OMA and NOMA, respectively. Numerical results are presented in Section 6. And Section 7 concludes the paper.

1.5. Notation

Bold upper- and lower-case letters describe matrices and column vectors, respectively. Letters $\mathbb{R}$, and $\mathbb{C}$ refer to real and complex numbers, respectively. We denote matrix inversion, transposition and conjugate transposition as $(\cdot)^{-1}, (\cdot)^{T}, (\cdot)^{H}$. Matrix $\mathbf{I}_n$ is an identity matrix of size n and $\mathbf{1}_n$ is a column vector made by n "1s". Symbol $\otimes$ denotes the Kronecker operator, $\text{vec}(\cdot)$ is the vectorization operator, and $\mathbb{E}[\cdot]$ is the statistical expectation. Notation $\text{diag}(A_1, A_2, ..., A_n)$ denotes a diagonal matrix with elements $A_1, A_2, ..., A_n$ on the main diagonal. The Q-function $\mathcal{Q}(\cdot)$ is the complementary cumulative distribution function of the standardized normal random variable, and $\mathcal{Q}^{-1}(\cdot)$ is its inverse.

2. System Model

The C-RAN architecture under study is illustrated in Figure 2. In this system, the BBUs communicate with multiple user equipments (UEs) belonging to M cells through M single-antenna Edge Nodes (ENs). The BBUs are co-located in the so-called BBU pool so that joint processing can be performed, while the ENs are geographically distributed. In particular, we assume here that cells are arranged in a line following the conventional circulant Wyner model ([35], Chapter 2), and each cell contains two single-antenna UEs with different service constraints: one eMBB user and one URLLC user.

Due to the strict latency constraints of URLLC traffic, the signal for the URLLC UEs is decoded on-site at the EN, while the eMBB signals are forwarded to the BBU through a multi-channel analog fronthaul. In this hybrid cloud-edge architecture, the mobile operator equips the EN with edge computing capabilities in order to provision the services required by the URLLC user directly from the EN. Following the A-RoX concept, the end-to-end channel from the eMBB UEs and the BBU pool is assumed to be fully analog: the EN performs only signal amplification and frequency translation to comply with fronthaul capabilities and forwards the signals to the BBU, where centralized decoding is performed. In practice, as detailed later in the paper (see Sections 4 and 5), we assume that each EN hosts a digital module, responsible for URLLC signal decoding, and an analog module, responsible for the mapping of received radio signal over the analog fronthauling, which is identified as Analog-to-Analog (A/A) mapping.

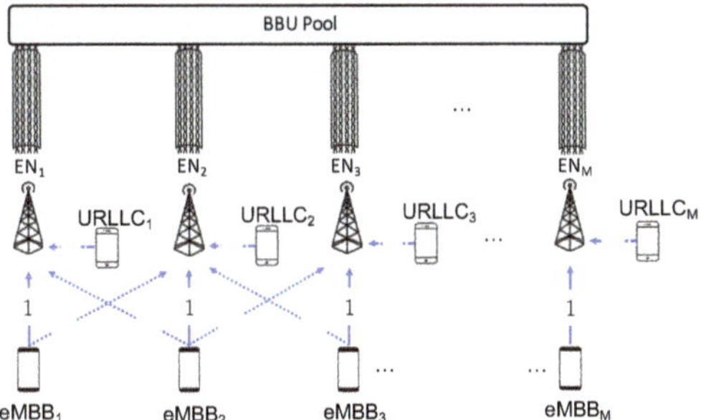

Figure 2. Model of the uplink of C-RAN system based on Analog Radio-over-Copper (A-RoC) fronthauling.

While the technology used for the analog fronthauling can be either fiber-optics (A-RoF), wireless (A-RoR) or cable (A-RoC), the system model proposed in this paper reflects mainly the last two solutions. Furthermore, we will focus on A-RoC, and we will use the corresponding terminology to fix the ideas (see Figure 2).

2.1. RAN Model

We consider the same Wyner-type radio access model of [32], which is described in this subsection. The Wyner model is an abstraction of cellular systems that captures one of the main aspects of such settings, namely the locality of inter-cell interference. The advantage of employing such a simple model is the possibility to obtain analytical insights, which is a first mandatory step for the performance assessment under more realistic operating conditions [35]. As illustrated in Figure 3, the direct channel gain from the eMBB UE and the EN belonging to the same cell is set to one, while the inter-cell eMBB channel gain is equal to $\alpha \in [0, 1]$. Furthermore, the URLLC UEs have a channel gain equal to $\beta > 0$. The URLLC user is assumed to be in the proximity of the EN, and thus it does not interfere with the neighboring cells. The eMBB user, instead, is assumed to be located at the edges of the cell in order to consider worst-case performance guarantees. As a result, each eMBB user interferes with both left and right neighboring cells, following the standard Wyner model [35]. All channel gains are assumed to be constant over the considered radio resources shown Figure 3, and known to all UEs and ENs.

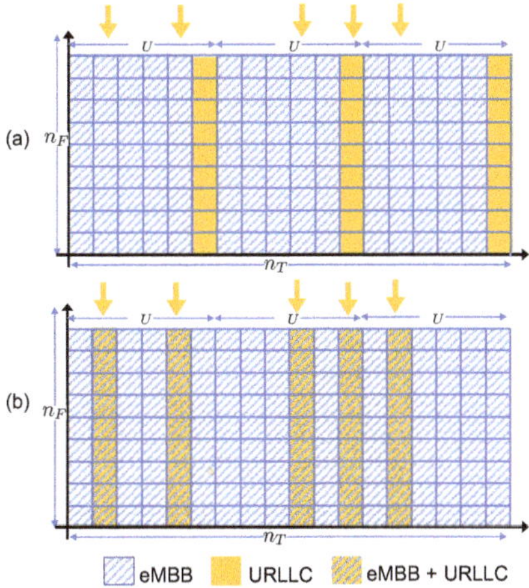

Figure 3. Time-frequency resource allocation: (**a**) Orthogonal Multiple Access (OMA) and (**b**) Non-Orthogonal Multiple Access (NOMA). Downwards arrows denote arrival of URLLC packets.

As illustrated in Figure 3, we assume that the time-frequency plane is divided in n_T minislots, indexed as $t \in [1, n_T]$, where each minislot is composed of n_F frequency channels, indexed as $f \in [1, n_F]$, for a total of $n_F n_T$ time-frequency radio resources. Each radio resource accommodates the transmission of a single symbol, although generalizations are straightforward. The eMBB UEs transmit over the entire $n_F \times n_T$ time-frequency frame. In contrast, due to the latency constraints of the URLLC traffic, each URLLC transmission is limited to the n_F frequency channels of a single minislot, and URLLC packets are generally small compared to the eMBB frame, which requires the condition $n_T \gg 1$. As illustrated in Figure 3, each URLLC UE generates an independent packet in each minislot with probability q. This packet is transmitted at the next available transmission opportunity in a grant-free manner.

In the case of OMA, one minislot is exclusively allocated for URLLC transmission every L_U minislots. Parameter L_U is considered here as the worst-case access latency. Accordingly, if more than one packet is generated within the L_U minislots between two transmission opportunities, only one of those packets is randomly selected for transmission and all the remaining are discarded. The signal $Y_k^f(t)$ received at the k-th EN at the f-th frequency under OMA is

$$Y_k^f(t) = \begin{cases} \beta A_k(t) U_k^f(t) + Z_k^f(t), & \text{if } t = L_U, 2L_U, \dots \\ X_k^f(t) + \alpha X_{k-1}^f(t) + \alpha X_{k+1}^f(t) + Z_k^f(t), & \text{otherwise} \end{cases} \tag{1}$$

where $X_k^f(t)$ and $U_k^f(t)$ are the signals transmitted at time t and subcarrier f by the k-th eMBB UE and URLLC UE, respectively; $Z_k^f(t) \sim \mathcal{CN}(0,1)$ is the unit-power zero-mean additive white Gaussian noise; and $A_k(t) \in \{0,1\}$ is a binary variable indicating whether or not the URLLC UE is transmitting at time t.

In case of NOMA, the URLLC UE transmits its packet in the same slot where it is generated by the application layer, so that the access latency is always minimal, i.e., $L_U = 1$ minislot. Under NOMA, the signal $Y_k^f(t)$ received at the k-th EN at the f-th frequency is

$$Y_k^f(t) = X_k^f(t) + \alpha X_{k-1}^f(t) + \alpha X_{k+1}^f(t) + \beta A_k(t) U_k^f(t) + Z_k^f(t). \tag{2}$$

According to the circulant Wyner model, in (1) and (2), we assume that $[k-1] = M$ for $k = 1$ and $[k+1] = 1$ for $k = M$, in order to guarantee symmetry.

For both OMA and NOMA, the power constraints for the k-th eMBB and URLLC users are defined within each radio resource frame respectively as

$$\frac{1}{n_F n_T} \sum_{t=1}^{n_T} \sum_{f=1}^{n_F} \mathbb{E}\left[\left| X_k^f(t) \right|^2 \right] \leq P_B, \tag{3}$$

and

$$\frac{1}{n_F} \sum_{f=1}^{n_F} \mathbb{E}\left[\left| U_k^f(t) \right|^2 \right] \leq P_U, \tag{4}$$

where the temporal average are taken over all symbols within a codeword.

Models (1) and (2) can be written in matrix form as

$$\mathbf{Y}(t) = \mathbf{X}(t)\mathbf{H} + \beta \mathbf{U}(t)\mathbf{A}(t) + \mathbf{Z}(t), \tag{5}$$

where matrix $\mathbf{Y}(t) = [\mathbf{y}_1(t), \mathbf{y}_2(t), \dots, \mathbf{y}_M(t)] \in \mathbb{C}^{n_F \times M}$ gathers all the signals received at all the M ENs over all the n_F frequencies, and the k-th column $\mathbf{y}_k(t) \in \mathbb{C}^{n_F \times 1}$ denotes the signal received across all the radio frequencies at the k-th EN. The channel matrix $\mathbf{H} \in \mathbb{R}^{M \times M}$ is circulant with the first column given by vector $[1, \alpha, 0, \dots, 0, \alpha]^T$; matrices $\mathbf{U}(t) \in \mathbb{C}^{n_F \times M}$ and $\mathbf{X}(t) \in \mathbb{C}^{n_F \times M}$ collect the signals transmitted by URLLC and eMBB UEs, respectively; and $\mathbf{Z}(t) \in \mathbb{C}^{n_F \times M}$ is the overall noise matrix. Finally, $\mathbf{A}(t)$ is a diagonal matrix whose k-th diagonal element is a Bernoulli random variable distributed as $A_k(t) \sim \mathcal{B}(q)$, $\forall k = 1, 2, ..., M$.

2.2. Space-Frequency Analog Fronthaul Channel

In the considered analog fronthaul architecture, the k-th EN forwards the signal $\mathbf{y}_k(t)$ received by the UEs to the BBU over a wired-access link in a fully analog fashion. As depicted in Figure 2, we focus our attention on a multichannel link that is possibly affected by inter-channel interference. While the model considered here can apply also to wireless multichannel links, as in Figure 1, we adopt here the terminology of Analog Radio-over-Copper (A-RoC) as an important example in which A-RoX is affected by fronthauling inter-link interference. Accordingly, each of the cables employed for the fronthaul contains l_S twisted-pairs, i.e., l_S space-separated channels, indexed as $c \in [1, l_S]$. Each pair carries a bandwidth equal to $l_F \leq n_F$ frequency channels of the RAN, indexed as $f' \in [1, l_F]$, so that a total of $l_S l_F$ space-frequency resource blocks are available over each cable, as shown in Figure 4. Furthermore, we assume that each analog fronthaul link has enough resources to accommodate the transmission of the whole radio signal at each EN, i.e., $l_S l_F \geq n_F$.

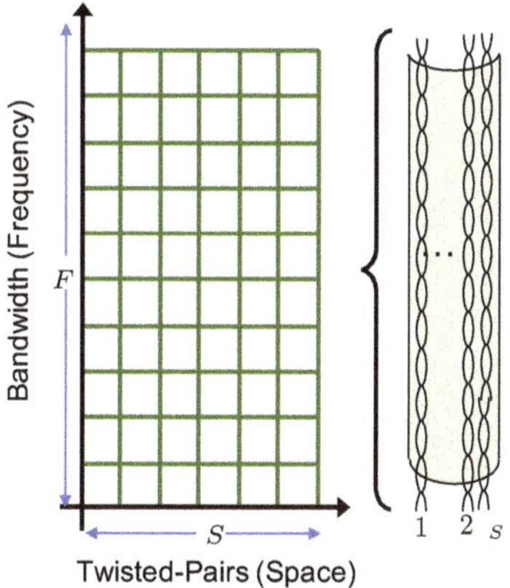

Figure 4. Space-frequency cable resource allocation.

The fronthaul channel between each EN and the BBU is described by the matrix $\mathbf{H}_c \in \mathbb{R}^{l_S \times l_S}$, which accounts for direct channel gains on each cable, given by the diagonal elements $[\mathbf{H}_c]_{ii}$, and for the intra-cable crosstalk, described by the off-diagonal elements $[\mathbf{H}_c]_{ij}$, with $i \neq j$. We assume here that the channel coefficients in $\mathbf{H}_c$ do not depend on frequency f'. Furthermore, in keeping the modeling assumptions of the Wyner model, we posit that the direct channel gains for all the pairs are normalized to 1, while the crosstalk coefficients between any pair of twisted-pairs are given by a coupling parameter $\gamma \geq 0$. It follows that the fronthaul channel matrix can be written as

$$\mathbf{H}_c = \gamma \mathbf{1}_{l_S} \mathbf{1}_{l_S}^T + (1 - \gamma)\mathbf{I}_{l_S}, \tag{6}$$

where $\mathbf{1}_n$ denotes a column vector of size n of all ones and $\mathbf{I}_n$ is the identity matrix of size n. We note that, in case of wireless fronthaul links such as in A-RoR, the coefficient γ accounts for the mutual interference between spatially separated radio links. As a result, one typically has $\gamma > 0$ when considering sub-6 GHz frequency bands, while the condition $\gamma = 0$ may be reasonable in the mmWave or THz bands, in which communication is mainly noise-limited due to the highly directive beams [36].

For a given time t, the symbols $\mathbf{y}_k(t)$ received at EN k-th over all the n_F radio frequency channels are transported to the BBU over the $l_S l_F$ cable resource blocks, where the mapping between radio and cable resources is to be designed (see Section 3.1) and depends on the bandwidth l_F available at each twisted-pair.

In this regards, we define the fraction $\mu \in [1/l_S, 1]$ of the radio bandwidth n_F that can be carried by each pair, referred to as *normalized cable bandwidth*, as

$$\mu = \frac{l_F}{n_F}. \tag{7}$$

As a result, the quantity

$$\eta = \mu \cdot l_S \tag{8}$$

expresses the bandwidth amplification factor (or redundancy) over the cable fronthaul, as $\eta \geq 1$. To simplify, we assume here that both $1/\mu$ and η are integer numbers. The two extreme situations with $\mu = 1$, or $\eta = l_S$, and $\mu = 1/l_S$, or $\eta = 1$, are shown in Figure 5 for $l_S = 4$ twisted-pairs and $n_F = 8$ subcarriers. For the first case, one replica of the whole radio signal $\mathbf{y}_k(t)$ can be transmitted over all of the l_S pairs, and the bandwidth amplification over cable is $\eta = l_S$. In the second case, disjoint fractions of the received bandwidth can be forwarded on each pair and the bandwidth amplification factor is $\eta = 1$.

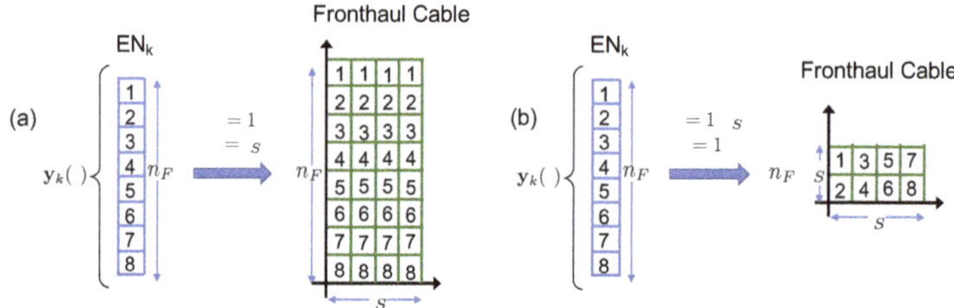

Figure 5. Mapping of radio resources over cable resources: (**a**) maximum normalized cable bandwidth (full redundancy), $\mu = 1$, or $\eta = l_S$; (**b**) minimal normalized cable bandwidth (no redundancy), $\mu = 1/l_S$, or $\eta = 1$.

We now detail the signal model for fronthaul transmission. To this end, let us define the $l_F \times l_S$ matrix $\tilde{\mathbf{Y}}_k$ containing the signal to be transmitted by the EN k-th to the BBU over the copper cable as

$$\tilde{\mathbf{Y}}_k = [\tilde{\mathbf{y}}_k^1(t), \tilde{\mathbf{y}}_k^2(t), ..., \tilde{\mathbf{y}}_k^{l_S}(t)], \tag{9}$$

where the k-th column $\tilde{\mathbf{y}}_k^c(t) \in \mathbb{C}^{l_F \times 1}$ denotes the signal transmitted on twisted-pair c across all the l_F cable frequency resources. The signal $\tilde{\mathbf{R}}_k \in \mathbb{C}^{l_F \times l_S}$ received at the BBU from the k-th EN across all the cable space-frequency resources is then computed as

$$\tilde{\mathbf{R}}_k(t) = \tilde{\mathbf{Y}}_k(t)\mathbf{H}_c + \tilde{\mathbf{W}}_k(t), \tag{10}$$

where $\tilde{\mathbf{W}}_k(t) = [\tilde{\mathbf{w}}_k^1(t), \tilde{\mathbf{w}}_k^2(t), ..., \tilde{\mathbf{w}}_k^{l_S}(t)] \in \mathbb{C}^{l_F \times l_S}$ is the additive white Gaussian cable noise uncorrelated over cable pairs and frequencies, i.e., $\tilde{\mathbf{w}}_k^c(t) \sim \mathcal{CN}(\mathbf{0}, \mathbf{I}_{l_F})$ for all pairs $c = 1, 2, ..., l_S$.

As commonly assumed in wireline communications to control cable radiations [37], the power of the cable symbol $[\tilde{\mathbf{y}}_k^c(t)]_{f'}$ transmitted from EN$_k$ over twisted-pair c at frequency f' is constrained to P_c by the (short-term: One can also consider the "long-term" power constraint

$n_T^{-1} \sum_t \mathbb{E}[|[\tilde{\mathbf{y}}_k^c(t)]_{f'}|^2] \leq P_c, \forall c \in [1, l_S], f' \in [1, l_F]$ with minor modifications to the analysis and final results) power constraint

$$\mathbb{E}\left[\left|[\tilde{\mathbf{y}}_k^c(t)]_{f'}\right|^2\right] \leq P_c \quad \forall c \in [1, l_S], f' \in [1, l_F], t \in [1, n_T]. \tag{11}$$

In the following, we will omit the time index t, when no confusion arises.

2.3. Performance Metrics

The performance metrics used to evaluate the interaction between eMBB and URLLC services in the considered A-RoC-based C-RAN architecture are detailed in the following.

2.3.1. eMBB

Capacity enhancement is the main goal of the eMBB service, which is envisioned to provide very high-rate communication to all the UEs. Therefore, for eMBB UEs, we are interested in the per-UE rate defined as

$$R_B = \frac{\log_2(M_B)}{n_T n_F}, \tag{12}$$

where M_B is the number of codewords in the codebook of each eMBB UE.

2.3.2. URLLC

Differently from eMBB, URLLC service is mainly focused on low-latency and reliability aspects. Due to the short length of URLLC packets, in order to guarantee ultra reliable communications, we need to ensure that the error probability for each URLLC UE, denoted as $\Pr[E_U]$, is bounded by a predefined value ϵ_U (typically smaller than 10^{-3}) as

$$\Pr[E_U] \leq \epsilon_U. \tag{13}$$

Concerning latency, we define the maximum access latency L_U as the maximum number of minislots that an URLLC UE has to wait before transmitting a packet. Finally, although rate enhancement is not one of the goals of URLLC service, it is still important to evaluate the per-UE URLLC rate that can be guaranteed while satisfying the aforementioned latency and reliability constraints. Similarly to (12), the per-UE URLLC rate is defined as

$$R_U = \frac{\log_2(M_U)}{n_F}, \tag{14}$$

where M_U is the number of URLLC codewords in the codebook used by the URLLC UE for each information packet.

3. Analog Fronthaul Signal Processing

The analog fronthaul links employed in the C-RAN system under study pose several challenges in the system design, which are addressed in this section. Firstly, the radio signal received at each EN needs to be mapped over the corresponding fronthaul resources in both frequency and space dimension. Secondly, the signal at the output of each fronthaul link needs to be processed in order to maximize the Signal-to-Noise Ratio (SNR) for all UE signals. Finally, the power constraints in (11) must be properly enforced. All these requirements are to be addressed by all-analog processing in order to meet the low-complexity and latency constraints of the analog fronthaul. In the rest of this section, we discuss each of these problems in turn.

3.1. Radio Resource Mapping over Fronthaul Channels

To maximize the SNRs for all the signals forwarded over the fronthaul by symmetry, we need to ensure that: (i) all the received signals are replicated η times across the cable twisted-pairs, where we recall that η is the bandwidth amplification factor defined in (8); and (ii) cable cross-talk interference among different radio frequency bands is minimized. In fact, as the transmitted power at the cable input is limited by the constraints in (11), a simple and effective way to cope with the impairments of the analog fronthaul links using the only analog-processing capability is by introducing redundancy in the fronthaul transmission. To this end, without loss of generality, we assume the following mapping rule between the n_F radio signals at each EN and the $l_S l_F$ cable resources.

Let us consider the radio signal $\mathbf{y}_k = \left[Y_k^1, Y_k^2, ..., Y_k^{n_F} \right]^T$ received at the k-th EN. For a given normalized cable bandwidth μ, the n_F frequency channels of the radio signal $\mathbf{y}_k$ can be split into $1/\mu$ sub-vectors of size $l_F = \mu n_F$ as

$$
\mathbf{y}_k = \begin{bmatrix} \mathbf{y}_k^1 \\ \mathbf{y}_k^2 \\ \vdots \\ \mathbf{y}_k^{\frac{1}{\mu}} \end{bmatrix}, \tag{15}
$$

where each vector $\mathbf{y}_k^j$ contains a disjoint fraction of the radio signal bandwidth n_F. Each vector $\mathbf{y}_k^j$ can be transmitted over η twisted-pairs, where we recall that η is the fronthaul redundancy factor. To this end, each signal $\mathbf{y}_k^j$ in (16) is transmitted over η consecutive cable twisted-pairs.

To formalize the described mapping, the first step consists in reorganizing the signal $\mathbf{y}_k$ into a $l_F \times \frac{1}{\mu}$ matrix as

$$
\mathbf{Y}_k = \mathrm{vec}_{\frac{1}{\mu}}^{-1}(\mathbf{y}_k) = \begin{bmatrix} \mathbf{y}_k^1 & \mathbf{y}_k^2 & \cdots & \mathbf{y}_k^{\frac{1}{\mu}} \end{bmatrix}, \tag{16}
$$

where the operator $\mathrm{vec}_{\frac{1}{\mu}}^{-1}(\cdot) : \mathbb{C}^{n_F} \to \mathbb{C}^{n_F \cdot \mu \times \frac{1}{\mu}}$ acts as the inverse of the vectorization operator $\mathrm{vec}(\cdot)$, with the subindex $1/\mu$ denoting the number of columns of the resulting matrix. Then, the overall cable signal $\tilde{\mathbf{Y}}_k$ transmitted by EN k-th in (9) can be equivalently written as

$$
\tilde{\mathbf{Y}}_k = \left[\underbrace{\mathbf{y}_k^1 \ \mathbf{y}_k^1 \ \cdots \ \mathbf{y}_k^1}_{\eta}, \ \underbrace{\mathbf{y}_k^2 \ \mathbf{y}_k^2 \ \cdots \ \mathbf{y}_k^2}_{\eta} \ \cdots \ \underbrace{\mathbf{y}_k^{\frac{1}{\mu}} \ \mathbf{y}_k^{\frac{1}{\mu}} \ \cdots \ \mathbf{y}_k^{\frac{1}{\mu}}}_{\eta} \right], \tag{17}
$$

or, in compact form, as

$$
\tilde{\mathbf{Y}}_k = \mathbf{Y}_k \otimes \mathbf{1}_\eta^T, \tag{18}
$$

where $\otimes$ denotes the Kronecker product (for a review of Kronecker product properties in signal processing we refer the reader to [38]). Notice that in case of full normalized cable bandwidth, i.e., $\mu = 1$ (corresponding to $\eta = l_S$), the signal $\tilde{\mathbf{Y}}_k$ transmitted over the fronthaul cable simplifies to $\tilde{\mathbf{Y}}_k = \mathbf{y}_k \otimes \mathbf{1}_{l_S}^T$, which implies that the radio signal $\mathbf{y}_k$ is replicated over all the l_S twisted-pairs. On the contrary, when the normalized cable bandwidth is minimal, i.e., $\mu = 1/l_S$ (corresponding to $\eta = 1$), the signal $\tilde{\mathbf{Y}}_k$ does not contain any redundancy, and disjoints signals are transmitted over all pairs, so that the cable signal $\tilde{\mathbf{Y}}_k$ equals the matrix radio signal in (16) as $\tilde{\mathbf{Y}}_k = \mathbf{Y}_k$.

Remark—Practical Implementation Issues. *The easiest practical implementation of the proposed analog radio resource mapping at the EN is by grouping the subcarriers onto a specific frequency portion of the cable, as described in [18–20,22]. As an example, let us assume that the EN is equipped with 5 antennas, that each antenna receives a 20-MHz radio signal, and that the analog fronthauling disposes of 4 links with 100 MHz bandwidth each. In this case, the above references have shown that it is possible to freely map, or to replicate, in an all-analog fashion the 5×20 MHz bands onto the overall 4×100 MHz = 400 MHz fronthaul bandwidth. This example corresponds to a special case of the model studied in this paper, obtained by setting $\mu = 1$,* *i.e., the whole radio signal bandwidth received at the ENs is mapped/replicated over the analog fronthauling. More generally, this paper posits the possibility to carry out the fronthaul mapping at a finer granularity, i.e., at a subcarrier level. In this case, filtering operations would in practice be mandatory in order to extract groups of subcarriers. This operation can be implemented in principle still by analog filters, whose design is left as future works.*

3.2. Signal Combining at the Fronthaul Output

As discussed, depending on the fronthaul bandwidth, a number η of noisy replicas of the radio signals received at each EN are relayed to the BBU over η different twisted-pairs. Hence, in order to maximize the SNRs for all signals, Maximum Ratio Combining (MRC) [39] is applied at the cable output as

$$\mathbf{R}_k = \tilde{\mathbf{R}}_k \mathbf{G}, \tag{19}$$

where $\mathbf{R}_k \in \mathbb{C}^{l_F \times \frac{1}{\mu}}$ is the signal received at BBU from the k-th EN after the combiner and $\mathbf{G} \in \mathbb{R}^{l_S \times \frac{1}{\mu}}$ is the MRC matrix. Under the assumptions here, MRC coincides with equal ratio combining and hence matrix $\mathbf{G}$ can be written as

$$\mathbf{G} = \frac{1}{\eta}\left(\mathbf{I}_{\frac{1}{\mu}} \otimes \mathbf{1}_\eta\right). \tag{20}$$

As an example, in the case of maximum redundancy, i.e., $\eta = l_S$, the MRC matrix $\mathbf{G} = l_S^{-1}\mathbf{1}_{l_S}$ combines the analog signals received over all pairs, since they carry the same information signal. On the contrary, in the case of minimal normalized bandwidth $\mu = 1/l_S$ or $\eta = 1$, the matrix $\mathbf{G}$ equals the identity matrix as $\mathbf{G} = \mathbf{I}_{l_S}$, since no combining is possible.

The signal $\mathbf{r}_k \in \mathbb{C}^{n_F}$ received at the BBU from EN_k across the n_F subcarriers is thus obtained by vectorizing matrix $\mathbf{R}_k$ in (19) as

$$\mathbf{r}_k = \mathrm{vec}(\mathbf{R}_k). \tag{21}$$

The relationship between the signal $\mathbf{r}_k$ (21) obtained at the output of the combiner and the radio received signal $\mathbf{y}_k$ in (15) is summarized by the block-scheme in Figure 6, and it is

$$\mathbf{r}_k = \mathrm{vec}\left[\left(\left(\mathrm{vec}_{\frac{1}{\mu}}^{-1}(\mathbf{y}_k) \otimes \mathbf{1}_\eta^T\right)\mathbf{H}_c + \tilde{\mathbf{W}}_k\right)\mathbf{G}_k\right]. \tag{22}$$

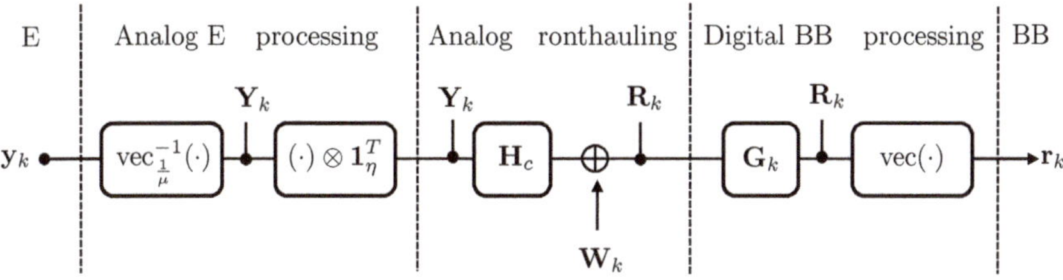

Figure 6. Relationship between the signal $\mathbf{r}_k$ (21) obtained at the output of the combiner and the radio received signal $\mathbf{y}_k$ in (15).

Finally, we collect the overall signal $\mathbf{R} \in \mathbb{C}^{n_F \times M}$ received at the BBU from all ENs across all frequencies in the matrix

$$\mathbf{R} = [\mathbf{r}_1, \mathbf{r}_2, ..., \mathbf{r}_M]. \tag{23}$$

After some algebraic manipulations, it is possible to express Equation (23) in a more compact form, which is reported in the following Lemma 1.

Lemma 1. *In the given C-RAN architecture with analog fronthaul links, for a given bandwidth amplification factor $\eta \geq 1$, the signal $\mathbf{R} \in \mathbb{C}^{n_F \times M}$ received at the BBU from all ENs across all radio frequencies after MRC can be written as*

$$\mathbf{R} = \left(\mathbf{H}_c^{\eta} \otimes \mathbf{I}_{l_F} \right) \mathbf{Y} + \mathbf{W}, \tag{24}$$

where

$$\mathbf{H}_c^{\eta} = \gamma \eta \mathbf{1}_{\frac{1}{\mu}} \mathbf{1}_{\frac{1}{\mu}}^{T} + (1 - \gamma) \mathbf{I}_{\frac{1}{\mu}} \tag{25}$$

is the equivalent fronthaul channel matrix; $\mathbf{Y}$ is the signal received at all ENs across all frequency radio channels in (5); and $\mathbf{W} = [\mathbf{w}_1, \mathbf{w}_2, ..., \mathbf{w}_M]$ is the equivalent cable noise at the BBU after MRC, with the k-th column distributed as $\mathbf{w}_k \sim \mathcal{CN} \left(\mathbf{0}, \frac{1}{\eta} \mathbf{I}_{n_F} \right)$ for all $k = 1, 2, ..., M$.

Proof. see Appendix A. $\square$

To gain some insights, it is useful again to consider the two extreme cases of maximum redundancy, i.e., $\eta = l_S$, and no redundancy, i.e., $\eta = 1$. In the former case, the equivalent channel (25) equals the scalar $\mathbf{H}_c^{\eta} = 1 + \gamma(l_S - 1)$. This demonstrates the effect of transmitting a replica of the whole radio signal over all pairs. In fact, the useful signal is received at the BBU not only through the direct path, which has unit gain, but also from the remaining $l_S - 1$ interfering paths, each with gain γ, which constructively contribute to the overall SNR after the combiner. More precisely, it can be observed that, in case of full redundancy, the SNR of the radio signal $\mathbf{Y}$ at the BBU is increased by the analog fronthaul links by a factor of $(1 + \gamma(l_S - 1))^2 / (1/\eta) = l_S(1 + \gamma(l_S - 1))^2$. As a result, in this case, for a coupling factor $\gamma > 0$, the SNR at the BBU increases with the cube of the number of fronthaul links l_S. In contrast, for $\mu = 1/l_S$, the equivalent fronthaul channel reflects the fact that signals forwarded over the different pairs interfere with each other, and is equal to $\mathbf{H}_c^{\eta} = \mathbf{H}_c$. The beneficial effect of redundantly transmitting radio signals over different pairs is reflected also in the power of the noise after the combiner, which is reduced proportionally to the bandwidth amplification factor η.

3.3. Fronthaul Power Constraints

To enforce the cable power constraints in (11), it is necessary to scale the radio signal $\mathbf{Y}$ in (24) by a factor of λ prior to the transmission over the fronthaul. This is given as

$$\lambda = \sqrt{\frac{P_c}{\delta P_B (1 + 2\alpha^2) + 1}}, \tag{26}$$

where δ is equal to $\delta = \left(1 - L_U^{-1} \right)^{-1}$ for OMA, accounting for the fact that only $L_U - 1$ minislots are devoted to the eMBB UE, while it equals $\delta = 1$ for NOMA, since the eMBB transmission spreads over all L_U minislots. To simplify the notation, in the following we will account for the gain λ by scaling the noise over the cable after MRC in Equation (25) accordingly as

$$\mathbf{w}_k(t) \sim \mathcal{CN} \left(\mathbf{0}, \frac{1}{\eta \lambda^2} \mathbf{I} \right). \tag{27}$$

4. Orthogonal Multiple Access (OMA)

As described in Section 2.1, under OMA over the radio channel, one minislot every L_U is exclusively allocated to URLLC UEs, while eMBB UEs transmit over the remaining minislots. In this way, URLLC UEs never interfere with eMBB transmissions. If more than one URLLC packet is generated at a user between two URLLC transmission opportunities, only one of such packets (randomly selected) is transmitted, while the others are discarded, causing a blockage error. Due to the latency constraints, URLLC signals are digitized and decoded locally at the ENs, while the eMB signals are first mapped over the fronthaul lines, and then analogically forwarded to the BBU, as mathematically summarized in Figure 7. In this section, we derive the expressions for the eMBB and URLLC rates under OMA for a given URLLC access latency L_U and, in the case of URLLC, for a fixed URLLC target error probability ϵ_U.

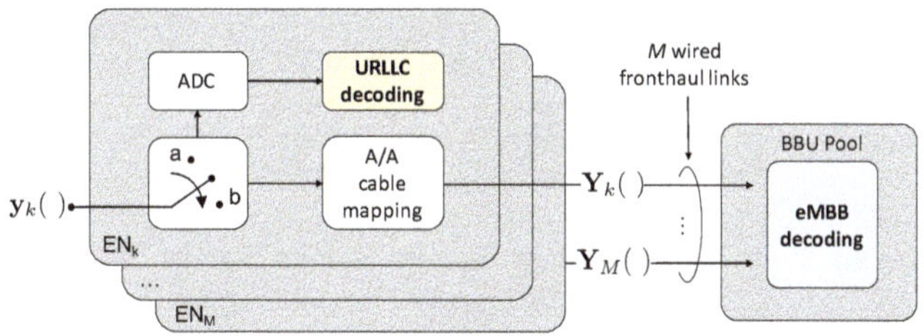

Figure 7. Block diagram of the operation of the ENs and BBU for Orthogonal Multiple Access (OMA). A/A stands for Analog-to-Analog.

4.1. URLLC Rate

To evaluate the per-UE URLLC rate under OMA and for a given URLLC target error probability ϵ_U, we follow the approach in [32], which is reviewed here. URLLC packets are generally short due to the strict latency constraints, and the maximum achievable rate can be computed by leveraging results from finite blocklength information theory. To this end, fix a given blocklength n_F and URLLC error decoding probability ϵ_U^D. Notice that this probability is different from the general URLLC target error probability ϵ_U, as detailed later in this section. According to [40], the URLLC rate can be well approximated by

$$R_U = \log_2(1 + \beta^2 P_U) - \sqrt{\frac{V}{n_F}} Q^{-1}(\epsilon_U^D), \tag{28}$$

where

$$V = \frac{\beta^2 P_U}{1 + \beta^2 P_U} \tag{29}$$

is the channel dispersion and $Q^{-1}(\cdot)$ is the inverse Q-function (see Section 1.5).

The error probability for URLLC packets is the sum of two contributions. The first represents the probability that an URLLC packet is discarded due to blockage, given that only one URLLC packet can be transmitted within the required L_U worst-case latency; while the second is the probability that the packet is transmitted but not successfully decoded. Accordingly, the overall error probability can be computed as

$$\Pr[E_U] = \sum_{n}^{L_U - 1} p(n) \frac{n}{n+1} + \sum_{n}^{L_U - 1} p(n) \frac{1}{n+1} \epsilon_U^D, \tag{30}$$

where $p(n) = \Pr[N_U(L_U) = n]$ is the distribution of the binomial random variable $N_U(L_U) \sim \text{Bin}(L_U - 1, q)$ representing the number of additional packets generated by the URLLC UE during the remaining minislots between two transmission opportunities. The decoding error probability ϵ_U^D in (28) can be obtained from the URLLC reliability constraint in (13), i.e., $\Pr[E_U] = \epsilon_U$.

4.2. eMBB Rate

The eMBB signals received at the ENs are forwarded over the analog fronthaul to the BBU, where centralized digital signal processing and decoding are performed. In the case of OMA, the eMBB signal is free from URLLC interference, hence signal $\mathbf{Y}$ received by all ENs over all radio channels in (5) can be written as

$$\mathbf{Y} = \mathbf{XH} + \mathbf{Z}. \tag{31}$$

By substituting (31) in (24), it is possible to compute the expression for the eMBB per-UE rate under OMA, as shown in Lemma 2. Notice that, unlike the case of URLLC packets, the $n_F n_T$ blocklength of eMBB packets allows for the use of standard asymptotic Shannon theory in the computation of eMBB information rate.

Lemma 2. *In the given C-RAN architecture with analog fronthaul links, for a given bandwidth amplification factor $\eta \geq 1$, the eMBB user rate under OMA is given as*

$$R_B = \mu \frac{1 - L_U^{-1}}{M} \log \left(\det \left(\mathbf{I} + \bar{P}_B \mathbf{R}_{z_{eq}}^{-1} \mathbf{H}_{eq} \mathbf{H}_{eq}^T \right) \right), \tag{32}$$

where $\bar{P}_B = P_B \left(1 - L_U^{-1} \right)^{-1}$ is the transmission power of eMBB users under OMA, $\mathbf{H}_{eq} = \mathbf{H} \otimes \mathbf{H}_c^\eta$ is the overall channel matrix comprising both the radio channel $\mathbf{H}$ and the equivalent cable channel $\mathbf{H}_c^\eta$ defined in Lemma 1, and $\mathbf{R}_{z_{eq}} = \mathbf{I}_M \otimes \mathbf{H}_c^\eta \mathbf{H}_c^\eta + \frac{1}{\lambda^2 \eta} \mathbf{I}_{\frac{M}{\mu}}$ is the overall wireless plus cable noise at the BBU.

Proof. see Appendix B. $\square$

As a first observation, the eMBB rate (32) linearly scales with the normalized bandwidth μ. This shows that a potential performance degradation in terms of spectral efficiency can be incurred in the presence of fronthaul channels with bandwidth limitations, i.e., with $\mu < 1$. This loss is pronounced in the presence of significant inter-channel interference, i.e., for large γ. In fact, a large γ increases the effective noise power as per expression of matrix $\mathbf{R}_{z_{eq}}$. It is also important to point out that in the considered C-RAN system based on the analog relaying of radio signals, the overall noise at the BBU is no longer white as it accounts both for the white cable noise and the wireless noise, where the latter is correlated when there is some bandwidth redundancy, i.e., when $\mu \geq 1/l_S$ or $\eta > 1$.

5. Non-Orthogonal Multiple Access (NOMA)

In NOMA, URLLC UEs transmit in the same minislot where the packet is generated, and hence the access latency is minimal and limited to $L_U = 1$ minislot. However, the URLLC signals mutually interfere with the eMBB transmission, which spans the whole time-frequency resource plane. Due to URLLC latency constraints, the eMBB signals necessarily need to be treated as noise while decoding URLLC packets at the ENs. On the contrary, several strategies can be adapted in order to deal with the interfering URLLC signal. Beside puncturing, considered for 5G NR standardization [33,34], this work considers two other techniques, namely Treating Interference as Noise (TIN) and Successive Interference Cancellation (SIC), as detailed in the rest of this section.

5.1. URLLC Rate under NOMA

The URLLC per-UE rate for NOMA can be computed by leveraging results from finite blocklength information theory similarly to the OMA case, but accounting for the additional eMBB interference [41]. The URLLC per-UE rate under NOMA is thus well approximated by [32]

$$R_U = \log_2(1 + S_U) - \sqrt{\frac{V}{n_F}} Q^{-1}(\epsilon_U^D),$$

where

$$S_U = \frac{\beta^2 P_U}{1 + (1 + 2\alpha^2) P_B} \tag{33}$$

is the Signal-to-Interference-plus-Noise Ratio (SINR) for the URLLC UE, and the channel dispersion V is given as

$$V = \frac{S_U}{1 + S_U}. \tag{34}$$

Notice that in NOMA the incoming URLLC packet is always transmitted, and hence an URLLC error occurs only if the decoding of such packet fails, which happens with probability ϵ_U^D. This implies that under NOMA, the probability of URLLC error is given by

$$\Pr[E_U] = \epsilon_U^D, \tag{35}$$

hence imposing the condition $\epsilon_U^D \leq \epsilon_U$ by the requirement (13).

5.2. eMBB Rate by Puncturing

To carry out joint decoding at the BBU of the eMBB signals under NOMA, the standard approach is to simply discard at the eMBB decoder those signals that are interfered by URLLC. As shown in Figure 8, this technique, referred to as puncturing, is based on the detection of URLLC transmissions at the BBU: if a URLLC transmission is detected in the signal received from EN_k, such signal is discarded. Otherwise, the interference-free eMBB signals are jointly decoded at the BBU.

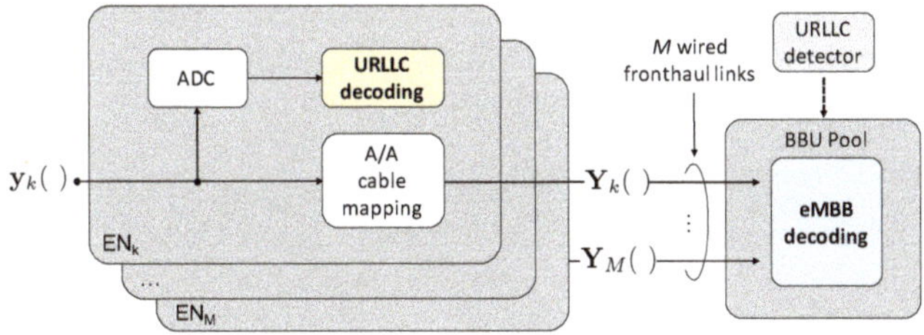

Figure 8. Block diagram of the operation of the ENs and BBU for Non-Orthogonal Multiple Access (NOMA) by puncturing and Treating Interference as Noise (TIN). A/A stands for Analog-to-Analog.

Considering the aforementioned assumptions, the signal model for puncturing can be equivalently described by assuming that if the signal Y_k^f received at the EN_k on frequency f is interfered by an URLLC transmission, then such signal is discarded, and the BBU receives only noise. This is mathematical described by

$$Y_k^f = B_k(X_k^f + \alpha X_{k+1}^f + \alpha X_{k-1}^f) + Z_k^f, \tag{36}$$

where the Bernoulli variable $B_k = 1 - A_k \sim \mathcal{B}(1-q)$ indicates the absence ($B_k = 1$) or presence ($B_k = 0$) of URLLC transmissions in the given minislot. The signal in (36) received across all ENs and frequencies can be written in matrix form as

$$\mathbf{Y} = \mathbf{XHB} + \mathbf{Z}, \tag{37}$$

with definitions given in Section 2 and with $\mathbf{B} = \mathrm{diag}(B_1, B_2, ..., B_M)$. The rate for the eMBB UE under NOMA by puncturing is reported in Lemma 3 and can be derived by substituting signal (37) in Equation (24).

Lemma 3. *In the given C-RAN architecture with analog fronthaul links, for a given bandwidth amplification factor $\eta \geq 1$, the eMBB user rate under NOMA by puncturing yields*

$$R_B = \frac{\mu}{M}\mathbb{E}_{\mathbf{B}}\left[log\left(det\left(\mathbf{I} + P_B \mathbf{R}_{z_{eq}}^{-1}\mathbf{H}_{B,eq}\mathbf{H}_{B,eq}^T\right)\right)\right], \tag{38}$$

where $\mathbf{R}_{z_{eq}}$ is defined as in Lemma 2; $\mathbf{H}_{B,eq} = \mathbf{BH} \otimes \mathbf{H}_c^\eta$ is the equivalent wireless plus cable channel in case of puncturing; and we have $\mathbf{B} = diag(B_1, B_2, ..., B_M)$, with B_k being i.i.d. $\mathcal{B}(1-q)$ variables .

Proof. Lemma 3 can be proved by following similar steps as for the proof of Lemma 2 with two minor differences: *(i)* the radio channel matrix $\mathbf{H}$ is right multiplied by the random matrix $\mathbf{B}$ and *(ii)* the capacity is computed by averaging over the distribution of $\mathbf{B}$. $\square$

Differently from the rate (32) achieved by OMA, under NOMA, the eMBB transmission spreads over all the minislots, so that there is no scaling factor $1 - L_U^{-1}$ in front of the rate expression (38) to account for the resulting loss in spectral efficiency. In case of NOMA by puncturing, the noise covariance is exactly as the one in the eMBB OMA rate in Equation (32), since the eMBB signal, if not discarded at the BBU, is guaranteed to be URLLC interference-free. The overall rate is computed by averaging over all the possible realizations of the random matrix $\mathbf{B}$, which left-multiplies the radio channel matrix $\mathbf{H}$ and accounts for the probability that the entire signal is discarded due to an incoming URLLC packet.

In the case of C-RAN with digital limited-capacity fronthaul links, as discussed in [32], it is advantageous to carry out the operation of detecting and, eventually, discarding the eMBB signal at the ENs. In fact, with a digital fronthaul, only the undiscarded minislots can be quantized, hence devoting the limited fronthaul resources to increase the resolution of interference-free eMBB samples [32]. The same does not apply to the analog fronthaul considered here, as signals are directly relayed to the BBU without any digitization.

5.3. eMBB Rate by Treating Interference as Noise

In the case of analog fronthaul, an enhanced strategy to jointly decode the eMBB signals under NOMA at the BBU is to treat the URLLC interfering transmissions as noise at the eMBB decoder, instead of discarding the corresponding minislot as in puncturing. The block diagram is the same as for puncturing and shown in Figure 8. Accordingly, based on the signals received over the fronthaul links, the BBU first detects the presence of URLLC transmission so as to properly select the decoding metric. Then, based on this knowledge, joint decoding is performed by TIN.

Lemma 4. *In the given C-RAN architecture with analog fronthaul links, for a given bandwidth amplification factor $\eta \geq 1$, the eMBB user rate under NOMA by treating URLLC interference as noise yields*

$$R_B = \frac{\mu}{M} \mathbb{E}_{\mathbf{A}} \left[log \left(det \left(\mathbf{I} + P_B \mathbf{R}_{A,z_{eq}}^{-1} \mathbf{H}_{eq} \mathbf{H}_{eq}^T \right) \right) \right], \tag{39}$$

where $\mathbf{R}_{A,z_{eq}} = \mathbf{R}_{z_{eq}} + \beta^2 P_U \left(\mathbf{A} \otimes \mathbf{H}_c^\eta \mathbf{H}_c^\eta \right)$ is the overall noise plus URLLC interference at the BBU; matrix $\mathbf{A}$ is as in (5); and matrices $\mathbf{R}_{z_{eq}}$ and $\mathbf{H}_{eq}$ are the same as in Lemma 2.

Proof. see Appendix C. $\square$

Differently from the two previous cases, in the case of NOMA under TIN, the noise covariance matrix $\mathbf{R}_{A,z_{eq}}$ needs to account also for the interfering URLLC transmissions, whose packet arrival probability is described by matrix $\mathbf{A}$. The achievable rate is then computed by taking the average over the random matrix $\mathbf{A}$. This average reflects the long-blocklength transmissions of the eMBB users.

5.4. eMBB Rate by Successive Interference Cancellation

Finally, a more complex receiver architecture can be considered at the BBU, whereby interference is cancelled out from the useful signal. This technique, referred to as Successive Interference Cancellation (SIC), is based on the idea that, if an URLLC signal is successfully decoded at the EN_k, it can be cancelled from the overall received signal $\mathbf{y}_k$ prior to the relaying over the cable, so that an ideally interference-free eMBB signal is forwarded to the BBU. We also assume that, if the URLLC signal is not successfully decoded, signal $\mathbf{y}_k$ is discarded.

As a practical note, SIC must be performed in the analog domain, thus complicating the system design. Practical complications are not considered in the analysis here. As shown in Figure 9, if the

URLLC signal is successfully decoded at EN_k, this needs first to be Digital-to-Analog Converted (DAC) and then cancelled from the analog signal $\mathbf{y}_k$. Therefore, signal $\mathbf{y}_k$ needs to be suitably delayed in order to wait for the cascade of ADC, decoding, and DAC operations to be completed at the URLLC decoder. Being latency not an issue for eMBB traffic, in this work we assume to employ ideal ADC/DAC, so that the delay in Figure 9 is assumed as ideally zero.

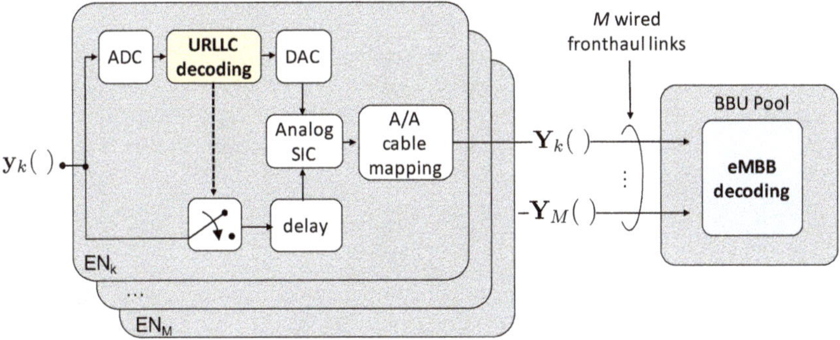

Figure 9. Block diagram of the operation of the ENs and BBU for Non-Orthogonal Multiple Access (NOMA) by Successive Interference Cancellation (SIC). A/A stands for Analog-to-Analog.

To account for imperfect SIC, the amplitude of the residual URLLC interference on eMBB signal is assumed to be proportional to a factor $\rho \in [0, 1]$. Accordingly, perfect SIC corresponds to $\rho = 0$, and no SIC to $\rho = 1$.

The signal received at the BBU from EN_k can be thus written as

$$Y_k^f = (1 - A_k E_k)(X_k^f + \alpha X_{k+1}^f + \alpha X_{k-1}^f) + \rho \beta A_k (1 - E_k) U_k^f + Z_k^f, \tag{40}$$

where the Bernoulli variable $E_k \sim \mathcal{B}(q\epsilon_U^D)$ indicates whether there has been an error in decoding the URLLC packet (i.e., $E_k = 1$), or it has been successfully decoded (i.e., $E_k = 0$), and A_k is the same as above. It is easy to show that the factor $(1 - A_k E_k)$ multiplying the eMBB signal indicates that the eMBB signal (40) is discarded only when the two following events simultaneously happen: (i) there is a URLLC transmission (i.e., $A_k = 1$, whose probability is q), and (ii) such URLLC transmission is not successfully decoded ($E_k = 1$, whose probability is ϵ_U^D). In turn, the factor $A_k(1 - E_k)$ multiplying the URLLC signal implies that, if there is a URLLC transmission (i.e., $A_k = 1$) and such transmission is successfully decoded at the EN_k (i.e., $E_k = 0$), then the URLLC signal is mitigated by analog SIC so that only a ρ-fraction of it is forwarded to the BBU and impairs the eMBB transmission.

The signal in (40) received across all ENs and frequencies in the case of NOMA by SIC can be equivalently written in matrix form as

$$\mathbf{Y} = \mathbf{XH}\,(\mathbf{I} - \mathbf{AE}) + \rho\beta\mathbf{UA}\,(\mathbf{I} - \mathbf{E}) + \mathbf{Z}, \tag{41}$$

where $\mathbf{E} = \mathrm{diag}(E_1, E_2, ..., E_M)$. The eMBB UE rate for NOMA by SIC can thus be computed by substituting signal (41) in (24) and the final result is in Lemma 5.

Lemma 5. *In the given C-RAN architecture with analog fronthaul links, for a given bandwidth amplification factor $\eta \geq 1$, the eMBB user rate under NOMA by SIC yields*

$$R_B = \frac{\mu}{M}\mathbb{E}_{\mathbf{A},\mathbf{E}}\left[\log\left(\det\left(\mathbf{I} + P_B \mathbf{R}_{AE,z_{eq}}^{-1}\mathbf{H}_{AE}\mathbf{H}_{AE}^T\right)\right)\right], \tag{42}$$

where $\mathbf{H}_{AE} = ((\mathbf{I} - \mathbf{AE})\mathbf{H}) \otimes \mathbf{H}_c^{\eta}$ *is the equivalent wireless plus cable channel in case of SIC;* $\mathbf{E} = diag(E_1, E_2, ..., E_M)$ *is a diagonal matrix whose k-th entry* $E_k \sim \mathcal{B}(q\epsilon_U^D)$ *accounts for the probability that the URLLC signal is not successfully decoded at* EN_k; *and* $\mathbf{R}_{AE,z_{eq}} = \mathbf{R}_{z_{eq}} + \rho^2\beta^2 P_U\left((\mathbf{A}(\mathbf{I} - \mathbf{E})) \otimes \mathbf{H}_c^{\eta}\mathbf{H}_c^{\eta}\right)$ *is the overall noise plus residual URLLC interference.*

Proof. Lemma 5 can be proved by following similar steps as for the proofs of the previous Lemmas. $\square$

SIC describes a more complex ENs architecture in which the URLLC signals, if successfully decoded, are successively canceled from the eMBB signals at the EN. However, in the case of imperfect interference cancellation, i.e., $\rho > 0$, the eMBB signal is still impaired by some residual URLLC interference, which is accounted for by the overall noise covariance $\mathbf{R}_{AE,z_{eq}}$ in (42), similarly to TIN. The URLLC arrival probability and the probability of successful decoding of the URLLC packets are reflected by random matrices $\mathbf{A}$ and $\mathbf{E}$, respectively.

6. Numerical Results

Numerical results based on the previous theoretical discussion are shown in this section with the aim of providing some useful intuitions about the performance of C-RAN systems based on analog RoC in the presence of both URLLC and eMBB services. Unless otherwise stated, we consider the following settings: $M = 6$ ENs, $n_F = 60$ subcarriers (This choice is motivated by the fact that, while still resembling the properties of URLLC short-packet transmissions, $n_F = 60$ is a sufficiently long packet size to ensure tight lower and upper bounds for the limited-blocklength channel capacity [5]), $P_B = 7$ dB, $P_U = 10$ dB, URLLC channel gain $\beta^2 = 1$, $P_C = 7$ dB, $l_S = 4$, and, conventionally, $\epsilon_U = 10^{-3}$. In the case of OMA, the worst-case access latency for URLLC users is set to $L_U = 2$ minislots.

Figure 10 shows URLLC and eMBB per-UE rates for both OMA and NOMA by varying the fronthaul crosstalk interference power γ^2. We consider two values for the normalized bandwidth μ of each copper cable, namely $\mu = 1/l_S = 1/4$ and $\mu = 1$. Please note that the first value corresponds to the minimal bandwidth, while the latter enables each twisted-pair to carry the whole signal bandwidth. For reference, the eMBB rates obtained in the case of ideal fronthaul are shown for both OMA and NOMA. The inter-cell interference power is set to $\alpha^2 = 0.2$, and the URLLC arrival probability to $q = 10^{-3}$. For NOMA, we consider here only puncturing.

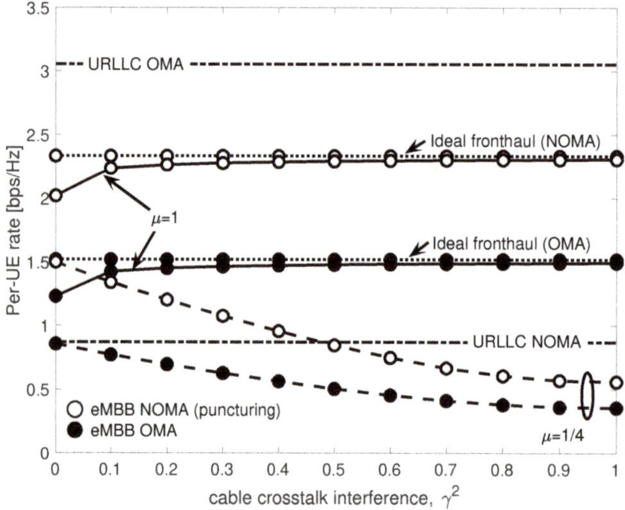

Figure 10. URLLC and eMBB per-UE rates as a function of fronthaul crosstalk interference power γ^2 for OMA and NOMA with puncturing.

The URLLC rates do not depend on cable crosstalk interference γ^2, since URLLC packets are decoded at the EN and thus never forwarded to the BBU over the fronthaul. Furthermore, with NOMA, the access latency of URLLC is minimum, i.e., $L_U = 1$, while for OMA it equals $L_U = 2$. However, Figure 10 shows that the price to pay for this reduced latency is in terms of transmission rate, which is lower than in the OMA case. On the contrary, in the case of eMBB, NOMA allows for a communication at higher rates than those achieved by OMA, thanks to the larger available bandwidth when q is small enough.

On the subject of eMBB rates, it is interesting to discuss the interplay between the normalized cable bandwidth μ and crosstalk power γ^2. For $\mu = 1$, the same signal is transmitted over all the $l_S = 4$ fronthaul twisted-pairs, and hence the four spatial paths sum coherently over the cable, thus turning crosstalk into a benefit. Hence, the eMBB rates under both OMA and NOMA increase with γ^2 and ultimately converge to those achieved over the ideal fronthaul (In this work, we consider the same interference gain γ for all fronthaul links. In practice, the performance boost shown in Figure 10 for increasing γ and for $\mu = 1$ would still be present, albeit to a different extent dependent on the channel realization, even if considering complex channel gain. However, this would require the use of more complex precoding techniques, such as Tomlinson-Harashima [37], which require an estimate of the fronthaul channel. Since for wired fronthauling the channel is nearly static and time-invariant, channel state information can be easily obtained [42]). For $\mu = 1/4$ instead, disjoint portions of the radio signal are transmitted over different interfering twisted-pairs, and the performance progressively decrease with γ^2. The leftmost portion of Figure 10 suggests that for mild cable interference, even when the cable bandwidth is small (i.e., $\mu = 1/4$), it is still possible to provide communication with acceptable performance degradation, i.e., with a ≈ 1 bps/Hz loss from the ideal fronthaul case for NOMA, and an even smaller loss for OMA. However, when the cable crosstalk increases, the rate degradation is severe, and fair performance are achieved only if the cable bandwidth is large enough to accommodate the redundant transmission of radio signals over all pairs, i.e., $\mu = 1$.

Figure 11 shows URLLC and eMBB rates as a function of the URLLC packet arrival probability q. eMBB rates under OMA are compared with those achieved by NOMA under puncturing, TIN and SIC. We consider here full cable bandwidth availability $\mu = 1$ and $\gamma^2 = 1$, so that, as in Figure 10, the rates achieved for both OMA and NOMA for low q (say, $q < 10^{-2}$) coincide with those achieved over the ideal fronthaul. Inter-cell interference is set to $\alpha^2 = 0.2$. As noted in [32], under OMA, when q increases, the probability of an URLLC packet to be dropped due to blockage becomes very high, preventing URLLC transmission from meeting the strict reliability constraints, and results in a vanishing URLLC rate. This is unlike in NOMA, whereby the URLLC rate is not affected by q, and the access latency is minimal, i.e., $L_U = 1$. For eMBB under NOMA, TIN always outperforms puncturing. This is because TIN does not discard any received minislot, thus contributing to the overall eMBB rate. The result is in contrast with the conventional digital capacity-constrained fronthaul considered in [32]. In fact, in the latter, for sufficient low q, it is preferable not to waste fronthaul capacity resources by quantizing samples received in minislots affected by URLLC interference in order to increase the resolution of the interference-free samples. Additional gains are achieved by SIC, which takes advantages of the high reliability, and thus high probability to be cancelled, of the URLLC signal at the EN.

Implementing SIC in a fully analog fashion is practically not trivial, and there is generally some residual URLLC interference. The effect of residual interference on the achievable eMBB rate is investigated in Figure 12 for $q = 0.3$, $\alpha^2 = 0.4$, $\gamma^2 = 0.5$, $\mu = 1$ and for different power of URLLC UE P_U. Once again, in case of perfect interference cancellation, i.e., $\rho = 0$, SIC approaches the ideal fronthaul performance, while for more severe values of the residual interference power ρ, the eMBB performance progressively decreases. Nevertheless, even in the worst-case SIC scenario, i.e., $\rho = 1$, the achievable rates are never worse than those achieved by TIN irrespective of the value of P_U. This is once again due to the high reliability of URLLC transmission. It is in fact easy to prove that for $\rho = 1$ and low values of ϵ_U^D, the SIC eMBB rate in (42) converges to the one of TIN in (39).

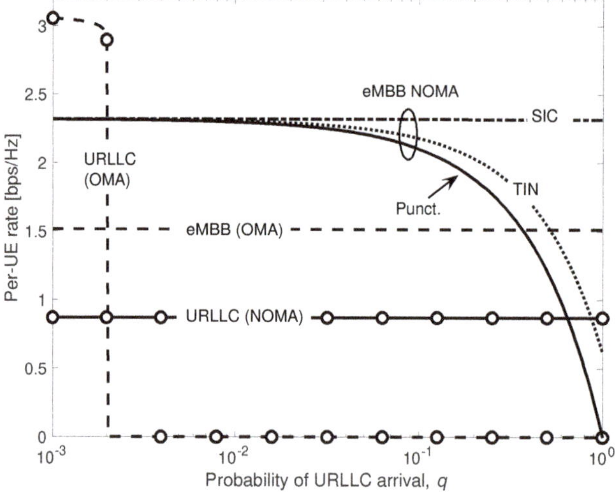

Figure 11. URLLC and eMBB rates vs. probability of URLLC arrival q for OMA and NOMA by puncturing, treating interference as noise (TIN), and successive interference cancellation (SIC).

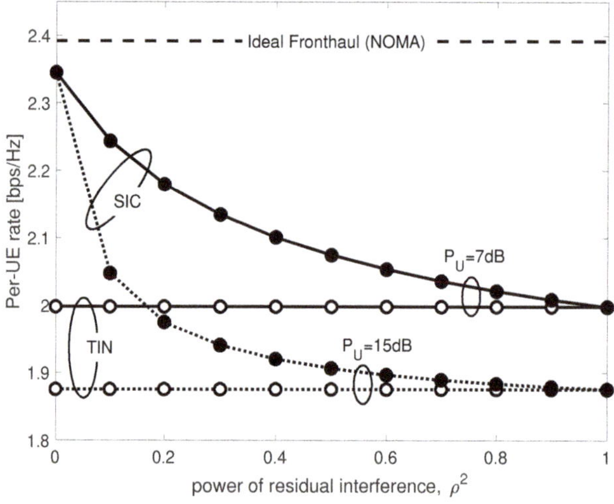

Figure 12. eMBB rates for NOMA by SIC vs. power of residual URLLC interference ρ^2.

For completeness, Figure 13 shows the trade-off between eMBB and URLLC per-UE rates as a function of the access latency L_U for OMA and NOMA with puncturing, and considering $q = 10^{-3}$, $\alpha^2 = 0.2$ and $\gamma^2 = 0.5$. The behavior of the RoC-based C-RAN system versus the access latency L_U is similar to the one observed for digital capacity-constrained fronthaul [32] for both $\mu = 1$ and $\mu = 1/4$. While under OMA it is not possible to achieve a non-zero URLLC rate at even relatively low access latency such as $L_U > 3$, NOMA provides a reliable communication with constant minimal $L_U = 1$ access latency, but with lower rate. For eMBB, NOMA achieves an higher per-UE rate regardless of the value of the normalized bandwidth μ.

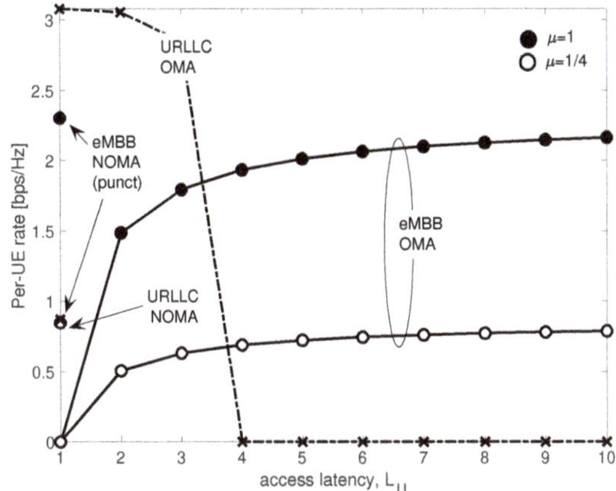

Figure 13. URLLC and eMBB per-UE rates as a function of access latency L_U for OMA and NOMA with puncturing.

7. Conclusions

This paper considers the coexistence of eMBB and URLLC services in the uplink of an analog C-RAN architecture from an information theoretic perspective. The rate expressions for URLLC and eMBB users under Orthogonal and Non-Orthogonal Multiple Access (OMA and NOMA, respectively) have been derived considering Analog Radio-over-Copper (A-RoC) as a sample scenario, although the proposed model can be easily adapted to other analog fronthaul technologies. For eMBB signals, performance have been evaluated in terms of information rate, while for URLLC we also took into account worst-case access latency and reliability. In case of NOMA, different decoding strategies have been considered in order to mitigate the impact of URLLC transmission on eMBB information rate. In particular, the performance achieved by puncturing, considered for 5G standardization, have been compared with those achieved by Treating URLLC Interference as Noise (TIN), and by Successive URLLC Interference Cancellation (SIC).

The analysis showed that NOMA allows for higher eMBB information rates with respect to OMA, while guaranteeing a reliable low-rate URLLC communication with minimal access latency. Furthermore, numerical results demonstrated that, differently from the digital C-RAN architecture based on limited-capacity fronthaul links, for analog C-RAN, TIN always outperforms puncturing, and SIC achieves the best performance at the price of an higher decoder complexity.

As work in progress, the theoretical model can be extended to account for fading channels or geometric mmWave-link channel models.

Similarly, a frequency-dependent cable channel can be considered by making the cable crosstalk coefficient γ increase with cable frequency [22]. Another interesting research direction is to consider the case in which the BBU has no knowledge about the incoming signal, i.e., it is not able to detect the URLLC transmissions, so that it is impossible for the BBU to choose the proper metric for joint signal decoding [43,44]. Finally, the overall system can be extended to the case of multiple users per-cell, where both ENs and users are equipped with multiple antennas.

Author Contributions: Conceptualization, A.M., O.S. and U.S.; Formal analysis, A.M., O.S. and U.S.; Methodology, A.M.; Software, A.M.; Supervision, O.S. and U.S.; Validation, R.K.; Writing—original draft, A.M.; Writing—review & editing, A.M., R.K., O.S. and U.S.

Funding: R.K. and O.S. have received funding from the European Research Council (ERC) under the European Union Horizon 2020 Research and Innovation Programme (Grant Agreement No. 725731).

Conflicts of Interest: The authors declare no conflict of interest.

Appendix A. Proof of Lemma 1

The proof of Lemma 1 is structured in two main steps detailed in the following: (i) MRC at the cable output, (ii) cable signal vectorization.

Appendix A.1. Maximum Ratio Combining at the Cable Output

The signal $\mathbf{R}_k \in \mathbb{C}^{l_F \times \frac{1}{\mu}}$ after the MRC at the cable output is

$$
\begin{aligned}
\mathbf{R}_k &= \tilde{\mathbf{R}}_k \mathbf{G} \\
&= \left(\tilde{\mathbf{Y}}_k \mathbf{H}_c + \tilde{\mathbf{W}}_k \right) \mathbf{G},
\end{aligned}
\tag{A1}
$$

where the definitions of $\tilde{\mathbf{R}}_k$, $\mathbf{G}$, $\tilde{\mathbf{Y}}_k$ and $\mathbf{H}_c$ follow from and (10), (20), (18) and (6), respectively. Thus, Equation (A1) can be rewritten as

$$
\begin{aligned}
\mathbf{R}_k &= \left(\mathbf{Y}_k \otimes \mathbf{1}_\eta^T \right) \left(\gamma \mathbf{1}_{l_S} \mathbf{1}_{l_S}^T + (1-\gamma)\mathbf{I}_{l_S} \right) \mathbf{G} + \tilde{\mathbf{W}}_k \mathbf{G} \\
&= \underbrace{\gamma \left(\mathbf{Y}_k \otimes \mathbf{1}_\eta^T \right) \mathbf{1}_{l_S} \mathbf{1}_{l_S}^T \mathbf{G}}_{\mathbf{R}_k'} + \underbrace{(1-\gamma) \left(\mathbf{Y}_k \otimes \mathbf{1}_\eta^T \right) \mathbf{G}}_{\mathbf{R}_k''} + \underbrace{\tilde{\mathbf{W}}_k \mathbf{G}}_{\mathbf{W}_k},
\end{aligned}
\tag{A2}
$$

where $\mathbf{W}_k = \tilde{\mathbf{W}}_k \mathbf{G}$ is the noise post MRC. The first term in (A2) can be rewritten as

$$
\begin{aligned}
\mathbf{R}_k' &= \frac{\gamma}{\eta} \left(\mathbf{Y}_k \otimes \mathbf{1}_\eta^T \right) \mathbf{1}_{l_S} \mathbf{1}_{l_S}^T \left(\mathbf{I}_{\frac{1}{\mu}} \otimes \mathbf{1}_\eta \right) \\
&\overset{(a)}{=} \gamma \left(\mathbf{Y}_k \otimes \mathbf{1}_\eta^T \right) \mathbf{1}_{l_S} \mathbf{1}_{\frac{1}{\mu}}^T \\
&\overset{(b)}{=} \gamma\eta \mathbf{Y}_k \mathbf{1}_{\frac{1}{\mu}} \mathbf{1}_{\frac{1}{\mu}}^T,
\end{aligned}
\tag{A3}
$$

where $\overset{(a)}{=}$ comes from the fact that $\mathbf{1}_{l_S}^T \left(\mathbf{I}_{\frac{1}{\mu}} \otimes \mathbf{1}_\eta \right) = \left(\mathbf{1}_{\frac{1}{\mu}}^T \otimes \mathbf{1}_\eta^T \right) \left(\mathbf{I}_{\frac{1}{\mu}} \otimes \mathbf{1}_\eta \right) = \eta \mathbf{1}_{\frac{1}{\mu}}^T$ due to the mixed-product property of Kronecker product operator $(\mathbf{A} \otimes \mathbf{B})(\mathbf{C} \otimes \mathbf{D}) = \mathbf{AC} \otimes \mathbf{BD}$ (see [38], Chapter 2), and, similarly, $\overset{(b)}{=}$ is obtained by $\left(\mathbf{Y}_k \otimes \mathbf{1}_\eta^T \right) \mathbf{1}_{l_S} \mathbf{1}_{\frac{1}{\mu}}^T = \left(\mathbf{Y}_k \otimes \mathbf{1}_\eta^T \right) \left(\mathbf{1}_{\frac{1}{\mu}} \mathbf{1}_{\frac{1}{\mu}}^T \otimes \mathbf{1}_\eta \right) = \eta \mathbf{Y}_k \mathbf{1}_{\frac{1}{\mu}} \mathbf{1}_{\frac{1}{\mu}}^T$. Using similar arguments, the second term in (A2) simplifies to

$$
\begin{aligned}
\mathbf{R}_k'' &= \frac{1-\gamma}{\eta} \left(\mathbf{Y}_k \otimes \mathbf{1}_\eta^T \right) \left(\mathbf{I}_{\frac{1}{\mu}} \otimes \mathbf{1}_\eta \right) \\
&= (1-\gamma)\mathbf{Y}_k.
\end{aligned}
\tag{A4}
$$

Finally, substituting (A3) and (A4) in (A2) we obtain

$$
\mathbf{R}_k = \mathbf{Y}_k \mathbf{H}_c^\eta + \mathbf{W}_k,
\tag{A5}
$$

where

$$
\mathbf{H}_c^\eta = \gamma\eta \mathbf{1}_{\frac{1}{\mu}} \mathbf{1}_{\frac{1}{\mu}}^T + (1-\gamma)\mathbf{I}_{\frac{1}{\mu}}
\tag{A6}
$$

is the equivalent cable channel matrix accounting for the bandwidth amplification factor over cable η, and $\mathbf{Y}_k$ is the radio signal reorganized in matrix form as in (16).

Appendix A.2. Cable Signal Vectorization

The vector signal $\mathbf{r}_k \in \mathbb{C}^{n_F \times 1}$ received at the BBU from the k-th EN over all the radio frequency channels can be obtained by vectorizing matrix $\mathbf{R}_k$ in (A5) as

$$
\begin{aligned}
\mathbf{r}_k &= \mathrm{vec}(\mathbf{Y}_k \mathbf{H}_c^{\eta} + \mathbf{W}_k) \\
&= \left(\mathbf{H}_c^{\eta} \otimes \mathbf{I}_{l_F} \right) \mathrm{vec}(\mathbf{Y}_k) + \mathrm{vec}(\mathbf{W}_k) \\
&= \left(\mathbf{H}_c^{\eta} \otimes \mathbf{I}_{l_F} \right) \mathbf{y}_k + \mathbf{w}_k,
\end{aligned}
\tag{A7}
$$

where $\mathbf{y}_k$ is the radio signal received at EN k-th. The overall cable noise vector $\mathbf{w}_k$ can be rewritten as

$$
\begin{aligned}
\mathbf{w}_k &= \mathrm{vec}(\mathbf{W}_k) \\
&= \mathrm{vec}(\tilde{\mathbf{W}}_k \mathbf{G}) \\
&= \left(\mathbf{G}^T \otimes \mathbf{I}_{l_F} \right) \mathrm{vec}(\tilde{\mathbf{W}}_k) \\
&= \frac{1}{\eta} \left(\mathbf{I}_{\frac{1}{\mu}} \otimes \mathbf{1}_{\eta}^T \otimes \mathbf{I}_{l_F} \right) \tilde{\mathbf{w}}_k.
\end{aligned}
\tag{A8}
$$

It is important to notice that since the cable noise $\tilde{\mathbf{W}}_k$ is white Gaussian and uncorrelated over cable pairs (see (10)), $\tilde{\mathbf{w}}_k = \mathrm{vec}(\tilde{\mathbf{W}}_k)$ is also white Gaussian distributed as $\tilde{\mathbf{w}}_k \sim \mathcal{CN}(\mathbf{0}, \mathbf{I}_{l_S l_F})$. Hence, the covariance $\mathbf{R}_w$ of the overall cable noise vector $\mathbf{w}_k$ yields

$$
\begin{aligned}
\mathbf{R}_w &= \mathbb{E}\left[\mathbf{w}_k \mathbf{w}_k^H \right] \\
&= \frac{1}{\eta^2} \left(\mathbf{I}_{\frac{1}{\mu}} \otimes \mathbf{1}_{\eta}^T \otimes \mathbf{I}_{l_F} \right) \left(\mathbf{I}_{\frac{1}{\mu}} \otimes \mathbf{1}_{\eta}^T \otimes \mathbf{I}_{l_F} \right)^T \\
&\overset{(a)}{=} \frac{1}{\eta} \left(\mathbf{I}_{\frac{1}{\mu}} \otimes \mathbf{I}_{l_F} \right) \\
&= \frac{1}{\eta} \mathbf{I}_{n_F},
\end{aligned}
\tag{A9}
$$

where the equality (a) comes again from the mixed-product property of Kronecker product. Equation (A9) shows that the MRC allows to take advantages from the signal redundancy over the cable, which results in a reduction of the cable noise power by a factor of η.

The proof is completed by gathering the signals $\mathbf{r}_k$ (A7) received at the BBU from all ENs as

$$
\begin{aligned}
\mathbf{R} &= [\mathbf{r}_1, \mathbf{r}_2, ..., \mathbf{r}_M] \\
&= \left(\mathbf{H}_c^{\eta} \otimes \mathbf{I}_{l_F} \right) [\mathbf{y}_1, \mathbf{y}_2, ..., \mathbf{y}_M] + [\mathbf{w}_1, \mathbf{w}_2, ..., \mathbf{w}_M] \\
&= \left(\mathbf{H}_c^{\eta} \otimes \mathbf{I}_{l_F} \right) \mathbf{Y} + \mathbf{W},
\end{aligned}
\tag{A10}
$$

where $\mathbf{Y}$ is the signal received by all ENs over all radio channels in (5).

Appendix B. Proof of Lemma 2

By substituting signal $\mathbf{Y}$ in (31) received by all ENs over all radio channels in case of OMA in Equation (24), we obtain

$$
\mathbf{R} = \left(\mathbf{H}_c^{\eta} \otimes \mathbf{I}_{l_F} \right) \mathbf{X} \mathbf{H} + \left(\mathbf{H}_c^{\eta} \otimes \mathbf{I}_{l_F} \right) \mathbf{Z} + \mathbf{W}.
\tag{A11}
$$

To compute the per-UE eMBB rate, a further vectorization is needed, leading to

$$
\begin{aligned}
\mathbf{r} &= \text{vec}(\mathbf{R}) \\
&= \text{vec}\left(\left(\mathbf{H}_c^\eta \otimes \mathbf{I}_{l_F}\right)\mathbf{XH}\right) + \text{vec}\left(\left(\mathbf{H}_c^\eta \otimes \mathbf{I}_{l_F}\right)\mathbf{Z}\right) + \text{vec}(\mathbf{W}) \\
&= \left(\mathbf{H}^T \otimes \mathbf{H}_c^\eta \otimes \mathbf{I}_{l_F}\right)\mathbf{x} + \left(\mathbf{I}_M \otimes \mathbf{H}_c^\eta \otimes \mathbf{I}_{l_F}\right)\mathbf{z} + \mathbf{w} \\
&= \bar{\mathbf{H}}_{\text{eq}}\mathbf{x} + \bar{\mathbf{z}}_{\text{eq}},
\end{aligned}
\tag{A12}
$$

where $\bar{\mathbf{H}}_{\text{eq}} = \mathbf{H}^T \otimes \mathbf{H}_c^\eta \otimes \mathbf{I}_{l_F}$ is the overall equivalent channel comprising both cable and radio channels over all ENs, and $\bar{\mathbf{z}}_{\text{eq}} = \left(\mathbf{I}_M \otimes \mathbf{H}_c^\eta \otimes \mathbf{I}_{l_F}\right)\mathbf{z} + \mathbf{w}$ is the overall noise vector comprising both the vectorized radio noise $\mathbf{z} = \text{vec}(\mathbf{Z}) \sim \mathcal{CN}(\mathbf{0}, \mathbf{I}_{n_F M})$ and the vectorized cable noise $\mathbf{w} = \text{vec}(\mathbf{W}) \sim \mathcal{CN}(\mathbf{0}, \frac{1}{\lambda^2 \eta}\mathbf{I}_{n_F M})$, where we recall that the scaling λ is due to the cable power constraints. Hence, the covariance of the equivalent noise $\bar{\mathbf{z}}_{\text{eq}}$ yields

$$
\begin{aligned}
\bar{\mathbf{R}}_{z_{\text{eq}}} &= \mathbb{E}\left[\bar{\mathbf{z}}_{\text{eq}}\bar{\mathbf{z}}_{\text{eq}}^H\right] \\
&= \left(\mathbf{I}_M \otimes \mathbf{H}_c^\eta \otimes \mathbf{I}_{l_F}\right)\left(\mathbf{I}_M \otimes \mathbf{H}_c^\eta \otimes \mathbf{I}_{l_F}\right)^H + \frac{1}{\lambda^2 \eta}\mathbf{I}_{n_F M} \\
&= \mathbf{I}_M \otimes \mathbf{H}_c^\eta \mathbf{H}_c^\eta \otimes \mathbf{I}_{l_F} + \frac{1}{\lambda^2 \eta}\mathbf{I}_{n_F M}.
\end{aligned}
\tag{A13}
$$

The eMBB per-UE rate under OMA is computed by

$$
\begin{aligned}
R_B &= \frac{(1 - L_U^{-1})}{n_F M}I(\mathbf{r}, \mathbf{x}) \\
&= \frac{(1 - L_U^{-1})}{n_F M}\log\left(\det\left(\mathbf{I} + \bar{P}_B \bar{\mathbf{R}}_{z_{\text{eq}}}^{-1}\bar{\mathbf{H}}_{\text{eq}}\bar{\mathbf{H}}_{\text{eq}}^T\right)\right),
\end{aligned}
\tag{A14}
$$

where $\bar{\mathbf{H}}_{\text{eq}}\bar{\mathbf{H}}_{\text{eq}}^T = \mathbf{HH} \otimes \mathbf{H}_c^\eta \mathbf{H}_c^\eta \otimes \mathbf{I}_{l_F}$. Finally, using the determinant property of the Kronecker product $|\mathbf{A} \otimes \mathbf{B}| = |\mathbf{A}|^m |\mathbf{B}|^n$, Equation (A14) simplifies to

$$
\begin{aligned}
R_B &= \frac{l_F(1 - L_U^{-1})}{n_F M}\log\left(\det\left(\mathbf{I} + \bar{P}_B \mathbf{R}_{z_{\text{eq}}}^{-1}\mathbf{H}_{\text{eq}}\mathbf{H}_{\text{eq}}^T\right)\right) \\
&= \mu\frac{1 - L_U^{-1}}{M}\log\left(\det\left(\mathbf{I} + \bar{P}_B \mathbf{R}_{z_{\text{eq}}}^{-1}\mathbf{H}_{\text{eq}}\mathbf{H}_{\text{eq}}^T\right)\right),
\end{aligned}
\tag{A15}
$$

where $\mathbf{H}_{\text{eq}} = \mathbf{H} \otimes \mathbf{H}_c^\eta$ and $\mathbf{R}_{z_{\text{eq}}} = \mathbf{I}_M \otimes \mathbf{H}_c^\eta \mathbf{H}_c^\eta + \frac{1}{\lambda^2 \eta}\mathbf{I}_{\frac{M}{\mu}}$, thus concluding the proof.

Appendix C. Proof of Lemma 3

By substituting the signal (5) received at the k-th EN under NOMA into (A10) and by following similar steps as for the proof of Lemma 2, the overall vector signal received at the BBU comprising both cable and radio channels over all ENs under NOMA by treating interference as noise yields

$$
\mathbf{r} = \bar{\mathbf{H}}_{\text{eq}}\mathbf{x} + \beta\bar{\mathbf{A}}_{\text{eq}}\mathbf{u} + \bar{\mathbf{z}}_{\text{eq}},
\tag{A16}
$$

where the definitions of $\bar{\mathbf{H}}_{\text{eq}}$ and $\bar{\mathbf{z}}_{\text{eq}}$ are the same as in (A12), $\bar{\mathbf{A}}_{\text{eq}} = \mathbf{A} \otimes \mathbf{H}_c^\eta \otimes \mathbf{I}_{l_F}$ accounts for the relay of URLLC signal over the cable, and $\mathbf{u}$ is the vectorization of the URLLC signal matrix $\mathbf{U}$ in (5). Similarly to the proof of Lemma 2, the eMBB per-UE rate under NOMA by treating URLLC interference as noise can be computed by

$$R_B = \frac{1}{n_F M} I\left(\mathbf{r}, \mathbf{x} \,|\, \mathbf{A}\right)$$
$$= \frac{\mu}{M} \mathbb{E}_{\mathbf{A}} \left[\log \left(\det \left(\mathbf{I} + P_B \mathbf{R}_{A,z_{eq}}^{-1} \mathbf{H}_{eq} \mathbf{H}_{eq}^{T} \right) \right) \right] \tag{A17}$$

where the average is taken over all the possible values of matrix $\mathbf{A}$, P_B is the power of the eMBB user under NOMA and $\mathbf{H}_{eq}$ is defined as in (A15). Finally, the covariance $\mathbf{R}_{A,z_{eq}}$ of the noise plus URLLC interference yields

$$\mathbf{R}_{A,z_{eq}} = \beta^2 P_U \quad \mathbf{A}_{eq} \mathbf{A}_{eq}^{T} + \mathbf{R}_{z_{eq}},$$

where $\mathbf{A}_{eq} = \mathbf{A} \otimes \mathbf{H}_c^{\eta}$ and $\mathbf{R}_{z_{eq}}$ is the same as in (A15). The proof is completed by noticing that $\mathbf{A}_{eq} \mathbf{A}_{eq}^{T} = (\mathbf{A} \otimes \mathbf{H}_c^{\eta})(\mathbf{A} \otimes \mathbf{H}_c^{\eta})^{H} = \mathbf{A} \otimes \mathbf{H}_c^{\eta} \mathbf{H}_c^{\eta}$, since matrix $\mathbf{A}$ is idempotent and $\mathbf{H}_c^{\eta}$ symmetric.

References

1. Shafi, M.; Molisch, A.F.; Smith, P.J.; Haustein, T.; Zhu, P.; Silva, P.D.; Tufvesson, F.; Benjebbour, A.; Wunder, G. 5G: A tutorial overview of standards, trials, challenges, deployment, and practice. *IEEE J. Sel. Areas Commun.* **2017**, *35*, 1201–1221. [CrossRef]
2. 5G PPP Architecture Working Group. View on 5G Architecture. 2016. Available online: https://5g-ppp.eu/wp-content/uploads/2014/02/5G-PPP-5G-Architecture-WP-July-2016.pdf (accessed on 31 August 2018).
3. Minimum Requirements Related to Technical Performance for IMT-2020 Radio Interface(s). 2017. Available online: https://www.itu.int/dms_pub/itu-r/opb/rep/R-REP-M.2410-2017-PDF-E.pdf (accessed on 31 August 2018).
4. Takeda, F. Study on New Radio (NR) Access Technology—Physical Layer Aspects. 2017. Available online: http://www.tech-invite.com/3m38/tinv-3gpp-38-802.html#e-3-1 (accessed on 31 August 2018).
5. Durisi, G.; Koch, T.; Popovski, P. Toward massive, ultrareliable, and low-latency wireless communication with short packets. *Proc. IEEE* **2016**, *104*, 1711–1726. [CrossRef]
6. Zhang, H.; Liu, N.; Chu, X.; Long, K.; Aghvami, A.H.; Leung, V.C.M. Network slicing based 5G and future mobile networks: mobility, resource management, and challenges. *IEEE Commun. Mag.* **2017**, *55*, 138–145, [CrossRef]
7. Popovski, P.; Nielsen, J.J.; Stefanovic, C.; de Carvalho, E.; Strom, E.; Trillingsgaard, K.F.; Bana, A.S.; Kim, D.M.; Kotaba, R.; Park, J.; et al. Wireless access for ultra-reliable low-latency communication: Principles and building blocks. *IEEE Netw.* **2018**, *32*, 16–23. [CrossRef]
8. Boccardi, F.; Heath, R.W.; Lozano, A.; Marzetta, T.L.; Popovski, P. Five disruptive technology directions for 5G. *IEEE Commun. Mag.* **2014**, *52*, 74–80. [CrossRef]
9. Wong, V.; Schober, R.; Ng, D.; Wang, L.E. *Key Technologies for 5G Wireless Systems*; Cambridge University Press: Cambridge, UK, 2017.
10. CPRI Specification V.6.1 (2014-07-01). 2014. Available online: http://www.cpri.info/downloads/CPRI_v_6_1_2014-07-01.pdf (accessed on 31 August 2018).
11. Bartelt, J.; Rost, P.; Wubben, D.; Lessmann, J.; Melis, B.; Fettweis, G. Fronthaul and backhaul requirements of flexibly centralized radio access networks. *IEEE Wirel. Commun.* **2015**, *22*, 105–111. [CrossRef]
12. Wake, D.; Nkansah, A.; Gomes, N.J. Radio over fiber link design for next generation wireless systems. *J. Lightw. Technol.* **2010**, *28*, 2456–2464. [CrossRef]
13. Gambini, J.; Spagnolini, U. Radio over telephone lines in femtocell systems. In Proceedings of the 21st Annual IEEE International Symposium on Personal, Indoor and Mobile Radio Communications, Istanbul, Turkey, 26–30 September 2010; pp. 1544–1549.
14. Gambini, J.; Spagnolini, U. Wireless over cable for femtocell systems. *IEEE Commun. Mag.* **2013**, *51*, 178–185. [CrossRef]
15. Bartelt, J.; Fettweis, G. Radio-over-radio: I/Q-stream backhauling for cloud-based networks via millimeter wave links. In Proceedings of the 2013 IEEE Globecom Workshops (GC Wkshps), Atlanta, GA, USA, 9–13 December 2013; pp. 772–777.

16. Combi, L.; Gatto, A.; Martinelli, M.; Parolari, P.; Spagnolini, U. Pulse-Width optical modulation for CRAN front-hauling. In Proceedings of the 2015 IEEE Globecom Workshops (GC Wkshps), San Diego, CA, USA, 6–10 December 2015; pp. 1–5.

17. Dat, P.T.; Kanno, A.; Yamamoto, N.; Kawanishi, T. 5G transport networks: The need for new technologies and standards. *IEEE Commun. Mag.* **2016**, *54*, 18–26. [CrossRef]

18. Matera, A.; Spagnolini, U. On the optimal Space-Frequency to Frequency mapping in indoor single-pair RoC fronthaul. In Proceedings of the 2017 European Conference on Networks and Communications (EuCNC), Oulu, Finland, 12–15 June 2017; pp. 1–5.

19. Matera, A.; Combi, L.; Naqvi, S.H.R.; Spagnolini, U. Space-frequency to space-frequency for MIMO radio over copper. In Proceedings of the 2017 IEEE International Conference on Communications (ICC), Paris, France, 21–25 May 2017; pp. 1–6.

20. Matera, A.; Spagnolini, U. Analog MIMO-RoC Downlink with SF2SF. *IEEE Wirel. Commun. Lett.* **2018**. [CrossRef]

21. Tonini, F.; Fiorani, M.; Furdek, M.; Raffaelli, C.; Wosinska, L.; Monti, P. Radio and transport planning of centralized radio architectures in 5G indoor scenarios. *IEEE J. Sel. Areas Commun.* **2017**, *35*, 1837–1848. [CrossRef]

22. Naqvi, S.H.R.; Matera, A.; Combi, L.; Spagnolini, U. On the transport capability of LAN cables in all-analog MIMO-RoC fronthaul. In Proceedings of the 2017 IEEE Wireless Communications and Networking Conference (WCNC), San Francisco, CA, USA, 19–22 March 2017; pp. 1–6.

23. Lu, C.; Berg, M.; Trojer, E.; Eriksson, P.E.; Laraqui, K.; Tridblad, O.V.; Almeida, H. Connecting the dots: Small cells shape up for high-performance indoor radio. *Ericsson Rev.* **2014**, *91*, 38–45. [CrossRef] [PubMed]

24. Cover, T.M.; Thomas, J.A. *Elements of Information Theory*; John Wiley & Sons: New York, NY, USA, 2012.

25. Saito, Y.; Kishiyama, Y.; Benjebbour, A.; Nakamura, T.; Li, A.; Higuchi, K. Non-Orthogonal Multiple Access (NOMA) for Cellular Future Radio Access. In Proceedings of the 2013 IEEE 77th Vehicular Technology Conference (VTC Spring), Dresden, Germany, 2–5 June 2013; pp. 1–5.

26. Shin, W.; Vaezi, M.; Lee, B.; Love, D.J.; Lee, J.; Poor, H.V. Non-orthogonal multiple access in multi-cell networks: Theory, performance, and practical challenges. *IEEE Commun. Mag.* **2017**, *55*, 176–183. [CrossRef]

27. Ding, Z.; Lei, X.; Karagiannidis, G.K.; Schober, R.; Yuan, J.; Bhargava, V.K. A Survey on Non-Orthogonal Multiple Access for 5G Networks: Research Challenges and Future Trends. *IEEE J. Sel. Areas Commun.* **2017**, *35*, 2181–2195. [CrossRef]

28. Anand, A.; de Veciana, G.; Shakkottai, S. Joint scheduling of URLLC and eMBB traffic in 5G wireless networks. *arXiv* **2017**, arXiv:1712.05344.

29. Popovski, P.; Trillingsgaard, K.F.; Simeone, O.; Durisi, G. 5G Wireless Network Slicing for eMBB, URLLC, and mMTC: A Communication-Theoretic View. *arXiv* **2018**, arXiv:1804.05057.

30. Vaezi, M.; Ding, Z.; Poor, H. *Multiple Access Techniques for 5G Wireless Networks and Beyond*; Springer: New York, NY, USA, 2018.

31. Dai, L.; Wang, B.; Yuan, Y.; Han, S.; I, C.-L.; Wang, Z. Non-orthogonal multiple access for 5G: solutions, challenges, opportunities, and future research trends. *IEEE Commun. Mag.* **2015**, *53*, 74–81. [CrossRef]

32. Kassab, R.; Simeone, O.; Popovski, P. Coexistence of URLLC and eMBB services in the C-RAN Uplink: An Information-Theoretic Study. In Proceedings of the 2018 IEEE Global Communications Conference (GLOBECOM), Abu Dhabi, UAE, 9–13 December 2018, pp. 1–6.

33. Making 5G NR a Reality. 2016. Available online: https://www.qualcomm.com/media/documents/files/whitepaper-making-5g-nr-a-reality.pdf (accessed on 31 August 2018).

34. eMBB and URLLC Multiplexing for NR. R1-1705051. 2017. Available online: https://portal.3gpp.org/ngppapp/CreateTdoc.aspx?mode=view&contributionId=775991 (accessed on 31 August 2018).

35. Simeone, O.; Levy, N.; Sanderovich, A.; Somekh, O.; Zaidel, B.M.; Poor, H.V.; Shamai, S. Cooperative wireless cellular systems: An information-theoretic view. *Found. Trends Commun. Inf. Theory* **2012**, *8*, 1–177. [CrossRef]

36. Rappaport, T.S.; Sun, S.; Mayzus, R.; Zhao, H.; Azar, Y.; Wang, K.; Wong, G.N.; Schulz, J.K.; Samimi, M.; Gutierrez, F., Jr. Millimeter wave mobile communications for 5G cellular: It will work! *IEEE Access* **2013**, *1*, 335–349. [CrossRef]

37. Hekrdla, M.; Matera, A.; Wang, W.; Wei, D.; Spagnolini, U. Ordered Tomlinson-Harashima Precoding in G.fast Downstream. In Proceedings of the 2015 IEEE Global Communications Conference (GLOBECOM), San Diego, CA, USA, 6–10 December 2015; pp. 1–6.

38. Spagnolini, U. *Statistical Signal Processing in Engineering*; John Wiley & Sons: New York, NY, USA, 2018.

39. Jakes, W.C.; Cox, D.C. *Microwave Mobile Communications*; John Wiley & Sons: New York, NY, USA, 1994.

40. Polyanskiy, Y.; Poor, H.; Verdu, S. Channel Coding Rate in the Finite Blocklength Regime. *IEEE Trans. Inf. Theory* **2010**, *56*, 2307–2359. [CrossRef]

41. Scarlett, J.; Tan, V.Y.; Durisi, G. The dispersion of nearest-neighbor decoding for additive non-Gaussian channels. *IEEE Trans. Inf. Theory* **2017**, *63*, 81–92. [CrossRef]

42. Zafaruddin, S.; Bergel, I.; Leshem, A. Signal processing for gigabit-rate wireline communications: An overview of the state of the art and research challenges. *IEEE Signal Process. Mag.* **2017**, *34*, 141–164. [CrossRef]

43. Zhang, W. A general framework for transmission with transceiver distortion and some applications. *IEEE Trans. Commun.* **2012**, *60*, 384–399. [CrossRef]

44. Merhav, N.; Kaplan, G.; Lapidoth, A.; Shitz, S.S. On information rates for mismatched decoders. *IEEE Trans. Inf. Theory* **1994**, *40*, 1953–1967. [CrossRef]

Article

Robust Baseband Compression Against Congestion in Packet-Based Fronthaul Networks Using Multiple Description Coding

Seok-Hwan Park [1,*], Osvaldo Simeone [2] and Shlomo Shamai (Shitz) [3]

[1] Division of Electronic Engineering, Chonbuk National University, Jeonju 54896, Korea
[2] Department of Informatics, King's College London, London WC2R2NA, UK; osvaldo.simeone@kcl.ac.uk
[3] Department of Electrical Engineering, Technion, Haifa 32000, Israel; sshlomo@ee.technion.ac.il
* Correspondence: seokhwan@jbnu.ac.kr; Tel.: +82-63-270-2357

Received: 4 February 2019; Accepted: 22 April 2019; Published: 24 April 2019

Abstract: In modern implementations of Cloud Radio Access Network (C-RAN), the fronthaul transport network will often be packet-based and it will have a multi-hop architecture built with general-purpose switches using network function virtualization (NFV) and software-defined networking (SDN). This paper studies the joint design of uplink radio and fronthaul transmission strategies for a C-RAN with a packet-based fronthaul network. To make an efficient use of multiple routes that carry fronthaul packets from remote radio heads (RRHs) to cloud, as an alternative to more conventional packet-based multi-route reception or coding, a multiple description coding (MDC) strategy is introduced that operates directly at the level of baseband signals. MDC ensures an improved quality of the signal received at the cloud in conditions of low network congestion, i.e., when more fronthaul packets are received within a tolerated deadline. The advantages of the proposed MDC approach as compared to the traditional path diversity scheme are validated via extensive numerical results.

Keywords: robust compression; congestion; packet-based fronthaul; multiple description coding; cloud radio access network; broadcast coding; eCPRI

---

1. Introduction

In a Cloud Radio Access Network (C-RAN) architecture, a cloud unit, or baseband processing unit (BBU), carries out baseband signal processing on behalf of a number of radio units, or remote radio heads (RRHs), that are connected to the cloud through an interface referred to as fronthaul links [1]. The C-RAN technology is recognized as one of the dominant architectural solutions for future wireless networks due to the promised reduction in capital and operational expenditures and the capability of large-scale interference management [2]. A major challenge of C-RAN deployment is that high-rate baseband in-phase and quadrature (IQ) samples need to be carried on the fronthaul links of limited data rate. The design of signal processing strategies, including fronthaul compression techniques, for C-RAN was widely studied in the literature [3–6].

The mentioned works [3–6] and references therein assume a conventional fronthaul topology, whereby there are dedicated point-to-point fronthaul links from the cloud to each RRH as in Common Public Radio Interface (CPRI) specification [7]. However, in modern implementations of C-RAN, as illustrated in Figure 1, the fronthaul transport network will often be packet-based and it will have a multi-hop architecture built with general-purpose switches using network function virtualization (NFV) and software-defined networking (SDN) [8,9]. Packet-based fronthaul network can leverage the wide deployment of Ethernet infrastructure [10].

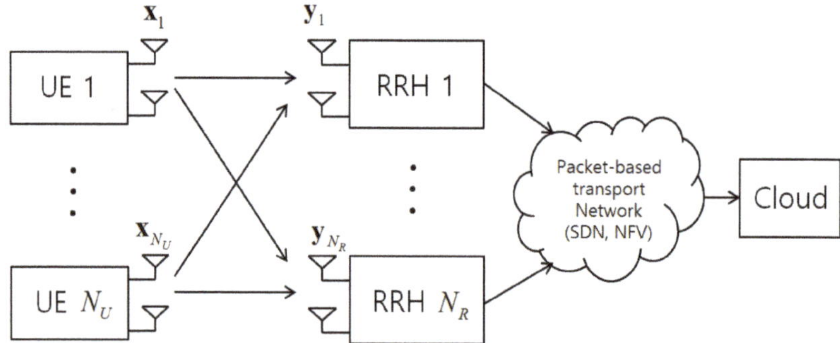

Figure 1. Illustration of the uplink of a Cloud Radio Access Network (C-RAN) with a packet-based fronthaul transport network.

Packet-based multi-hop networks are subject to congestion and packet losses. The traditional path diversity approach repeats the same packet on the multiple routes in order to mitigate these issues [11,12]. This approach can successfully reduce the packet loss probability at the cost of increasing the overhead in the fronthaul network. A limitation of these traditional schemes is that, when multiple packets arrive at the cloud within the tolerated delay, the signal quality utilized for channel decoding at the cloud is the same as if a single packet is received. To make a more efficient use of the multiple routes, in this paper, we propose a multiple description coding (MDC) scheme that operates directly on the baseband signals. Thanks to MDC, a better distortion level is obtained as more packets arrive at the cloud within the deadline. We refer to [13] for an overview and for a discussion on applications of MDC. In addition, the work [14] proposed the use of MDC to improve the achievable rate of a multicast cognitive interference channel.

Since, thanks to MDC, the signal quality varies depending on the number of packets arriving at the cloud, we propose that user equipments (UEs) leverage the broadcast approach in order to enable the adaptation of the transmission rate to the effective received signal-to-noise ratio (SNR) [15,16]. The broadcast approach defines a variable-to-fixed channel code [17] that enables the achievable rate to adapt to the channel state when the latter is known only at the receiving end. The broadcast approach splits the message of each UE into multiple submessages that are encoded independently, and transmitted as a superposition of the encoded signals. With the proposed MDC-based solution, based on the packets received within a given deadline, the cloud performs successive interference cancellation (SIC) decoding of the UEs' submessages with a given order so that the achievable rate can be adapted to the number of delivered packets. Therefore, the number of received packets determines the quality of the channel state known only at the receiver. Related methods were introduced in [18] and [19], where broadcast coding with layered compression [20] was applied to the uplink of C-RAN systems with distributed channel state information [18] and with uncertain fronthaul capacity [19].

More specifically, in this work, we study joint radio and fronthaul transmission for the uplink of a C-RAN with a packet-based fronthaul network. In the system, the uplink received baseband signal of each RRH is quantized and compressed producing a bit stream. The output bits are then packetized and transmitted on the fronthaul network. Following a standard approach to increase robustness to network losses and random delays (see, e.g., [11,12]), we assume that the packets are sent over multiple paths towards the cloud as seen in Figure 2. This can be done by using either conventional packet-based duplication [11,12] or the proposed MDC approach. The packets may be lost due to network delays or congestion when they are not received within a tolerable fronthaul delay dependent on the application. Based on the packets that have arrived within the delay, the cloud carries out decompression and channel decoding.

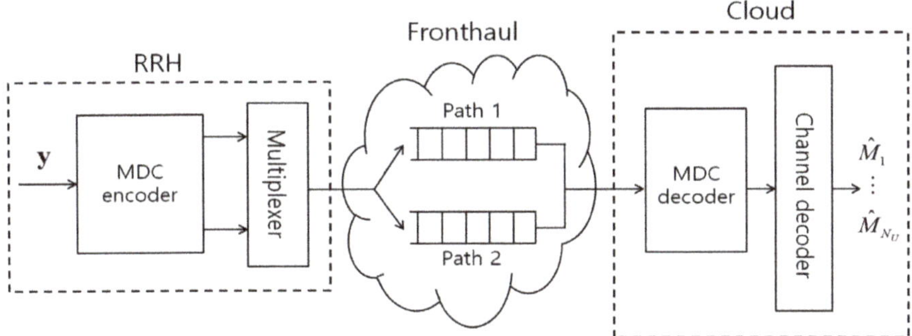

Figure 2. Illustration of the multiple description coding (MDC) and packet-based transport network.

The rest of the paper is organized as follows. In Section 2, we describe the system model for the uplink of a C-RAN with packet-based fronthaul network. In Section 3, we present the proposed MDC scheme which operates in a combination with the broadcast coding. The optimization of the proposed scheme is discussed in Section 4, and the advantages of the proposed scheme are validated with extensive numerical results in Section 5. We discuss extension to general cases in Section 6, and the paper is concluded in Section 7.

We summarize some notations used throughout the paper as follows. The mutual information between random variables X and Y conditioned on Z is denoted as $I(X;Y|Z)$, and $h(X)$ denotes the differential entropy of X. We define $\mathcal{CN}(\boldsymbol{\mu}, \boldsymbol{\Sigma})$ as the circularly symmetric complex Gaussian distribution with mean $\boldsymbol{\mu}$ and covariance $\boldsymbol{\Sigma}$. The expectation, trace, determinant and Hermitian transpose operations are denoted by $\mathrm{E}(\cdot)$, $\mathrm{tr}(\cdot)$, $\det(\cdot)$ and $(\cdot)^H$, respectively, and $\mathbb{C}^{M \times N}$ represents the set of all $M \times N$ complex matrices. We denote as $\mathbf{I}_N$ an identity matrix of size N, and $\otimes$ represents the Kronecker product. $\mathbf{A} \succeq \mathbf{0}$ indicates that the matrix $\mathbf{A}$ is positive semidefinite.

2. System Model

We consider the uplink of a C-RAN in which N_U UEs communicate with a cloud unit through N_R RRHs. To emphasize the main idea, we first focus on the case of $N_R = 1$ and discuss extension to a general number of RRHs in Section 6. Also, for convenience, we define the set $\mathcal{N}_U \triangleq \{1, \ldots, N_U\}$ of UEs, and denote the numbers of antennas of UE k and of the RRH by $n_{U,k}$ and n_R, respectively. The key novel aspect as compared to the prior work reviewed above is the assumption of packet-based fronthaul connecting between RRH and cloud.

2.1. Uplink Wireless Channel

Each UE k encodes its message to be decoded at the cloud and obtains an encoded baseband signal $\mathbf{x}_k \sim \mathcal{CN}(\mathbf{0}, \boldsymbol{\Sigma}_{\mathbf{x}_k}) \in \mathbb{C}^{n_{U,k} \times 1}$ which is transmitted on the uplink channel toward the RRH. Assuming flat-fading channel, the signal $\mathbf{y} \in \mathbb{C}^{n_R \times 1}$ received by the RRH is given as

$$\mathbf{y} = \sum_{k \in \mathcal{N}_U} \mathbf{H}_k \mathbf{x}_k + \mathbf{z} = \mathbf{H}\mathbf{x} + \mathbf{z}, \tag{1}$$

where $\mathbf{H}_k \in \mathbb{C}^{n_R \times n_{U,k}}$ is the channel transfer matrix from UE k to the RRH, $\mathbf{z} \sim \mathcal{CN}(\mathbf{0}, \boldsymbol{\Sigma}_{\mathbf{z}})$ is the additive noise vector, $\mathbf{H} = [\mathbf{H}_1 \cdots \mathbf{H}_{N_U}]$ is the channel matrix from all the UEs to the RRH, and $\mathbf{x} = [\mathbf{x}_1^H \cdots \mathbf{x}_{N_U}^H]^H \sim \mathcal{CN}(\mathbf{0}, \boldsymbol{\Sigma}_{\mathbf{x}})$ is the signal transmitted by all the UEs with $\boldsymbol{\Sigma}_{\mathbf{x}} = \mathrm{diag}(\{\boldsymbol{\Sigma}_{\mathbf{x}_k}\}_{k \in \mathcal{N}_U})$. We define the covariance matrix $\boldsymbol{\Sigma}_{\mathbf{y}} = \mathbf{H}\boldsymbol{\Sigma}_{\mathbf{x}}\mathbf{H}^H + \boldsymbol{\Sigma}_{\mathbf{z}}$ of $\mathbf{y}$.

The RRH quantizes and compresses the received signal $\mathbf{y}$ producing a number of packets. As detailed next, these packets are sent to the cloud on a packet-based fronthaul network, and the cloud jointly decodes the messages sent by the UEs based on the signals received within some maximum allowed fronthaul delay.

2.2. Packet-Based Fronthaul Transport Network

As discussed in Section 1, in modern implementations of C-RAN, the fronthaul transport network is expected to be packet-based and to have a multi-hop architecture built with general-purpose switches using NFV and SDN [8,9]. As a result, upon compression, the received signals need to be packetized, and the packets to be transmitted on the fronthaul network to the cloud. Packets may be lost due to network delays or congestion when they are not received within a tolerated fronthaul delay dependent on the application.

A standard approach to increase robustness to network losses and random delays is to send packets over multiple paths towards the destination (see, e.g., [11,12]). As seen in Figure 2, following [11], we model transmission on each such path, or route, as a queue. Furthermore, as seen in Figure 3, transmission on the fronthaul transport network is slotted, with each slot carrying a payload of B_F bits. The duration of each wireless frame, of L_W symbols, encompasses T_F fronthaul slots. Due to congestion, each fronthaul packet sent by the RRH on route j takes a geometrically distributed number of time slots to be delivered. Accordingly, transmission is successful independently in each slot with probability $1 - \epsilon_{F,j}$ on the jth route.

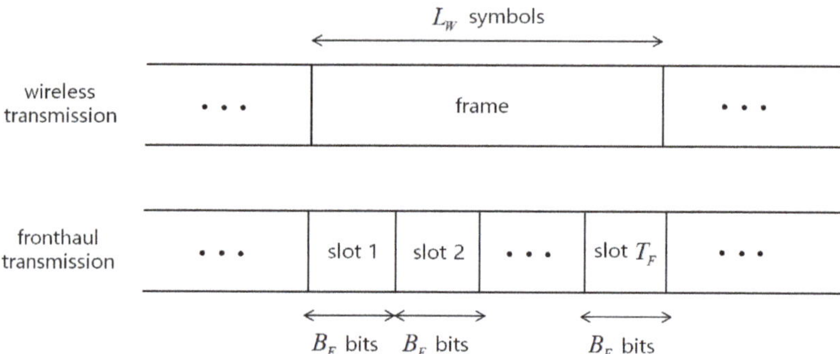

Figure 3. Illustration of wireless frame and fronthaul slotted transmission.

3. Robust Compression Based on Multiple Description Coding

In this section, we propose a robust compression technique based on MDC, which, in combination with broadcast coding, enables the achievable rate to be adapted to the number of packets collected by the cloud, and hence to the current network congestion level. To highlight the idea, we assume that the RRH has available two paths to the cloud. Extensions will be discussed in Section 6. The traditional path diversity approach repeats the same packet on the two routes [11]. More sophisticated forms of packet-based encoding, such as erasure coding studied in [12], are not applicable to the case of two paths. Accordingly, if one or two packets are received by the fronthaul deadline of T_F slots, the signal is decompressed and decoding is carried out at the cloud. Note that, if both packets are received, the signal quality is the same as if one packet is received. In contrast, we propose to adopt MDC as seen in Figure 2. With MDC, if one packet is received by the deadline T_F, we obtain a certain distortion level, while we obtain a better distortion level if both packets are received ([21] Ch. 14).

In the MDC approach, the RRH first quantizes and compresses the received signal $\mathbf{y}$ to produce quantized signals $\hat{\mathbf{y}}_0$, $\hat{\mathbf{y}}_1$ and $\hat{\mathbf{y}}_2$. Packets $\hat{\mathbf{y}}_1$ and $\hat{\mathbf{y}}_2$ are sent on two separate paths to the cloud, with $\hat{\mathbf{y}}_l$ sent on route l. By the properties of MDC, if only a single packet $l \in \{1, 2\}$ arrives at the cloud within deadline T_F, the MDC decoder can recover the quantized signal $\hat{\mathbf{y}}_l$, while the signal $\hat{\mathbf{y}}_0$ can be recovered if the both packets are received in time.

Denote as R_F the number of bits per symbol used to represent the signal for each of the quantized packets $\hat{\mathbf{y}}_1$ and $\hat{\mathbf{y}}_2$. We refer to R_F as the compression output rate. As shown in ([21] Ch. 14), the rate R_F should satisfy the conditions

$$R_F \geq I\left(\mathbf{y}; \hat{\mathbf{y}}_1\right), \tag{2}$$

$$R_F \geq I\left(\mathbf{y}; \hat{\mathbf{y}}_2\right), \tag{3}$$

$$\text{and } 2R_F \geq I\left(\mathbf{y}; \{\hat{\mathbf{y}}_l\}_{l\in\{0,1,2\}}\right) + I\left(\hat{\mathbf{y}}_1; \hat{\mathbf{y}}_2\right). \tag{4}$$

To evaluate (2)–(4), as in, e.g., [3–6], we assume standard Gaussian quantization codebooks, so that the quantized signals can be modeled as

$$\hat{\mathbf{y}}_l = \mathbf{y} + \mathbf{q}_l, \tag{5}$$

for $l \in \{0,1,2\}$, where the quantization noise $\mathbf{q}_l$ is independent of the signal $\mathbf{y}$ and distributed as $\mathbf{q}_l \sim \mathcal{CN}(\mathbf{0}, \boldsymbol{\Omega})$ for $l \in \{1,2\}$ and $\mathbf{q}_0 \sim \mathcal{CN}(\mathbf{0}, \boldsymbol{\Omega}_0)$. The right-hand sides (RHSs) of (2)–(4) can hence be written as

$$
\begin{aligned}
g_l\left(\boldsymbol{\Omega}, \boldsymbol{\Omega}_0\right) =& I\left(\mathbf{y}; \hat{\mathbf{y}}_l\right) \\
=& \log_2 \det\left(\boldsymbol{\Sigma}_\mathbf{y} + \boldsymbol{\Omega}\right) - \log_2 \det\left(\boldsymbol{\Omega}\right), \ l \in \{1,2\},
\end{aligned}
\tag{6}
$$

$$
\begin{aligned}
\text{and } g_{\text{sum}}\left(\boldsymbol{\Omega}, \boldsymbol{\Omega}_0\right) =& I\left(\mathbf{y}; \{\hat{\mathbf{y}}_l\}_{l\in\{0,1,2\}}\right) + I\left(\hat{\mathbf{y}}_1; \hat{\mathbf{y}}_2\right) \\
=& h\left(\mathbf{y}\right) + h\left(\{\hat{\mathbf{y}}_l\}_{l\in\{0,1,2\}}\right) - h\left(\mathbf{y}, \{\hat{\mathbf{y}}_l\}_{l\in\{0,1,2\}}\right) \\
& + h\left(\hat{\mathbf{y}}_1\right) + h\left(\hat{\mathbf{y}}_2\right) - h\left(\hat{\mathbf{y}}_1, \hat{\mathbf{y}}_2\right) \\
=& \log_2 \det\left(\boldsymbol{\Sigma}_\mathbf{y}\right) + \log_2 \det\left(\mathbf{A}_3 \boldsymbol{\Sigma}_\mathbf{y} \mathbf{A}_3^H + \bar{\boldsymbol{\Omega}}\right) \\
& - \log_2 \det\left(\mathbf{A}_4 \boldsymbol{\Sigma}_\mathbf{y} \mathbf{A}_4^H + \text{diag}(\mathbf{0}_{n_R}, \bar{\boldsymbol{\Omega}})\right) \\
& + 2\log_2 \det\left(\boldsymbol{\Sigma}_\mathbf{y} + \boldsymbol{\Omega}\right) - \log_2 \det\left(\mathbf{A}_2 \boldsymbol{\Sigma}_\mathbf{y} \mathbf{A}_2^H + \mathbf{I}_2 \otimes \boldsymbol{\Omega}\right),
\end{aligned}
\tag{7}
$$

where we have defined the notations $\bar{\boldsymbol{\Omega}} = \text{diag}(\boldsymbol{\Omega}_0, \mathbf{I}_2 \otimes \boldsymbol{\Omega})$ and $\mathbf{A}_m = \mathbf{1}_m \otimes \mathbf{I}_{n_R}$ with $\mathbf{1}_m \in \mathbb{C}^{m\times 1}$ denoting a column vector of all ones.

We now discuss the derivation of the probability that a packet l is delivered to the cloud within the given deadline T_F. The number N_F of fronthaul packets that need to be delivered within the time T_F to the cloud for the lth description is given as

$$N_F = \left\lceil \frac{L_W R_F}{B_F} \right\rceil, \tag{8}$$

since $L_W R_F$ is the number of bits per description and B_F is the number of available bits per frame. Note that N_F increases with the compression output rate R_F and decreases with the size of the fronthaul packet B_F. Then, the probability that description $l \in \{1,2\}$ sent on route l is received at the cloud within the deadline T_F is given as

$$P_l^c(T_F) = \Pr\left[\sum_{m=1}^{N_F} T_{l,m} \leq T_F\right], \tag{9}$$

where $\{T_{l,m}\}_{m=1}^{N_F}$ are independent and geometrically distributed random variables with parameter $1 - \epsilon_{F,l}$ such that the sum $\sum_{m=1}^{N_F} T_{l,m}$ is a negative binomial random variable with parameters $1 - \epsilon_{F,l}$ and N_F ([22] Ch. 3). Therefore, the probability (9) can be written as

$$P_l^c(T_F) = 1 - I_{\epsilon_{F,l}}\left(T_F - N_F + 1, \ N_F\right), \tag{10}$$

where $I_x(a,b)$ is the regularized incomplete beta function defined as

$$I_x(a,b) = \frac{B(x;a,b)}{B(1;a,b)}, \tag{11}$$

with $B(x;a,b) = \int_0^x t^{a-1}(1-t)^{b-1}dt$. For simplicity of notation, we also define the probabilities $P_\varnothing^c(T_F) = (1 - P_1^c(T_F))(1 - P_2^c(T_F))$ and $P_{\text{all}}^c(T_F) = P_1^c(T_F)P_2^c(T_F)$ that no or both descriptions arrive at the cloud within the deadline.

Define as $M \in \{0,1,2\}$ the number of descriptions that arrive at the cloud within the given deadline T_F. The probability distribution $p_M(m) = \Pr[M = m]$ can then be written as

$$p_M(m) = \begin{cases} P_\varnothing^c(T_F), & m = 0 \\ \sum_{l=1}^2 P_l^c(T_F)\left(1 - P_{\bar{l}}^c(T_F)\right), & m = 1 \, , \\ P_{\text{all}}^c(T_F), & m = 2 \end{cases} \tag{12}$$

with the notation $\bar{1} = 2$ and $\bar{2} = 1$.

Broadcast Coding

With MDC, the quality of the information available at the cloud for decoding the transmitted signals $\{x_k\}_{k\in\mathcal{N}_U}$ is determined by the number M of descriptions that arrive at the cloud. Since the state $M \in \{0,1,2\}$ is not known to the UEs, the rate cannot be a priori adapted by the UEs depending on the congestion level. To handle this issue, we propose that each UE k adopts a broadcast coding strategy [15–18] as

$$\mathbf{x}_k = \mathbf{x}_{k,1} + \mathbf{x}_{k,2}, \tag{13}$$

where the signals $\mathbf{x}_{k,1}$ and $\mathbf{x}_{k,2}$ encode independent messages of UE k, and the decoder at the cloud is required to reliably recover only the signals $\{\mathbf{x}_{k,j}\}_{k\in\mathcal{N}_U}$ with $j \leq m$ when $M = m$ descriptions arrive at the cloud. We denote the rate of the signal $\mathbf{x}_{k,m}$ as $R_{k,m}$ for $k \in \mathcal{N}_U$ and $m \in \{1,2\}$. We make the standard assumption that the jth signal $\mathbf{x}_{k,j}$ of each UE k is distributed as $\mathbf{x}_{k,j} \sim \mathcal{CN}(\mathbf{0}, P_{k,j}\mathbf{I}_{n_{U,k}})$, where the powers $P_{k,j}$ need to satisfy the power constraint $P_{k,1} + P_{k,2} = P$. Under the described assumption, the covariance matrix $\mathbf{\Sigma_x}$ of all the transmitted signals $\mathbf{x}$ is given as $\mathbf{\Sigma_x} = P\mathbf{I}_{n_U}$ with $n_U = \sum_{k\in\mathcal{N}_U} n_{U,k}$.

The signal $\mathbf{r}_m$ collected at the cloud when $M = m$ descriptions have arrived at the cloud is given as

$$\mathbf{r}_m = \begin{cases} \mathbf{0}, & m = 0 \\ \hat{\mathbf{y}}_1, & m = 1 \, . \\ \hat{\mathbf{y}}_0, & m = 2 \end{cases} \tag{14}$$

For the case of $m = 1$, the cloud receives $\mathbf{r}_1 = \hat{\mathbf{y}}_1$ or $\mathbf{r}_1 = \hat{\mathbf{y}}_2$. In (14), we set $\mathbf{r}_1 = \hat{\mathbf{y}}_1$ without loss of generality, since $\hat{\mathbf{y}}_1$ and $\hat{\mathbf{y}}_2$ are statistically equivalent.

When no description arrives at the cloud (i.e., $M = 0$), the cloud has no information received from the RRH, and none of the signals $\{\mathbf{x}_{k,0}, \mathbf{x}_{k,1}\}_{k\in\mathcal{N}_U}$ can be decoded by the cloud. When only a single description arrives at the cloud ($M = 1$), the cloud jointly decodes the first-layer signals $\{\mathbf{x}_{k,1}\}_{k\in\mathcal{N}_U}$ based on the received quantized signal $\mathbf{r}_1$. Therefore, the achievable sum-rate $R_{\Sigma,1} = \sum_{k\in\mathcal{N}_U} R_{k,1}$ of the first-layer signals is given as

$$R_{\Sigma,1} = f_1(\mathbf{P}, \mathbf{\Omega}, \mathbf{\Omega}_0) = I(\bar{\mathbf{x}}_1; \mathbf{r}_1) \tag{15}$$

$$= \log_2 \det\left(\mathbf{H}\mathbf{\Sigma_x}\mathbf{H}^H + \mathbf{\Sigma_z} + \mathbf{\Omega}\right) - \log_2 \det\left(\mathbf{H}\bar{\mathbf{P}}_2\mathbf{H}^H + \mathbf{\Sigma_z} + \mathbf{\Omega}\right),$$

where we have defined the vector $\bar{\mathbf{x}}_m = [\mathbf{x}_{1,m}^H \cdots \mathbf{x}_{N_U,m}^H]^H \sim \mathcal{CN}(\mathbf{0}, \bar{\mathbf{P}}_m)$ that stacks the layer-m signals of all the UEs, and the notations $\mathbf{P} = \{P_{k,j}\}_{k\in\mathcal{N}_U, j\in\{1,2\}}$ and $\bar{\mathbf{P}}_m = \text{diag}(\{P_{k,m}\}_{k\in\mathcal{N}_U})$.

If both descriptions arrive at the cloud (i.e., $M = 2$), the cloud first jointly decodes the first-layer signals $\{\mathbf{x}_{k,1}\}_{k \in \mathcal{N}_U}$ from the recovered quantized signal $\mathbf{r}_2$, and cancels the impact of the decoded signals from $\mathbf{r}_2$, i.e., $\tilde{\mathbf{r}}_2 \leftarrow \mathbf{r}_2 - \sum_{k \in \mathcal{N}_U} \mathbf{H}_k \mathbf{x}_{k,1}$. Then, the cloud decodes the second-layer signals $\{\mathbf{x}_{k,2}\}_{k \in \mathcal{N}_U}$ based on $\tilde{\mathbf{r}}_2$. Thus, the achievable sum-rate $R_{\Sigma,2} = \sum_{k \in \mathcal{N}_U} R_{k,2}$ of the second-layer signals is given as

$$R_{\Sigma,2} = f_2\left(\mathbf{P}, \boldsymbol{\Omega}, \boldsymbol{\Omega}_0\right) = I\left(\bar{\mathbf{x}}_2; \mathbf{r}_2 | \bar{\mathbf{x}}_1\right) \tag{16}$$
$$= \log_2 \det\left(\mathbf{H}\bar{\mathbf{P}}_2\mathbf{H}^H + \boldsymbol{\Sigma}_{\mathbf{z}} + \boldsymbol{\Omega}_0\right) - \log_2 \det\left(\boldsymbol{\Sigma}_{\mathbf{z}} + \boldsymbol{\Omega}_0\right).$$

In summary, the whole system operates as follows. The cloud first obtains the channel state information and optimizes the variables related to broadcast coding and MDC coding. The optimization will be discussed in Section 4. After the optimization algorithm is finished, the cloud informs the UEs and the RRH of the optimized variables. The UEs perform broadcast coding and uplink transmission, and the RRH compresses the received signal obtaining two descriptions which are packetized and sent on fronthaul paths to the cloud. Based on the received packets, the cloud performs MDC decoding of the quantized signals and SIC decoding of the UEs' messages. We provide a flowchart that illustrates the described operations of the proposed system in Figure 4.

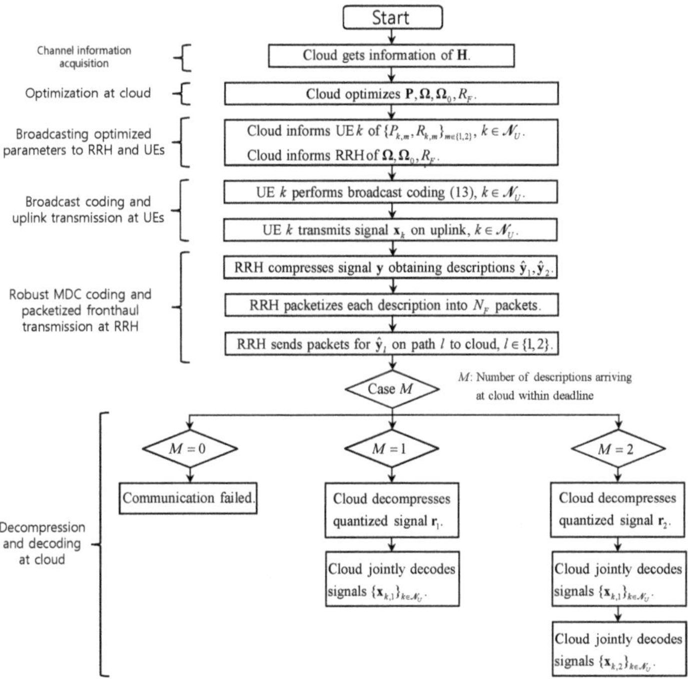

Figure 4. A flowchart that illustrates the operations of the proposed uplink system based on broadcast coding and multiple description coding (MDC).

4. Problem Definition and Optimization

For fixed instantaneous channel states $\{\mathbf{H}_k\}_{k \in \mathcal{N}_U}$, we aim at jointly optimizing the compression output rate R_F, the power allocation variables $\mathbf{P}$ and the quantization noise covariance matrices

$\{\Omega, \Omega_0\}$ with the goal of maximizing the expected sum-rate denoted as $\bar{R}_\Sigma$. Here the expectation is taken with respect to the random variables $\{T_{l,m}\}_{l \in \{1,2\}, m \in \mathcal{N}_F}$ with $\mathcal{N}_F = \{1, 2, \ldots, N_F\}$, which depend on the current congestion level of the packet network. The expected sum-rate $\bar{R}_\Sigma$ is hence given as

$$\bar{R}_\Sigma = p_M(1)R_{\Sigma,1} + p_M(2)\left(R_{\Sigma,1} + R_{\Sigma,2}\right) \tag{17}$$
$$= \bar{p}_M(1)R_{\Sigma,1} + \bar{p}_M(2)R_{\Sigma,2},$$

with the notations $\bar{p}_M(1) = p_M(1) + p_M(2)$ and $\bar{p}_M(2) = p_M(2)$. The expected sum-rate $\bar{R}_\Sigma$ can be expressed as a function of R_F, $\mathbf{P}$ and $\{\Omega, \Omega_0\}$:

$$\bar{R}_\Sigma = f_\Sigma\left(R_F, \mathbf{P}, \Omega, \Omega_0\right) \tag{18}$$
$$= \bar{p}_M(1)f_1\left(\mathbf{P}, \Omega, \Omega_0\right) + \bar{p}_M(2)f_2\left(\mathbf{P}, \Omega, \Omega_0\right).$$

We note that increasing the compression output rate R_F has conflicting effects on the expected sum-rate $\bar{R}_\Sigma$. On the one hand, the probability of timely reception of all fronthaul packets decreases with R_F due to the increased number N_F of packets in (8). On the other hand, once the packets have arrived at the cloud, a better sum-rate can be achieved with larger R_F, since the quantization noise signals have smaller powers.

The problem mentioned above can be stated as

$$\underset{R_F, \mathbf{P}, \Omega, \Omega_0}{\text{maximize}} \ f_\Sigma\left(R_F, \mathbf{P}, \Omega, \Omega_0\right) \tag{19a}$$

$$\text{s.t.} \ R_F \geq g_1\left(\Omega, \Omega_0\right), \tag{19b}$$

$$2R_F \geq g_{\text{sum}}\left(\Omega, \Omega_0\right), \tag{19c}$$

$$\Omega \succeq 0, \ \Omega_0 \succeq 0, \tag{19d}$$

$$P_{k,1} + P_{k,2} = P, \ k \in \mathcal{N}_U, \tag{19e}$$

$$P_{k,1} \geq 0, \ P_{k,2} \geq 0, \ k \in \mathcal{N}_U. \tag{19f}$$

To tackle the problem (19), we first note that, if we fix the compression output rate variable R_F, the problem becomes a difference-of-convex (DC) problem as in [23]. Therefore, we can find an efficient solution by adopting the concave convex procedure (CCCP) approach (see, e.g., [24,25]). The detailed algorithm that tackles (19) with the CCCP approach is described in Algorithm 1, where we have defined the functions $\tilde{f}_\Sigma(R_F, \mathbf{P}, \Omega, \Omega_0, \mathbf{P}^{(t)}, \Omega^{(t)}, \Omega_0^{(t)})$, $\tilde{g}_1(\Omega, \Omega_0, \Omega^{(t)}, \Omega_0^{(t)})$ and $\tilde{g}_{\text{sum}}(\Omega, \Omega_0, \Omega^{(t)}, \Omega_0^{(t)})$ as

$$\tilde{f}_\Sigma\left(R_F, \mathbf{P}, \Omega, \Omega_0, \mathbf{P}^{(t)}, \Omega^{(t)}, \Omega_0^{(t)}\right) = \bar{p}_M(1)\left(\begin{array}{c} \log_2 \det\left(\mathbf{H}\Sigma_\mathbf{x}\mathbf{H}^H + \Sigma_\mathbf{z} + \Omega\right) \\ -\phi\left(\begin{array}{c} \mathbf{H}\bar{\mathbf{P}}_2\mathbf{H}^H + \Sigma_\mathbf{z} + \Omega, \\ \mathbf{H}\bar{\mathbf{P}}_2^{(t)}\mathbf{H}^H + \Sigma_\mathbf{z} + \Omega^{(t)} \end{array}\right) \end{array}\right)$$

$$+ \bar{p}_M(2)\left(\begin{array}{c} \log_2 \det\left(\mathbf{H}\bar{\mathbf{P}}_2\mathbf{H}^H + \Sigma_\mathbf{z} + \Omega_0\right) \\ -\phi\left(\Sigma_\mathbf{z} + \Omega_0, \Sigma_\mathbf{z} + \Omega_0^{(t)}\right) \end{array}\right),$$

$$\tilde{g}_1\left(\Omega, \Omega_0, \Omega^{(t)}, \Omega_0^{(t)}\right) = \phi\left(\Sigma_\mathbf{y} + \Omega, \Sigma_\mathbf{y} + \Omega^{(t)}\right) - \log_2 \det\left(\Omega\right),$$

$$\text{and} \ \tilde{g}_{\text{sum}}\left(\Omega, \Omega_0, \Omega^{(t)}, \Omega_0^{(t)}\right) = \log_2 \det\left(\Sigma_\mathbf{y}\right) + \phi\left(\mathbf{A}_3\Sigma_\mathbf{y}\mathbf{A}_3^H + \bar{\Omega}, \mathbf{A}_3\Sigma_\mathbf{y}\mathbf{A}_3^H + \bar{\Omega}^{(t)}\right)$$

$$- \log_2 \det\left(\mathbf{A}_4\Sigma_\mathbf{y}\mathbf{A}_4^H + \text{diag}(\mathbf{0}_{n_R}, \bar{\Omega})\right)$$

$$+ 2\phi\left(\Sigma_\mathbf{y} + \Omega, \Sigma_\mathbf{y} + \Omega^{(t)}\right) - \log_2 \det\left(\mathbf{A}_2\Sigma_\mathbf{y}\mathbf{A}_2^H + \mathbf{I}_2 \otimes \Omega\right),$$

with the function $\phi\left(\mathbf{A},\mathbf{B}\right)$ defined as

$$\phi\left(\mathbf{A},\mathbf{B}\right) = \log_2 \det(\mathbf{B}) + \frac{1}{\ln 2}\mathrm{tr}\left(\mathbf{B}^{-1}(\mathbf{A}-\mathbf{B})\right).$$

Algorithm 1 CCCP algorithm for problem (19) for fixed R_F

---

1. Initialize the variables $\mathbf{P}^{(1)},\mathbf{\Omega}^{(1)},\mathbf{\Omega}_0^{(1)}$ to arbitrary matrices that satisfy the constraints (19b), (19c) and (19d), and set $t \leftarrow 1$.

2. Update the variables $\mathbf{P}^{(t+1)},\mathbf{\Omega}^{(t+1)},\mathbf{\Omega}_0^{(t+1)}$ as a solution of the convex problem:

$$\underset{\mathbf{P},\mathbf{\Omega},\mathbf{\Omega}_0}{\text{maximize}}\ \tilde{f}_\Sigma\left(R_F,\mathbf{P},\mathbf{\Omega},\mathbf{\Omega}_0,\mathbf{P}^{(t)},\mathbf{\Omega}^{(t)},\mathbf{\Omega}_0^{(t)}\right) \tag{20a}$$

$$\text{s.t.}\ R_F \geq \tilde{g}_1\left(\mathbf{\Omega},\mathbf{\Omega}_0,\mathbf{\Omega}^{(t)},\mathbf{\Omega}_0^{(t)}\right), \tag{20b}$$

$$2R_F \geq \tilde{g}_{\text{sum}}\left(\mathbf{\Omega},\mathbf{\Omega}_0,\mathbf{\Omega}^{(t)},\mathbf{\Omega}_0^{(t)}\right), \tag{20c}$$

$$\mathbf{\Omega} \succeq \mathbf{0},\ \mathbf{\Omega}_0 \succeq \mathbf{0}, \tag{20d}$$

$$P_1 + P_2 = P, \tag{20e}$$

$$P_1 \geq 0,\ P_2 \geq 0. \tag{20f}$$

3. Stop if a convergence criterion is satisfied. Otherwise, set $t \leftarrow t+1$ and go back to Step 2.

We have discussed the optimization of the power allocation variables $\mathbf{P}$ and the quantization noise covariance matrices $\{\mathbf{\Omega},\mathbf{\Omega}_0\}$ for fixed compression output rate R_F. For the optimization of R_F, we propose to perform a 1-dimensional discrete search over $R_F \in \mathcal{R} = \{\Delta_{R_F}, 2\Delta_{R_F}, \ldots, N_{F,\max}\Delta_{R_F}\}$ with $\Delta_{R_F} = B_F/L_W$ and $N_{F,\max} = T_F + 1$. Here we have excluded the values $\tau\Delta_{R_F}$ with non-integer τ from the search space $\mathcal{R}$. This does not cause a loss of optimality, since we can increase the compression output rate, hence improving the compression fidelity to $\lceil \tau \rceil \Delta_{R_F}$ without increasing the number N_F of packets in (8) that needs to be delivered to the cloud.

Optimization of Traditional Path-Diversity Scheme

In this subsection, we discuss the optimization of the traditional path-diversity (PD) scheme, in which the RRH repeats to send the same packet on the available two routes [11]. Accordingly, the RRH produces only a single quantized signal $\hat{\mathbf{y}} = \mathbf{y} + \mathbf{q}$, where the quantization noise $\mathbf{q}$ is independent of $\mathbf{y}$ and distributed as $\mathbf{q} \sim \mathcal{CN}(\mathbf{0},\mathbf{\Omega})$ under the assumption of standard Gaussian quantization codebooks. Denoting as R_F the compression output rate for the quantized signal $\hat{\mathbf{y}}$, the rate R_F should satisfy the condition

$$R_F \geq g\left(\mathbf{\Omega}\right) = I\left(\mathbf{y};\hat{\mathbf{y}}\right) \tag{21}$$
$$= \log_2 \det\left(\mathbf{\Sigma_y} + \mathbf{\Omega}\right) - \log_2 \det\left(\mathbf{\Omega}\right).$$

To evaluate the achievable sum-rate, we define the binary variable $D \in \{0,1\}$, which takes 1 if at least one packet arrives at the cloud, and 0 otherwise. The probability distribution of D can be written as

$$\Pr\left[D=d\right] = \begin{cases} P_\varnothing^c(T_F), & d=0 \\ 1 - P_\varnothing^c(T_F), & d=1 \end{cases}. \tag{22}$$

If both packets sent on two routes are lost (i.e., $D=0$), the cloud cannot decode the signals sent by the UEs. If the cloud receives at least one packet ($D=1$), the cloud can perform decoding of the signals $\mathbf{x}$ based on the received quantized signal $\hat{\mathbf{y}}$, and the achievable sum-rate can be written as

$$R_\Sigma = f_\Sigma(\mathbf{\Omega}) = I\left(\mathbf{x}; \hat{\mathbf{y}}\right) \tag{23}$$

$$= \log_2 \det\left(\mathbf{H}\mathbf{\Sigma}_\mathbf{x}\mathbf{H}^H + \mathbf{\Sigma}_\mathbf{z} + \mathbf{\Omega}\right) - \log_2 \det\left(\mathbf{\Sigma}_\mathbf{z} + \mathbf{\Omega}\right).$$

The expected sum-rate $\bar{R}_\Sigma$ can be expressed as

$$\bar{R}_\Sigma = f_\Sigma\left(R_F, \mathbf{\Omega}\right) = \Pr\left[D = 1\right] f_\Sigma(\mathbf{\Omega}). \tag{24}$$

The problem of maximizing the expected sum-rate $\bar{R}_\Sigma$ with the traditional PD scheme can hence be stated as

$$\underset{R_F, \mathbf{\Omega}}{\text{maximize}} \ f_\Sigma\left(R_F, \mathbf{\Omega}\right) \tag{25a}$$

$$\text{s.t.} \ R_F \geq g\left(\mathbf{\Omega}\right), \tag{25b}$$

$$\mathbf{\Omega} \succeq \mathbf{0}. \tag{25c}$$

We can tackle the problem (25) in a similar approach to that proposed for addressing (19).

5. Numerical Results

In this section, we provide numerical results that validate the advantages of the proposed robust baseband compression technique based on MDC coding scheme. We consider a system bandwidth of 100 MHz and assume that each wireless frame consists of $L_W = 5000$ channel uses. We also assume that each fronthaul packet has $B_F = 6000$ bits (i.e., 750 bytes) which corresponds to a half of the maximum payload size per frame defined in Ethernet [10]. Denoting as C_F the fronthaul capacity in bit/s, each fronthaul packet has the duration of B_F/C_F. If we define the maximum tolerable delay on fronthaul network as $T_{\max}$ s, the deadline T_F in packet duration is given as $T_F = \lfloor T_{\max}/(B_F/C_F)\rfloor$. In the simulation, we set $T_{\max} = 1$ ms. For simplicity, we assume that all paths have the same error probability $\epsilon_{F,l} = \epsilon_F$ for all $l \in \{1,2\}$. Regarding the channel statistics, we assume that the positions of the UEs and the RRH are uniformly distributed within a circular area of radius 100 m. The elements of the channel matrix $\mathbf{H}_k$ are independent and identically distributed (i.i.d.) as $\mathcal{CN}(0, \rho_k)$. Here the path-loss ρ_k is modeled as $\rho_k = 1/(1 + (d_k/d_0)^3)$, where d_k represents the distance between the RRH and UE k, and d_0 is the reference distance set to $d_0 = 30$ m. We set the noise covariance to $\mathbf{\Sigma}_\mathbf{z} = N_0 \mathbf{I}_{n_R}$, and the SNR is defined as P/N_0.

5.1. Fixed Compression Output Rate R_F

We first evaluate the expected sum-rate performance $\mathrm{E}[R_{\text{sum}}]$ when only the power allocation variables $\mathbf{P}$ and the quantization noise covariance matrices $\mathbf{\Omega}$ are optimized according to Algorithm 1 for fixed compression output rate R_F. In Figure 5, we plot the expected sum-rate $\mathrm{E}[R_{\text{sum}}]$ versus the compression output rate R_F for various values of path error probability ϵ_F with $N_U = 2$, $n_R = 2$, $n_{U,k} = 1$, $C_F = 100$ Mbit/s and 25 dB SNR. We observe that, for both the MDC and PD schemes, the optimal compression output rate R_F increases as the fronthaul error probability ϵ_F decreases. This suggests that, with smaller ϵ_F, the packet networks become more reliable and hence more packets can be reliably delivered to the cloud within the deadline. Furthermore, the figure shows that, with MDC, it is optimal to choose a lower compression output rate with respect to PD. This is because, as the fronthaul quality improves in terms of the error probability ϵ_F, the PD scheme can only increase the sum-rate by increasing the quality, or the compression output rate R_F, of each individual description, since it cannot benefit from reception of both descriptions. In contrast, the MDC scheme can operate at a lower R_F, since the quality of the compressed signal is improved by reception of both descriptions. Receiving both descriptions tends to be more likely if the compression output rate is lower and hence the number of fronthaul packets per frame is reduced.

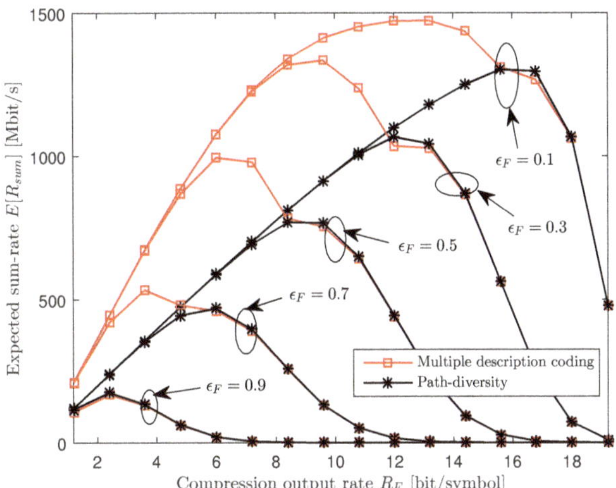

Figure 5. Expected sum-rate $E[R_{sum}]$ versus the compression output rate R_F for various values of $\epsilon_F \in \{0.1, 0.3, 0.5, 0.7, 0.9\}$ ($N_U = 2$, $n_R = 2$, $n_{U,k} = 1$, $C_F = 100$ Mbit/s and 25 dB signal-to-noise ratio (SNR)).

In Figure 6, we depict the expected sum-rate $E[R_{sum}]$ versus the compression output rate R_F for various SNR levels with $\epsilon_F = 0.4$, $N_U = 3$, $n_R = 3$, $n_{U,k} = 1$ and $C_F = 100$ Mbit/s. The figure shows that, as the SNR increases, the optimal compression output rate R_F slightly increases for both MDC and PD. This is because, while the SNR level does not affect the reliability of the packet fronthaul network, it is desirable for the RRH to report better descriptions of the uplink received signals to the cloud when the received signals carry more information on the UEs' messages.

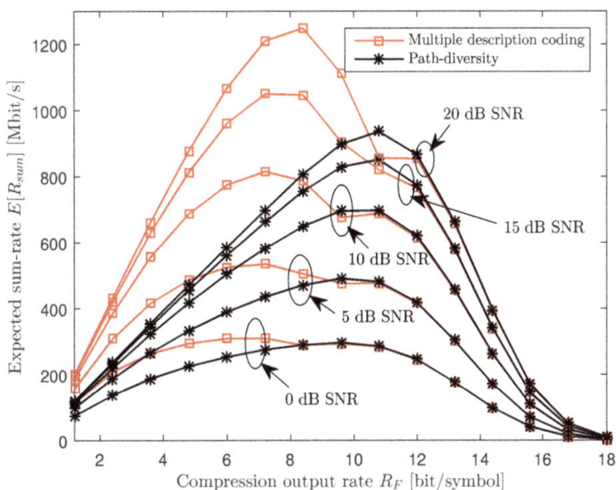

Figure 6. Expected sum-rate $E[R_{sum}]$ versus the compression output rate R_F for various SNR levels ($\epsilon_F = 0.4$, $N_U = 3$, $n_R = 3$, $n_{U,k} = 1$ and $C_F = 100$ Mbit/s).

Figure 7 plots the expected sum-rate $E[R_{sum}]$ with respect to the compression output rate R_F for various fronthaul capacity C_F with $\epsilon_F = 0.6$, $N_U = 2$, $n_R = 2$, $n_{U,k} = 1$ and 25 dB SNR. Since more packets, and hence more bits, can be transferred to the cloud within the deadline T_F with increased fronthaul capacity C_F, the optimal compression output rate R_F grows with C_F for both the MDC and PD schemes.

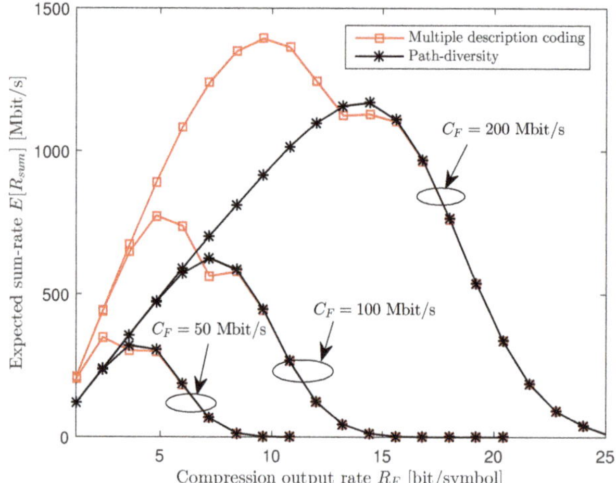

Figure 7. Expected sum-rate $\mathrm{E}[R_{\mathrm{sum}}]$ versus the compression output rate R_F for various fronthaul capacity C_F ($\epsilon_F = 0.6$, $N_U = 2$, $n_R = 2$, $n_{U,k} = 1$ and 25 dB SNR).

5.2. Optimized Compression Output Rate R_F

In this subsection, we present the expected sum-rate $\mathrm{E}[R_{\mathrm{sum}}]$ achieved when the power allocation variables $\mathbf{P}$, the quantization noise covariance matrices $\boldsymbol{\Omega}$ and the compression output rate R_F are jointly optimized as discussed in Section 4. In Figure 8, we plot the expected sum-rate $\mathrm{E}[R_{\mathrm{sum}}]$ versus the SNR for $N_U = 3$, $n_R = 3$, $n_{U,k} = 1$, $\epsilon_F \in \{0.4, 0.6\}$ and $C_F = 100$ Mbit/s. We observe from the figure that the MDC scheme shows a larger gain at a higher SNR level. This suggests that, as the SNR increases, the overall performance becomes limited by the quantization distortion which is smaller for the MDC scheme than for PD.

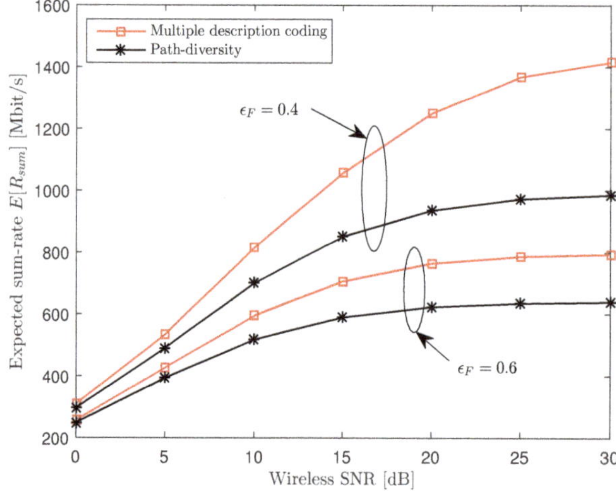

Figure 8. Expected sum-rate $\mathrm{E}[R_{\mathrm{sum}}]$ versus the SNR ($N_U = 3$, $n_R = 3$, $n_{U,k} = 1$, $\epsilon_F \in \{0.4, 0.6\}$ and $C_F = 100$ Mbit/s).

In Figure 9, we plot the expected sum-rate $E[R_{\text{sum}}]$ versus the fronthaul capacity C_F for $N_U = 3$, $n_R = 3$, $n_{U,k} = 1$, $\epsilon_F \in \{0.4, 0.6\}$ and 25 dB SNR. The figure illustrates that the MDC scheme shows relevant gains over the PD scheme in the intermediate regime of C_F. This is because, when the fronthaul capacity C_F is sufficiently large, the whole system has a performance bottleneck in the wireless uplink rather than in the fronthaul network, and the sum-rate converges to 0 as C_F approaches 0.

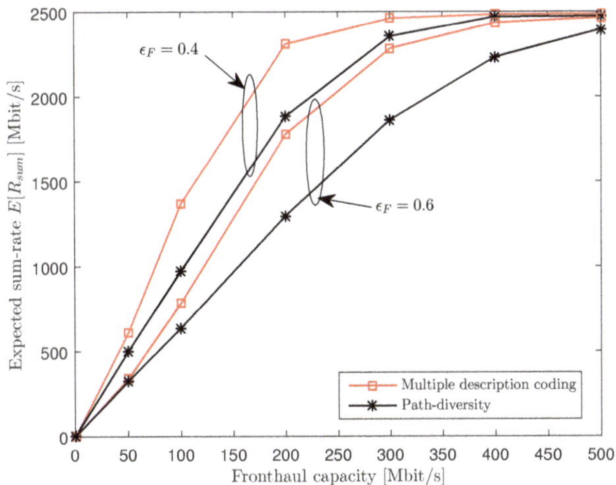

Figure 9. Expected sum-rate $E[R_{\text{sum}}]$ versus the fronthaul capacity C_F ($N_U = 3$, $n_R = 3$, $n_{U,k} = 1$, $\epsilon_F \in \{0.4, 0.6\}$ and 25 dB SNR).

6. Extension to General Numbers of RRHs and Fronthaul Paths

In this section, we briefly discuss the application of MDC to the case of general number N_R of RRHs and N_P fronthaul paths. Each RRH i sends N_P descriptions $\hat{\mathbf{y}}_{i,l}$, $l \in \{1, \ldots, N_P\}$, one on each of the routes to the cloud, where $\hat{\mathbf{y}}_{i,l}$ is a quantized version of the received signal $\mathbf{y}_i$ defined as

$$\hat{\mathbf{y}}_{i,l} = \mathbf{y}_i + \mathbf{q}_{i,l}. \tag{26}$$

As in (5), under Gaussian quantization codebook, the quantization noise $\mathbf{q}_{i,l}$ is independent of $\mathbf{y}_i$ and is distributed as $\mathbf{q}_{i,l} \sim \mathcal{CN}(\mathbf{0}, \mathbf{\Omega}_i)$. With MDC, the cloud can recover the signal $\hat{\mathbf{y}}_{i,l}$ if only the packets for the lth description $\hat{\mathbf{y}}_{i,l}$ arrive at the cloud within the deadline. If a subset of descriptions from RRH i arrive in time, the cloud can obtain a better signal from RRH i, whose quality increases with the size of the subset. Generalizing (2)–(4), conditions relating the resulting quantization noise covariance matrices and the output compression rate R_F can be found in [26].

We define as $M_i \in \{0, 1, \ldots, N_P\}$ the number of descriptions of RRH i that arrive at the cloud within the deadline T_F. The probability distribution $p_{M_i}(m) = \Pr[M_i = m]$ of M_i is then given as

$$p_{M_i}(m) = \sum_{(c_1,\ldots,c_{N_P}) \in \{0,1\}^{N_P}} 1\left(\sum_{l=1}^{N_P} c_l = m\right) \prod_{l=1}^{N_P} \tilde{P}_l(T_F), \tag{27}$$

where $1(\cdot)$ is an indicator function that outputs 1 if the input statement is true and 0 otherwise; and the probability $\tilde{P}_l(T_F)$ is defined as $\tilde{P}_l(T_F) = 1(c_l = 1)P_l^c(T_F) + 1(c_l = 0)(1 - P_l^c(T_F))$.

As discussed, with MDC, the quality of the information available at the cloud depends on the numbers of descriptions that arrive at the cloud. This means that there are $(N_P + 1)^{N_R}$ distinct states depending on the current congestion level of the packet network. In principle, the broadcast coding can be applied in such a way that each UE k sends a superposition of $(N_P + 1)^{N_R}$ layers. However, this approach does not scale well with respect to N_R, and it is not straightforward to rank all the $(N_P + 1)^{N_R}$ states.

To adopt a broadcast coding strategy with a scalable complexity, a possible option is to fix the number of layers, denoted as L, as in, e.g., [27]. Accordingly, the transmit signal $\mathbf{x}_k$ of each UE k is given by a superposition of L independent signals $\mathbf{x}_{k,l} \sim \mathcal{CN}(\mathbf{0}, P_{k,l}\mathbf{I}_{n_{U,k}})$, $l \in \mathcal{L} = \{1, \ldots, L\}$, i.e., $\mathbf{x}_k = \sum_{l \in \mathcal{L}} \mathbf{x}_{k,l}$ with the power constraint $\sum_{l \in \mathcal{L}} P_{k,l} = P$. We then partition the $(N_P + 1)^{N_R}$ congestion states into L groups, denoted as $\mathcal{S}_1, \ldots, \mathcal{S}_L$, so that the layer-$l$ signals $\{\mathbf{x}_{k,l}\}_{k \in \mathcal{N}_U}$ can be decoded by the cloud for all congestion states in $\mathcal{S}_j$ with $j \geq l$. Since we can evaluate the probability of all the states using (27), the expected sum-rate can be expressed as a function of the compression output rate, the power allocation variables and the quantization covariance matrices. Therefore, we can tackle the problem of jointly optimizing these variables in a similar approach to that proposed in Section 4. We leave the evaluation of the impacts of the numbers of RRHs N_R and fronthaul paths N_P to future work.

7. Conclusions

In this paper, we have studied the joint design of uplink radio and fronthaul packet transmission strategies for the uplink of C-RAN with a packet-based fronthaul network. To efficiently use multiple fronthaul paths that carry fronthaul packets from RRHs to cloud, we have proposed an MDC scheme that operates directly on the baseband signals. Since the signal quality available at the cloud depends on the current network congestion level, a broadcast coding strategy has been investigated with MDC in order to enable variable-rate transmission. The advantages of the proposed MDC scheme compared to the traditional PD technique have been validated through extensive numerical results. Among open problems, we mention the analysis in the presence of imperfect channel state information [28], the impact of joint decompression of the signals received from multiple RRHs at the cloud [3,23], and design of downlink transmission for C-RAN systems with packet-based fronthaul network.

Author Contributions: S.-H.P. is the primary author; O.S. and S.S. contributed in terms of problem formulation, key theoretical ideas, and writing.

Funding: S.-H. Park was supported by the National Research Foundation of Korea (NRF) grant funded by the Korea government [NRF-2018R1D1A1B07040322]. The work of O. Simeone was partially supported by the U.S. NSF through grant 1525629 and by the European Research Council (ERC) under the European Union's Horizon 2020 research and innovation programme (grant agreement No 725731). The work of S. Shamai has been supported by the European Union's Horizon 2020 research and innovation programme, grant agreement no. 694630.

Conflicts of Interest: The authors declare no conflict of interest.

References

1. Checko, A.; Christiansen, H.L.; Yan, Y.; Scolari, L.; Kardaras, G.; Berger, M.S.; Dittmann, L. Cloud RAN Mob. Networks—A Technol. Overview. *IEEE Commun. Surv. Tutors* **2015**, *17*, 405–426. [CrossRef]
2. Simeone, O.; Maeder, A.; Peng, M.; Sahin, O.; Yu, W. Cloud radio access network: Virtualizing wireless access for dense heterogeneous systems. *J. Commun. Netw.* **2016**, *18*, 135–149. [CrossRef]
3. Park, S.-H.; Simeone, O.; Sahin, O.; Shamai, S. Robust and efficient distributed compression for cloud radio access networks. *IEEE Trans. Veh. Technol.* **2013**, *62*, 692–703. [CrossRef]
4. Park, S.-H.; Simeone, O.; Sahin, O.; Shamai, S. Joint precoding and multivariate backhaul compression for the downlink of cloud radio access networks. *IEEE Trans. Signal Process.* **2013**, *61*, 5646–5658. [CrossRef]
5. Zhou, Y.; Yu, W. Fronthaul compression and transmit beamforming optimization for multi-antenna uplink C-RAN. *IEEE Trans. Signal Process.* **2016**, *64*, 4138–4151. [CrossRef]
6. Patil, P.; Dai, B.; Yu, W. Hybrid data-sharing and compression strategy for downlink cloud radio access network. *IEEE Trans. Commun.* **2018**, *66*, 5370–5384. [CrossRef]
7. Ericsson AB; Huawei Technologies Co. Ltd.; NEC Corporation. *Alcatel Lucent and Nokia Networks, Common Public Radio Interface (CPRI): Interface Specification*; CPRI Specification V7.0; Huawei Technologies Co. Ltd.: Shenzhen, China; NEC Corporation: Tokyo, Japan, 2015.

8. De la Oliva, A.; Hernandez, J.A.; Larrabeiti, D.; Azcorra, A. An overview of the CPRI specification and its application to C-RAN-based LTE scenarios. *IEEE Commun. Mag.* **2016**, *54*, 152–159. [CrossRef]

9. Ericsson AB; Huawei Technologies Co. Ltd.; NEC Corporation and Nokia. *Common Public Radio Interface: ECPRI Interface Specification*. eCPRI Specification V1.0; Huawei Technologies Co. Ltd.: Shenzhen, China; NEC Corporation: Tokyo, Japan, 2017.

10. *The Ethernet—A Local Area Network—Data Link Layer and Physical Layer Specifications*, Ver. 2.0; The ACM Digital Library: New York, NY, USA, 1982.

11. Alasti, M.; Sayrafian-Pour, K.; Emphremides, A.; Farvardin, N. Multiple description coding in networks with congestion problem. *IEEE Trans. Inf. Theory* **2001**, *47*, 891–902. [CrossRef]

12. Mountaser, G.; Mahmoodi, T.; Simeone, O. Reliable and low-latency fronthaul for tactile internet applications. *IEEE J. Sel. Areas Commun.* **2018**, *36*, 2455–2463. [CrossRef]

13. Goyal, V.K. Multiple description coding: Compression meets the network. *IEEE Signal Process. Mag.* **2001**, *18*, 74–93. [CrossRef]

14. Benammar, M.; Piantanida, P.; Shamai, S. Capacity Results for the Multicast Cognitive Interference Channel. *IEEE Trans. Inf. Theory* **2017**, *63*, 4119–4136. [CrossRef]

15. Cover, T.M. Comments on broadcast channels. *IEEE Trans. Inf. Theory* **1998**, *44*, 2524–2530. [CrossRef]

16. Shamai, S.; Steiner, A. A broadcast approach for a single-user slowly fading MIMO channel. *IEEE Trans. Inf. Theory* **2003**, *49*, 2617–2635. [CrossRef]

17. Verdú, S.; Shamai, S. Variable-rate channel capacity. *IEEE Trans. Inf. Theory* **2010**, *56*, 2651–2667. [CrossRef]

18. Park, S.-H.; Simeone, O.; Sahin, O.; Shamai, S. Robust layered transmission and compression for distributed uplink reception in cloud radio access networks. *IEEE Trans. Veh. Technol.* **2014**, *63*, 204–216. [CrossRef]

19. Karasik, R.; Simeone, O.; Shamai, S. Robust Uplink Communications over Fading Channels with Variable Backhaul Connectivity. *IEEE Trans. Wirel. Commun.* **2013**, *12*, 5788–5799. [CrossRef]

20. Ng, C.T.K.; Tian, C.; Goldsmith, A.J.; Shamai, S. Minimum Expected Distortion in Gaussian Source Coding With Fading Side Information. *IEEE Trans. Inf. Theory* **2012**, *58*, 5725–5739. [CrossRef]

21. Gamal, A.E.; Kim, Y.-H. *Network Information Theory*; Cambridge University Press: Cambridge, UK, 2011.

22. Leon-Garcia, A. *Probability and Random Processes for Electrical Engineering*; Addison Wesley: Boston, MA, USA, 1994.

23. Park, S.-H.; Simeone, O.; Sahin, O.; Shamai, S. Joint decompression and decoding for cloud radio access networks. *IEEE Signal Process. Lett.* **2013**, *20*, 503–506. [CrossRef]

24. Tao, M.; Chen, E.; Zhou, H.; Yu, W. Content-centric sparse multicast beamforming for cache-enabled cloud RAN. *IEEE Trans. Wirel. Commun.* **2016**, *15*, 6118–6131. [CrossRef]

25. Park, S.-H.; Simeone, O.; Shamai, S. Joint optimization of cloud and edge processing for fog radio access networks. *IEEE Trans. Wirel. Commun.* **2016**, *15*, 7621–7632. [CrossRef]

26. Venkataramani, R.; Kramer G.; Goyal, V.K. Multiple description coding with many channels. *IEEE Trans. Inf. Theory* **2003**, *49*, 2106–2114. [CrossRef]

27. Steiner, A.; Shamai, S. Multi-Layer Broadcasting over a Block Fading MIMO Channel. *IEEE Trans. Wirel. Commun.* **2007**, *6*, 3937–3945. [CrossRef]

28. Kang, J.; Simeone, O.; Kang, J.; Shamai, S. Joint Signal and Channel State Information Compression for the Backhaul of Uplink Network MIMO Systems. *IEEE Trans. Wirel. Commun.* **2014**, *13*, 1555–1567. [CrossRef]

Article

Amplitude Constrained MIMO Channels: Properties of Optimal Input Distributions and Bounds on the Capacity [†]

Alex Dytso [1,*], Mario Goldenbaum [1,*], H. Vincent Poor [1,*] and Shlomo Shamai (Shitz) [2,*]

[1]   Department of Electrical Engineering, Princeton University, Princeton, NJ 08544, USA
[2]   Department of Electrical Engineering, Technion-Israel Institute of Technology,
    Technion City, Haifa 32000, Israel
*   Correspondence: adytso@princeton.edu (A.D.); goldenbaum@princeton.edu (M.G.);
    poor@princeton.edu (H.V.P.); sshlomo@ee.technion.ac.il (S.S.)
†   Parts of the material in this paper were presented at the 2017 IEEE Global Communications Conference
    (Singapore, 4–8 December 2017).

Received: 21 January 2019; Accepted: 13 February 2019; Published: 19 February 2019

Abstract: In this work, the capacity of multiple-input multiple-output channels that are subject to constraints on the support of the input is studied. The paper consists of two parts. The first part focuses on the general structure of capacity-achieving input distributions. Known results are surveyed and several new results are provided. With regard to the latter, it is shown that the support of a capacity-achieving input distribution is a small set in both a topological and a measure theoretical sense. Moreover, explicit conditions on the channel input space and the channel matrix are found such that the support of a capacity-achieving input distribution is concentrated on the boundary of the input space only. The second part of this paper surveys known bounds on the capacity and provides several novel upper and lower bounds for channels with arbitrary constraints on the support of the channel input symbols. As an immediate practical application, the special case of multiple-input multiple-output channels with *amplitude constraints* is considered. The bounds are shown to be within a constant gap to the capacity if the channel matrix is invertible and are tight in the high amplitude regime for arbitrary channel matrices. Moreover, in the regime of high amplitudes, it is shown that the capacity scales linearly with the minimum between the number of transmit and receive antennas, similar to the case of average power-constrained inputs.

Keywords: MIMO; channel capacity; amplitude constraint; input distrbution; capacity bounds

---

1. Introduction

While the capacity of a multiple-input multiple-output (MIMO) channel with an average power constraint is well understood [1], there is surprisingly little known about the capacity of the more practically relevant case in which the channel inputs are subject to *amplitude constraints*. Shannon was the first who considered a channel that is constrained in its amplitude [2]. In that paper, he derived corresponding upper and lower bounds and showed that in the low-amplitude regime, the capacity behaves as that of channel with an average power constraint. The next major contribution to this problem was a seminal paper of Smith [3] published in 1971. Smith showed that, for the single-input single-output (SISO) Gaussian noise channel with an amplitude-constrained input, the capacity-achieving inputs are discrete with finite support. In [4], this result is extended to peak-power-constrained quadrature Gaussian channels. Using the approach of Shamai [4], it is shown in [5] that the input distribution that achieves the capacity of a MIMO channel with an identity channel matrix and a Euclidian norm constraint on the input vector is discrete. Even though the optimal input distribution is known to be discrete, very little is known about the number or the optimal positions of the corresponding constellation points. To the best of our knowledge, the only case for which the input distribution is precisely known is considered in [6], where it is shown for the Gaussian SISO channel with an amplitude constraint that two point masses are optimal if amplitude values are smaller than 1.665 and three for amplitude values of up to 2.786. Finally, it has been shown very recently that the number of mass points in the support of the capacity-achieving input distribution of a SISO channel is of the order $O(A^2)$ with A the amplitude constraint.

Based on a dual capacity expression, in [7], McKellips derived an upper bound on the capacity of a SISO channel that is subject to an amplitude constraint. The bound is asymptotically tight; that is, for amplitude values that tend to infinity. By cleverly choosing an auxiliary channel output distribution in the dual capacity expression, the authors of [8] sharpened McKellips' upper bound and extended it to parallel MIMO channels with a Euclidian norm constraint on the input. The SISO version of the upper bound in [8] has been further sharpened in [9] by yet another choice of auxiliary output distribution. In [10], asymptotic lower and upper bounds for a 2×2 MIMO channel are presented and the gap between the bounds is specified.

In this work, we make progress on this open problem by deriving several new upper and lower bounds that hold for channels with *arbitrary constraints* on the support of the channel input distribution and then apply them to the practically relevant special case of MIMO channels that are subject to amplitude-constraints.

1.1. Contributions and Paper Organization

The remainder of the paper is organized as follows. The problem is stated in Section 2. In Section 3, we study properties of input distributions that achieve the capacity of input-constrained MIMO channels. The section reviews known results on the structure of optimal input distributions and presents several new results. In particular, Theorem 3 shows that the support of a capacity-achieving input distribution must necessarily be a small set both topologically and measure theoretically. Moreover, Theorem 8 characterizes conditions on the channel input space as well as on the channel matrix such that the support of the optimal input distribution is concentrated on the boundary of the channel input space.

In Section 4, we derive novel upper and lower bounds on the capacity of a MIMO channel that is subject to an arbitrary constraint on the support of the input. In particular, three families of upper bounds are proposed, which are based on: (i) the maximum entropy principle (see the bound in Theorem 9); (ii) the dual capacity characterization (see the bound in Theorem 10); and (iii) a relationship between mutual information and the minimum mean square error that is known as the I-MMSE relationship (see the bound in Theorem 11). On the other hand, Section 4 provides three different lower bounds. The first one is given in Theorem 12 and is based on the entropy power inequality. The second one (see Theorem 13) is based on a generalization of the celebrated Ozarow–Wyner bound [11] to the MIMO case. The third upper bound (see Theorem 14) is based on Jensen's inequality and depends on the characteristic function of the channel input distribution.

In Section 5, we evaluate the performance of our bounds by studying MIMO channels with invertible channel matrices. In particular, Theorem 17 states that our upper and lower bounds are within $n \log_2(\rho)$ bits, where ρ is the packing efficiency and n the number of transmit and receive antennas. For diagonal channel matrices, it is then shown (see Theorem 18) that the Cartesian product of simple pulse-amplitude modulation (PAM) constellations achieves the capacity to within $1.64n$ bits.

Section 6 is devoted to MIMO channels with arbitrary channel matrices. It is shown that, in the regime of high amplitudes, similar to the case of average power-constrained channel inputs, the capacity scales linearly with the minimum of the number of transmit and receive antennas.

In Section 7, our upper and lower bounds are applied to the SISO case, which are then compared with bounds known from the literature. Finally, Section 8 concludes the paper. Note that parts of the results in this paper were also published in [12].

1.2. Notation

Vectors are denoted as bold lowercase letters, random vectors as bold uppercase letters, and matrices as bold uppercase sans serif letters (e.g., $\mathbf{x}$, $\mathbf{X}$, X). For any deterministic vector $\mathbf{x} \in \mathbb{R}^n$, $n \in \mathbb{N}$, we denote the Euclidian norm of $\mathbf{x}$ by $\|\mathbf{x}\|$. For some random $\mathbf{X} \in \mathrm{supp}(\mathbf{X}) \subseteq \mathbb{R}^n$ and any $p > 0$, we define

$$\|\mathbf{X}\|_p := \left(\frac{1}{n} \mathbb{E}[\|\mathbf{X}\|^p] \right)^{\frac{1}{p}}, \tag{1}$$

where $\mathrm{supp}(\mathbf{X})$ denotes the support of $\mathbf{X}$. Note that for $p \geq 1$, the quantity in Equation (1) defines a norm and for $n = 1$ we simply have $\|X\|_p^p = \mathbb{E}[|X|^p]$.

The norm of a matrix $\mathbf{H} \in \mathbb{R}^{n \times n}$ is defined as

$$\|\mathbf{H}\| := \sup_{\mathbf{x}: \mathbf{x} \neq 0} \frac{\|\mathbf{H}\mathbf{x}\|}{\|\mathbf{x}\|},$$

whereas $\mathrm{Tr}(\mathbf{H})$ is denoting its trace. The $n \times n$ identity matrix is represented as I_n.

Let $\mathcal{S}$ be a subset of $\mathbb{R}^n$. Then,

$$\mathrm{Vol}(\mathcal{S}) := \int_{\mathcal{S}} \mathrm{d}\mathbf{x}$$

denotes its volume. Moreover, the boundary of $\mathcal{S}$ is denoted as $\partial \mathcal{S}$.

Let $\mathbb{R}_+ := \{x \in \mathbb{R} : x \geq 0\}$. We define an n-dimensional ball or radius $r \in \mathbb{R}_+$ centered at $\mathbf{x} \in \mathbb{R}^n$ as the set

$$\mathcal{B}_{\mathbf{x}}(r) := \{\mathbf{y} \in \mathbb{R}^n : \|\mathbf{x} - \mathbf{y}\| \leq r\}.$$

Recall that, for any $\mathbf{x} \in \mathbb{R}^n$ and $r \in \mathbb{R}_+$,

$$\mathrm{Vol}(\mathcal{B}_{\mathbf{x}}(r)) = \frac{\pi^{\frac{n}{2}}}{\Gamma(\frac{n}{2} + 1)} r^n,$$

where $\Gamma(z)$ denotes the gamma function.

For any matrix $\mathbf{H} \in \mathbb{R}^{k \times n}$ and some $\mathcal{S} \subset \mathbb{R}^n$, we define

$$\mathbf{H}\mathcal{S} := \{\mathbf{y} \in \mathbb{R}^k : \mathbf{y} = \mathbf{H}\mathbf{x}, \mathbf{x} \in \mathcal{S}\}.$$

Note that for an invertible $\mathbf{H} \in \mathbb{R}^{n \times n}$, we have

$$\mathrm{Vol}(\mathbf{H}\mathcal{S}) = |\det(\mathbf{H})| \mathrm{Vol}(\mathcal{S})$$

with $\det(\mathbf{H})$ the determinant of $\mathbf{H}$. We define the maximum and minimum radius of a set $\mathcal{S} \subset \mathbb{R}^n$ that contains the origin as

$$r_{\max}(\mathcal{S}) := \min\{r \in \mathbb{R}_+ : \mathcal{S} \subset \mathcal{B}_0(r)\},$$
$$r_{\min}(\mathcal{S}) := \max\{r \in \mathbb{R}_+ : \mathcal{B}_0(r) \subseteq \mathcal{S}\}.$$

For a given vector $\mathbf{a} = (a_1, \ldots, a_n) \in \mathbb{R}_+^n$, we define

$$\mathrm{Box}(\mathbf{a}) := \{\mathbf{x} \in \mathbb{R}^n : |x_i| \leq a_i, i = 1, \ldots, n\}$$

and the smallest box containing a given set $\mathcal{S} \subset \mathbb{R}^n$ as

$$\mathrm{Box}(\mathcal{S}) := \inf\{\mathrm{Box}(\mathbf{a}) : \mathcal{S} \subseteq \mathrm{Box}(\mathbf{a})\},$$

respectively.

The entropy of any discrete random object $\mathbf{X}$ is denoted as $H(\mathbf{X})$, whereas $h(\mathbf{X})$ (i.e., the differential entropy) is used whenever $\mathbf{X}$ is continuous. The mutual information between two random objects $\mathbf{X}$ and $\mathbf{Y}$ is denoted as $I(\mathbf{X};\mathbf{Y})$ and $\mathcal{N}(\mathbf{m},\mathbf{C})$ denotes the multivariate normal distribution with mean vector $\mathbf{m}$ and covariance matrix $\mathbf{C}$. Finally, $\log_a^+(x) := \max\{\log_a(x),0\}$, for any base $a > 0$, $Q(x)$, $x \in \mathbb{R}$, denotes the Q-function, and $\delta_{\mathbf{x}}(\mathbf{y})$ the Kronecker delta, which is one for $\mathbf{x} = \mathbf{y}$ and zero otherwise.

2. Problem Statement

Consider a MIMO system with $n_t \in \mathbb{N}$ transmit and $n_r \in \mathbb{N}$ receive antennas. The corresponding n_r-dimensional channel output for a single channel use is of the form

$$\mathbf{Y} = \mathbf{H}\mathbf{X} + \mathbf{Z},$$

for some fixed channel matrix $\mathbf{H} \in \mathbb{R}^{n_r \times n_t}$ (considering a real-valued channel model is without loss of generality). Here and hereafter, we assume $\mathbf{Z} \sim \mathcal{N}(\mathbf{0}, \mathbf{I}_{n_r})$ is independent of the channel input $\mathbf{X} \in \mathbb{R}^{n_t}$ and $\mathbf{H}$ is known to both the transmitter and the receiver.

Now, let $\mathcal{X} \subset \mathbb{R}^{n_t}$ be a *convex and compact* channel input space that contains the origin (i.e., the length-n_t zero vector) and let $F_{\mathbf{X}}$ denote the cumulative distribution function of $\mathbf{X}$. As of the writing of this paper, the capacity

$$C(\mathcal{X},\mathbf{H}) := \max_{F_{\mathbf{X}}:\mathbf{X}\in\mathcal{X}} I(\mathbf{X};\mathbf{Y}) = \max_{F_{\mathbf{X}}:\mathbf{X}\in\mathcal{X}} I(\mathbf{X};\mathbf{H}\mathbf{X} + \mathbf{Z}), \tag{2}$$

of this channel is unknown and we are interested in finding novel lower and upper bounds. Even though most of the results in this paper hold for arbitrary convex and compact $\mathcal{X}$, we are mainly interested in the two important special cases:

(i) per-antenna amplitude constraints, i.e., $\mathcal{X} = \mathrm{Box}(\mathbf{a})$ for some given $\mathbf{a} = (A_1,\ldots,A_{n_t}) \in \mathbb{R}_+^{n_t}$; and

(ii) n_t-dimensional amplitude constraint, i.e., $\mathcal{X} = \mathcal{B}_0(A)$ for some given $A \in \mathbb{R}_+$.

Remark 1. *Note that determining the capacity of a MIMO channel with average per-antenna power constraints is also still an open problem and has been solved for some special cases only [13–17].*

3. Properties of an Optimal Input Distribution

Unlike the special cases of real and complex-valued SISO channels (i.e., $n_t = n_r = 1$), the structure of the capacity-achieving input distribution, denoted as $F_{\mathbf{X}}^\star$, is in general not known. To motivate why in this paper we are seeking for novel upper and lower bounds on the capacity (Equation (2)) instead of trying to solve the optimization problem directly, in this section we first summarize properties optimal input distributions must posses, which demonstrate how complicated the optimization problem actually is. Note that, whereas an optimal input distribution always exists, it does not necessarily need to be unique.

3.1. Necessary and Sufficient Conditions for Optimality

To study properties of an optimal input distribution, we need the notion of a point of increase of probability distribution.

Definition 1. *(Points of Increase of a Distribution) A point $\mathbf{x} \in \mathbb{R}^n$, $n \in \mathbb{N}$, is said to be a point of increase of a given probability distribution $F_{\mathbf{X}}$ if for any open set $\mathcal{A} \subset \mathbb{R}^n$ containing $\mathbf{x}$, $F_{\mathbf{X}}(\mathcal{A}) > 0$.*

The following result provides necessary and sufficient conditions for the optimality of a channel input distribution.

Theorem 1. *Let $F_{\mathbf{X}}$ be some given channel input distribution and let $\mathcal{E}(F_{\mathbf{X}}) \subset \mathcal{X}$ denote the set of points of increase of $F_{\mathbf{X}}$. Then, the following holds:*

- *$F_{\mathbf{X}}$ is capacity-achieving if and only if the Karush–Kuhn–Tucker (KKT) conditions*

$$h(\mathbf{x}; F_{\mathbf{X}}) \le h(\mathbf{HX} + \mathbf{Z}), \ \mathbf{x} \in \mathcal{X}, \tag{3a}$$

$$h(\mathbf{x}; F_{\mathbf{X}}) = h(\mathbf{HX} + \mathbf{Z}), \ \mathbf{x} \in \mathcal{E}(F_{\mathbf{X}}) \subset \mathcal{X}, \tag{3b}$$

are satisfied [3,18], where

$$h(\mathbf{x}; F_{\mathbf{X}}) := - \int_{\mathbb{R}^{n_r}} \frac{1}{(2\pi)^{\frac{n_r}{2}}} e^{-\frac{\|\mathbf{y} - \mathbf{Hx}\|^2}{2}} \log_2\left(f_{\mathbf{Y}}(\mathbf{y})\right) d\mathbf{y}$$

with $f_{\mathbf{Y}}$ being the probability density of the channel output induced by the channel input $\mathbf{X} \sim F_{\mathbf{X}}$.
- *$F_{\mathbf{X}}$ is unique and symmetric if $\mathbf{H}$ is left invertible [18].*
- *$F_{\mathbf{Y}}$ (i.e., the channel output distribution) is unique [18,19].*

3.2. General Structure of Capacity-Achieving Input Distributions

Theorem 1 can be used to find general properties of the support of a capacity-achieving input distribution, which we will do in this subsection.

Remark 2. *Fully characterizing an input distribution that achieves the capacity of a general MIMO channel with per-antenna or an n_t-dimensional amplitude constraint is still an open problem. To the best of our knowledge, the only general available for showing that discrete channel inputs are optimal was developed by Smith in [3] for the amplitude and variance-constrained Gaussian SISO channel. Since then, it has been useful to also characterize the optimal input distribution of several other SISO channels (see, for instance, [4,20–24]). The method relies on the following series of steps:*

1. *Towards a contradiction, it is assumed that the set of points of increase $\mathcal{E}(F_{\mathbf{X}})$ is infinite.*
2. *The assumption in Step 1 is then used to establish a certain property of the function $h(\mathbf{x}; F_{\mathbf{X}})$ on the input space $\mathcal{X}$. For example, by showing that $h(\mathbf{x}; F_{\mathbf{X}})$ has an analytic continuation to $\mathbb{C}$. Then, by means of the Identity Theorem of complex analysis and the Bolzano–Weierstrass Theorem [25], Smith was able to show that $h(\mathbf{x}; F_{\mathbf{X}})$ must be constant.*
3. *By using either the Fourier or Laplace transform of $h(\mathbf{x}; F_{\mathbf{X}})$ together with the property of $h(\mathbf{x}; F_{\mathbf{X}})$ established in Step 2, a new a property of the channel output distribution $F_{\mathbf{Y}}$ is established. For example, Smith was able to show that $F_{\mathbf{Y}}$ must be constant.*
4. *A conclusion out of Step 3 is used to reach a contradiction. The contradiction implies that $\mathcal{E}(F_{\mathbf{X}})$ must be finite. For example, to reach a contradiction, Smith was using the fact that the channel output distribution $F_{\mathbf{Y}}$ results from a convolution with a Gaussian probability density, which cannot be constant.*

Remark 3. *Under the restriction that the output space, $\mathcal{Y}$, of a Gaussian SISO channel is finite and the channel input space, $\mathcal{X}$, is subject to an amplitude constraint, Witsenhausen has shown in [26] that the capacity-achieving input distribution is discrete with the number of mass points bounded as $|\mathcal{X}| \le |\mathcal{Y}|$. The approach of Witsenhausen, however, does not use the variational technique of Smith and relies on arguments from convex analysis instead.*

According to Remark 2, assuming in the MIMO case that $\mathcal{E}(F_{\mathbf{X}})$ is of infinite cardinality does not help (or at least it is not clear how this assumption should be used) in showing that the capacity-achieving input distribution is discrete and finite. However, by using the weaker assumption that $\mathcal{E}(F_{\mathbf{X}})$ contains a non-empty open subset in conjunction with the following version of the Identity Theorem, we can show that the support of the optimal input distribution is a small set in a certain topological sense.

Theorem 2. (Identity Theorem for Real-Analytic Functions [27]) *For some $n \in \mathbb{N}$ let $\mathcal{U}$ be a subset of $\mathbb{R}^n$ and $f, g : \mathcal{U} \to \mathbb{R}$ be two real-analytic functions that agree on a set $\mathcal{A} \subseteq \mathcal{U}$. Then, f and g agree on $\mathbb{R}^n$ if one of the following two conditions is satisfied:*

(i) $\mathcal{A}$ is an open set.

(ii) $\mathcal{A}$ is a set of positive Lebesgue measure.

Furthermore, for $n = 1$, it suffices for $\mathcal{A}$ to be an arbitrary set with an accumulation point.

We also need the definitions of a dense and a nowhere dense set.

Definition 2. *(Dense and Nowhere Dense Sets) A subset $\mathcal{A} \subset \mathcal{X}$ is said to be* dense *in the set $\mathcal{X}$ if every element $\mathbf{x} \in \mathcal{X}$ either belongs to $\mathcal{A}$ or is an accumulation point of $\mathcal{A}$. A subset $\mathcal{A} \subset \mathcal{X}$ is said to be* nowhere dense *if for every nonempty open subset $\mathcal{U} \subset \mathcal{X}$, the intersection $\mathcal{A} \cap \mathcal{U}$ is not dense in $\mathcal{U}$.*

With Theorem 2 at our disposal, we are now able to prove the following result on the structure of the support of the optimal input distribution.

Theorem 3. *The set of points of increase $\mathcal{E}(F_{\mathbf{X}}^{\star})$ of an optimal input distribution $F_{\mathbf{X}}^{\star}$ is a nowhere dense subset of $\mathcal{X}$ that is of Lebesgue measure zero.*

Proof. It is not difficult to show that $h(\mathbf{x}; F_{\mathbf{X}}^{\star})$ is a real-analytic function on $\mathbb{R}^{n_t}$ ([18] Proposition 5). Now, in order to prove the result, we follow a series of steps similar to those outlined in Remark 2. Towards a contradiction, assume that the set of points of increase $\mathcal{E}(F_{\mathbf{X}}^{\star})$ of $F_{\mathbf{X}}^{\star}$ is *not* a nowhere dense subset of $\mathcal{X}$. Then, according to Definition 2, there exists an open set $\mathcal{U} \subset \mathcal{X}$ such that $\mathcal{E}(F_{\mathbf{X}}^{\star}) \cap \mathcal{U}$ is dense in $\mathcal{U}$.

By using the KKT condition in Equation (3b), we have that $h(\mathbf{x}; F_{\mathbf{X}}^{\star})$ is constant on the intersection $\mathcal{E}(F_{\mathbf{X}}^{\star}) \cap \mathcal{U}$. Thus, as $\mathcal{E}(F_{\mathbf{X}}^{\star}) \cap \mathcal{U}$ is dense in $\mathcal{U}$, it follows by the properties of continuous functions (real-analytic functions are continuous) that $h(\mathbf{x}; F_{\mathbf{X}}^{\star})$ is also constant on $\mathcal{U}$. Moreover, as $\mathcal{U}$ is an open set, Theorem 2 implies that $h(\mathbf{x}; F_{\mathbf{X}}^{\star})$ must also be constant on $\mathbb{R}^{n_t}$. This, however, leads to a contradiction as $h(\mathbf{x}; F_{\mathbf{X}}^{\star})$ cannot be constant on all of $\mathbb{R}^{n_t}$, which can be shown by taking the Fourier transform of $h(\mathbf{x}; F_{\mathbf{X}}^{\star})$ and solving for the probability density $f_{\mathbf{Y}}(\mathbf{y})$ of the channel output (the reader is referred to [3] for details). Therefore, we conclude that $\mathcal{E}(F_{\mathbf{X}}^{\star})$ is a nowhere dense subset of $\mathcal{X}$.

Showing that $\mathcal{E}(F_{\mathbf{X}}^{\star})$ has Lebesgue measure zero follows along similar lines by assuming that $\mathcal{E}(F_{\mathbf{X}}^{\star})$ is a set of positive measure. Then, Property (ii) of Theorem 2 can be used to conclude that $h(\mathbf{x}; F_{\mathbf{X}}^{\star})$ must be zero on all of $\mathcal{U}$. This again leads to a contradiction, which implies that $\mathcal{E}(F_{\mathbf{X}}^{\star})$ must be of Lebesgue measure zero. $\square$

Remark 4. *Note that if $\mathcal{X} = \mathcal{B}_0(A)$ for some $A \in \mathbb{R}_+$ and $h(\mathbf{x}; F_{\mathbf{X}}^{\star})$ is orthogonally equivariant (i.e., it only depends on $\|\mathbf{x}\|$), then $\mathcal{E}(F_{\mathbf{X}}^{\star})$ can be written as a union of concentric spheres. That is,*

$$\mathcal{E}(F_{\mathbf{X}}^{\star}) = \bigcup_{j} \mathcal{C}(A_j) \tag{4}$$

with $\mathcal{C}(A_j) := \{\mathbf{x} \in \mathbb{R}^{n_t} : \|\mathbf{x}\| = A_j\}$ for some $A_j \in \mathbb{R}_+$. To see this, let

$$g(\mathbf{x}) := h(\mathbf{x}; F_{\mathbf{X}}^{\star}) - h(\mathbf{H}\mathbf{X} + \mathbf{Z})$$

and observe that if $\mathbf{x} \in \mathcal{E}(F_{\mathbf{X}}^{\star})$, then $g(\mathbf{x}) = 0$. Combining this with the symmetry of the function $\|\mathbf{x}\| \mapsto g(\|\mathbf{x}\|)$, we have that (We know that it is abuse of notation to use the same letter for the functions $\mathbf{x} \mapsto g(\mathbf{x})$ and $\|\mathbf{x}\| \mapsto g(\|\mathbf{x}\|)$ even if they are different. It is an attempt to say in a compact way that g is orthogonally equivariant.)

$$\forall \|\mathbf{x}\| = \|\mathbf{y}\| : \mathbf{x} \in \mathcal{E}(F_{\mathbf{X}}^{\star}) \Rightarrow \mathbf{y} \in \mathcal{E}(F_{\mathbf{X}}^{\star}).$$

Moreover, this implies that

$$\mathcal{E}(F_{\mathbf{X}}^{\star}) = \bigcup_{j \in \mathcal{I}} \mathcal{C}(A_j),$$

where $\mathcal{I}$ is possibly of infinite cardinality. In fact, $\mathcal{I}$ has finite cardinality. To see this, note that, if $g(\mathbf{x})$ is real-analytic, then so is $g(\|\mathbf{x}\|)$. However, as $\|\mathbf{x}\| \mapsto g(\|\mathbf{x}\|)$ is a non-zero real-analytic function on $\mathbb{R}$, it can have at most finitely many zeros on an interval.

As an example consider the special case $n_r = n_t = n$ with $\mathbf{H} = \mathbf{I}_n$. Then, the union in Equation (4) implies that the cardinality of $\mathcal{E}(F_{\mathbf{X}}^{\star})$ is uncountable and that discrete inputs are in general not optimal. Therefore, Theorem 3 can generally not be improved in the sense that for $n > 1$, statements about the cardinality of $\mathcal{E}(F_{\mathbf{X}}^{\star})$ cannot be made. Note, however, that the magnitude of $\mathbf{X}$ is discrete. An example of the corresponding optimal input distribution for the case of $n = 2$ is given in Figure 1.

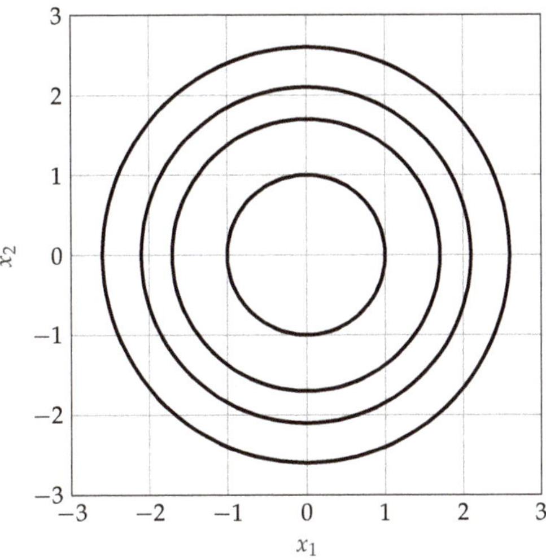

Figure 1. An example of a support of an optimal input distribution for the special case $n_t = n_r = n = 2$.

Even though Theorem 3 does not allow us to conclude that the optimal input distribution of an arbitrary MIMO channel is discrete and finite, for the special case of a SISO channel we have the following partial result.

Theorem 4. (Optimal Input Distribution of a SISO Channel [3,6,28]) *For some fixed $h \in \mathbb{R}$ and $A \in \mathbb{R}_+$, consider the SISO channel $Y = hX + Z$ with input space $\mathcal{X} = [-A, A]$. Let $F_X^{\star}$ be an input distribution that achieves the capacity, $C(\mathcal{X}, h)$, of that channel. Then, $F_X^{\star}$ satisfies the following properties:*

- $F_X^{\star}$ *is unique.*
- $F_X^{\star}$ *is symmetric.*
- $F_X^{\star}$ *is discrete with the number of mass points being of the order $O(A^2)$.*
- $F_X^{\star}$ *contains probability mass points at $\{-A, A\}$.*

Moreover, binary communication with mass points at $\{-A, A\}$ is optimal if and only if $A \leq \bar{A}$, where $\bar{A} \approx 1.665$.

Theorem 4 can now be used to also address the special cases of multiple-input single output (MISO) and single-input multiple output (SIMO) channels.

Theorem 5. *Let $Y = \mathbf{h}^T \mathbf{X}^\star + Z$ be a MISO channel with channel matrix $\mathbf{h}^T \in \mathbb{R}^{n_t}$ and some optimal input $\mathbf{X}^\star \in \mathcal{X} \subset \mathbb{R}^{n_t}$. Then, the distribution of $\mathbf{h}^T \mathbf{X}^*$ is discrete with finitely many mass points. On the other hand, let $\mathbf{Y} = \mathbf{h}X^\star + \mathbf{Z}$ be a SIMO channel with channel matrix $\mathbf{h} \in \mathbb{R}^{n_r}$. Then, the optimal input $X^\star \in \mathcal{X} \subset \mathbb{R}$ has a discrete distribution with finitely many mass points.*

Proof. For the MISO case, the capacity can expressed as

$$
\begin{aligned}
C(\mathcal{X}, \mathbf{h}^T) &= \max_{F_\mathbf{X}: \mathbf{X} \in \mathcal{X}} I(\mathbf{X}; \mathbf{h}^T \mathbf{X} + Z) \\
&= \max_{F_\mathbf{X}: \mathbf{X} \in \mathcal{X}} I(\mathbf{h}^T \mathbf{X}; \mathbf{h}^T \mathbf{X} + Z) \\
&= \max_{F_S: S \in \mathbf{h}^T \mathcal{X}} I(S; S + Z) \\
&= \max_{F_S: |S| \leq r_{\max}(\mathbf{h}^T \mathcal{X})} I(S; S + Z).
\end{aligned}
\tag{5}
$$

Using Theorem 5 we have that the maximizing distribution in Equation (5) $F_S^\star$, where $S := \mathbf{h}^T \mathbf{X}$, is discrete with finitely many mass points.

For the SIMO case, the channel input distribution is discrete as a SIMO channel can be transformed into a SISO channel. Thus, let $A \in \mathbb{R}_+$ be finite. Then, the capacity of the SIMO channel can be expressed as

$$
\begin{aligned}
C(\mathcal{X}, \mathbf{h}) &= \max_{F_X: |X| \leq A} I(X; \mathbf{h}X + \mathbf{Z}) \\
&= \max_{F_X: |X| \leq A} I(X; \|\mathbf{h}\|X + Z).
\end{aligned}
\tag{6}
$$

Again, it follows from Theorem 5 that the mutual information in Equation (6) is maximized by a channel input distribution, $F_X^\star$, that is discrete with finitely many mass points. This concludes the proof. $\square$

Remark 5. *Note that in the MISO case, we do not claim $F_X^\star$ to be discrete with finitely many points. To illustrate the difficulty, let $\mathbf{h}^T = [1, -1]$ so that*

$$
\mathbf{h}^T \mathbf{X}^\star = \mathbf{h}^T \begin{pmatrix} X_1^\star \\ X_2^\star \end{pmatrix} = X_1^\star - X_2^\star.
$$

As $X_1^\star$ and $X_2^\star$ can be arbitrarily correlated, we cannot rule out cases in which $X_1^\star = X - D$ and $X_2^\star = X - 2D$, with D a discrete random variable and X of arbitrary distribution. Clearly the distribution of $X^\star$ is not discrete.

Note that in general it can be shown that the capacity-achieving input distribution is discrete if the optimization problem in Equation (2) can be reformulated as an optimization over one dimensional distributions. This, for example, has been done in [5] for parallel channels with a total amplitude constraint.

3.3. Properties of Capacity-Achieving Input Distributions in the Small (But Not Vanishing) Amplitude Regime

In this subsection, we study properties of capacity-achieving input distribution in the regime of small amplitudes. To that end, we will need the notion of a subharmonic function.

Definition 3. *(Subharmonic Function) Let f be a real-valued function that is twice continuously differentiable on an open set $\mathcal{G} \subset \mathbb{R}^n$. Then, f is subharmonic if $\nabla^2 f \geq 0$ on $\mathcal{G}$, where ∇^2 denotes the Laplacian (if f is twice differentiable, its Laplacian is given by $\nabla^2 f(x_1, \ldots, x_n) = \sum_{i=1}^{n} \frac{\partial^2 f}{\partial x_i^2}$).*

We use the following Theorem, which states that a subharmonic function always attains its maximum on the boundary of its domain.

Theorem 6. (Maximum Principle of Subharmonic Functions [29]) *Let $\mathcal{G} \subset \mathbb{R}^n$ be a connected open set. If $f : \mathcal{G} \to \mathbb{R}$ is subharmonic and attains a global maximum in the interior of $\mathcal{G}$, then f is constant on $\mathcal{G}$.*

In addition to Theorem 6, we need the following result that has been proven in [30].

Lemma 1. *Let the likelihood function of the output of a MIMO channel be defined as*

$$\ell : \mathbb{R}^{n_r} \to \mathbb{R}, \; \ell(\mathbf{y}) = \frac{f_{\mathbf{Y}}(\mathbf{y})}{\frac{1}{(2\pi)^{\frac{n_r}{2}}} e^{-\frac{\|\mathbf{y}\|^2}{2}}}$$

and let $\mathbf{A_y}$ denote the Hessian matrix of $\log_e(\ell(\mathbf{y}))$. Then, the Laplacian (or the trace of $\mathbf{A_y}$) is given by

$$\nabla^2 \log_e(\ell(\mathbf{y})) = \mathrm{Tr}(\mathbf{A_y}) = \mathbb{E}\big[\|\mathbf{HX}\|^2 | \mathbf{Y} = \mathbf{y}\big] - \big\|\mathbb{E}[\mathbf{HX} | \mathbf{Y} = \mathbf{y}]\big\|^2. \tag{7}$$

Theorem 7. *Suppose that $r_{\max}^2(\mathbf{H}\mathcal{X}) \leq \log_2(e)$. Then, $\mathbf{x} \mapsto h(\mathbf{x}; F_{\mathbf{X}})$ is a subharmonic function for every $F_{\mathbf{X}}$.*

Proof. Let $F_{\mathbf{X}}$ be arbitrary and observe that

$$
\begin{aligned}
h(\mathbf{x}; F_{\mathbf{X}}) &= -\mathbb{E}\big[\log_2\big(f_{\mathbf{Y}}(\mathbf{Hx} + \mathbf{Z})\big)\big] \\
&= -\mathbb{E}\big[\ell(\mathbf{Hx} + \mathbf{Z})\big] - \mathbb{E}\left[\log_2\left(\frac{1}{(2\pi)^{\frac{n_r}{2}}} e^{-\frac{\|\mathbf{Hx}+\mathbf{Z}\|^2}{2}}\right)\right] \\
&= -\mathbb{E}\big[\ell(\mathbf{Hx} + \mathbf{Z})\big] + \mathbb{E}\left[\frac{\|\mathbf{Hx} + \mathbf{Z}\|^2}{2}\right]\log_2(e) + \log_2\left((2\pi)^{\frac{n_r}{2}}\right) \\
&= -\mathbb{E}\big[\ell(\mathbf{Hx} + \mathbf{Z})\big] + \frac{\|\mathbf{Hx}\|^2 + n_r}{2}\log_2(e) + \log_2\left((2\pi)^{\frac{n_r}{2}}\right).
\end{aligned}
$$

With this expression in hand, the Laplacian of $h(\mathbf{x}; F_{\mathbf{X}})$ with respect to $\mathbf{x}$ can be bounded from below as follows:

$$
\begin{aligned}
\nabla^2 h(\mathbf{x}; F_{\mathbf{X}}) &= \nabla^2\left(-\mathbb{E}\big[\ell(\mathbf{Hx} + \mathbf{Z})\big] + \frac{\|\mathbf{Hx}\|^2 + n_r}{2}\log_2(e) + \log_2\left((2\pi)^{\frac{n_r}{2}}\right)\right) \\
&= -\mathbb{E}\big[\nabla^2 \ell(\mathbf{Hx} + \mathbf{Z})\big] + \nabla^2 \frac{\|\mathbf{Hx}\|^2}{2}\log_2(e) \\
&\overset{(a)}{=} -\mathbb{E}\big[\mathrm{Tr}(\mathbf{HH}^T \mathbf{A_{Hx+Z}})\big] + \nabla^2 \frac{\|\mathbf{Hx}\|^2}{2}\log_2(e) \\
&= -\mathbb{E}\big[\mathrm{Tr}(\mathbf{HH}^T \mathbf{A_{Hx+Z}})\big] + \mathrm{Tr}(\mathbf{HH}^T)\log_2(e) \\
&\overset{(b)}{\geq} -\mathbb{E}\big[\mathrm{Tr}(\mathbf{HH}^T)\mathrm{Tr}(\mathbf{A_{Hx+Z}})\big] + \mathrm{Tr}(\mathbf{HH}^T)\log_2(e) \\
&= \mathrm{Tr}(\mathbf{HH}^T)\big(-\mathbb{E}\big[\mathrm{Tr}(\mathbf{A_{Hx+Z}})\big] + \log(e)\big) \\
&\overset{(c)}{\geq} \mathrm{Tr}(\mathbf{HH}^T)\big(-r_{\max}^2(\mathbf{H}\mathcal{X}) + \log(e)\big), \tag{8}
\end{aligned}
$$

where (a) follows from Equation (7) and the chain rule for the Hessian; (b) from using the well-known inequality

$$\mathrm{Tr}(\mathbf{CD})^2 \leq \mathrm{Tr}(\mathbf{C})^2 \mathrm{Tr}(\mathbf{D})^2$$

that holds for **C** and **D** positive semi-definite; and (c) from using the inequality

$$\mathrm{Tr}(\mathbf{A_y}) = \mathbb{E}\big[\|\mathbf{HX}\|^2|\mathbf{Y} = \mathbf{y}\big] - \big\|\mathbb{E}[\mathbf{HX}|\mathbf{Y} = \mathbf{y}]\big\|^2$$
$$\leq \mathbb{E}\big[\|\mathbf{HX}\|^2|\mathbf{Y} = \mathbf{y}\big]$$
$$\leq r_{\mathsf{max}}^2(\mathbf{H}\mathcal{X}).$$

Thus, according to the assumption that $r_{\mathsf{max}}^2(\mathbf{H}\mathcal{X}) \leq \log_2(\mathrm{e})$, the right-hand side of Equation (8) is nonnegative, which proves the result. $\square$

Now, knowing that $h(\mathbf{x}; F_{\mathbf{X}})$ is a subharmonic function allows us to characterize the support of an optimal input distribution of a MIMO channel provided that the radius of the channel input space, $\mathcal{X}$, is sufficiently small.

Theorem 8. *Let $F_{\mathbf{X}}^{\star}$ be a capacity-achieving input distribution and $r_{\mathsf{max}}^2(\mathbf{H}\mathcal{X}) \leq \log_2(\mathrm{e})$. Then, $\mathcal{E}(F_{\mathbf{X}}^{\star}) \subseteq \partial\mathcal{X}$.*

Proof. From the KKT conditions in Equation (3), we know that, if $\mathbf{x} \in \mathcal{E}(F_{\mathbf{X}}^{\star})$, then $\mathbf{x}$ is a maximizer of $h(\mathbf{x}; F_{\mathbf{X}}^{\star})$. According to Theorem 7, we also know that $h(\mathbf{x}; F_{\mathbf{X}}^{\star})$ is subharmonic. Hence, from the Maximum Principle of Subharmonic Functions (i.e., Theorem 6), it follows $\mathcal{E}(F_{\mathbf{X}}^{\star}) \subseteq \partial\mathcal{X}$. $\square$

Combining Theorem 8 with the observations made in Remark 4 leads to the following corollary.

Corollary 1. *Let $\mathcal{X} = \mathcal{B}_0(A)$ and $A \leq \frac{1}{\|\mathbf{H}\|}\log_2(\mathrm{e})$. Then,*

$$\mathcal{E}(F_{\mathbf{X}}^{\star}) \subseteq \mathcal{C}(A),$$

where $\mathcal{C}(A) := \{\mathbf{x} \in \mathbb{R}^n : \|\mathbf{x}\| = A\}$ denotes a sphere of radius A.

We conclude this section by noting that for the special case $n_t = n_r = n$ with $\mathbf{H} = \mathbf{I}_n$, the exact value of A such that $\mathcal{E}(F_{\mathbf{X}}^{\star}) = \mathcal{C}(A)$ has been characterized in terms of an integral equation in [31], which is approximately equal to $1.5\sqrt{n}$.

4. Upper and Lower Bounds on the Capacity

The considerations in the previous section have shown that characterizing the structure of an optimal channel-input distribution is a challenging question in itself that we could only partially answer. A full characterization, however, is a necessary prerequisite to narrow down the search space in Equation (2) to one that is tractable. Except for some special cases (i.e., special choices of $\mathcal{X}$), optimizing over the most general space of input distributions that consists of *all* continuous n_t-dimensional probability distributions $F_{\mathbf{X}}$ with $\mathbf{X} \in \mathcal{X}$, is prohibitive (Note that Dytso et al. [32] summarized methods of how to optimize functionals over the space of probability distributions that are constrained in there support). Thus, up to the writing of this paper, there is little hope in being able to solve the problem in Equation (2) in full generality so that in this section we are proposing novel lower and upper bounds on the capacity $C(\mathcal{X}, \mathbf{H})$. Nevertheless, these bounds will allow us to better understand how the capacity of such MIMO channels behaves.

Towards this end, in Section 4.1, we provide four upper bounds. The first is based on an upper bound on the differential entropy of a random vector that is constraint in its pth moment, the second and third bounds are based on duality arguments, and the fourth on the relationship between mutual information and the minimum mean square error (MMSE), I-MMSE relationship for short, known from [33]. The three lower bounds proposed in Section 4.2, on the other hand, are based on the celebrated entropy power inequality, a generalization of the Ozarow–Wyner capacity bound taken from [11], and on Jensen's inequality.

4.1. Upper Bounds

To establish our first upper bound on Equation (2), we need the following result ([11] Th. 1).

Lemma 2. (Maximum Entropy Under *p*th Moment Constraint) *Let $n \in \mathbb{N}$ and $p \in (0, \infty)$ be arbitrary. Then, for any $\mathbf{U} \in \mathbb{R}^n$ such that $h(\mathbf{U}) < \infty$ and $\|\mathbf{U}\|_p < \infty$, we have*

$$h(\mathbf{U}) \leq n \log_2 \left(k_{n,p} \, n^{\frac{1}{p}} \, \|\mathbf{U}\|_p \right),$$

where

$$k_{n,p} := \frac{\sqrt{\pi} \, \mathrm{e}^{\frac{1}{p}} \left(\frac{p}{n}\right)^{\frac{1}{p}} \Gamma\left(\frac{n}{p}+1\right)^{\frac{1}{n}}}{\Gamma\left(\frac{n}{2}+1\right)^{\frac{1}{n}}}.$$

Theorem 9. (Moment Upper Bound) *For any channel input space $\mathcal{X}$ and any fixed channel matrix $\mathbf{H}$, we have*

$$C(\mathcal{X}, \mathbf{H}) \leq \bar{C}_{\mathsf{M}}(\mathcal{X}, \mathbf{H}) := \inf_{p>0} n_r \log_2 \left(\frac{k_{n_r,p}}{(2\pi e)^{\frac{1}{2}}} \, n_r^{\frac{1}{p}} \, \|\tilde{\mathbf{x}} + \mathbf{Z}\|_p \right),$$

where $\tilde{\mathbf{x}} \in \mathbf{H}\mathcal{X}$ is chosen such that $\|\tilde{\mathbf{x}}\| = r_{\mathsf{max}}(\mathbf{H}\mathcal{X})$.

Proof. Expressing Equation (2) in terms of differential entropies results in

$$
\begin{aligned}
C(\mathcal{X}, \mathbf{H}) &= \max_{F_{\mathbf{X}}: \mathbf{X} \in \mathcal{X}} h(\mathbf{HX} + \mathbf{Z}) - h(\mathbf{Z}) \\
&\overset{(a)}{\leq} \max_{F_{\mathbf{X}}: \mathbf{X} \in \mathcal{X}} n_r \log_2 \left(\frac{k_{n_r,p}}{(2\pi e)^{\frac{1}{2}}} \, n_r^{\frac{1}{p}} \, \|\mathbf{HX} + \mathbf{Z}\|_p \right) \\
&\overset{(b)}{=} n_r \log_2 \left(\frac{k_{n_r,p}}{(2\pi e)^{\frac{1}{2}}} \, n_r^{\frac{1}{p}} \max_{F_{\mathbf{X}}: \mathbf{X} \in \mathcal{X}} \|\mathbf{HX} + \mathbf{Z}\|_p \right),
\end{aligned}
\tag{9}
$$

where (a) follows from Lemma 2 with the fact that $h(\mathbf{Z}) = \frac{n_r}{2} \log_2(2\pi e)$; and (b) from the monotonicity of the logarithm.

Now, notice that $\|\mathbf{HX} + \mathbf{Z}\|_p$ is linear in $F_{\mathbf{X}}$ and therefore it attains its maximum at an extreme point of the set $\mathcal{F}_{\mathbf{X}} := \{F_{\mathbf{X}} : \mathbf{X} \in \mathcal{X}\}$ (i.e., the set of all cumulative distribution functions of $\mathbf{X}$). As a matter of fact [26], the extreme points of $\mathcal{F}_{\mathbf{X}}$ are given by the set of degenerate distributions on $\mathcal{X}$; that is, $\{F_{\mathbf{X}}(\mathbf{y}) = \delta_{\mathbf{x}}(\mathbf{y}), \mathbf{y} \in \mathcal{X}\}_{\mathbf{x} \in \mathcal{X}}$. This allows us to conclude

$$\max_{F_{\mathbf{X}}: \mathbf{X} \in \mathcal{X}} \|\mathbf{HX} + \mathbf{Z}\|_p = \max_{\mathbf{x} \in \mathcal{X}} \|\mathbf{Hx} + \mathbf{Z}\|_p.$$

Observe that the Euclidian norm is a convex function, which is therefore maximized at the boundary of the set $\mathbf{H}\mathcal{X}$. Combining this with Equation (9) and taking the infimum over $p > 0$ completes the proof. $\square$

The following Theorem provides two alternative upper bounds that are based on duality arguments.

Theorem 10. (Duality Upper Bounds) *For any channel input space $\mathcal{X}$ and any fixed channel matrix $\mathbf{H}$*

$$C(\mathcal{X}, \mathbf{H}) \leq \bar{C}_{\mathsf{Dual},1}(\mathcal{X}, \mathbf{H}) := \log_2 \left(c_{n_r}(d) + \frac{\mathrm{Vol}(\mathcal{B}_0(d))}{(2\pi e)^{\frac{n_r}{2}}} \right),
\tag{10}$$

where

$$d := r_{\max}(\mathbf{H}\mathcal{X}), \; c_{n_r}(d) := \sum_{i=1}^{n_r-1} \binom{n_r-1}{i} \frac{\Gamma\left(\frac{n_r-1}{2}\right)}{2^{\frac{n_r}{2}} \Gamma\left(\frac{n_r}{2}\right)} d^i,$$

and

$$C(\mathcal{X}, \mathbf{H}) \le \bar{C}_{\mathsf{Dual},2}(\mathcal{X}, \mathbf{H}) := \sum_{i=1}^{n_r} \log_2\left(1 + \frac{2A_i}{\sqrt{2\pi e}}\right), \tag{11}$$

where $\mathbf{a} = (A_1, \ldots, A_{n_r})$ *such that* $\mathrm{Box}(\mathbf{a}) = \mathrm{Box}(\mathbf{H}\mathcal{X})$.

Proof. Using duality bounds, it has been shown in [8] that for any centered n-dimensional ball of radius $r \in \mathbb{R}_+$

$$\max_{F_\mathbf{X}: \mathbf{X} \in \mathcal{B}_0(r)} I(\mathbf{X}; \mathbf{X} + \mathbf{Z}) \le \log_2\left(c_n(r) + \frac{\mathrm{Vol}(\mathcal{B}_0(r))}{(2\pi e)^{\frac{n}{2}}}\right), \tag{12}$$

where $c_n(r) := \sum_{i=1}^{n-1} \binom{n-1}{i} \frac{\Gamma\left(\frac{n-1}{2}\right)}{2^{\frac{n}{2}} \Gamma\left(\frac{n}{2}\right)} r^i$.

Now, observe that

$$\begin{aligned}
C(\mathcal{X}, \mathbf{H}) &= \max_{F_\mathbf{X}: \mathbf{X} \in \mathcal{X}} h(\mathbf{H}\mathbf{X} + \mathbf{Z}) - h(\mathbf{H}\mathbf{X} + \mathbf{Z} | \mathbf{H}\mathbf{X}) \\
&= \max_{F_\mathbf{X}: \mathbf{X} \in \mathcal{X}} I(\mathbf{H}\mathbf{X}; \mathbf{H}\mathbf{X} + \mathbf{Z}) \\
&= \max_{F_{\tilde{\mathbf{X}}}: \tilde{\mathbf{X}} \in \mathbf{H}\mathcal{X}} I(\tilde{\mathbf{X}}; \tilde{\mathbf{X}} + \mathbf{Z}) \\
&\overset{(a)}{\le} \max_{F_{\tilde{\mathbf{X}}}: \tilde{\mathbf{X}} \in \mathcal{B}_0(d), d := r_{\max}(\mathbf{H}\mathcal{X})} I(\tilde{\mathbf{X}}; \tilde{\mathbf{X}} + \mathbf{Z}) \\
&\overset{(b)}{\le} \log_2\left(c_{n_r}(d) + \frac{\mathrm{Vol}(\mathcal{B}_0(d))}{(2\pi e)^{\frac{n_r}{2}}}\right).
\end{aligned} \tag{13}$$

where (a) follows from enlarging the optimization domain; and (b) from using the upper bound in Equation (12). This proves Equation (10).

To show the upper bound in Equation (11), we proceed with an alternative upper bound to Equation (13):

$$\begin{aligned}
C(\mathcal{X}, \mathbf{H}) &= \max_{F_{\tilde{\mathbf{X}}}: \tilde{\mathbf{X}} \in \mathbf{H}\mathcal{X}} I(\tilde{\mathbf{X}}; \tilde{\mathbf{X}} + \mathbf{Z}) \\
&\overset{(a)}{\le} \max_{F_{\tilde{\mathbf{X}}}: \tilde{\mathbf{X}} \in \mathrm{Box}(\mathbf{H}\mathcal{X})} I(\tilde{\mathbf{X}}; \tilde{\mathbf{X}} + \mathbf{Z}) \\
&\overset{(b)}{\le} \max_{F_{\tilde{\mathbf{X}}}: \tilde{\mathbf{X}} \in \mathrm{Box}(\mathbf{H}\mathcal{X})} \sum_{i=1}^{n_r} I(\tilde{X}_i; \tilde{X}_i + Z_i) \\
&\overset{(c)}{=} \sum_{i=1}^{n_r} \max_{F_{\tilde{X}_i}: |\tilde{X}_i| \le A_i} I(\tilde{X}_i; \tilde{X}_i + Z_i) \\
&\overset{(d)}{\le} \sum_{i=1}^{n_r} \log_2\left(1 + \frac{2A_i}{\sqrt{2\pi e}}\right),
\end{aligned}$$

where the (in)equalities follow from: (a) enlarging the optimization domain; (b) single-letterizing the mutual information; (c) choosing individual amplitude constraints $(A_1, \ldots, A_{n_r}) =: \mathbf{a} \in \mathbb{R}_+^{n_r}$ such that $\mathrm{Box}(\mathbf{a}) = \mathrm{Box}(\mathbf{H}\mathcal{X})$; and (d) using the upper bound in Equation (12) for $n = 1$. This concludes the proof. $\square$

As mentioned at the beginning of the section, another simple technique for deriving upper bounds on the capacity is to use the I-MMSE relationship [33]

$$I(\mathbf{X}; \mathbf{X} + \mathbf{Z}) = \frac{\log_2(e)}{2} \int_0^1 \mathbb{E}\left[\|\mathbf{X} - \mathbb{E}[\mathbf{X}\,|\,\sqrt{\gamma}\mathbf{X} + \mathbf{Z}]\|^2\right] d\gamma. \tag{14}$$

For any $\gamma \geq 0$, the quantity $\mathbb{E}\left[\|\mathbf{X} - \mathbb{E}[\mathbf{X}\,|\,\sqrt{\gamma}\mathbf{X} + \mathbf{Z}]\|^2\right]$ is known as the MMSE of estimating $\mathbf{X}$ from the noisy observation $\sqrt{\gamma}\mathbf{X} + \mathbf{Z}$. An important fact that will be useful is that the conditional expected value $\mathbb{E}[\mathbf{X}\,|\,\sqrt{\gamma}\mathbf{X} + \mathbf{Z}]$ is the best estimator in the sense that it minimizes the mean square error over all measurable functions $f : \mathbb{R}^{n_r} \to \mathbb{R}^{n_t}$; that is, for any $\mathbf{Y} \in \mathbb{R}^{n_r}$ and $\mathbf{X} \in \mathbb{R}^{n_t}$

$$\mathbb{E}\left[\|\mathbf{X} - \mathbb{E}[\mathbf{X}|\mathbf{Y}]\|^2\right] = \inf_{f \text{ is measurable}} \mathbb{E}\left[\|\mathbf{X} - f(\mathbf{Y})\|^2\right]. \tag{15}$$

Theorem 11. (I-MMSE Upper Bound) *For any channel input space* $\mathcal{X}$ *and any fixed channel matrix* $\mathbf{H}$

$$C(\mathcal{X}, \mathbf{H}) \leq \bar{C}_{\text{I-MMSE}}(\mathcal{X}, \mathbf{H}) = \log_2(e) \begin{cases} \frac{n_r}{2} + \frac{n_r}{2}\log_2\left(\frac{r_{\max}^2(\mathbf{H}\mathcal{X})}{n_r}\right), & r_{\max}^2(\mathbf{H}\mathcal{X}) \geq n_r \\ \frac{r_{\max}^2(\mathbf{H}\mathcal{X})}{2}, & r_{\max}^2(\mathbf{H}\mathcal{X}) \leq n_r \end{cases}.$$

Proof. Fix some $\epsilon \in [0, 1]$ and observe that

$$\frac{2}{\log_2(e)} I(\mathbf{X}; \mathbf{H}\mathbf{X} + \mathbf{Z}) \overset{(a)}{=} \frac{2}{\log_2(e)} I(\mathbf{H}\mathbf{X}; \mathbf{H}\mathbf{X} + \mathbf{Z})$$

$$\overset{(b)}{=} \int_0^1 \mathbb{E}\left[\|\mathbf{H}\mathbf{X} - \mathbb{E}[\mathbf{H}\mathbf{X}|\sqrt{\gamma}\mathbf{H}\mathbf{X} + \mathbf{Z}]\|^2\right] d\gamma$$

$$= \int_0^\epsilon \mathbb{E}\left[\|\mathbf{H}\mathbf{X} - \mathbb{E}[\mathbf{H}\mathbf{X}|\sqrt{\gamma}\mathbf{H}\mathbf{X} + \mathbf{Z}]\|^2\right] d\gamma$$

$$+ \int_\epsilon^1 \mathbb{E}\left[\|\mathbf{H}\mathbf{X} - \mathbb{E}[\mathbf{H}\mathbf{X}|\sqrt{\gamma}\mathbf{H}\mathbf{X} + \mathbf{Z}]\|^2\right] d\gamma$$

$$\overset{(c)}{\leq} \int_0^\epsilon \mathbb{E}\left[\|\mathbf{H}\mathbf{X} - \mathbf{0}\|^2\right] d\gamma$$

$$+ \int_\epsilon^1 \mathbb{E}\left[\left\|\mathbf{H}\mathbf{X} - \frac{1}{\sqrt{\gamma}}(\sqrt{\gamma}\mathbf{H}\mathbf{X} + \mathbf{Z})\right\|^2\right] d\gamma$$

$$= \epsilon\, \mathbb{E}\left[\|\mathbf{H}\mathbf{X}\|^2\right] + \int_\epsilon^1 \frac{1}{\gamma}\mathbb{E}\left[\|\mathbf{Z}\|^2\right] d\gamma$$

$$= \epsilon\, \mathbb{E}\left[\|\mathbf{H}\mathbf{X}\|^2\right] + \mathbb{E}\left[\|\mathbf{Z}\|^2\right] \log_e\left(\frac{1}{\epsilon}\right)$$

$$= \epsilon\, \mathbb{E}\left[\|\mathbf{H}\mathbf{X}\|^2\right] + n_r \log_e\left(\frac{1}{\epsilon}\right),$$

where the (in)equalities follow from: (a) using that $I(\mathbf{H}\mathbf{X}; \mathbf{H}\mathbf{X} + \mathbf{Z}) = I(\mathbf{X}; \mathbf{H}\mathbf{X} + \mathbf{Z})$ for any fixed $\mathbf{H}$; (b) using the I-MMSE relationship in Equation (14); and (c) using the property that conditional expectation minimizes mean square error (i.e., (15)).

Now, notice that

$$\max_{F_{\mathbf{X}}:\mathbf{X}\in\mathcal{X}} I(\mathbf{X}; \mathbf{H}\mathbf{X} + \mathbf{Z}) \leq \max_{F_{\mathbf{X}}:\mathbf{X}\in\mathcal{X}} \frac{\log_2(e)}{2}\left(\epsilon\, \mathbb{E}\left[\|\mathbf{H}\mathbf{X}\|^2\right] + n_r \log_e\left(\frac{1}{\epsilon}\right)\right)$$

$$\overset{(a)}{=} \frac{1}{2}\left(\epsilon \max_{\tilde{\mathbf{x}}\in\mathcal{X}} \|\mathbf{H}\tilde{\mathbf{x}}\|^2 + n_r \log_e\left(\frac{1}{\epsilon}\right)\right)$$

$$\overset{(b)}{=} \frac{1}{2}\left(\epsilon\, r_{\max}^2(\mathbf{H}\mathcal{X}) + n_r \log_e\left(\frac{1}{\epsilon}\right)\right), \tag{16}$$

where (a) follows from $\max_{F_X:X\in\mathcal{X}} \mathbb{E}\left[\|\mathbf{H}X\|^2\right] = \max_{\tilde{x}\in\mathcal{X}} \|\mathbf{H}\tilde{x}\|^2$ (the same argument was used in the proof of Theorem 9); and (b) from the definition of $r^2_{\max}(\mathbf{H}\mathcal{X})$.

Since ϵ is arbitrary, we can choose it to minimize the upper bound in Equation (16). Towards this end, we need the following optimization result

$$\min_{\epsilon\in[0,1]}\left(\epsilon\, a + b\log\left(\frac{1}{\epsilon}\right)\right) = \begin{cases} b + b\log(\frac{a}{b}), & a \geq b \\ a, & a \leq b \end{cases}, \tag{17}$$

which is easy to show. Combining Equation (16) with Equation (17), we obtain the following upper bound on the capacity

$$\max_{F_X:X\in\mathcal{X}} I(X; \mathbf{H}X + Z) \leq \begin{cases} \frac{n_r}{2} + \frac{n_r}{2}\log\left(\frac{r^2_{\max}(\mathbf{H}\mathcal{X})}{n_r}\right), & r^2_{\max}(\mathbf{H}\mathcal{X}) \geq n_r \\ r^2_{\max}(\mathbf{H}\mathcal{X}), & r^2_{\max}(\mathbf{H}\mathcal{X}) \leq n_r \end{cases}.$$

This concludes the proof. $\square$

Corollary 2. *For any channel input space $\mathcal{X}$ and any fixed channel matrix $\mathbf{H}$*

$$\bar{C}_{\text{l-MMSE}}(\mathcal{X},\mathbf{H}) \leq \log_2(e) \begin{cases} \frac{n_r}{2} + \frac{n_r}{2}\log_e\left(\frac{\|\mathbf{H}\|^2 r^2_{\max}(\mathcal{X})}{n_r}\right), & \|\mathbf{H}\|^2 r^2_{\max}(\mathcal{X}) \geq n_r \\ \frac{\|\mathbf{H}\|^2 r^2_{\max}(\mathcal{X})}{2}, & \|\mathbf{H}\|^2 r^2_{\max}(\mathcal{X}) \leq n_r \end{cases}.$$

Proof. The corollary follows by upper bounding Equation (16) using the fact that $r^2_{\max}(\mathbf{H}\mathcal{X}) \leq \|\mathbf{H}\| r^2_{\max}(\mathcal{X})$. $\square$

Remark 6. *In the proof of Theorem 11, instead of using sub-optimal estimators $f(\mathbf{Y}) = 0$ and $f(\mathbf{Y}) = \frac{1}{\gamma}\mathbf{Y}$, we could have used an optimal linear estimator of the form $f(\mathbf{Y}) = \mathbf{K}_{XY}\mathbf{K}_Y^{-1}\mathbf{Y}$, where $\mathbf{K}_{XY}$ denotes the cross-covariance matrix between $\mathbf{X}$ and $\mathbf{Y}$ and $\mathbf{K}_Y$ the covariance matrix of $\mathbf{Y}$. This choice would result in the capacity upper bound*

$$C(\mathcal{X},\mathbf{H}) \leq \max_{\mathbf{K}_X:X\in\mathcal{X}} \frac{1}{2}\log_2\left(\det\left(\mathbf{I}_{n_r} + \mathbf{H}\mathbf{K}_X\mathbf{H}^T\right)\right) \tag{18}$$

with $\mathbf{K}_X$ the covariance matrix of the channel input. While Equation (18) is a valid upper bound, as of the writing of this paper, it is not clear how to perform an optimization over covariance matrices of random variables with bounded support. One possibility to avoid this is to use the inequality between arithmetic and geometric mean and bound the determinant by the trace:

$$\det\left(\mathbf{I}_{n_r} + \mathbf{H}\mathbf{K}_X\mathbf{H}^T\right) \leq \left(\frac{\text{Tr}\left(\mathbf{I}_{n_r} + \mathbf{H}\mathbf{K}_X\mathbf{H}^T\right)}{n_r}\right)^{n_r} = \left(\frac{\|\mathbf{Z} + \mathbf{H}X\|_2^2}{n_r}\right)^{n_r}. \tag{19}$$

However, combining Equation (19) with Equation (18) is merely a special case of the moment upper bound of Theorem 9 for $p = 2$. Therefore, the estimators in Theorem 11 are chosen to obtain a non-trivial upper bound avoiding the optimization over covariance matrices.

In Section 5, we present a comparison of the upper bounds of Theorems 9–11 by means of a simple example.

4.2. Lower Bounds

A classical approach to bound a mutual information from below is to use the entropy power inequality (EPI).

Theorem 12. (EPI Lower Bounds) *For any fixed channel matrix* $\mathbf{H}$ *and any channel input space* $\mathcal{X}$ *with* $\mathbf{X}$ *absolutely continuous, we have*

$$C(\mathcal{X}, \mathbf{H}) \geq \underline{C}_{\mathsf{EPI}}(\mathcal{X}, \mathbf{H}) := \max_{F_{\mathbf{X}}: \mathbf{X} \in \mathcal{X}} \frac{n_r}{2} \log_2\left(1 + \frac{2^{\frac{2}{n_r} h(\mathbf{HX})}}{2\pi e}\right). \tag{20}$$

Moreover, if $n_t = n_r = n$, $\mathbf{H} \in \mathbb{R}^{n \times n}$ *invertible, and* $\mathbf{X}$ *uniformly distributed over* $\mathcal{X}$, *then*

$$C(\mathcal{X}, \mathbf{H}) \geq \underline{C}_{\mathsf{EPI}}(\mathcal{X}, \mathbf{H}) := \frac{n}{2} \log_2\left(1 + \frac{|\det(\mathbf{H})|^{\frac{2}{n}} \mathrm{Vol}(\mathcal{X})^{\frac{2}{n}}}{2\pi e}\right). \tag{21}$$

Proof. By means of the EPI

$$2^{\frac{2}{n_r} h(\mathbf{HX}+\mathbf{Z})} \geq 2^{\frac{2}{n_r} h(\mathbf{HX})} + 2^{\frac{2}{n_r} h(\mathbf{Z})},$$

we conclude

$$2^{\frac{2}{n_r} C(\mathcal{X}, \mathbf{H})} \geq 1 + (2\pi e)^{-1} 2^{\frac{2}{n_r} \max_{F_{\mathbf{X}}: \mathbf{X} \in \mathcal{X}} h(\mathbf{HX})},$$

which finalizes the proof of the lower bound in Equation (20).

To show the lower bound in Equation (21), all we need is to recall that

$$h(\mathbf{HX}) = h(\mathbf{X}) + \log_2 |\det(\mathbf{H})|,$$

which is maximized for $\mathbf{X}$ uniformly distributed over $\mathcal{X}$. However, if $\mathbf{X}$ is uniformly drawn from $\mathcal{X}$, we have

$$2^{\frac{2}{n} h(\mathbf{HX})} = \mathrm{Vol}(\mathbf{H}\mathcal{X})^{\frac{2}{n}} = |\det(\mathbf{H})|^{\frac{2}{n}} \mathrm{Vol}(\mathcal{X})^{\frac{2}{n}},$$

which completes the proof. $\square$

The considerations in Section 3 suggest that a channel input distribution that maximizes Equation (2) might be discrete. Therefore, there is a need for lower bounds that unlike the bounds in Theorem 12 rely on discrete inputs.

Theorem 13. (Ozarow–Wyner Type Lower Bound) *Let* $\mathbf{X}_D \in \mathrm{supp}(\mathbf{X}_D) \subset \mathbb{R}^{n_t}$ *be a discrete random vector of finite entropy,* $g : \mathbb{R}^{n_r} \to \mathbb{R}^{n_t}$ *a measurable function, and* $p > 0$. *Furthermore, let* $\mathcal{K}_p$ *be a set of continuous random vectors, independent of* $\mathbf{X}_D$, *such that for every* $\mathbf{U} \in \mathcal{K}_p$ *we have* $h(\mathbf{U}) < \infty$, $\|\mathbf{U}\|_p < \infty$, *and*

$$\mathrm{supp}(\mathbf{U} + \mathbf{x}_i) \cap \mathrm{supp}(\mathbf{U} + \mathbf{x}_j) = \varnothing \tag{22}$$

for all $\mathbf{x}_i, \mathbf{x}_j \in \mathrm{supp}(\mathbf{X}_D)$, $i \neq j$. *Then,*

$$C(\mathcal{X}, \mathbf{H}) \geq \underline{C}_{\mathsf{OW}}(\mathcal{X}, \mathbf{H}) := [H(\mathbf{X}_D) - \mathrm{gap}]^+,$$

where

$$\mathrm{gap} := \inf_{\substack{\mathbf{U} \in \mathcal{K}_p \\ g \text{ measurable} \\ p > 0}} \left(G_{1,p}(\mathbf{U}, \mathbf{X}_D, g) + G_{2,p}(\mathbf{U})\right)$$

with

$$G_{1,p}(\mathbf{U}, \mathbf{X}_D, g) := n_t \log_2\left(\frac{\|\mathbf{U} + \mathbf{X}_D - g(\mathbf{Y})\|_p}{\|\mathbf{U}\|_p}\right), \tag{23}$$

$$G_{2,p}(\mathbf{U}) := n_t \log_2\left(\frac{k_{n_t, p}\, n_t^{\frac{1}{p}}\, \|\mathbf{U}\|_p}{2^{\frac{1}{n_t} h(\mathbf{U})}}\right), \tag{24}$$

and $k_{n_t,p}$ as defined in Lemma 2, respectively.

Proof. The proof is identical to ([11] Theorem 2). To make the manuscript more self-contained, we repeat it here.

Let $\mathbf{U}$ and $\mathbf{X}_D$ be statistically independent. Then, the mutual information $I(\mathbf{X}_D; \mathbf{Y})$ can be lower bounded as

$$
\begin{aligned}
I(\mathbf{X}_D; \mathbf{Y}) &\overset{(a)}{\geq} I(\mathbf{X}_D + \mathbf{U}; \mathbf{Y}) \\
&= h(\mathbf{X}_D + \mathbf{U}) - h(\mathbf{X}_D + \mathbf{U}|\mathbf{Y}) \\
&\overset{(b)}{=} H(\mathbf{X}_D) + h(\mathbf{U}) - h(\mathbf{X}_D + \mathbf{U}|\mathbf{Y}).
\end{aligned}
\tag{25}
$$

Here, (a) follows from the data processing inequality as $\mathbf{X}_D + \mathbf{U} \rightarrow \mathbf{X}_D \rightarrow \mathbf{Y}$ forms a Markov chain in that order; and (b) from the assumption in Equation (22). By using Lemma 2, we have that the last term in Equation (25) can be bounded from above as

$$
h(\mathbf{X}_D + \mathbf{U}|\mathbf{Y}) \leq n_t \log_2\left(k_{n_t,p}\, n_t^{\frac{1}{p}}\, \|\mathbf{X}_D + \mathbf{U} - g(\mathbf{Y})\|_p \right).
$$

Combining this expression with Equation (25) results in

$$
I(\mathbf{X}_D; \mathbf{Y}) \geq H(\mathbf{X}_D) - \left(G_{1,p}(\mathbf{U}, \mathbf{X}_D, g) + G_{2,p}(\mathbf{U}) \right),
$$

with $G_{1,p}$ and $G_{2,p}$ as defined in Equations (23) and (24), respectively. Maximizing the right-hand side over all $\mathbf{U} \in \mathcal{K}_p$, measurable functions $g : \mathbb{R}^{n_r} \rightarrow \mathbb{R}^{n_t}$, and $p > 0$ provides the bound. $\quad\square$

Interestingly, the bound of Theorem 13 holds for arbitrary channels and is therefore not restricted to MIMO channels. The interested reader is referred to [11] for details.

We conclude the section by providing a lower bound that is based on Jensen's inequality and holds for arbitrary inputs.

Theorem 14. (Jensen's Inequality Lower Bound) *For any channel input space $\mathcal{X}$ and any fixed channel matrix $\mathbf{H}$, we have*

$$
C(\mathcal{X}, \mathbf{H}) \geq C_{\mathrm{Jensen}}(\mathcal{X}, \mathbf{H}) := \max_{F_{\mathbf{X}}: \mathbf{X} \in \mathcal{X}} \log_2^+\left(\left(\frac{2}{e}\right)^{\frac{n_r}{2}} \psi^{-1}(\mathbf{X}, \mathbf{H}) \right)
\tag{26}
$$

with

$$
\psi(\mathbf{X}, \mathbf{H}) := \mathbb{E}\left[\exp\left(-\frac{\|\mathbf{H}(\mathbf{X} - \mathbf{X}')\|^2}{4} \right) \right] = \mathbb{E}\left[\left| \phi_{\mathbf{X}}\left(\frac{\mathbf{H}^T \mathbf{Z}}{\sqrt{2}} \right) \right|^2 \right],
$$

where $\mathbf{X}'$ is an independent copy of $\mathbf{X}$ and $\phi_{\mathbf{X}}$ denotes the characteristic function of $\mathbf{X}$.

Proof. To show the lower bound, we follow an approach of Dytso et al. [34]. Note that by Jensen's inequality

$$
h(\mathbf{Y}) = -\mathbb{E}[\log_2 f_{\mathbf{Y}}(\mathbf{Y})] \geq -\log_2 \mathbb{E}[f_{\mathbf{Y}}(\mathbf{Y})] = -\log_2 \int_{\mathbb{R}^{n_r}} f_{\mathbf{Y}}(\mathbf{y}) f_{\mathbf{Y}}(\mathbf{y})\, \mathrm{d}\mathbf{y}.
\tag{27}
$$

Now, evaluating the integral in Equation (27) results in

$$
\int_{\mathbb{R}^{n_r}} f_{\mathbf{Y}}(\mathbf{y}) f_{\mathbf{Y}}(\mathbf{y})\, \mathrm{d}\mathbf{y} = \frac{1}{(2\pi)^{n_r}} \int_{\mathbb{R}^{n_r}} \mathbb{E}\left[e^{-\frac{\|\mathbf{y} - \mathbf{H}\mathbf{X}\|^2}{2}} \right] \mathbb{E}\left[e^{-\frac{\|\mathbf{y} - \mathbf{H}\mathbf{X}'\|^2}{2}} \right] \mathrm{d}\mathbf{y}
$$

$$
\stackrel{(a)}{=} \frac{1}{(2\pi)^{n_r}} \mathbb{E}\left[\int_{\mathbb{R}^{n_r}} e^{-\frac{\|\mathbf{y}-\mathbf{HX}\|^2+\|\mathbf{y}-\mathbf{HX}'\|^2}{2}} d\mathbf{y}\right]
$$

$$
\stackrel{(b)}{=} \frac{1}{(2\pi)^{n_r}} \mathbb{E}\left[e^{-\frac{\|\mathbf{HX}-\mathbf{HX}'\|^2}{4}} \int_{\mathbb{R}^{n_r}} e^{-\|\mathbf{y}-\frac{\mathbf{H(X-X')}}{2}\|^2} d\mathbf{y}\right]
$$

$$
\stackrel{(c)}{=} \frac{1}{2^{n_r}\pi^{\frac{n_r}{2}}} \mathbb{E}\left[e^{-\frac{\|\mathbf{H(X-X')}\|^2}{4}}\right], \tag{28}
$$

where (a) follows from the independence of $\mathbf{X}$ and $\mathbf{X}'$ and Tonelli's Theorem ([35] Chapter 5.9); (b) from completing a square; and (c) from the fact that $\int_{\mathbb{R}^{n_r}} e^{-\|\mathbf{y}-\frac{\mathbf{H(X-X')}}{2}\|^2} d\mathbf{y} = \int_{\mathbb{R}^{n_r}} e^{-\|\mathbf{y}\|^2} d\mathbf{y} = \pi^{\frac{n_r}{2}}$. Combining Equation (27) with Equation (28) and subtracting $h(\mathbf{Z}) = \frac{n_r}{2}\log_2(2\pi e)$ completes the proof of the first version of the bound.

To show the second version, observe that

$$
\mathbb{E}\left[e^{-\frac{\|\mathbf{H(X-X')}\|^2}{4}}\right] \stackrel{(d)}{=} \mathbb{E}\left[\phi_{\frac{\mathbf{H(X-X')}}{\sqrt{2}}}(\mathbf{Z})\right]
$$

$$
\stackrel{(e)}{=} \mathbb{E}\left[\phi_{\frac{\mathbf{HX}}{\sqrt{2}}}(\mathbf{Z})\phi_{-\frac{\mathbf{HX}'}{\sqrt{2}}}(\mathbf{Z})\right]
$$

$$
\stackrel{(f)}{=} \mathbb{E}\left[\phi_{\mathbf{X}}\left(\frac{\mathbf{H}^T\mathbf{Z}}{\sqrt{2}}\right)\phi_{\mathbf{X}'}\left(-\frac{\mathbf{H}^T\mathbf{Z}}{\sqrt{2}}\right)\right]
$$

$$
\stackrel{(g)}{=} \mathbb{E}\left[\phi_{\mathbf{X}}\left(\frac{\mathbf{H}^T\mathbf{Z}}{\sqrt{2}}\right)\phi_{\mathbf{X}}\left(-\frac{\mathbf{H}^T\mathbf{Z}}{\sqrt{2}}\right)\right]
$$

$$
\stackrel{(h)}{=} \mathbb{E}\left[\phi_{\mathbf{X}}\left(\frac{\mathbf{H}^T\mathbf{Z}}{\sqrt{2}}\right)\phi_{\mathbf{X}}^*\left(\frac{\mathbf{H}^T\mathbf{Z}}{\sqrt{2}}\right)\right]
$$

$$
= \mathbb{E}\left[\left|\phi_{\mathbf{X}}\left(\frac{\mathbf{H}^T\mathbf{Z}}{\sqrt{2}}\right)\right|^2\right],
$$

where (d) follows from Parseval's identity ([35] Chapter 9.5) by noting that $\exp(-\|\cdot\|^2/2)$ is a characteristic function of $\mathbf{Z}$ and $\phi_{\frac{\mathbf{H(X-X')}}{\sqrt{2}}}(\cdot)$ is a characteristic function of $\frac{\mathbf{H(X-X')}}{\sqrt{2}}$; (e) from using the property that the characteristic function of a sum of random vectors is equal to the product of its characteristic functions; (f) from using the fact that a characteristic function is a linear transformation; (g) from using that $\mathbf{X}$ and $\mathbf{X}'$ have the same characteristic function; and (h) from the fact that the characteristic function is Hermitian. This completes the proof. $\square$

Remark 7. *As is evident from our examples in the following sections, in many cases, the Jensen's inequality lower bound of Theorem 14 performs remarkably well. The bound, however, is also useful for MIMO channels that are subject to an average power constraint. For example, evaluating Equation (26) with $\mathbf{X} \sim \mathcal{N}(\mathbf{0}, \mathbf{K_X})$ results in*

$$
I(\mathbf{X}; \mathbf{HX}+\mathbf{Z}) \geq \frac{1}{2}\log_2^+\left(\left(\frac{2}{e}\right)^{\min(n_r,n_t)} \det\left(\mathbf{I}_{n_r}+\mathbf{HK_X}\mathbf{H}^T\right)\right).
$$

Note that this bound is within $\frac{\min(n_r,n_t)}{2}\log_2\left(\frac{2}{e}\right)$ bits of the capacity of the power-constrained channel.

In Section 3, we discuss that the distributions that maximize mutual information in n_t-dimensions are typically singular, which means that they are concentrated on a set of Lebesgue measure zero. Singular distributions generally do not have a probability density, whereas the characteristic function always exists. This is why the version of Jensen's inequality lower bound in Theorem 14 that is based on the characteristic function of the channel input is especially useful for amplitude-constrained MIMO channels.

5. Invertible Channel Matrices

Consider the symmetric case of $n_t = n_r = n$ antennas with $\mathbf{H} \in \mathbb{R}^{n \times n}$ being invertible. In this section, we evaluate some of the lower and upper bounds proposed in the previous section for the special case of $\mathbf{H}$ being also diagonal and then characterize the gap to the capacity for arbitrary invertible channel matrices.

5.1. Diagonal Channel Matrices

Suppose the channel inputs are subject to per-antenna or an n-dimensional amplitude constraint. Then, the duality upper bound $\bar{C}_{\text{Dual},2}(\mathcal{X}, \mathbf{H})$ of Theorem 10 takes on the following form.

Theorem 15. (Upper Bounds) *Let* $\mathbf{H} = \text{diag}(h_{11}, \ldots, h_{nn}) \in \mathbb{R}^{n \times n}$ *be fixed. If* $\mathcal{X} = \text{Box}(\mathbf{a})$ *for some* $\mathbf{a} = (A_1, \ldots, A_n) \in \mathbb{R}_+^n$, *then*

$$\bar{C}_{\text{Dual},2}(\text{Box}(\mathbf{a}), \mathbf{H}) = \sum_{i=1}^{n} \log_2\left(1 + \frac{2|h_{ii}|A_i}{\sqrt{2\pi e}}\right). \tag{29}$$

Moreover, if $\mathcal{X} = \mathcal{B}_0(A)$ *for some* $A \in \mathbb{R}_+$, *then*

$$\bar{C}_{\text{Dual},2}(\mathcal{B}_0(A), \mathbf{H}) = \sum_{i=1}^{n} \log_2\left(1 + \frac{2|h_{ii}|A}{\sqrt{n}\sqrt{2\pi e}}\right). \tag{30}$$

Proof. The bound in Equation (29) immediately follows from Theorem 10 by observing that $\text{Box}(\mathbf{H}\text{Box}(\mathbf{a})) = \text{Box}(\mathbf{H}\mathbf{a})$. The bound in Equation (30) follows from Theorem 10 by the fact that

$$\text{Box}\big(\mathbf{H}\mathcal{B}_0(A)\big) \subset \text{Box}\big(\mathbf{H}\text{Box}(\mathcal{B}_0(A))\big) = \text{Box}(\mathbf{h}),$$

where $\mathbf{h} := \frac{A}{\sqrt{n}}(|h_{11}|, \ldots, |h_{nn}|)$. This concludes the proof. $\square$

For an arbitrary channel input space $\mathcal{X}$, the EPI lower bound of Theorem 12 and Jensen's inequality lower bound of Theorem 14 take on the following form.

Theorem 16. (Lower Bounds) *Let* $\mathbf{H} = \text{diag}(h_{11}, \ldots, h_{nn}) \in \mathbb{R}^{n \times n}$ *be fixed and* $\mathcal{X}$ *arbitrary. Then,*

$$\underline{C}_{\text{Jensen}}(\mathcal{X}, \mathbf{H}) = \log_2^+\left(\left(\frac{2}{e}\right)^{\frac{n}{2}} \frac{1}{\psi(\mathbf{H}, \mathbf{b}^\star)}\right) \tag{31}$$

with

$$\psi(\mathbf{H}, \mathbf{b}^\star) := \min_{\mathbf{b} \in \mathcal{X}} \prod_{i=1}^{n} \varphi(|h_{ii}|B_i),$$

where $\mathbf{b} := (B_1, \ldots, B_n)$ *and* $\varphi : \mathbb{R}_+ \to \mathbb{R}_+$,

$$\varphi(x) := \frac{1}{x^2}\left(e^{-x^2} - 1 + \sqrt{\pi}x(1 - 2Q(\sqrt{2}x))\right). \tag{32}$$

Moreover,

$$\underline{C}_{\text{EPI}}(\mathcal{X}, \mathbf{H}) = \frac{n}{2} \log_2\left(1 + \text{Vol}(\mathcal{X})^{\frac{2}{n}} \frac{|\prod_{i=1}^{n} h_{ii}|^{\frac{2}{n}}}{2\pi e}\right). \tag{33}$$

Proof. For some given values $B_i \in \mathbb{R}_+$, $i = 1, \ldots, n$, let the ith component of $\mathbf{X} = (X_1, \ldots, X_n)$ be independent and uniformly distributed over the interval $[-B_i, B_i]$. Thus, the expected value appearing in the bound of Theorem 14 can be written as

$$\mathbb{E}\left[e^{-\frac{\|\mathbf{H}(\mathbf{X}-\mathbf{X}')\|^2}{4}}\right] = \mathbb{E}\left[e^{-\frac{\sum_{i=1}^{n} h_{ii}^2 (X_i - X_i')^2}{4}}\right] = \mathbb{E}\left[\prod_{i=1}^{n} e^{-\frac{h_{ii}^2 (X_i - X_i')^2}{4}}\right] = \prod_{i=1}^{n} \mathbb{E}\left[e^{-\frac{h_{ii}^2 (X_i - X_i')^2}{4}}\right]. \tag{34}$$

Now, if $\mathbf{X}'$ is an independent copy of $\mathbf{X}$, it can be shown that the expected value at the right-hand side of Equation (34) is of the explicit form

$$\mathbb{E}\left[e^{-\frac{h_{ii}^2 (x_i - x_i')^2}{4}}\right] = \varphi(|h_{ii}| B_i)$$

with φ as defined in Equation (32). Finally, optimizing over all $\mathbf{b} = (B_1, \ldots, B_n) \in \mathcal{X}$ results in the bound (31). The bound in Equation (33) follows by inserting $|\det(\mathbf{H})| = |\prod_{i=1}^{n} h_{ii}|$ into Equation (21), which concludes the proof. $\square$

In Figure 2, the upper bounds of Theorems 9 and 15 and the lower bounds of Theorem 16 are depicted for a diagonal 2×2 MIMO channel with per-antenna amplitude constraints. It turns out that the moment upper bound and the EPI lower bound perform well in the small amplitude regime while the duality upper bound and Jensen's inequality lower bound perform well in the high amplitude regime. Interestingly, for this specific example, the duality upper bound and Jensen's lower bound are asymptotically tight.

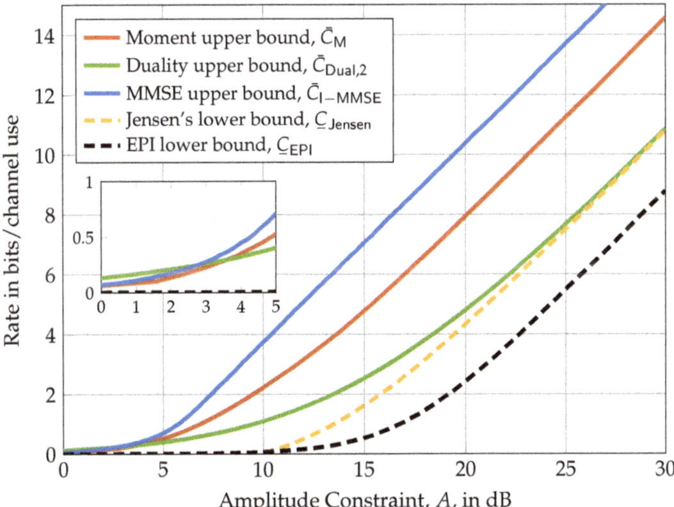

Figure 2. Comparison of the upper and lower bounds of Theorems 9, 11, 15, and 16 evaluated for a 2×2 MIMO system with per-antenna amplitude constraints $A_1 = A_2 = A$ (i.e., $\mathbf{a} = (A, A)$) and channel matrix $\mathbf{H} = \left(\begin{smallmatrix} 0.3 & 0 \\ 0 & 0.1 \end{smallmatrix}\right)$. The nested figure represents a zoom into the region $0\,\text{dB} \leq A \leq 5\,\text{dB}$ to visualize the differences between the bounds at small amplitude constraints.

5.2. Gap to the Capacity

Our first result provides and upper bound to the gap between the capacity in Equation (2) and the lower bound in Equation (21).

Theorem 17. *Let* $\mathbf{H} \in \mathbb{R}^{n \times n}$ *be of full rank and*

$$\rho(\mathcal{X}, \mathbf{H}) := \frac{\mathrm{Vol}\big(\mathcal{B}_0\left(r_{\max}(\mathbf{H}\mathcal{X})\right)\big)}{\mathrm{Vol}(\mathbf{H}\mathcal{X})}.$$

Then,

$$C(\mathcal{X}, \mathbf{H}) - \underline{C}_{\mathsf{EPI}}(\mathcal{X}, \mathbf{H}) \leq \frac{n}{2} \log_2\left((\pi n)^{\frac{1}{n}} \rho(\mathcal{X}, \mathbf{H})^{\frac{2}{n}} \right).$$

Proof. For notational convenience, let the volume of an n-dimensional ball of radius $r > 0$ be denoted as

$$V_n(r) := \mathrm{Vol}\big(\mathcal{B}_0(r)\big) = V_n(1)r^n = \frac{\pi^{\frac{n}{2}} r^n}{\Gamma\left(\frac{n}{2} + 1\right)}.$$

Now, observe that, by choosing $p = 2$, the upper bound of Theorem 9 can further be upper bounded as

$$\bar{C}_{\mathsf{M}}(\mathcal{X}, \mathbf{H}) \leq n \log_2\left(\frac{k_{n,2}}{(2\pi e)^{\frac{1}{2}}} n^{\frac{1}{2}} \|\tilde{\mathbf{x}} + \mathbf{Z}\|_2 \right)$$

$$\overset{(a)}{=} \frac{n}{2} \log_2\left(\frac{1}{n} \mathbb{E}\big[\|\tilde{\mathbf{x}} + \mathbf{Z}\|^2 \big] \right)$$

$$\overset{(b)}{=} \frac{n}{2} \log_2\left(1 + \frac{1}{n} \|\tilde{\mathbf{x}}\|^2 \right),$$

where (a) follows since $k_{n,2} = \sqrt{\frac{2\pi e}{n}}$; and (b) since $\mathbb{E}[\|\mathbf{Z}\|^2] = n$. Therefore, the gap between Equation (21) and the moment upper bound of Theorem 9 can be upper bounded as follows:

$$\bar{C}_{\mathsf{M}}(\mathcal{X}, \mathbf{H}) - \underline{C}_{\mathsf{EPI}}(\mathcal{X}, \mathbf{H}) = \frac{n}{2} \log_2\left(\frac{1 + \frac{1}{n}\|\tilde{\mathbf{x}}\|^2}{1 + \frac{\mathrm{Vol}(\mathbf{H}\mathcal{X})^{\frac{2}{n}}}{2\pi e}} \right)$$

$$\overset{a)}{=} \frac{n}{2} \log_2\left(\frac{1 + \frac{1}{n}\left(\frac{V_n(\|\tilde{\mathbf{x}}\|)}{V_n(1)} \right)^{\frac{2}{n}}}{1 + \frac{\mathrm{Vol}(\mathbf{H}\mathcal{X})^{\frac{2}{n}}}{2\pi e}} \right)$$

$$= \frac{n}{2} \log_2\left(\frac{1 + \frac{1}{n}\left(\frac{\rho(\mathcal{X},\mathbf{H})\mathrm{Vol}(\mathbf{H}\mathcal{X})}{V_n(1)} \right)^{\frac{2}{n}}}{1 + \frac{\mathrm{Vol}(\mathbf{H}\mathcal{X})^{\frac{2}{n}}}{2\pi e}} \right)$$

$$\overset{b)}{\leq} \frac{n}{2} \log_2\left(\frac{1}{n} 2\pi e \left(\frac{\rho(\mathcal{X},\mathbf{H})}{V_n(1)} \right)^{\frac{2}{n}} \right)$$

$$\overset{c)}{\leq} \frac{n}{2} \log_2\left((\pi n)^{\frac{1}{n}} \rho(\mathcal{X}, \mathbf{H})^{\frac{2}{n}} \right).$$

where (a) is due to the fact that $\|\tilde{\mathbf{x}}\|$ is the radius of an n-dimensional ball; (b) follows from the inequality $\frac{1+cx}{1+x} \leq c$ for $c \geq 1$ and $x \in \mathbb{R}_+$; and (c) follows from using Stirling's approximation to obtain $\left(\frac{1}{V_n(1)} \right)^{\frac{2}{n}} \leq \frac{1}{2e\pi^{1-\frac{1}{n}}} n^{1+\frac{1}{n}}$. $\quad\square$

The term $\rho(\mathcal{X}, \mathbf{H})$ is referred to as the *packing efficiency* of the set $\mathbf{H}\mathcal{X}$. In the following proposition, we present the packing efficiencies for important special cases.

Proposition 1. (Packing Efficiencies) *Let* $\mathbf{H} \in \mathbb{R}^{n \times n}$ *be of full rank,* $A \in \mathbb{R}_+$, *and* $\mathbf{a} := (A_1, \ldots, A_n) \in \mathbb{R}_+^n$. *Then,*

$$\rho(\mathcal{B}_0(A), \mathbf{I}_n) = 1, \tag{35}$$

$$\rho(\mathcal{B}_0(A), \mathbf{H}) = \frac{\|\mathbf{H}\|^n}{|\det(\mathbf{H})|}, \tag{36}$$

$$\rho(\mathrm{Box}(\mathbf{a}), \mathbf{I}_n) = \frac{\pi^{\frac{n}{2}}}{\Gamma(\frac{n}{2}+1)} \frac{\|\mathbf{a}\|^n}{\prod_{i=1}^n A_i}, \tag{37}$$

$$\rho(\mathrm{Box}(\mathbf{a}), \mathbf{H}) \leq \frac{\pi^{\frac{n}{2}}}{\Gamma(\frac{n}{2}+1)} \frac{\|\mathbf{H}\|^n \|\mathbf{a}\|^n}{|\det(\mathbf{H})| \prod_{i=1}^n A_i}. \tag{38}$$

Proof. The packing efficiency in Equation (35) follows immediately. Note that

$$r_{\max}(\mathbf{H}\mathcal{B}_0(A)) = \max_{\mathbf{x} \in \mathcal{B}_0(A)} \|\mathbf{H}\mathbf{x}\| = \|\mathbf{H}\|A.$$

Thus, as $\mathbf{H}$ is assumed to be invertible, we have $\mathrm{Vol}(\mathbf{H}\mathcal{B}_0(A)) = |\det(\mathbf{H})|\mathrm{Vol}(\mathcal{B}_0(A))$, which results in Equation (36). To show Equation (37), observe that

$$\mathrm{Vol}\big(\mathcal{B}_0\big(r_{\max}(\mathbf{I}_n \mathrm{Box}(\mathbf{a}))\big)\big) = \mathrm{Vol}\big(\mathcal{B}_0(\|\mathbf{a}\|)\big) = \frac{\pi^{\frac{n}{2}}}{\Gamma(\frac{n}{2}+1)} \|\mathbf{a}\|^n.$$

The proof of Equation (37) is concluded by observing that $\mathrm{Vol}(\mathbf{I}_n \mathrm{Box}(\mathbf{a})) = \prod_{i=1}^n A_i$. Finally, observe that $\mathrm{Box}(\mathbf{a}) \subset \mathcal{B}_0(\|\mathbf{a}\|)$ implies $r_{\max}(\mathbf{H}\mathrm{Box}(\mathbf{a})) \leq r_{\max}(\mathbf{H}\mathcal{B}_0(\|\mathbf{a}\|))$ so that

$$\rho(\mathbf{H}, \mathrm{Box}(\mathbf{a})) \leq \frac{\mathrm{Vol}(\mathcal{B}_0(\|\mathbf{H}\|\|\mathbf{a}\|))}{\mathrm{Vol}(\mathbf{H}\mathrm{Box}(\mathbf{a}))} = \frac{\pi^{\frac{n}{2}}}{\Gamma(\frac{n}{2}+1)} \frac{\|\mathbf{H}\|^n \|\mathbf{a}\|^n}{|\det(\mathbf{H})| \prod_{i=1}^n A_i},$$

which is the bound in Equation (38). $\quad\square$

We conclude this section by characterizing the gap to the capacity when $\mathbf{H}$ is diagonal and the channel input space is the Cartesian product of n PAM constellations. In this context, $\mathrm{PAM}(N, A)$ refers to the set of $N \in \mathbb{N}$ equidistant PAM-constellation points with amplitude constraint $A \in \mathbb{R}_+$ (see Figure 3 for an illustration), whereas $X \sim \mathrm{PAM}(N, A)$ means that X is uniformly distributed over $\mathrm{PAM}(N, A)$ [11].

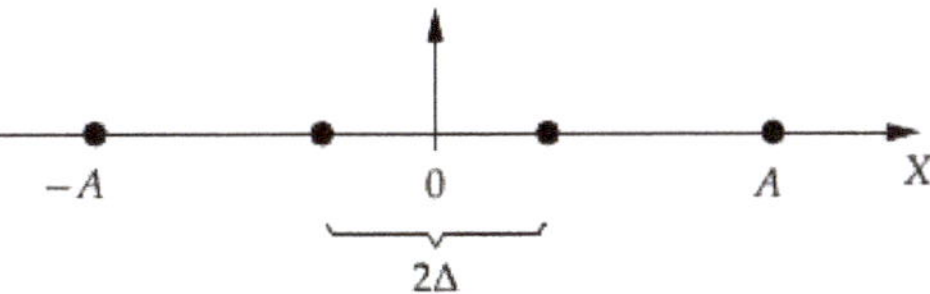

Figure 3. Example of a pulse-amplitude modulation constellation with $N = 4$ points and amplitude constraint A (i.e., $\mathrm{PAM}(4, A)$), where $\Delta := A/(N-1)$ denotes half the Euclidean distance between two adjacent constellation points. In the case N is odd, 0 is a constellation point.

Theorem 18. *Let* $\mathbf{H} = \mathrm{diag}(h_{11}, \ldots, h_{nn}) \in \mathbb{R}^{n \times n}$ *be fixed and* $\mathbf{X} = (X_1, \ldots, X_n)$. *Then, if* $X_i \sim \mathrm{PAM}(N_i, A_i)$, $i = 1, \ldots, n$, *for some given* $\mathbf{a} = (A_1, \ldots, A_n) \in \mathbb{R}_+^n$, *it holds that*

$$\bar{C}_{\mathrm{Dual},2}(\mathrm{Box}(\mathbf{a}), \mathbf{H}) - \underline{C}_{\mathrm{OW}}(\mathrm{Box}(\mathbf{a}), \mathbf{H}) \leq c \cdot n \,\mathrm{bits}, \tag{39}$$

where $N_i := \left\lfloor 1 + \frac{2A_i|h_{ii}|}{\sqrt{2\pi e}} \right\rfloor$ and

$$c := 1 + \frac{1}{2}\log_2\left(\frac{\pi e}{6}\right) + \frac{1}{2}\log_2\left(1 + \frac{6}{\pi e}\right) \approx 1.64.$$

Moreover, if $X_i \sim \mathrm{PAM}(N_i, A)$, $i = 1, \ldots, n$, for some given $A \in \mathbb{R}_+$, it holds that

$$\bar{C}_{\mathsf{Dual},2}(\mathcal{B}_0(A), \mathbf{H}) - \underline{C}_{\mathsf{OW}}(\mathcal{B}_0(A), \mathbf{H}) \leq c \cdot n \text{ bits,} \tag{40}$$

where $N_i := \left\lfloor 1 + \frac{2A|h_{ii}|}{\sqrt{n}\sqrt{2\pi e}} \right\rfloor$.

Proof. Since the channel matrix is diagonal, letting the channel input $\mathbf{X}$ be such that its elements X_i, $i = 1, \ldots, n$, are independent, we have that

$$I(\mathbf{X}; \mathbf{H}\mathbf{X} + \mathbf{Z}) = \sum_{i=1}^{n} I(X_i; h_{ii}X_i + Z_i).$$

Let $X_i \sim \mathrm{PAM}(N_i, A_i)$ with $N_i := \left\lfloor 1 + \frac{2A_i|h_{ii}|}{\sqrt{2\pi e}} \right\rfloor$ and observe that half the Euclidean distance between any pair of adjacent points in $\mathrm{PAM}(N_i, A_i)$ is equal to $\Delta_i := A_i/(N_i - 1)$ (see Figure 3), $i = 1, \ldots, n$. To lower bound the mutual information $I(X_i; h_{ii}X_i + Z_i)$, we use the bound of Theorem 13 for $p = 2$ and $n_t = 1$. Thus, for some continuous random variable U that is uniformly distributed over the interval $[-\Delta_i, \Delta_i)$ and independent of X_i, we have that

$$I(X_i; h_{ii}X_i + Z_i) \geq H(X_i) - \frac{1}{2}\log_2\left(\frac{\pi e}{6}\right) - \frac{1}{2}\log_2\left(\frac{\mathbb{E}\left[(U + X_i - g(Y_i))^2\right]}{\mathbb{E}[U^2]}\right). \tag{41}$$

Now, note that the entropy term in Equation (41) can be lower bounded as

$$H(X_i) = \log_2\left(\left\lfloor 1 + \frac{2A_i|h_{ii}|}{\sqrt{2\pi e}} \right\rfloor\right) \geq \log_2\left(1 + \frac{2A_i|h_{ii}|}{\sqrt{2\pi e}}\right) + \log_2(2), \tag{42}$$

where we have used that $\lfloor x \rfloor \geq \frac{x}{2}$ for every $x \geq 1$. On the other hand, the last term in Equation (41) can be upper bounded by upper bounding its argument as follows:

$$\begin{aligned}
\frac{\mathbb{E}\left[(U + X_i - g(Y_i))^2\right]}{\mathbb{E}[U^2]} &\overset{a)}{=} 1 + \frac{3\mathbb{E}\left[(X_i - g(Y_i))^2\right]}{\Delta_i^2} \\
&\overset{b)}{\leq} 1 + \frac{3\mathbb{E}[Z_i^2](N_i - 1)^2}{A_i^2|h_{ii}|^2} \\
&= 1 + \frac{3(N_i - 1)^2}{A_i^2|h_{ii}|^2} \\
&\overset{c)}{\leq} 1 + \frac{3\left(\frac{2A_i|h_{ii}|}{\sqrt{2\pi e}}\right)^2}{A_i^2|h_{ii}|^2} \\
&= 1 + \frac{6}{\pi e}.
\end{aligned} \tag{43}$$

where (a) follows from using that X_i and U are independent and $\mathbb{E}[U^2] = \frac{\Delta_i^2}{3}$; (b) from using the estimator $g(Y_i) = \frac{1}{h_{ii}}Y_i$; and (c) from $N_i = \left\lfloor 1 + \frac{2A_i|h_{ii}|}{\sqrt{2\pi e}} \right\rfloor \leq 1 + \frac{2A_i|h_{ii}|}{\sqrt{2\pi e}}$. Combining Equations (41), (42), and (43) results in the gap in (39).

The proof of the capacity gap in Equation (40) follows along similar lines, which concludes the proof. $\square$

We are also able to determine the gap to the capacity for a general invertible channel matrix.

Theorem 19. *For any $\mathcal{X}$ and any invertible* $\mathbf{H}$

$$C(\mathcal{X}, \mathbf{H}) - \underline{C}_{\mathrm{OW}}(\mathcal{X}, \mathbf{H}) \leq \log_2(\pi n) + \frac{n}{2}\log_2\left(1 + 4n(4 + 4n)\left(\frac{\|\mathbf{H}^{-1}\|^2 r_{\max}^2(\mathbf{H}\mathcal{X})}{r_{\min}^2(\mathcal{X})} + \frac{n\|\mathbf{H}^{-1}\|^2}{r_{\min}^2(\mathcal{X})}\right)\right).$$

Proof. Let $\mathbf{X}$ be uniformly distributed over a set constructed from an n-dimensional cubic lattice with the number of points equal to $N = \lfloor \|\tilde{\mathbf{x}} + \mathbf{Z}\|_2^n \rfloor$, where $\tilde{\mathbf{x}} \in \mathbf{H}\mathcal{X}$ is chosen such that $\|\tilde{\mathbf{x}}\| = r_{\max}(\mathbf{H}\mathcal{X})$, and scaled such that it is contained in the input space $\mathcal{X}$. Note that the minimum distance between point in $\mathbf{X}$ are given by

$$d_{\min}(\mathrm{supp}(\mathbf{X})) := \frac{r_{\min}(\mathcal{X})}{N^{\frac{1}{n}}}.$$

Now, we compute the difference between the moment upper bound of Theorem 9 and the Ozarow–Wyner lower bound of Theorem 13:

$$
\begin{aligned}
\bar{C}_{\mathrm{M}}(\mathcal{X}, \mathbf{H}) - \underline{C}_{\mathrm{OW}}(\mathcal{X}, \mathbf{H}) &\overset{(a)}{\leq} \log_2(\|\tilde{\mathbf{x}} + \mathbf{Z}\|_2) - H(\mathbf{X}) + \mathrm{gap} \\
&= n\log_2(\|\tilde{\mathbf{x}} + \mathbf{Z}\|_2) - \log_2(\lfloor\|\tilde{\mathbf{x}} + \mathbf{Z}\|_2^n\rfloor) + \mathrm{gap} \\
&\overset{(b)}{\leq} \log_2(2) + \mathrm{gap},
\end{aligned}
\tag{44}
$$

where (a) follows from Theorem 9 by choosing $p = 2$; and (b) by using the bound $\lfloor x \rfloor \geq \frac{x}{2}$ for $x > 1$. The next step in the proof consists in bounding the gap term, which requires to upper bound the terms in Equations (23) and (24) individually. Towards this end, choose $p = 2$ and let $\mathbf{U}$ be a random vector that is uniformly distributed over a ball of radius $d_{\min}(\mathbf{X})$. Thus, for (23) it follows

$$
\begin{aligned}
G_{1,2}(\mathbf{U}, \mathbf{X}, g) &= n\log_2\left(\frac{\|\mathbf{U} + \mathbf{X} - g(\mathbf{Y})\|_2}{\|\mathbf{U}\|_2}\right) \\
&\overset{(a)}{=} n\log_2\left(\frac{\|\mathbf{U} - \mathbf{H}^{-1}\mathbf{Z}\|_2}{\|\mathbf{U}\|_2}\right) \\
&= \frac{n}{2}\log_2\left(1 + \frac{\|\mathbf{H}^{-1}\mathbf{Z}\|_2^2}{\|\mathbf{U}\|_2^2}\right) \\
&\overset{(b)}{=} \frac{n}{2}\log_2\left(1 + \frac{4(4 + 4n)\|\mathbf{H}^{-1}\mathbf{Z}\|_2^2}{d_{\min}^2(\mathrm{supp}(\mathbf{X}))}\right) \\
&\overset{(c)}{\leq} \frac{n}{2}\log_2\left(1 + \frac{(4 + 4n)\|\mathbf{H}^{-1}\mathbf{Z}\|_2^2\,\|\tilde{\mathbf{x}} + \mathbf{Z}\|_2^2}{r_{\min}^2(\mathcal{X})}\right) \\
&\overset{(d)}{=} \frac{n}{2}\log_2\left(1 + \frac{(4 + 4n)\|\mathbf{H}^{-1}\mathbf{Z}\|_2^2\,(r_{\max}^2(\mathbf{H}\mathcal{X}) + n)}{r_{\min}^2(\mathcal{X})}\right) \\
&\overset{(e)}{\leq} \frac{n}{2}\log_2\left(1 + \frac{4(4 + 4n)\|\mathbf{H}^{-1}\|^2\|\mathbf{Z}\|_2^2\,(r_{\max}^2(\mathbf{H}\mathcal{X}) + n)}{r_{\min}^2(\mathcal{X})}\right) \\
&= \frac{n}{2}\log_2\left(1 + \frac{4(4 + 4n)\|\mathbf{H}^{-1}\|^2 n\,(r_{\max}^2(\mathbf{H}\mathcal{X}) + n)}{r_{\min}^2(\mathcal{X})}\right).
\end{aligned}
$$

where (a) follows by choosing $g(\mathbf{Y}) = \mathbf{H}^{-1}\mathbf{Y}$; ($b$) by using $\|\mathbf{U}\|_2^2 = \frac{r^2}{4+2n}$, where $r = \frac{d_{\min}(\mathrm{supp}(\mathbf{X}))}{2}$ is the radius of an n-dimensional ball; (c) from dropping the floor function in the expression for the minimum distance, i.e.,

$$d_{\min}^{-1}(\mathrm{supp}(\mathbf{X})) = \frac{N^{\frac{1}{n}}}{r_{\min}(\mathcal{X})} = \frac{\left\lfloor \|\tilde{\mathbf{x}} + \mathbf{Z}\|_2^n \right\rfloor^{\frac{1}{n}}}{r_{\min}(\mathcal{X})} \leq \frac{\|\tilde{\mathbf{x}} + \mathbf{Z}\|_2}{r_{\min}(\mathcal{X})};$$

(d) follows by expanding $\|\tilde{\mathbf{x}} + \mathbf{Z}\|_2^2$ using that $\|\tilde{\mathbf{x}}\| = r_{\max}(\mathbf{H}\mathcal{X})$; and ($e$) from using the bound $\|\mathbf{H}^{-1}\mathbf{Z}\|_2 \leq \|\mathbf{H}^{-1}\|\|\mathbf{Z}\|_2$.

On the other hand, the term $G_{2,p}(\mathbf{U})$ can be bounded from above as follows ([36] Appendix L):

$$G_{2,p}(\mathbf{U}) = n\log_2\left(\frac{k_{n,p}\, n^{\frac{1}{p}}\|\mathbf{U}\|_p}{2^{\frac{1}{n}h(\mathbf{U})}}\right) \leq n\log_2\left((\pi n)^{\frac{1}{n}}\right).$$

Combining these two bounds with the one in (44) provides the result. $\square$

6. Arbitrary Channel Matrices

For an arbitrary MIMO channel with an average power constraint, it is well known that the capacity is achieved by a singular value decomposition (SVD) of the channel matrix (i.e., $\mathbf{H} = \mathbf{U}\boldsymbol{\Lambda}\mathbf{V}^T$) along with considering the equivalent channel model

$$\tilde{\mathbf{Y}} = \boldsymbol{\Lambda}\tilde{\mathbf{X}} + \tilde{\mathbf{Z}},$$

where $\tilde{\mathbf{Y}} := \mathbf{U}^T\mathbf{Y}$, $\tilde{\mathbf{X}} := \mathbf{V}^T\mathbf{X}$, and $\tilde{\mathbf{Z}} := \mathbf{U}^T\mathbf{Z}$, respectively.

To provide lower bounds for channels with amplitude constraints and SVD precoding, we need the following lemma.

Lemma 3. *For any given orthogonal matrix $\mathbf{V} \in \mathbb{R}^{n_t \times n_t}$ and constraint vector $\mathbf{a} = (A_1, \ldots, A_{n_t}) \in \mathbb{R}_+^{n_t}$, there exists a distribution $F_{\mathbf{X}}$ of $\mathbf{X}$ such that $\tilde{\mathbf{X}} = \mathbf{V}^T\mathbf{X}$ is uniformly distributed over $\mathrm{Box}(\mathbf{a})$. Moreover, the components $\tilde{X}_1, \ldots, \tilde{X}_{n_t}$ of $\tilde{\mathbf{X}}$ are mutually independent with $\tilde{X}_i$ uniformly distributed over $[-A_i, A_i]$, $i = 1, \ldots, n_t$.*

Proof. Suppose that $\tilde{\mathbf{X}}$ is uniformly distributed over $\mathrm{Box}(\mathbf{a})$; that is, the density of $\tilde{\mathbf{X}}$ is of the form

$$f_{\tilde{\mathbf{X}}}(\tilde{\mathbf{x}}) = \frac{1}{\mathrm{Vol}(\mathrm{Box}(\mathbf{a}))}, \ \tilde{\mathbf{x}} \in \mathrm{Box}(\mathbf{a}).$$

Since $\mathbf{V}$ is orthogonal, we have $\mathbf{V}\tilde{\mathbf{X}} = \mathbf{X}$ and by the change of variable Theorem for $\mathbf{x} \in \mathbf{V}\mathrm{Box}(\mathbf{a})$

$$f_{\mathbf{X}}(\mathbf{x}) = \frac{1}{|\det(\mathbf{V})|}f_{\tilde{\mathbf{X}}}(\mathbf{V}^T\mathbf{x}) = \frac{1}{|\det(\mathbf{V})|\mathrm{Vol}(\mathrm{Box}(\mathbf{a}))} = \frac{1}{\mathrm{Vol}(\mathrm{Box}(\mathbf{a}))}.$$

Therefore, such a distribution of $\mathbf{X}$ exists. $\square$

Theorem 20. (Lower Bounds with SVD Precoding) *Let $\mathbf{H} \in \mathbb{R}^{n_r \times n_t}$ be fixed, $n_{\min} := \min(n_r, n_t)$, and $\mathcal{X} = \mathrm{Box}(\mathbf{a})$ for some $\mathbf{a} = (A_1, \ldots, A_{n_t}) \in \mathbb{R}_+^{n_t}$. Furthermore, let σ_i, $i = 1, \ldots, n_{\min}$, be the ith singular value of $\mathbf{H}$. Then,*

$$\underline{C}_{\mathrm{Jensen}}(\mathrm{Box}(\mathbf{a}), \mathbf{H}) = \log_2^+\left(\left(\frac{2}{e}\right)^{\frac{n_{\min}}{2}}\frac{1}{\psi(\mathbf{H}, \mathbf{b}^\star)}\right) \tag{45}$$

and

$$\underline{C}_{\mathrm{EPI}}(\mathrm{Box}(\mathbf{a}), \mathbf{H}) = \frac{n_{\min}}{2}\log_2\left(1 + \frac{\left|\prod_{i=1}^{n_{\min}} A_i\sigma_i\right|^{\frac{2}{n_{\min}}}}{2\pi e}\right), \tag{46}$$

where

$$\psi(\mathbf{H}, \mathbf{b}^\star) := \min_{\mathbf{b} \in \mathrm{Box}(\mathbf{a})} \prod_{i=1}^{n_{\min}} \varphi(\sigma_i B_i)$$

with $\mathbf{b} := (B_1, \ldots, B_{n_t})$ *and* φ *as defined in Equation* (32).

Proof. Performing the SVD, the expected value in Theorem 14 can be written as

$$\mathbb{E}\left[e^{-\frac{\|\mathbf{H}(\mathbf{x}-\mathbf{x}')\|^2}{4}}\right] = \mathbb{E}\left[e^{-\frac{\|\mathbf{U}\mathbf{\Lambda}\mathbf{V}^T(\mathbf{x}-\mathbf{x}')\|^2}{4}}\right] = \mathbb{E}\left[e^{-\frac{\|\mathbf{\Lambda}\mathbf{V}^T(\mathbf{x}-\mathbf{x}')\|^2}{4}}\right] = \mathbb{E}\left[e^{-\frac{\|\mathbf{\Lambda}(\tilde{\mathbf{x}}-\tilde{\mathbf{x}}')\|^2}{4}}\right].$$

By Lemma 3, there exists a distribution $F_{\mathbf{X}}$ such that the components of $\tilde{\mathbf{X}}$ are independent and uniformly distributed. Since $\mathbf{\Lambda}$ is a diagonal matrix, we can use Theorem 16 to arrive at Equation (45).

Note that by Lemma 3 there exists a distribution on $\mathbf{X}$ such that $\tilde{\mathbf{X}}$ is uniform over $\mathrm{Box}(\mathbf{a}) \subset \mathbb{R}^{n_t}$ and $\mathbf{\Lambda}\tilde{\mathbf{X}}$ is uniform over $\mathbf{\Lambda}\mathrm{Box}(\mathbf{a}) \subset \mathbb{R}^{n_{\min}}$, respectively. Therefore, by the EPI lower bound given in Equation (20), we obtain

$$\begin{aligned}
\underline{C}_{\mathsf{EPI}}(\mathrm{Box}(\mathbf{a}), \mathbf{H}) &= \frac{n_{\min}}{2} \log_2\left(1 + \frac{2^{\frac{2}{n_{\min}}h(\mathbf{\Lambda}\tilde{\mathbf{X}})}}{2\pi e}\right) \\
&= \frac{n_{\min}}{2} \log_2\left(1 + \frac{\mathrm{Vol}\left(\mathbf{\Lambda}\mathrm{Box}(\mathbf{a})\right)^{\frac{2}{n_{\min}}}}{2\pi e}\right) \\
&= \frac{n_{\min}}{2} \log_2\left(1 + \frac{\left(\prod_{i=1}^{n_{\min}} A_i\right)^{\frac{2}{n_{\min}}} \left|\prod_{i=1}^{n_{\min}} \sigma_i\right|^{\frac{2}{n_{\min}}}}{2\pi e}\right),
\end{aligned}$$

which is exactly the expression in Equation (46). This concludes the proof. $\qquad\square$

Remark 8. *Notice that choosing the optimal* $\mathbf{b}$ *for the lower bound in Equation* (45) *is an amplitude allocation problem, which is reminiscent of waterfilling in the average power constraint case. It would be interesting to study whether the bound in Equation* (45) *is connected to what is called* mercury waterfilling *in* [37,38].

In Figure 4, the lower bounds of Theorem 20 are compared to the moment upper bound of Theorem 2 for the special case of a 3×1 MIMO channel. Similar to the example presented in Figure 2, the EPI lower bound performs well in the low amplitude regime, while Jensen's inequality lower bound performs well in the high amplitude regime.

We conclude this section by showing that for an arbitrary channel input space $\mathcal{X}$, in the large amplitude regime the capacity pre-log is given by $\min(n_r, n_t)$.

Theorem 21. *Let* $\mathcal{X}$ *be arbitrary and* $\mathbf{H} \in \mathbb{R}^{n_r \times n_t}$ *fixed. Then,*

$$\lim_{r_{\min}(\mathcal{X}) \to \infty} \frac{C(\mathcal{X}, \mathbf{H})}{\log_2\left(1 + \frac{2r_{\min}(\mathcal{X})}{\sqrt{2\pi e}}\right)} = \min(n_r, n_t).$$

Proof. Notice that there always exists $\mathbf{a} \in \mathbb{R}_+^{n_t}$ and $c \in \mathbb{R}_+$ such that $\mathrm{Box}(\mathbf{a}) \subseteq \mathcal{X} \subset c\mathrm{Box}(\mathbf{a})$. Thus, without loss of generality, we can consider $\mathcal{X} = \mathrm{Box}(\mathbf{a})$, $\mathbf{a} = (A, \ldots, A)$, for sufficiently large $A \in \mathbb{R}_+$. To prove the result, we therefore start with enlarging the constraint set of the bound in Equation (11):

$$\begin{aligned}
\mathrm{Box}\left(\mathbf{H}\mathrm{Box}(\mathbf{a})\right) &\subseteq \mathcal{B}_0\left(r_{\max}\left(\mathbf{H}\mathrm{Box}(\mathbf{a})\right)\right) \\
&\subseteq \mathcal{B}_0\left(r_{\max}\left(\mathbf{H}\mathcal{B}_0(\sqrt{n_t}A)\right)\right) \\
&= \mathcal{B}_0\left(r_{\max}\left(\mathbf{U}\mathbf{\Lambda}\mathbf{V}^T\mathcal{B}_0(\sqrt{n_t}A)\right)\right)
\end{aligned}$$

$$= \mathcal{B}_0\big(r_{\max}\big(\mathbf{U\Lambda}\mathcal{B}_0(\sqrt{n_t}A)\big)\big)$$
$$= \mathcal{B}_0\big(r_{\max}\big(\mathbf{\Lambda}\mathcal{B}_0(\sqrt{n_t}A)\big)\big)$$
$$\subseteq \mathcal{B}_0(r)$$
$$\subseteq \mathrm{Box}(\mathbf{a}'),$$

where $r := \sqrt{n_t}A\sqrt{\sum_{i=1}^{n_{\min}} \sigma_i^2}$ and $\mathbf{a}' := \big(\frac{r}{\sqrt{n_{\min}}}, \ldots, \frac{r}{\sqrt{n_{\min}}}\big) \in \mathbb{R}_+^{n_{\min}}$. Therefore, by using the upper bound in Equation (11), it follows that

$$C(\mathrm{Box}(\mathbf{a}), \mathbf{H}) \le \sum_{i=1}^{n_r} \log_2\left(1 + \frac{2A_i}{\sqrt{2\pi e}}\right) \le n_{\min} \log_2\left(1 + \frac{2}{\sqrt{2\pi e}} \frac{\sqrt{n_t}A\sqrt{\sum_{i=1}^{n_{\min}} \sigma_i^2}}{\sqrt{n_{\min}}}\right).$$

Moreover,

$$\lim_{A\to\infty} \frac{C(\mathrm{Box}(\mathbf{a}), \mathbf{H})}{\log_2\left(1 + \frac{2A}{\sqrt{2\pi e}}\right)} \le n_{\min} \lim_{A\to\infty} \frac{\log_2\left(1 + \frac{2}{\sqrt{2\pi e}} \frac{\sqrt{n_t}A\sqrt{\sum_{i=1}^{n_{\min}} \sigma_i^2}}{\sqrt{n_{\min}}}\right)}{\log_2\left(1 + \frac{2A}{\sqrt{2\pi e}}\right)} = n_{\min}.$$

Next, using the EPI lower bound in Equation (46), we have that

$$\lim_{A\to\infty} \frac{\underline{C}_{\mathrm{EPI}}(\mathrm{Box}(\mathbf{a}), \mathbf{\Lambda})}{\log_2\left(1 + \frac{2A}{\sqrt{2\pi e}}\right)} = n_{\min} \lim_{A\to\infty} \frac{\frac{1}{2}\log_2\left(1 + \frac{A\left|\Pi_{i=1}^{n_{\min}} \sigma_i\right|^{\frac{2}{n_{\min}}}}{2\pi e}\right)}{\log_2\left(1 + \frac{2A}{\sqrt{2\pi e}}\right)} = n_{\min}.$$

This concludes the proof. $\square$

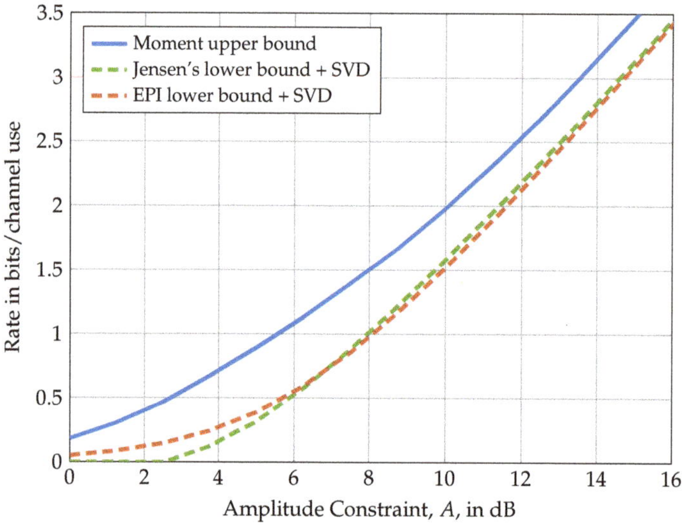

Figure 4. Comparison of the upper bound in Theorem 2 with the lower bounds of Theorem 20 for a 3×1 MIMO system with amplitude constraints $A_1 = A_2 = A_3 = A$ (i.e., $\mathbf{a} = (A, A, A)$) and channel matrix $\mathbf{h} = (0.6557, 0.0357, 0.8491)$.

7. The SISO Case

In this section, we apply the upper and lower bounds presented in the previous sections to the special case of a SISO channel that is subject to an amplitude constraint (i.e., $\mathcal{X} = [-A, A]$ for some $A \in \mathbb{R}_+$) and compare them with the state-of-the art. More precisely, we are interested in upper and lower bounds to the capacity

$$C([-A, A], h) := \max_{F_X : X \in [-A, A]} I(X; hX + Y). \tag{47}$$

Without loss of generality, we assume $h = 1$ in all that follows.

7.1. Upper and Lower Bounds

As a starting point for our comparisons, the following Theorem summarizes bounds on the capacity (47) that are known from the literature. The bounds are all based on the duality approach that we generalize in Section 4 to the MIMO case.

Theorem 22. (Known Duality Upper Bounds) *Let $A > 0$ be arbitrary. Then, the following are valid upper bounds to the capacity of the amplitude-constrained SISO channel defined in Equation* (47).

- *McKellips upper bound [7]:*

$$C([-A, A], 1) \leq \bar{C}_{\mathsf{McK}}([-A, A], 1) := \log_2\left(1 + \frac{2A}{\sqrt{2\pi e}}\right). \tag{48}$$

- *Thangaraj–Kramer–Böcherer upper bound ([8] Theorem 1):*

$$C([-A, A], 1) \leq \bar{C}_{\mathsf{TKB}}([-A, A], 1) := \begin{cases} \beta(A) \log_e\left(\sqrt{\frac{2}{\pi e}} A\right) + H_{\mathsf{b}}(\beta(A)), & A^2 \leq 6.304\,\mathrm{dB} \\ \bar{C}_{\mathsf{McK}}([-A, A], 1), & \text{else,} \end{cases} \tag{49}$$

where $\beta(A) := \frac{1}{2} - Q(2A)$ and H_{b} denotes the binary entropy function.
- *Rassouli–Clerckx upper bound [9]:*

$$C([-A, A], 1) \leq \bar{C}_{\mathsf{RC}}([-A, A], 1) := \bar{C}_{\mathsf{TKB}}([-A, A], 1) + W(A), \tag{50}$$

where

$$W(A) := \frac{1}{2}\left(\log_e(\sigma^2(A)) + \frac{1}{\sigma^2(A)} - 1\right)\left(\frac{1}{2} + Q(2A)\right) + \frac{g(2A)}{2\sigma^2(A)},$$

$$\sigma^2(A) := 1 + \frac{2g(2A)}{1 + 2Q(2A)},$$

and

$$g(x) := x^2 Q(x) - \frac{x}{\sqrt{2\pi}} e^{-\frac{x^2}{2}}.$$

Now, we apply the moment upper bound of Theorem 9 to the SISO case.

Theorem 23. (Moment Upper Bound) *Let $A > 0$ be arbitrary. Then,*

$$C([-A, A], 1) \leq \bar{C}_{\mathsf{M}}([-A, A], 1) = \inf_{p > 0} \log_2\left(\frac{k_{1,p}}{(2\pi e)^{\frac{1}{2}}} \mathbb{E}\big[|A + Z|^p\big]^{\frac{1}{p}}\right), \tag{51}$$

where the expected value is of the explicit form

$$\mathbb{E}\left[|A+Z|^p\right] = \frac{2^{\frac{p}{2}}\Gamma\left(\frac{p+1}{2}\right)}{\sqrt{\pi}} \, {}_1F_1\left(-\frac{p}{2};\frac{1}{2};-\frac{A^2}{2}\right)$$

with ${}_1F_1(a;b;z)$ being the confluent hypergeometric function of the first kind ([39] Chapter 13).

Proof. First, note that $r_{\max}([-A,A]) = A$. Then, by using the expression for the raw absolute moment of a Gaussian distribution given in [40], we have that

$$\max_{a\in[0,A]} \mathbb{E}\left[|a+Z|^p\right] = \max_{a\in[0,A]} \frac{2^{\frac{p}{2}}\Gamma\left(\frac{p+1}{2}\right)}{\sqrt{\pi}} \, {}_1F_1\left(-\frac{p}{2};\frac{1}{2};-\frac{a^2}{2}\right).$$

The proof is concluded by observing that $f(a) := {}_1F_1\left(-\frac{p}{2};\frac{1}{2};-\frac{a^2}{2}\right)$ is an increasing function in a. $\square$

The following theorem establishes the EPI and the Jensen lower bound of Section 4.2 assuming the channel input symbols are uniformly distributed.

Theorem 24. (Lower Bounds with Uniform Inputs) *Let $A > 0$ be arbitrary and the channel input X be uniformly distributed over $[-A, A]$. Then,*

$$C([-A,A],1) \geq \underline{C}_{\mathsf{EPI}}([-A,A],1) = \frac{1}{2}\log_2\left(1 + \frac{2A^2}{\pi e}\right) \tag{52}$$

and

$$C([-A,A],1) \geq \underline{C}_{\mathsf{Jensen}}([-A,A],1) = \log_2\left(\frac{\sqrt{2}A^2}{e^{\frac{1}{2}}\left(e^{-A^2} - 1 + \sqrt{\pi}A\left(1 - 2Q(\sqrt{2}A)\right)\right)}\right). \tag{53}$$

Proof. The lower bound in Equation (52) follows from Theorem 12 by observing that $\mathrm{Vol}(\mathcal{X}) = 2A$. To show the lower bound in Equation (53), consider Theorem 14 and let X and X' be independent and uniformly distributed over $[-A, A]$. Then, we have

$$\mathbb{E}\left[e^{-\frac{|X-X'|^2}{4}}\right] = \frac{1}{4A^2}\int_{-A}^{A}\int_{-A}^{A} e^{-\frac{(x-x')^2}{4}}\,dx\,dx'$$

$$= \frac{1}{A^2}\left(e^{-A^2} - 1 + \sqrt{\pi}A\left(1 - 2Q(\sqrt{2}A)\right)\right),$$

which concludes the proof. $\square$

Restricting the channel inputs to be discrete allows for another set of lower bounds on Equation (47).

Theorem 25. (Lower Bounds with Discrete Inputs) *Let $A > 1$ be arbitrary, $X_B \in \{-A, A\}$ equally likely, and $X_D \sim \mathrm{PAM}(N)$ with $N = \left\lceil 1 + \frac{A}{\sqrt{2\pi e}}\right\rceil$. Then,*

$$C([-A,A],1) \geq \underline{C}_{\mathsf{Binary}}([-A,A],1) := I(X_B; X_B + Z) \tag{54}$$

$$= \frac{1}{\log_e(2)}\left(A^2 - \int_{-\infty}^{\infty} \frac{e^{-\frac{y^2}{2}}}{\sqrt{2\pi}}\log_e\left(\cosh(A^2 - Ay)\right)dy\right), \tag{55}$$

$$C([-A, A], 1) \geq \underline{C}_{\text{Jensen}}([-A, A], 1) = -\log_2\left(\sqrt{\frac{e}{2}}\frac{1}{N^2}\sum_{(x_{D_i}, x_{D_j}) \in \text{PAM}(N)^2} e^{-\frac{(x_{D_i} - x_{D_j})^2}{4}}\right), \qquad (56)$$

and

$$C([-A, A], 1) \geq \underline{C}_{\text{OW}}([-A, A], 1) = \bar{C}_{\text{McK}}([-A, A], 1) - 2. \qquad (57)$$

Proof. The expression of the mutual information in Equation (54) for a uniform binary input $X_B \in \{-A, A\}$ is found in [41] by using the I-MMSE relationship. The bound in Equation (56) follows from using Theorem 14 and the bound in Equation (57) from Theorem 13, respectively. This concludes the proof. $\square$

Figure 5 compares the upper and lower bounds presented in this section in dependency of the amplitude constraint A. Observe that for values of A smaller than ≈ 1.665 (i.e., to the left of the gray vertical line), the lower bound (55) is in fact equal to the capacity. Up to constraints of $A \approx 1$, the moment upper bound in Equation (51) is the best after which the bound in Equation (50) becomes the tightest. The best lower bound for constraint values smaller than $A \approx 10$ is the bound in Equation (56) after which the lower bound in Equation (53) becomes the tightest. Note that all lower and upper bounds are asymptotically tight (i.e., for $A \to \infty$).

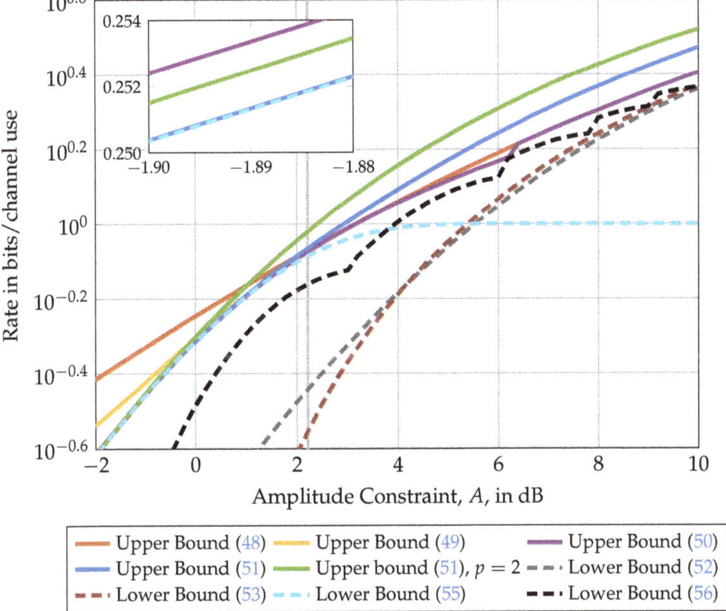

Figure 5. Comparison of upper and lower bounds on the capacity of a SISO channel with amplitude constraint A. The capacity of this channel is known for amplitudes smaller than $A \approx 10\log_{10}(1.665) = 2.214\,\text{dB}$ only (i.e., to the left of the gray vertical line) and unknown elsewhere. The nested figure represents a zoom into the region $-1.9\,\text{dB} \leq A \leq -1.88\,\text{dB}$ to highlight the differences between the Moment upper bound (51), the Rassouli–Clerckx upper bound in Equation (50), and the lower bound with binary inputs in Equation (56).

7.2. High and Low Amplitude Asymptotics

In this subsection, we study how the capacity in Equation (47) behaves in the high and low amplitude regimes. To this end, we need the following expression

$$\bar{C}_{\text{AWGN}}([-A, A], 1) := \frac{1}{2} \log_2\left(1 + A^2\right),$$

which is either the capacity of a SISO channel with an average power constraint A^2 or the moment bound in Equation (51) evaluated for $p = 2$.

Theorem 26. (SISO High and Low Amplitude Asymptotics) *It holds*

$$\lim_{A \to 0} \frac{C([-A, A], 1)}{\bar{C}_{\text{AWGN}}([-A, A], 1)} = 1, \tag{58}$$

$$\lim_{A \to \infty} \frac{C([-A, A], 1)}{\bar{C}_{\text{McK}}([-A, A], 1)} = 1, \tag{59}$$

and

$$\lim_{A \to \infty} C([-A, A], 1) - \bar{C}_{\text{AWGN}}([-A, A], 1) = \frac{1}{2} \log_2\left(\frac{\pi e}{2}\right) \approx 1.044. \tag{60}$$

Proof. The capacity of an amplitude-constrained SISO channel in the regime of low amplitudes (i.e., for amplitudes smaller than $A \approx 1.655$) was given by Guo et al. [41]

$$C([-A, A], 1) = \frac{1}{\log_e(2)} \left(A^2 - \int_{-\infty}^{\infty} \frac{e^{-\frac{y^2}{2}}}{\sqrt{2\pi}} \log_e\left(\cosh(A^2 - Ay)\right) dy\right).$$

Now, observe that

$$\lim_{A \to 0} \frac{1}{\frac{1}{2} \log_2(1 + A^2)} \int_{-\infty}^{\infty} \frac{e^{-\frac{y^2}{2}}}{\sqrt{2\pi}} \log_e\left(\cosh(A^2 - Ay)\right) dy$$

$$= \frac{2}{\log_2(e)} \int_{-\infty}^{\infty} \frac{e^{-\frac{y^2}{2}}}{\sqrt{2\pi}} \frac{\log_e\left(\cosh(A^2 - Ay)\right)}{\log_e(1 + A^2)} dy$$

$$= \frac{2}{\log_2(e)} \int_{-\infty}^{\infty} \frac{e^{-\frac{y^2}{2}}}{\sqrt{2\pi}} \lim_{A \to 0} \frac{\log_e\left(\cosh(A^2 - Ay)\right)}{\log_e(1 + A^2)} dy$$

$$= \frac{2}{\log_2(e)} \int_{-\infty}^{\infty} \frac{e^{-\frac{y^2}{2}}}{\sqrt{2\pi}} \frac{y^2}{2} dy$$

$$= \frac{1}{\log_2(e)}.$$

Therefore, the limit in Equation (58) is given by

$$\lim_{A \to 0} \frac{C([-A, A], 1)}{\bar{C}_{\text{AWGN}}([-A, A], 1)} = \lim_{A \to 0} \frac{A^2}{\log_e(2) \frac{1}{2} \log_2(1 + A^2)} - \frac{1}{\log_e(2) \log_2(e)}$$

$$= \frac{2 \log_e(2)}{\log_e(2)} - \frac{1}{\log_e(2) \log_2(e)}$$

$$= 2 - 1 = 1.$$

The limit in Equation (59) follows from comparing the EPI lower bound $\underline{C}_{\mathsf{EPI}}([-A,A],1) = \frac{1}{2}\log_2\left(1+\frac{2A^2}{\pi e}\right)$ in (52) with the McKellips upper bound $\bar{C}_{\mathsf{McK}}([-A,A],1) = \log_2\left(1+\frac{2A}{\sqrt{2\pi e}}\right)$ given in Equation (48).

Finally, to show Equation (60), observe that

$$
\lim_{A\to\infty} C([-A,A],1) - \bar{C}_{\mathsf{AWGN}}([-A,A],1) = \lim_{A\to\infty} \bar{C}_{\mathsf{McK}}([-A,A],1) - \bar{C}_{\mathsf{AWGN}}([-A,A],1)
$$
$$
= \lim_{A\to\infty} \log_2\left(1+\frac{2A}{\sqrt{2\pi e}}\right) - \frac{1}{2}\log_2(1+A^2)
$$
$$
= \frac{1}{2}\log_2\left(\frac{\pi e}{2}\right).
$$

This concludes the proof. $\square$

8. Conclusions

In this work, we studied the capacity of MIMO channels with bounded input spaces. Several new properties of input distributions that achieve the capacity of such channels have been provided. In particular, it is shown that the support of a capacity-achieving channel input distribution is a set that is small in a topological and measure theoretical sense. In addition to that, it is shown that, if the radius of the underlying channel input space, $\mathcal{X}$, is small enough, then the support of a corresponding capacity-achieving input distribution must necessarily be a subset of the boundary of $\mathcal{X}$. As the considerations on the input distribution have demonstrated that determining the capacity is a very challenging problem, we proposed several new upper and lower bounds that are shown to be tight in the high amplitude regime. An interesting future direction would be to study generalizations of our techniques to wireless optical MIMO channels [42] and other channels such as the wiretap channel [43].

Author Contributions: All authors contributed equally to this work.

Funding: This work was supported in part by the U. S. National Science Foundation under Grants CCF-093970 and CCF-1513915, by the German Research Foundation (DFG) under Grant GO 2669/1-1, and by the European Union's Horizon 2020 Research And Innovation Programme, grant agreement No. 694630.

Conflicts of Interest: The authors declare no conflict of interest.

References

1. Telatar, I.E. Capacity of multi-antenna Gaussian channels. *Eur. Trans. Telecommun.* **1999**, *10*, 585–595.
2. Shannon, C. A mathematical theory of communication. *Bell Syst. Tech. J.* **1948**, *27*.
3. Smith, J.G. The information capacity of amplitude-and variance-constrained scalar Gaussian channels. *Inf. Control* **1971**, *18*, 203–219.
4. Shamai (Shitz), S.; Bar-David, I. The capacity of average and peak-power-limited quadrature Gaussian channels. *IEEE Trans. Inf. Theory* **1995**, *41*, 1060–1071.
5. Rassouli, B.; Clerckx, B. On the capacity of vector Gaussian channels with bounded inputs. *IEEE Trans. Inf. Theory* **2016**, *62*, 6884–6903.
6. Sharma, N.; Shamai (Shitz), S. Transition points in the capacity-achieving distribution for the peak-power limited AWGN and free-space optical intensity channels. *Probl. Inf. Transm.* **2010**, *46*, 283–299.
7. McKellips, A.L. Simple tight bounds on capacity for the peak-limited discrete-time channel. In Proceedings of the 2004 IEEE International Symposium on Information Theory (ISIT), Chicago, IL, USA, 27 June–2 July 2004; pp. 348–348.
8. Thangaraj, A.; Kramer, G.; Böcherer, G. Capacity Bounds for Discrete-Time, Amplitude-Constrained, Additive White Gaussian Noise Channels. Available Online: https://arxiv.org/abs/1511.08742 (accessed on 15 February 2019).
9. Rassouli, B.; Clerckx, B. An Upper Bound for the Capacity of Amplitude-Constrained Scalar AWGN Channel. *IEEE Commun. Lett.* **2016**, *20*, 1924–1926.

10. ElMoslimany, A.; Duman, T.M. On the Capacity of Multiple-Antenna Systems and Parallel Gaussian Channels With Amplitude-Limited Inputs. *IEEE Trans. Commun.* **2016**, *64*, 2888–2899.

11. Dytso, A.; Goldenbaum, M.; Poor, H.V.; Shamai (Shitz), S. A Generalized Ozarow-Wyner Capacity Bound with Applications. In Proceedings of the 2017 IEEE International Symposium on Information Theory (ISIT), Aachen, Germany, 25–30 June 2017; pp. 1058–1062.

12. Dytso, A.; Goldenbaum, M.; Shamai (Shitz), S.; Poor, H.V. Upper and Lower Bounds on the Capacity of Amplitude-Constrained MIMO Channels. In Proceedings of the 2017 IEEE Global Communications Conference (GLOBECOM), Singapore, 4–8 December 2017; pp. 1–6.

13. Cao, P.L.; Oechtering, T.J. Optimal transmit strategy for MIMO channels with joint sum and per-antenna power constraints. In Proceedings of the 2017 IEEE International Conference on Acoustics, Speech and Signal Processing (ICASSP), New Orleans, LA, USA, 5–9 March 2017.

14. Loyka, S. The Capacity of Gaussian MIMO Channels Under Total and Per-Antenna Power Constraints. *IEEE Trans. Commun.* **2017**, *65*, 1035–1043.

15. Loyka, S. On the Capacity of Gaussian MIMO Channels under the Joint Power Constraints. Available online: https://arxiv.org/abs/1809.00056 (accessed on 15 February 2019).

16. Tuninetti, D. On the capacity of the AWGN MIMO channel under per-antenna power constraints. In Proceedings of the 2014 IEEE International Conference on Communications (ICC), Sydney, NSW, Australia, 10–14 June 2014, pp. 2153–2157.

17. Vu, M. MISO capacity with per-antenna power constraint. *IEEE Trans. Commun.* **2011**, *59*, 1268–1274.

18. Chan, T.H.; Hranilovic, S.; Kschischang, F.R. Capacity-achieving probability measure for conditionally Gaussian channels with bounded inputs. *IEEE Trans. Inf. Theory* **2005**, *51*, 2073–2088.

19. Csiszar, I.; Körner, J. *Information Theory: Coding Theorems for Discrete Memoryless Systems*; Cambridge University Press: Cambridge, UK, 2011.

20. Abou-Faycal, I.C.; Trott, M.D.; Shamai, S. The capacity of discrete-time memoryless Rayleigh-fading channels. *IEEE Trans. Inf. Theory* **2001**, *47*, 1290–1301.

21. Fahs, J.; Abou-Faycal, I. On properties of the support of capacity-achieving distributions for additive noise channel models with input cost constraints. *IEEE Trans. Inf. Theory* **2018**, *64*, 1178–1198.

22. Gursoy, M.C.; Poor, H.V.; Verdú, S. The noncoherent Rician fading channel-part I: Structure of the capacity-achieving input. *IEEE Trans. Wireless Commun.* **2015**, *4*, 2193–2206.

23. Katz, M.; Shamai, S. On the capacity-achieving distribution of the discrete-time noncoherent and partially coherent AWGN channels. *IEEE Trans. Inf. Theory* **2004**, *50*, 2257–2270.

24. Ozel, O.; Ekrem, E.; Ulukus, S. Gaussian wiretap channel with amplitude and variance constraints. *IEEE Trans. Inf. Theory* **2015**, *61*, 5553–5563.

25. Bak, J.; Newman, D.J. *Complex Analysis*; Springer Science & Business Media: Berlin, Germany, 2010.

26. Witsenhausen, H.S. Some Aspects of Convexity Useful in Information Theory. *IEEE Trans. Inf. Theory* **1980**, *26*, 265–271.

27. Krantz, S.G.; Parks, H.R. *A Primer of Real Analytic Functions*; Springer Science & Business Media: New York, NY, USA, 2002.

28. Dytso, A.; Yagli, S.; Poor, H.V.; Shamai (Shitz), S. Capacity Achieving Distribution for the Amplitude Constrained Additive Gaussian Channel: An Upper Bound on the Number of Mass Points. Available online: https://arxiv.org/abs/1901.03264 (accessed on 15 February 2019).

29. Ransford, T. *Potential Theory in the Complex Plane*; *London Mathematical Society Student Texts*; Cambridge University Press: Cambridge, UK, 1995; Volume 28.

30. Hatsell, C.; Nolte, L. Some geometric properties of the likelihood ratio (Corresp.). *IEEE Trans. Inf. Theory* **1971**, *17*, 616–618.

31. Dytso, A.; Poor, H.V.; Shamai (Shitz), S. On the Capacity of the Peak Power Constrained Vector Gaussian Channel: An Estimation Theoretic Perspective. Available online: https://arxiv.org/abs/1804.08524 (accessed on 15 February 2019).

32. Dytso, A.; Goldenbaum, M.; Poor, H.V.; Shamai (Shitz), S. When are Discrete Channel Inputs Optimal?—Optimization Techniques and Some New Results. In Proceedings of the 2018 Annual Conference on Information Sciences and Systems (CISS), Princeton, NJ, USA, 21–23 March 2018; pp. 1–6.

33. Guo, D.; Shamai, S.; Verdú, S. Mutual information and minimum mean-square error in Gaussian channels. *IEEE Trans. Inf. Theory* **2005**, *51*, 1261–1282.

34. Dytso, A.; Tuninetti, D.; Devroye, N. Interference as Noise: Friend or Foe? *IEEE Trans. Inf. Theory* **2016**, *62*, 3561–3596.

35. Resnick, S.I. *A Probability Path*; Springer Science & Business Media: New York, NY, USA, 2013.

36. Dytso, A.; Bustin, R.; Tuninetti, D.; Devroye, N.; Shamai (Shitz), S.; Poor, H.V. On the Minimum Mean p-th Error in Gaussian Noise Channels and its Applications. *IEEE Trans. Inf. Theory* **2018**, *64*, 2012–2037.

37. Lozano, A.; Tulino, A.M.; Verdú, S. Optimum power allocation for parallel Gaussian channels with arbitrary input distributions. *IEEE Trans. Inf. Theory* **2006**, *52*, 3033–3051.

38. Pérez-Cruz, F.; Rodrigues, M.R.; Verdú, S. MIMO Gaussian channels with arbitrary inputs: Optimal precoding and power allocation. *IEEE Trans. Inf. Theory* **2010**, *56*, 1070–1084.

39. Abramowitz, M.; Stegun, I.A. *Handbook of Mathematical Functions with Formulas, Graphs, and Mathematical Tables*, 9th dover printing ed.; Dover Publications: New York, NY, USA, 1964.

40. Winkelbauer, A. Moments and Absolute Moments of the Normal Distribution. Available online: https://arxiv.org/abs/1209.4340 (accessed on 15 February 2019).

41. Guo, D.; Wu, Y.; Shamai, S.; Verdú, S. Estimation in Gaussian Noise: Properties of the Minimum Mean-Square Error. *IEEE Trans. Inf. Theory* **2011**, *57*, 2371–2385.

42. Moser, S.M.; Mylonakis, M.; Wang, L.; Wigger, M. Asymptotic Capacity Results for MIMO Wireless Optical Communication. In Proceedings of the 2017 IEEE International Symposium on Information Theory (ISIT), Aachen, Germany, 25–30 June 2017; pp. 536–540.

43. Dytso, A.; Egan, M.; Perlaza, S.; Poor, H.V.; Shamai (Shitz), S. Optimal Inputs for Some Classes of Degraded Wiretap Channels. In Proceedings of the 2018 IEEE Information Theory Workshop (ITW), Guangzhou, China, 25–29 November 2018; pp. 1–5.

Article

Quasi-Concavity for Gaussian Multicast Relay Channels

Mohit Thakur [1] and Gerhard Kramer [2,*]

[1] Independent Researcher, Amalienstr. 49A, 80799 Munich, Germany; thakur.mohit@gmail.com
[2] Institute for Communications Engineering, Technical University of Munich, 80333 Munich, Germany
*   Correspondence: gerhard.kramer@tum.de

Received: 6 January 2019; Accepted: 22 January 2019; Published: 24 January 2019

Abstract: Standard upper and lower bounds on the capacity of relay channels are cut-set (CS), decode-forward (DF), and quantize-forward (QF) rates. For real additive white Gaussian noise (AWGN) multicast relay channels with one source node and one relay node, these bounds are shown to be quasi-concave in the receiver signal-to-noise ratios and the squared source-relay correlation coefficient. Furthermore, the CS rates are shown to be quasi-concave in the relay position for a fixed correlation coefficient, and the DF rates are shown to be quasi-concave in the relay position. The latter property characterizes the optimal relay position when using DF. The results extend to complex AWGN channels with random phase variations.

Keywords: capacity; decode-forward; multicast; relaying

---

1. Introduction

A multicast relay channel (MRC) is an information network with a source node, a relay node, and two or more destination nodes, and where one message originating at the source should be received reliably at the destinations. We consider additive white Gaussian noise (AWGN) MRCs and show that certain information rate expressions are quasi-concave in the receiver signal-to-noise ratios (SNRs), the squared source-relay correlation coefficient, and the relay position. In particular, we study cut-set (CS), decode-forward (DF), and quantize-forward (QF) rates. Quasi-concavity suggests that efficient algorithms can optimize signaling and the relay position. However, the main motivation of this work is not practicality, but simply to provide better understanding of the problem.

Relay positioning has been studied by many authors, with a focus on rate enhancement (e.g., [1,2]), range extension (e.g., [3,4]), and outage probability (e.g., [1,5,6]). We study the problem of placing a relay to maximize the multicast rate by extending results of [7–10]. A preliminary version of this paper without proofs appeared in [11]. Our focus is on real alphabet channels. However, our main results also apply to complex alphabet channels if there are random phase variations so that beamforming is not useful.

This paper is organized as follows. Section 2 presents the MRC model and reviews the CS, DF, and QF rates. Section 3 develops quasi-concavity results in the squared source-relay correlation coefficient ρ^2 and the channel SNRs. Section 4 introduces a distance dependence for the channel gains and shows that the CS rate is quasi-concave in the relay position when ρ is fixed. We further show that the DF rate is quasi-concave in the relay position. Section 5 illustrates quasi-concavity for one-, two-, and three-dimensional networks, and compares the performance of two DF strategies. Section 6 discusses complex AWGN channels and a sum (source plus relay) power constraint. Section 7 concludes the paper. Appendices A and B review useful results on concavity and quasi-concavity, and prove a few new results.

2. Model and Information Rates

2.1. Model

An MRC has three types of nodes:

- a source node s that generates a message W and transmits the symbols $X_s^n = X_{s,1}, X_{s,2}, \ldots, X_{s,n}$;
- a relay node r that receives and forwards symbols $Y_{r,k}$ and $X_{r,k}$, respectively, for $k = 1, 2, \cdots, n$;
- destination nodes $j = 1, 2, \ldots, N$ where node j receives $Y_j^n = Y_{j,1}, Y_{j,2}, \ldots, Y_{j,n}$ and estimates W as $\widehat{W}_j$.

We denote the destination node set as $\mathcal{T} = \{1, 2, \ldots, N\}$. The classic relay channel has $N = 1$ and Figure 1 shows an MRC with $N = 2$.

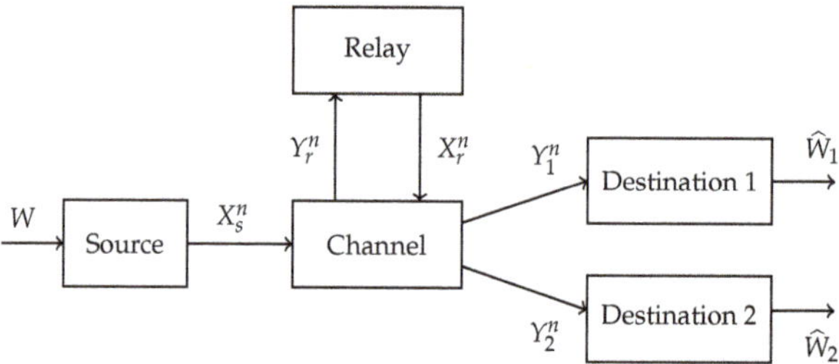

Figure 1. Multicast relay channel (MRC) with two destinations.

A memoryless MRC has a function $h(\cdot)$ and a noise random variable $\mathbf{Z}$ so that for every time instant the $N + 1$ channel outputs $\mathbf{Y} = (Y_r \ Y_1 \ \ldots \ Y_N)$ are given by

$$\mathbf{Y} = h(X_s, X_r, \mathbf{Z}).$$

The noise $\mathbf{Z}$ is statistically independent of X_s and X_r, and the noise variables at different times are statistically independent.

An encoding strategy for M messages has

- W uniformly distributed over $\{1, 2, \ldots, M\}$;
- an encoding function $e_s(\cdot)$ such that $X_s^n = e_s(W)$;
- relay functions $e_{r,k}(\cdot)$ with $X_{r,k} = e_{r,k}(Y_{r,1}, \ldots, Y_{r,k-1})$, where $k = 1, 2, \ldots, n$;
- decoding functions $d_j(\cdot)$ such that $d_j(Y_j^n) = \widehat{W}_j, j \in \mathcal{T}$.

The error probability at destination j is $P_{e,j} = \Pr\left[\widehat{W}_j \neq W\right]$. The multicast rate is $R = (\log_2 M)/n$ bits/use. The rate R is achievable if, for any $\epsilon > 0$ and sufficiently large n, there is an encoding strategy with $P_{e,j} \leq \epsilon$ for all $j \in \mathcal{T}$. The capacity C is the supremum of the achievable rates.

2.2. Information Rates

The following bounds were given in [12] for the relay channel ($N = 1$). Their extensions to MRCs are straightforward.

- CS Rate: $C \leq R_{CS}$ where

$$R_{CS} = \max \left\{ \min_{1 \leq j \leq N} \min \left(I(X_s X_r; Y_j), I(X_s; Y_r Y_j | X_r) \right) \right\} \tag{1}$$

and where the maximization is over all $X_s X_r$.

- Direct-Transmission (DT) Rate: $C \geq R_{DT}$ where

$$R_{DT} = \max\{\min_{1 \leq j \leq N} I(X_s; Y_j | X_r = x^*)\} \tag{2}$$

and where the maximization is over all x^* and X_s.

- DF Rate: $C \geq R_{DF}$ where

$$R_{DF} = \max \left\{ \min_{1 \leq j \leq N} \min \left(I(X_s X_r; Y_j), I(X_s; Y_r | X_r) \right) \right\} \tag{3}$$

and where the maximization is over all $X_s X_r$.

- QF Rate: $C \geq R_{QF}$ where

$$R_{QF} = \max \left\{ \min_{1 \leq j \leq N} \min \left(I(X_s X_r; Y_j) - I(Y_r; \widehat{Y}_r | X_s X_r Y_j), I(X_s; \widehat{Y}_r Y_j | X_r) \right) \right\} \tag{4}$$

where $\widehat{Y}_r$ is an auxiliary random variable, and where the maximization is over all $X_s X_r \widehat{Y}_r$ such that X_s and X_r are independent and $X_s - X_r Y_r - \widehat{Y}_r$ forms a Markov chain.

2.3. Real Alphabet AWGN MRC

The real alphabet AWGN MRC has real channel symbols and

$$Y_r = a_{s,r} X_s + Z_r \tag{5}$$
$$Y_j = a_{s,j} X_s + a_{r,j} X_r + Z_j \tag{6}$$

where $j \in \mathcal{T}$. The $a_{s,r}$, $a_{s,j}$, and $a_{r,j}$ are channel gains between the nodes (see Figure 2). We later relate these gains to distances between the nodes. The Z_r and Z_j, $j = 1, 2, \ldots, N$, are independent and identically distributed Gaussian random variables with zero mean and unit variance. We may alternatively write (5) and (6) in vector form as

$$\mathbf{Y}_j = \mathbf{A}_j \mathbf{X} + \mathbf{Z}_j \tag{7}$$

where $\mathbf{X} = (X_s \ X_r)^T$, $\mathbf{Y}_j = (Y_r \ Y_j)^T$, $\mathbf{Z} = (Z_r \ Z_j)^T$, and

$$\mathbf{A}_j = \begin{pmatrix} a_{s,r} & 0 \\ a_{s,j} & a_{r,j} \end{pmatrix}. \tag{8}$$

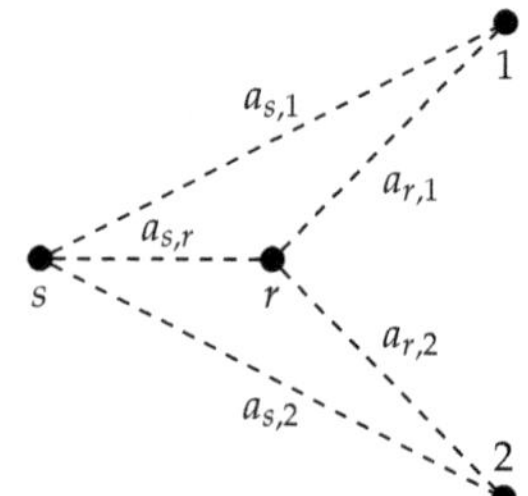

Figure 2. AWGN MRC with two destinations.

We consider individual average block power constraints

$$\mathrm{E}\left[\sum_{k=1}^{n} X_{s,k}^2\right] \le nP_s, \quad \mathrm{E}\left[\sum_{k=1}^{n} X_{r,k}^2\right] \le nP_r. \tag{9}$$

The SNR and the capacity of the link from node u (with transmit power P_u) to node v are the respective

$$\mathrm{SNR}_{u,v} = a_{u,v}^2 P_u \tag{10}$$

$$\mathrm{C}(\mathrm{SNR}_{u,v}) = \frac{1}{2}\log\left(1 + \mathrm{SNR}_{u,v}\right). \tag{11}$$

We simplify the above rate bounds for the AWGN MRC.

- CS Rate:

$$R_{CS} = \max_{\rho} \left[\min_{1 \le j \le N} \min\left(\mathrm{C}\left(\mathrm{SNR}_{s,j} + \mathrm{SNR}_{r,j} + 2\rho\sqrt{\mathrm{SNR}_{s,j}\mathrm{SNR}_{r,j}}\right), \right.\right.$$
$$\left.\left. \mathrm{C}\left((1-\rho^2)(\mathrm{SNR}_{s,j} + \mathrm{SNR}_{s,r})\right)\right)\right] \tag{12}$$

where the correlation coefficient ρ satisfies $|\rho| \le 1$. One can restrict attention to non-negative ρ.

- DT Rate:

$$R_{DT} = \min_{1 \le j \le N} \mathrm{C}(\mathrm{SNR}_{s,j}). \tag{13}$$

- DF Rate:

$$R_{DF} = \max_{\rho} \left[\min_{1 \le j \le N} \min\left(\mathrm{C}(\mathrm{SNR}_{s,j} + \mathrm{SNR}_{r,j} + 2\rho\sqrt{\mathrm{SNR}_{s,j}\mathrm{SNR}_{r,j}}), \mathrm{C}((1-\rho^2)\mathrm{SNR}_{s,r})\right)\right]. \tag{14}$$

One can again restrict attention to non-negative ρ.

- QF Rate: Optimizing $X_s X_r \widehat{Y}_r$ seems difficult. Instead, we choose X_s and X_r to be zero-mean Gaussian with variances P_s and P_r, respectively. We further choose $\widehat{Y}_r = Y_r + Z_r$ where Z_r is zero-mean Gaussian with variance N_r. Optimizing N_r gives (see [13], pp. 336–337)

$$\widetilde{R}_{QF} = \min_{1 \le j \le N} \mathrm{C}\left(\mathrm{SNR}_{s,j} + \frac{\mathrm{SNR}_{r,j}\mathrm{SNR}_{s,r}}{\mathrm{SNR}_{s,j} + \mathrm{SNR}_{r,j} + \mathrm{SNR}_{s,r} + 1}\right). \tag{15}$$

3. Quasi-Concavity in SNRs and ρ^2

3.1. CS Rate

We consider two characterizations of R_{CS}. First, let $\mathbf{a}_j^T = (a_{s,j}\ a_{r,j})$ be the second row of $\mathbf{A}_j$, let $\mathbf{Q_X}$ be the covariance matrix of $\mathbf{X}$ (see Appendix A), and let $\det \mathbf{M}$ be the determinant of the square matrix $\mathbf{M}$. The CS rate (12) can be expressed as the maximum of

$$R_{CS}(\mathbf{Q_X}) = \min_{1 \le j \le N} \min \left(\frac{1}{2} \log \left(\mathbf{a}_j^T \mathbf{Q_X}\, \mathbf{a}_j + 1 \right), \frac{1}{2} \log \left(\frac{\det \mathbf{Q}_{(\mathbf{Y}_j^T\, X_r)^T}}{P_r} \right) \right) \tag{16}$$

over the convex set of $\mathbf{Q_X}$ with diagonal entries P_s and P_r. The first logarithm in (16) is clearly concave in $\mathbf{Q_X}$. The second logarithm is concave in $\mathbf{Q}_{(\mathbf{Y}_j^T\, X_r)^T}$ (see Appendix A) and $\mathbf{Q}_{(\mathbf{Y}_j^T\, X_r)^T}$ is linear in $\mathbf{Q_X}$. To prove the latter claim, observe that

$$\mathbf{Q}_{(\mathbf{Y}_j^T\, X_r)^T} = \tilde{\mathbf{A}}_j \mathbf{Q_X} \tilde{\mathbf{A}}_j^T + \begin{pmatrix} \mathbf{I}_2 & \mathbf{0} \\ \mathbf{0} & 0 \end{pmatrix} \tag{17}$$

where $\tilde{\mathbf{A}}_j^T = \left(\mathbf{A}_j^T\ [0\ 1]^T \right)$ and $\mathbf{I}_2$ is the 2×2 identity matrix. Hence $R_{CS}(\mathbf{Q_X})$ is concave in (the convex set of) $\mathbf{Q_X}$ because it is the minimum of $2N$ concave functions.

Suppose next that we wish to consider ρ and the SNRs individually rather than via $\mathbf{Q_X}$. Define the vector

$$\mathbf{S} = (\mathrm{SNR}_{s,r}, \mathrm{SNR}_{s,1}, \cdots, \mathrm{SNR}_{s,N}, \mathrm{SNR}_{r,1}, \cdots, \mathrm{SNR}_{r,N}) \tag{18}$$

and the functions

$$f_j(\rho, \mathbf{S}) = \mathrm{SNR}_{s,j} + \mathrm{SNR}_{r,j} + 2\rho \sqrt{\mathrm{SNR}_{s,j}\mathrm{SNR}_{r,j}} \tag{19}$$

$$g_j(\rho, \mathbf{S}) = (1 - \rho^2)\left(\mathrm{SNR}_{s,j} + \mathrm{SNR}_{s,r} \right) \tag{20}$$

$$R_{CS}(\rho, \mathbf{S}) = \min_{1 \le j \le N} \min \left(C(f_j(\rho, \mathbf{S})), C(g_j(\rho, \mathbf{S})) \right). \tag{21}$$

We establish the following results. We restrict attention to $0 \le \rho \le 1$ and positive $\mathbf{S}$.

Lemma 1. $f_j(\rho, \mathbf{S})$ *and* $g_j(\rho, \mathbf{S})$ *are concave in* ρ*, concave in* $\mathbf{S}$*, and quasi-concave in* $(\rho^2, \mathbf{S})$*.*

Proof. Concavity with respect to ρ is established by observing that $f_j(\rho, \mathbf{S})$ is linear in ρ, and $g_j(\rho, \mathbf{S})$ is linear in $-\rho^2$ which is concave in ρ.

Consider next concavity with respect to $\mathbf{S}$. The Hessian of $f_j(\rho, \mathbf{S})$ with respect to $\mathbf{S}$ has only one non-zero eigenvalue

$$-\frac{\rho}{2} \cdot \frac{\mathrm{SNR}_{s,j}^2 + \mathrm{SNR}_{r,j}^2}{\mathrm{SNR}_{s,j}^{3/2}\mathrm{SNR}_{r,j}^{3/2}}. \tag{22}$$

Thus, $f_j(\rho, \mathbf{S})$ is concave in $\mathbf{S}$ for non-negative ρ and positive $\mathbf{S}$. The function $g_j(\rho, \mathbf{S})$ is linear in $\mathbf{S}$, and thus concave in $\mathbf{S}$.

Now consider quasi-concavity with respect to $(\rho^2, \mathbf{S})$. Substituting $a = \mathrm{SNR}_{s,j}, b = \mathrm{SNR}_{r,j}, c = \rho^2$ into the fifth function of Lemma A6 in Appendix B, we find that $f_j(\rho, \mathbf{S})$ is quasi-concave in $(\rho^2, \mathbf{S})$. For the $g_j(\rho, \mathbf{S})$, observe that ab is quasi-concave for non-negative (a, b), see the first function of Lemma A6. This implies (see (A7))

$$(\lambda a_1 + \bar{\lambda} a_2)(\lambda b_1 + \bar{\lambda} b_2) \geq \min(a_1 b_1, a_2 b_2) \tag{23}$$

for $0 \leq \lambda \leq 1$, and where $\bar{\lambda} = 1 - \lambda$. Substituting $a_i = 1 - \rho_i^2$ and $b_i = \mathrm{SNR}_{s,j,i} + \mathrm{SNR}_{s,r,i}$ for $i = 1, 2$, we find that $g_j(\rho, \mathbf{S})$ is quasi-concave in $(\rho^2, \mathbf{S})$. $\quad\square$

Theorem 1. *$R_{CS}(\rho, \mathbf{S})$ is concave in ρ, concave in $\mathbf{S}$, and quasi-concave in $(\rho^2, \mathbf{S})$.*

Proof. $R_{CS}(\rho, \mathbf{S})$ involves taking logarithms and minima of (quasi-) concave functions. The results thus follow by applying Lemma 1 above and Lemma A5, Parts 2 and 3, in Appendix B. $\quad\square$

Corollary 1. *Consider $\mathbf{S}$ as a function of $\mathbf{P} = (P_s, P_r)$. Then $R_{CS}(\rho, \mathbf{S}(\mathbf{P}))$ is quasi-concave in $(\rho^2, \mathbf{P})$.*

Proof. The proof follows from the proof of Theorem 1 and because $\mathbf{S}$ is a linear function of $\mathbf{P}$. $\quad\square$

To illustrate the quasi-concavity, consider one relay and the channel gains $a_{s,r} = 5/2$, $a_{s,1} = 1$, and $a_{r,1} = 5/3$. This scenario corresponds to the geometry in Section 5.1 with $\mathbf{r} = 0.4$. Figure 3 shows a contour plot of $R_{CS}(\rho, \mathbf{S}(\mathbf{P}))$ when $P_s = 1$. Observe that the contour lines form convex regions, as predicted by Corollary 1.

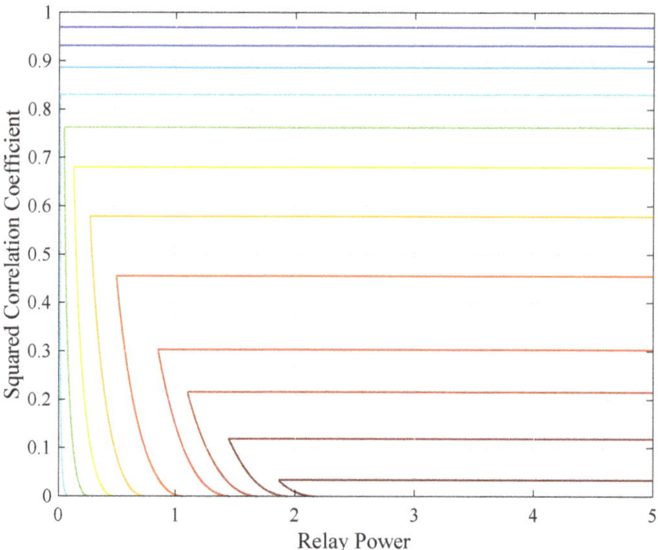

Figure 3. Contour plot of $R_{CS}(\rho, \mathbf{S}(\mathbf{P}))$ when $P_s = 1$.

3.2. DF Rate

Consider the functions

$$g_j^*(\rho, \mathbf{S}) = (1 - \rho^2)\text{SNR}_{s,r} \tag{24}$$

$$R_{DF}(\rho, \mathbf{S}) = \min_{1 \leq j \leq N} \min \left(C(f_j(\rho, \mathbf{S})), C(g_j^*(\rho, \mathbf{S})) \right). \tag{25}$$

As above, we restrict attention to $0 \leq \rho \leq 1$ and positive $\mathbf{S}$.

Theorem 2. *$R_{DF}(\rho, \mathbf{S})$ is concave in ρ, concave in $\mathbf{S}$, and quasi-concave in $(\rho^2, \mathbf{S})$.*

Proof. The proof is similar to that of Theorem 1. $\square$

Corollary 2. *$R_{DF}(\rho, \mathbf{S}(\mathbf{P}))$ is quasi-concave in $(\rho^2, \mathbf{P})$.*

Proof. See the proof of Corollary 1. $\square$

3.3. DT Rate

The DT rate (13) is clearly concave in $\mathbf{S}$ and $\mathbf{P}$.

3.4. QF Rate

Consider the functions

$$h_j(\mathbf{S}) = \text{SNR}_{s,j} + \frac{\text{SNR}_{r,j}\text{SNR}_{s,r}}{\text{SNR}_{s,j} + \text{SNR}_{r,j} + \text{SNR}_{s,r} + 1} \tag{26}$$

$$\widetilde{R}_{QF}(\mathbf{S}) = \min_{1 \leq j \leq N} C(h_j(\mathbf{S})). \tag{27}$$

We establish the following results. We restrict attention to non-negative $\mathbf{S}$.

Lemma 2. *$h_j(\mathbf{S})$ is quasi-concave in $(\text{SNR}_{r,j}, \text{SNR}_{s,r})$.*

Proof. Substitute $a = \text{SNR}_{r,j}, b = \text{SNR}_{s,r}, k = \text{SNR}_{s,j} + 1$ into the second function of Lemma A6 in Appendix B, and apply Lemma A5, Part 1. $\square$

Theorem 3. *$\widetilde{R}_{QF}(\mathbf{S})$ is quasi-concave in $\mathbf{S}$ if the $\text{SNR}_{s,j}, j = 1, 2, \ldots, n$, are held fixed.*

Proof. Apply Lemma 2 above and Lemma A5, Parts 2 and 3, in Appendix B. $\square$

4. Quasi-Concavity in Relay Position

Suppose the channel gain for the node pair (i, j) is

$$a_{i,j} = \sqrt{\xi_{i,j}} \Big/ D_{i,j}^{\alpha/2} \tag{28}$$

where $\xi_{i,j}$ is a "fading" gain, $D_{i,j} = \|\mathbf{i} - \mathbf{j}\|$ is the Euclidean distance between the positions $\mathbf{i}$ and $\mathbf{j}$ of nodes i and j, respectively, and $\alpha \geq 2$ is a path-loss exponent. We thus have

$$\text{SNR}_{i,j} = \frac{\xi_{i,j} P_i}{D_{i,j}^{\alpha}} = \frac{\xi_{i,j} P_i}{\|\mathbf{i} - \mathbf{j}\|^{\alpha}}.$$

We establish quasi-concavity results in ρ^2 and $\mathbf{r}$, where $\mathbf{r}$ is the position of the relay node.

4.1. CS Rate

Consider the functions (19)–(21) but relabeled as $f_j(\rho, \mathbf{r})$, $g_j(\rho, \mathbf{r})$, and $R_{CS}(\rho, \mathbf{r})$ to emphasize the dependence on the considered parameters. We again consider $0 \leq \rho \leq 1$ and positive $\mathbf{S}$.

Lemma 3. *$f_j(\rho, \mathbf{r})$ and $g_j(\rho, \mathbf{r})$ are quasi-concave in $\mathbf{r}$ for fixed ρ. Furthermore, $f_j(\rho, \mathbf{r})$ is quasi-concave in $(\rho^2, \mathbf{r})$.*

Proof. Consider the functions

$$\tilde{f}_j(\rho, D^{\alpha}) = \frac{\xi_{s,j} P_s}{D_{s,j}^{\alpha}} + \frac{\xi_{r,j} P_r}{D^{\alpha}} + 2\rho \sqrt{\frac{\xi_{s,j} P_s}{D_{s,j}^{\alpha}} \frac{\xi_{r,j} P_r}{D^{\alpha}}} \tag{29}$$

$$\tilde{g}_j(\rho, D^{\alpha}) = (1 - \rho^2) \left(\frac{\xi_{s,j} P_s}{D_{s,j}^{\alpha}} + \frac{\xi_{s,r} P_s}{D^{\alpha}} \right) \tag{30}$$

which are quasi-linear in D^{α} for fixed ρ since they are decreasing in D^{α}. However, $D_{r,j}^{\alpha}$ is a convex function of $\mathbf{r}$ for $\alpha \geq 1$, and thus Lemma A5, Part 5, in Appendix B establishes that $f_j(\rho, \mathbf{r})$ is quasi-concave in $\mathbf{r}$ for fixed ρ. Similarly, $D_{s,r}^{\alpha}$ is a convex function of $\mathbf{r}$ for $\alpha \geq 1$, and we find that $g_j(\rho, \mathbf{r})$ is quasi-concave in $\mathbf{r}$ for fixed ρ.

Next, substitute $a = D^{\alpha}$ and $b = \rho^2$ into the third function of Lemma A6, and use Lemma A5, Part 1, to show that $\tilde{f}_j(\rho, D^{\alpha})$ is quasi-concave in (ρ^2, D^{α}). However, $\tilde{f}_j$ is decreasing in D^{α} and $D_{r,j}^{\alpha}$ is convex in $\mathbf{r}$, so Lemma A5, Part 5, establishes that $f_j(\rho, \mathbf{r})$ is quasi-concave in $(\rho^2, \mathbf{r})$. $\square$

Unfortunately, $\tilde{g}_j$ is quasi-*convex* (and not quasi-concave) in (ρ^2, D^{α}). To see this, substitute $a = D^{\alpha}$ and $b = \rho^2$ into the fourth function of Lemma A6. Quasi-concavity would have been useful since it would have permitted using Lemma A5, Parts 2 and 4, to establish the quasi-concavity of

$$R_{CS}(\mathbf{r}) = \max_{\rho} \left[\min_{1 \leq j \leq N} \min \left(C(f_j(\rho, \mathbf{r})), C(g_j(\rho, \mathbf{r})) \right) \right]. \tag{31}$$

However, we have been unable to prove this, and our numerical results suggest that $R_{CS}(\rho, \mathbf{r})$ is not quasi-concave in $(\rho^2, \mathbf{r})$. Nevertheless, Lemma 3 suffices to establish an intermediate result which is useful in Section 5 when we study $\rho = 0$.

Theorem 4. *$R_{CS}(\rho, \mathbf{r})$ is quasi-concave in $\mathbf{r}$ for fixed ρ, $0 \leq \rho \leq 1$.*

Proof. $R_{CS}(\rho, \mathbf{r})$ is the minimum of functions that are quasi-concave in $\mathbf{r}$. Lemma A5, Part 2, thus establishes the theorem. $\square$

4.2. DF Rate

The quasi-convexity of $\tilde{g}_j(\rho, D^\alpha)$ relaxes for the DF rate (25). Consider the negative of the fourth function of Lemma A6 in Appendix B with $k_1 = 0$:

$$f(a,b) = (1-b)k_2/a. \tag{32}$$

This function is quasi-linear in (a,b) since both its superlevel and sublevel sets are convex. This result implies the following theorem. We again consider the functions (24)–(25) but relabeled as $g_j^*(\rho, \mathbf{r})$ and $R_{DF}(\rho, \mathbf{r})$. We further define

$$\tilde{g}_j^*(\rho, D^\alpha) = (1-\rho^2)\frac{\zeta_{s,r}P_s}{D^\alpha} \tag{33}$$

$$R_{DF}(\mathbf{r}) = \max_\rho \left[\min_{1 \le j \le N} \min\left(\mathsf{C}(f_j(\rho, \mathbf{r})), \mathsf{C}(g_j^*(\rho, \mathbf{r})) \right) \right]. \tag{34}$$

As above, we consider $0 \le \rho \le 1$ and positive $\mathbf{S}$.

Theorem 5. $R_{DF}(\rho, \mathbf{r})$ *is quasi-concave in* $(\rho^2, \mathbf{r})$*, and* $R_{DF}(\mathbf{r})$ *is quasi-concave in* $\mathbf{r}$*.*

Proof. $\tilde{g}_j^*(\rho, D^\alpha)$ is quasi-linear in (ρ^2, D^α) and decreasing in D^α. Furthermore, $D_{s,r}^\alpha$ is convex in $\mathbf{r}$, and thus Lemma A5, Part 5, in Appendix B establishes that $g_j^*(\rho, \mathbf{r})$ is quasi-concave in $(\rho^2, \mathbf{r})$. $R_{DF}(\rho, \mathbf{r})$ is therefore quasi-concave in $\mathbf{r}$, as it is the minimum of quasi-concave functions (see Lemma A5, Part 2). Furthermore, $R_{DF}(\mathbf{r})$ is concave in $\mathbf{r}$ by Lemma A5, Part 4. $\square$

5. DF Performance

This section presents numerical results for the DF strategy and compares them to results from [7–9]. We consider 1-, 2-, and 3-dimensional MRCs with different numbers N of destination nodes. For simplicity, we consider the low SNR or broadband regime where

$$\mathsf{C}(\mathrm{SNR}) = \frac{1}{2}\log(1 + \mathrm{SNR}) \to \frac{1}{2}\mathrm{SNR}. \tag{35}$$

In other words, we consider the CS and DF rates without the logarithms. This approach is valid not only in the limit of low SNR, but more generally because we proved our quasi-concavity results without taking logarithms. Furthermore, in the low SNR regime the rates of full-duplex and half-duplex transmission are the same under a block power constraint.

We choose $P_s = P_r = P = 1$, $\alpha = 2$, and $\zeta_{u,v} = 1$ for all node pairs (u,v). We study both coherent transmission where ρ is optimized and non-coherent transmission with $\rho = 0$. The rates are in nats/channel use. Alternatively, suppose we use sync pulses sampled at $2W$ samples per second, where W is the (one-sided) signal bandwidth. Suppose further that the (one-sided) noise power spectral density is 1 Watt/Hz. Then at low SNR the rates in nats/channel use are the same as the rates in nats/sec.

5.1. One Dimension

Consider a relay channel ($N = 1$) where the source is at the origin ($\mathbf{s} = 0$) and the destination is at point 1 ($\mathbf{1} = 1$). Figure 4 shows the low SNR CS rates, DF rates, and the routing-based DF (RDF) rates developed in [7], which are given by

$$R_{CS} \to \frac{1}{2} \min \left(\frac{\zeta_{s,1} P_s}{\|\mathbf{s} - \mathbf{1}\|^\alpha} + \frac{\zeta_{r,1} P_r}{\|\mathbf{r} - \mathbf{1}\|^\alpha} + \frac{2\rho \sqrt{\zeta_{s,1} \zeta_{r,1} P_s P_r}}{\|\mathbf{s} - \mathbf{1}\|^{\alpha/2} \|\mathbf{r} - \mathbf{1}\|^{\alpha/2}}, \ (1 - \rho^2) \left(\frac{\zeta_{s,1} P_s}{\|\mathbf{s} - \mathbf{1}\|^\alpha} + \frac{\zeta_{s,r} P_s}{\|\mathbf{s} - \mathbf{r}\|^\alpha} \right) \right) \tag{36}$$

$$R_{DF} \to \frac{1}{2} \min \left(\frac{\zeta_{s,1} P_s}{\|\mathbf{s} - \mathbf{1}\|^\alpha} + \frac{\zeta_{r,1} P_r}{\|\mathbf{r} - \mathbf{1}\|^\alpha} + \frac{2\rho \sqrt{\zeta_{s,1} \zeta_{r,1} P_s P_r}}{\|\mathbf{s} - \mathbf{1}\|^{\alpha/2} \|\mathbf{r} - \mathbf{1}\|^{\alpha/2}}, \ (1 - \rho^2) \frac{\zeta_{s,r} P_s}{\|\mathbf{s} - \mathbf{r}\|^\alpha} \right) \tag{37}$$

$$R_{RDF} \to \max_{0 \le \beta \le 1} \frac{1}{2} \left[\min \left(\frac{\zeta_{r,1} P_r}{\|\mathbf{r} - \mathbf{1}\|^\alpha}, \frac{\beta \zeta_{s,r} P_s}{\|\mathbf{s} - \mathbf{r}\|^\alpha} \right) + \frac{(1 - \beta) \zeta_{s,1} P_s}{\|\mathbf{s} - \mathbf{1}\|^\alpha} \right]. \tag{38}$$

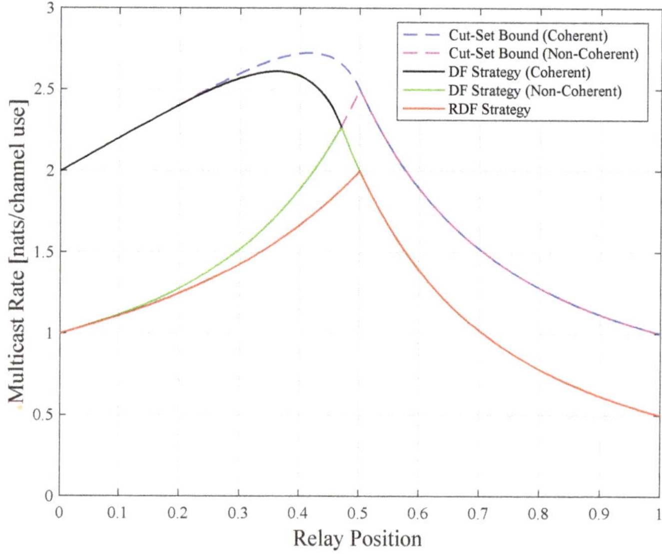

Figure 4. Relay channel rates for low signal-to-noise ratio (SNR) and $P = 1$.

Observe that all curves are quasi-concave (but not concave) in $\mathbf{r}$. Theorems 4 and 5 predict the quasi-concavity for all curves except for the coherent CS rates. Observe also that the curves for the coherent and non-coherent rates merge for relay positions exceeding a certain value ($r = 0.5$ and $r \approx 0.47$ for the respective CS and DF rates). The reason for this behavior is that $\rho = 0$ is optimal for the coherent CS and DF rates beyond these positions, see the ρ curve in [1] (Figure 16). Furthermore, the non-coherent CS rates coincide with the non-coherent DF rates for a large range of $\mathbf{r}$.

The best relay positions for the two strategies are different. For example, $\mathbf{r} = 0.5$ maximizes R_{RDF} while the $\mathbf{r}$ maximizing R_{DF} is closer to the source. This is because when the source transmits, the relay and the destination listen, and the destination "collects" information. The relay can thus be positioned closer to the source while maintaining the same information rate from the source to the relay, and from the source-relay pair to the destination. At the optimal positions, we compute $R_{DF} \approx 2.26P$ nats/sec and $R_{RDF} = 2P$ nats/sec, so the DF gain is $\approx 13\%$.

Finally, we illustrate that $R_{DF}(\rho, \mathbf{r})$ is quasi-concave in $(\rho^2, \mathbf{r})$ in Figure 5. The contour lines form convex regions, as predicted by Theorem 5.

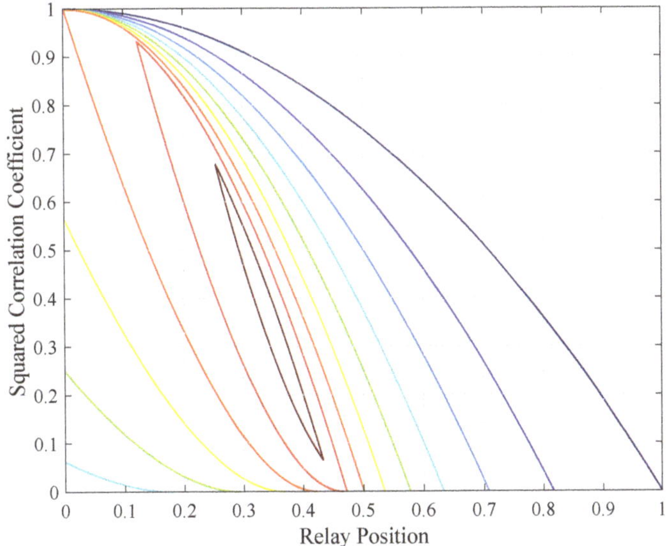

Figure 5. Contour plot of $R_{DF}(\rho, \mathbf{r})$ in (37).

5.2. Two Dimensions

Consider $N = 5$ destinations positioned on a square in the two-dimensional Euclidean plane with the source node at the origin. Figure 6a plots the node positions as circles, and the non-coherent R_{DF} as a function of the relay position. The best relay position is shown by a circle labeled $\mathbf{r}^*_{DF}$ and the corresponding rate is $R_{DF} \approx 0.011P$ nats/sec. Figure 6c plots the low SNR two-hop rate

$$R_{2H} \to \min_{1 \leq j \leq 5} \frac{1}{2} \min \left(\frac{\xi_{s,r} P_s}{\|\mathbf{s} - \mathbf{r}\|^{\alpha}}, \frac{\xi_{r,j} P_r}{\|\mathbf{r} - \mathbf{j}\|^{\alpha}} \right) \tag{39}$$

as a function of the relay position. The best relay position is shown by a circle labeled $\mathbf{r}^*_{2H}$ and the corresponding two-hop rate is $R_{2H} = 0.01P$ nats/sec. The non-coherent DF gain is thus $\approx 10\%$.

Figure 6b,d shows contour plots for R_{DF} and R_{2H}. The contours form convex regions, as predicted by Theorem 5. Again, the relay position maximizing R_{DF} lies closer to the source than the relay position maximizing R_{2H}.

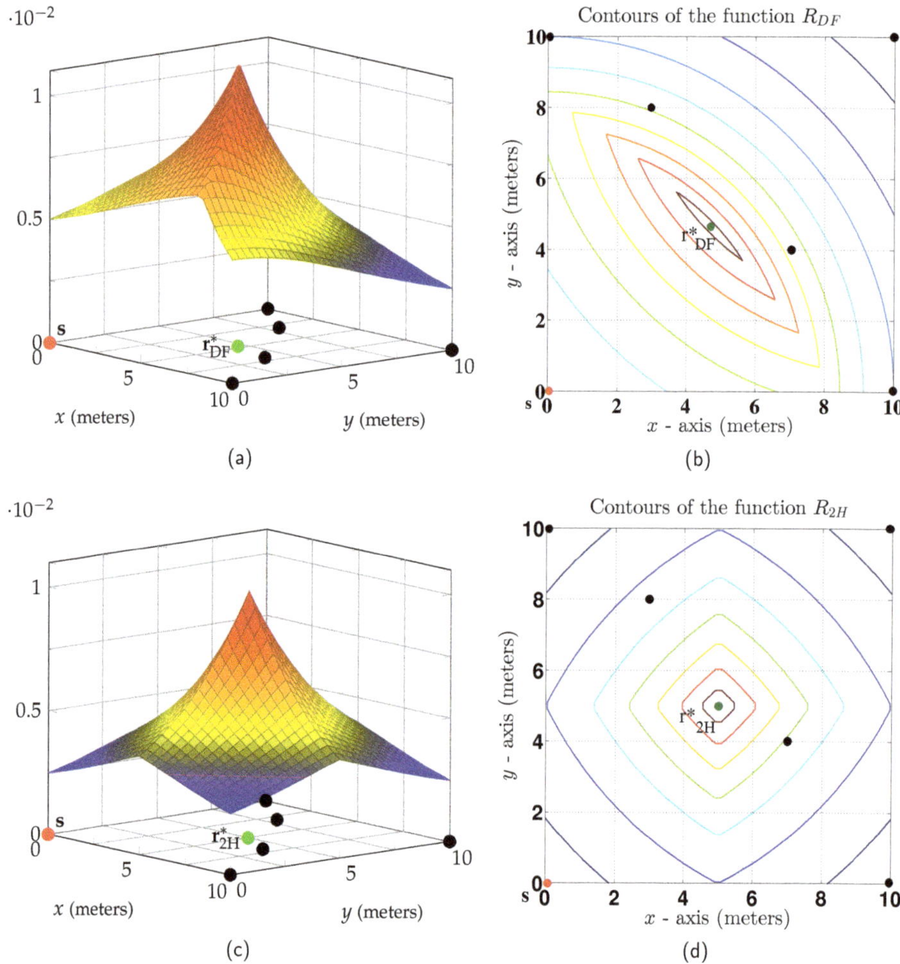

Figure 6. (a) R_{DF} for $N = 5$; (b) R_{DF} contour plot; (c) R_{2H} for the same network; (d) R_{2H} contour plot.

5.3. Three Dimensions

Consider $N = 5$ destinations positioned in 3-dimensional Euclidean space as in Figure 7. The figure also shows the convex hull (a polyhedron) of the points. The points $\mathbf{r}_{DF}^*$ and $\mathbf{r}_{2H}^*$ denote the relay positions that maximize the non-coherent R_{DF} and R_{2H}, respectively. We remark that $\mathbf{r}_{DF}^*$ and $\mathbf{r}_{2H}^*$ remain unchanged if more destinations are positioned inside the polyhedron. This is because the points in the polyhedron receive at least the same rate as the worst of the five nodes at the corner points.

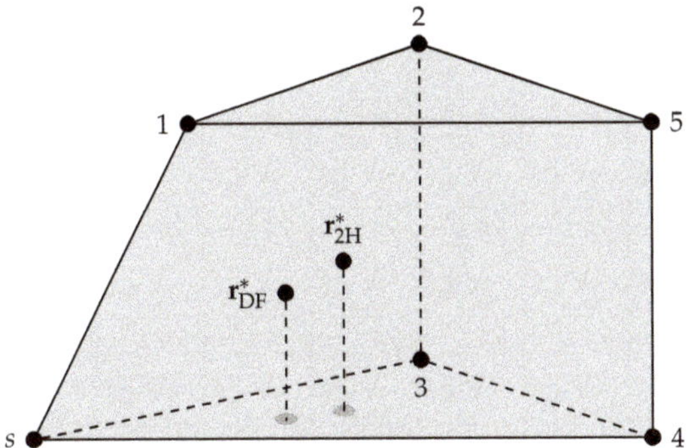

Figure 7. $N = 5$ destination geometry in three dimensions.

6. Discussion

6.1. Complex AWGN Channels

For complex-alphabet AWGN channels, we could replace (5) and (6) by adding phases $\phi_{i,j}$ for $i = s, r$ and $j = 1, 2, \ldots, N$ as follows:

$$Y_r = a_{s,r}\, e^{j\phi_{s,r}}\, X_s + Z_r \tag{40}$$

$$Y_j = a_{s,j}\, e^{j\phi_{s,j}}\, X_s + a_{r,j}\, e^{j\phi_{r,j}}\, X_r + Z_j \tag{41}$$

where the noise variables Z_r and Z_j, $j = 1, 2, \ldots, N$, are independent, identically distributed, circularly symmetric, complex, Gaussian random variables with zero mean and unit variance. The distance dependence of $a_{i,j}$ can be chosen as in (28) and the phase dependence as

$$\phi_{i,j} = 2\pi D_{i,j} / \lambda \tag{42}$$

where $\lambda = c/f_o$ is the wavelength, c is the speed of light, and f_o is the carrier frequency.

For example, the DF rate (14), normalized by the number of real dimensions, is

$$R_{DF} = \max_{\rho} \left[\min_{1 \leq j \leq N} \min \left(C(\mathsf{SNR}_{s,j} + \mathsf{SNR}_{r,j} + \right. \right.$$
$$\left. \left. 2\Re \left\{ \rho e^{j(\phi_{s,j} - \phi_{r,j})} \right\} \sqrt{\mathsf{SNR}_{s,j} \mathsf{SNR}_{r,j}} \right), C((1 - |\rho|^2)\mathsf{SNR}_{s,r}) \right) \right] \tag{43}$$

where the complex correlation coefficient ρ satisfies $0 \leq |\rho| \leq 1$. Observe that for a classic relay channel, with $N = 1$ destination, one can choose ρ to make $\Re \left\{ \rho e^{j(\phi_{s,1} - \phi_{r,1})} \right\}$ real and non-negative, as for real alphabet AWGN channels. However, for $N \geq 2$ one must choose complex ρ in general. Furthermore, the quasi-concavity in $\mathbf{r}$ will not be valid in general because the phases $\phi_{r,j}$ change with $\mathbf{r}$, and we cannot optimize ρ for each destination node separately. However, we remark that this effect is "local" in the sense that for large carrier frequencies the phase variations are sensitive to changes in $\mathbf{r}$. A pragmatic approach would then be to optimize $\mathbf{r}$ for non-coherent transmission ($\rho = 0$) even if beamforming is permitted. Furthermore, if the channel exhibits random phase variations, then the best approach is to choose $\rho = 0$ (see [1], Figure 18) in which case we have quasi-concavity for both the CS and DF rates. Finally, we remark that it might be interesting to consider quasi-concavity in the correlation coefficients for problems where the source and relay have sufficiently many antennas to overcome the problem outlined above.

6.2. Sum-Power Constraint

For some applications, it is interesting to consider a *sum*-power constraint

$$\mathrm{E}\left[\sum_{k=1}^{n} |X|_{s,k}^2 + |X|_{r,k}^2\right] \leq nP_T. \tag{44}$$

As is usually done, we set $P_r = P_T - P_s$ and consider $P_s, 0 \leq P_s \leq P_T$ as a new optimization parameter. One might now hope that $R_{CS}(P_s, \mathbf{r})$ or $R_{DF}(P_s, \mathbf{r})$ are quasi-concave in $(P_s, \mathbf{r})$ for fixed ρ, or at least for $\rho = 0$. Unfortunately, we have found counterexamples that show this is not the case. The rate functions do seem to have interesting properties, however, and these deserve further exploration.

7. Conclusions

Various quasi-concavity results were established for AWGN MRCs. In particular, the CS rates are quasi-concave in the relay position for a fixed correlation coefficient (Theorem 4) and the DF rates are quasi-concave in the relay position (Theorem 5).

Author Contributions: Both authors conceived the problem and solution and wrote the paper.

Funding: This research was funded by the German Ministry of Education and Research in the framework of an Alexander von Humboldt Professorship.

Acknowledgments: The authors thank the reviewers for their comments that helped to improve the paper.

Conflicts of Interest: The authors declare no conflicts of interest.

Appendix A. Covariance Matrices and Concavity

The covariance matrix of a real-valued random column vector $\mathbf{V}$ is

$$\mathbf{Q_V} = \mathrm{E}\left[(\mathbf{V} - \mathrm{E}\,[\mathbf{V}])(\mathbf{V} - \mathrm{E}\,[\mathbf{V}])^T\right]. \tag{A1}$$

A useful property of covariance matrices is as follows (see [14], p. 684). If $\mathbf{Q_V^*}$ is a principal minor of $\mathbf{Q_V}$, then the following function is concave in $\mathbf{Q_V}$:

$$f(\mathbf{Q_V}) = \log \frac{\det \mathbf{Q_V}}{\det \mathbf{Q_V^*}}. \tag{A2}$$

Appendix B. Concave and Quasi-Concave Functions

We review results on quasi-concavity, and then establish quasi-concavity for several functions.

Appendix B.1. Definitions

Consider the following sets. The *domain* $\mathcal{D}_f$ of a real-valued function $f : \mathbf{R}^n \to \mathbf{R}$ is the set of arguments for which f is defined. The *hypograph* and *hypergraph* of f are the respective

$$\mathcal{H}_f = \{(\mathbf{x}, y) \,|\, y \leq f(\mathbf{x})\}, \qquad \widehat{\mathcal{H}}_f = \{(\mathbf{x}, y) \,|\, y \geq f(\mathbf{x})\}. \tag{A3}$$

The *superlevel* and *sublevel* sets of f with respect to $\beta \in \mathbf{R}$ are the respective

$$\mathcal{S}_{f,\beta} = \{\mathbf{x}|f(\mathbf{x}) \geq \beta\}, \qquad \widehat{\mathcal{S}}_{f,\beta} = \{\mathbf{x}|f(\mathbf{x}) \leq \beta\}. \tag{A4}$$

Concave and quasi-concave functions can be defined via the convexity of these sets. Recall that a set $\mathcal{S}, \mathcal{S} \subseteq \mathbf{R}^n$, is *convex* if for any two points $\mathbf{x}_1$ and $\mathbf{x}_2$ in $\mathcal{S}$ and for any λ satisfying $0 \leq \lambda \leq 1$ we have

$$\lambda \mathbf{x}_1 + \bar{\lambda} \mathbf{x}_2 \in \mathcal{S} \tag{A5}$$

where $\bar{\lambda} = 1 - \lambda$. Suppose that $\mathcal{D}_f$ is convex. The function f is *concave* over $\mathcal{D}_f$ if and only if its hypograph $\mathcal{H}_f$ is convex. Similarly, f is *convex* over $\mathcal{D}_f$ if and only if $\widehat{\mathcal{H}}_f$ is convex. The function f is *quasi-concave* over $\mathcal{D}_f$ if and only if all its superlevel sets are convex, and f is *quasi-convex* over $\mathcal{D}_f$ if and only if all its sublevel sets are convex. A function that is quasi-convex and quasi-concave is called *quasi-linear*. For example, any non-increasing or non-decreasing function is quasi-linear.

Appendix B.2. Basic Properties

Two properties of concave and quasi-concave functions are as follows; these properties are often used as the definitions of such functions. Similar properties exist for convex and quasi-convex functions.

Lemma A1. *The function f is concave if and only if*

$$f(\lambda \mathbf{x}_1 + \bar{\lambda} \mathbf{x}_2) \geq \lambda f(\mathbf{x}_1) + \bar{\lambda} f(\mathbf{x}_2) \tag{A6}$$

for all $\mathbf{x}_1$ and $\mathbf{x}_2$ in $\mathcal{D}_f$ and for all $0 \leq \lambda \leq 1$.

Lemma A2. *The function f is quasi-concave if and only if*

$$f(\lambda \mathbf{x}_1 + \bar{\lambda} \mathbf{x}_2) \geq \min(f(\mathbf{x}_1), f(\mathbf{x}_2)) \tag{A7}$$

for all $\mathbf{x}_1$ and $\mathbf{x}_2$ in $\mathcal{D}_f$ and for all $0 \leq \lambda \leq 1$.

The next two properties assume that f is twice differentiable and that $\mathcal{D}_f$ is convex. Let $\mathbf{H}_f(\mathbf{x})$ and $\mathbf{B}_f(\mathbf{x})$ be the respective Hessian and bordered Hessian of f at $\mathbf{x}$.

Lemma A3. *(see [15], Section 3.1.4) f is concave if and only if $\mathbf{H}_f(\mathbf{x})$ is negative semidefinite for all $\mathbf{x} \in \mathcal{D}_f$.*

Lemma A4. *(see [16], p. 771) f is quasi-concave on the open and convex set $\mathcal{D}_f$ if the determinants $D_2, D_3, \ldots, D_n$ of the respective second to nth leading principal minors of $\mathbf{B}_f(\mathbf{x})$ satisfy $(-1)^k D_k < 0$ for $k = 2, 3, \ldots, n$ and for all $\mathbf{x} \in \mathcal{D}_f$.*

Appendix B.3. Compositions Preserving Quasi-Concavity

The following compositions preserve quasi-concavity.

Lemma A5. *Suppose f and f_i, $1 \leq i \leq n$, are quasi-concave, then so are the functions*

1. $h = k_1 f + k_2$, *where $k_1 \geq 0$ and $k_2 \in \mathbf{R}$;*
2. $h = \min_{1 \leq i \leq n} f_i$;
3. $h = g \circ f$ *where f is quasi-concave and g is non-decreasing;*

4. $h(\mathbf{a}) = \sup_{\mathbf{b} \in \mathcal{B}} f(\mathbf{a}, \mathbf{b})$ *where $\mathcal{B}$ is a convex set;*
5. $h(\mathbf{a}, \mathbf{b}) = f(g(\mathbf{a}), \mathbf{b})$ *where g is convex and $f(\tilde{\mathbf{a}}, \mathbf{b})$ is non-increasing in $\tilde{\mathbf{a}}$ for fixed $\mathbf{b}$.*

Proof. Properties 1)–4) are standard (see [15], Section 3.4). For property 5), observe that

$$
\begin{aligned}
&h(\lambda \mathbf{a}_1 + \bar{\lambda} \mathbf{a}_2, \lambda \mathbf{b}_1 + \bar{\lambda} \mathbf{b}_2) \\
&= f(g(\lambda \mathbf{a}_1 + \bar{\lambda} \mathbf{a}_2), \lambda \mathbf{b}_1 + \bar{\lambda} \mathbf{b}_2) \\
&\overset{(a)}{\geq} f(\lambda g(\mathbf{a}_1) + \bar{\lambda} g(\mathbf{a}_2), \lambda \mathbf{b}_1 + \bar{\lambda} \mathbf{b}_2) \\
&\overset{(b)}{\geq} \min\left(f(g(\mathbf{a}_1), \mathbf{b}_1), f(g(\mathbf{a}_2), \mathbf{b}_2)\right)
\end{aligned}
\tag{A8}
$$

where (a) follows because $g(\lambda \mathbf{a}_1 + \bar{\lambda} \mathbf{a}_2) \leq \lambda g(\mathbf{a}_1) + \bar{\lambda} g(\mathbf{a}_2)$ and $f(\tilde{\mathbf{a}}, \mathbf{b})$ is non-increasing in $\tilde{\mathbf{a}}$. Step (b) follows because f is quasi-concave. $\square$

Appendix B.4. Examples of Quasi-Concave Functions

We establish quasi-concavity for several useful functions.

Lemma A6. *The following functions are quasi-concave for $\mathbf{x} = (a\ b)$ with non-negative entries.*

1. $f(\mathbf{x}) = ab$
2. $f(\mathbf{x}) = \frac{ab}{a+b+k}$ *for a positive constant k*
3. $f(\mathbf{x}) = k_1/a + 2\sqrt{k_2 b/a}$ *for positive constants k_1, k_2*
4. $f(\mathbf{x}) = -(1-b)(k_1 + k_2/a)$ *for positive constants k_1, k_2, and $b \leq 1$*

Furthermore, the following function is quasi-concave for $\mathbf{x} = (a\ b\ c)$ with non-negative entries.

5. $f(\mathbf{x}) = a + b + 2\sqrt{abc}$

Proof. We consider positive $\mathbf{x}$, and we use bordered Hessians $\mathbf{B}_f(\mathbf{x})$ and the derivatives D_k of their kth leading principal minors, $k = 2, 3, \ldots, n$. The results extend to non-negative $\mathbf{x}$ by using continuity at zero values, except for the third and fourth functions where $a = 0$ makes the functions undefined.

1. We have $D_2 < 0$ and $D_3 > 0$ for

$$
\mathbf{B}_f(\mathbf{x}) = \begin{pmatrix} 0 & b & a \\ b & 0 & 1 \\ a & 1 & 0 \end{pmatrix}.
$$

2. We have $D_2 < 0$ and $D_3 > 0$ for

$$
\mathbf{B}_f(\mathbf{x}) = \begin{pmatrix} 0 & \frac{b(b+k)}{(a+b+k)^2} & \frac{a(a+k)}{(a+b+k)^2} \\ \frac{b(b+k)}{(a+b+k)^2} & \frac{-2b(b+k)}{(a+b+k)^3} & \frac{2ab+(a+b+k)k}{(a+b+k)^3} \\ \frac{a(a+k)}{(a+b+k)^2} & \frac{2ab+(a+b+k)k}{(a+b+k)^3} & \frac{-2a(a+k)}{(a+b+k)^3} \end{pmatrix}.
$$

3. We have $D_2 < 0$ and $D_3 > 0$ for

$$\mathbf{B}_f(\mathbf{x}) = \begin{pmatrix} 0 & -\dfrac{k_1+\sqrt{k_2 ab}}{a^2} & \sqrt{\dfrac{k_2}{ab}} \\ -\dfrac{k_1+\sqrt{k_2 ab}}{a^2} & \dfrac{4k_1+3\sqrt{k_2 ab}}{2a^3} & -\dfrac{\sqrt{k_2}}{2a^{3/2}\sqrt{b}} \\ \sqrt{\dfrac{k_2}{ab}} & -\dfrac{\sqrt{k_2}}{2a^{3/2}\sqrt{b}} & -\dfrac{\sqrt{k_2}}{2b^{3/2}\sqrt{a}} \end{pmatrix}.$$

4. If $b \leq 1$, we have $D_2 < 0$ and $D_3 > 0$ for

$$\mathbf{B}_f(\mathbf{x}) = \begin{pmatrix} 0 & \dfrac{(1-b)k_2}{a^2} & k_1 + \dfrac{k_2}{a} \\ \dfrac{(1-b)k_2}{a^2} & -\dfrac{2(1-b)k_2}{a^3} & -\dfrac{k_2}{a^2} \\ k_1 + \dfrac{k_2}{a} & -\dfrac{k_2}{a^2} & 0 \end{pmatrix}.$$

5. We have $D_2 < 0$, $D_3 > 0$ and $D_4 < 0$ for

$$\mathbf{B}_f(\mathbf{x}) = \begin{pmatrix} 0 & 1+\sqrt{\dfrac{bc}{a}} & 1+\sqrt{\dfrac{ac}{b}} & \sqrt{\dfrac{ab}{c}} \\ 1+\sqrt{\dfrac{bc}{a}} & -\dfrac{\sqrt{bc}}{2a^{3/2}} & \dfrac{1}{2}\sqrt{\dfrac{c}{ab}} & \dfrac{1}{2}\sqrt{\dfrac{b}{ac}} \\ 1+\sqrt{\dfrac{ac}{b}} & \dfrac{1}{2}\sqrt{\dfrac{c}{ab}} & -\dfrac{\sqrt{ac}}{2b^{3/2}} & \dfrac{1}{2}\sqrt{\dfrac{a}{bc}} \\ \sqrt{\dfrac{ab}{c}} & \dfrac{1}{2}\sqrt{\dfrac{b}{ac}} & \dfrac{1}{2}\sqrt{\dfrac{a}{bc}} & -\dfrac{\sqrt{ab}}{2c^{3/2}} \end{pmatrix}. \qquad (A9)$$

$\square$

References

1. Kramer, G.; Gastpar, M.; Gupta, P. Cooperative strategies and capacity theorems for relay networks. *IEEE Trans. Inf. Theory* **2005**, *51*, 3037–3063. [CrossRef]
2. Lin, B.; Ho, P.-H.; Xie, L.-L.; Shen, X.; Tapolcai, J. Optimal relay station placement in broadband wireless access networks. *IEEE Trans. Mob. Comput.* **2010**, *9*, 259–269. [CrossRef]
3. Aggarwal, V.; Bennatan, A.; Calderbank, A. R. Calderbank. On maximizing coverage in Gaussian relay channels. *IEEE Trans. Inf. Theory* **2009**, *55*, 2518–2536. [CrossRef]
4. Joshi, G.; Karandikar, A. Optimal relay placement for cellular coverage extension. In Proceedings of the Seventeenth National Conference on Communications, Bangalore, India, 28–30 January 2011; pp. 1196–1200.
5. Lee, J.; Wang, H.; Andrews, J.G.; Hong, D. Outage probability of cognitive relay networks with interference constraints. *IEEE Trans. Wirel. Commun.* **2011**, *10*, 390–395. [CrossRef]
6. Chen, X.; Song, S.H.; Letaief, K.B. Relay position optimization improves finite-SNR diversity gain of decode-and-forward MIMO relay systems. *IEEE Trans. Commun.* **2012**, *60*, 3311–3321. [CrossRef]
7. Thakur, M.; Fawaz, N.; Médard, M. Optimal relay location and power allocation for low-SNR broadcast relay channels. In Proceedings of the 30th IEEE International Conference on Computer Communications (INFOCOM 2011), Shanghai, China, 10–15 April 2011; pp. 2822–2830.
8. Thakur, M.; Fawaz, N.; Médard, M. On the geometry of wireless network multicast in 2-D. In Proceedings of the 2011 IEEE International Symposium on Information Theory, Saint Petersburg, Russian, 31 July–5 August 2011; pp. 1628–1632.
9. Thakur, M.; Fawaz, N.; Médard, M. Reducibility of joint relay positioning and flow optimization problem. In Proceedings of the 2012 IEEE International Symposium on Information Theory, Boston, MA, USA, 1–6 July 2012; pp. 1117–1121.
10. Thakur, M.; Kramer, G. Relay Positioning for Multicast Relay Networks. In Proceedings of the 2013 IEEE International Symposium on Information Theory, Istanbul, Turkey, 7–12 July 2013; pp. 1954–1958.

11. Thakur, M.; Kramer, G. Quasi-concavity for Gaussian multicast relay channels. In Proceedings of the 2015 IEEE International Symposium on Information Theory, Hong Kong, China, 14–19 June 2015; pp. 2867–2869.

12. Cover, T.M.; El Gamal, A. Capacity theorems for the relay channel. *IEEE Trans. Inf. Theory* **1979**, *25*, 572–584. [CrossRef]

13. Kramer, G.; Maric, I.; Yates, R.D. Cooperative Communications. *Found. Trends Netw.* **2006**, *1*, 271–425. [CrossRef]

14. Cover, T.M.; Thomas, J.A. *Elements of Information Theory*, 2nd ed.; John Wiley & Sons: New York, NY, USA, 2006.

15. Boyd, S.; Vandenberghe, L. *Convex Optimization*; Cambridge University Press: New York, NY, USA, 2004.

16. Bazaraa, M.S.; Sherali, H.D.; Shetty, C.M. *Nonlinear Programming*; Wiley: Hoboken, NJ, USA, 2006

Article

Gaussian Multiple Access Channels with One-Bit Quantizer at the Receiver [†,‡]

Borzoo Rassouli [1,*], Morteza Varasteh [2,*] and Deniz Gündüz [2,*]

[1]   School of Computer Science and Electronic Engineering, University of Essex, Colchester CO4 3SQ, UK
[2]   The Intelligent Systems and Networks group of Department of Electrical and Electronics,
      Imperial College London, London SW7 2AZ, UK
*   Correspondence: b.rassouli@essex.ac.uk (B.R.); m.varasteh12@imperial.ac.uk (M.V.);
     d.gunduz@imperial.ac.uk (D.G.); Tel.: +44-(0)-207-594-6218 (D.G.)
†   This paper is an extended version of our paper published in the 2017 IEEE International Symposium on
     Information Theory (ISIT), Aachen, Germany, 25–30 June 2017.
‡   This work has been carried out when the first author was with the Information Processing and
     Communications Laboratory at Imperial College London.

Received: 21 May 2018; Accepted: 5 September 2018; Published: 7 September 2018

Abstract: The capacity region of a two-transmitter Gaussian multiple access channel (MAC) under average input power constraints is studied, when the receiver employs a zero-threshold one-bit analogue-to-digital converter (ADC). It is proven that the input distributions of the two transmitters that achieve the boundary points of the capacity region are discrete. Based on the position of a boundary point, upper bounds on the number of the mass points of the corresponding distributions are derived. Furthermore, a lower bound on the sum capacity is proposed that can be achieved by time division with power control. Finally, inspired by the numerical results, the proposed lower bound is conjectured to be tight.

Keywords: Gaussian multiple access channel; one-bit quantizer; capacity region

1. Introduction

The energy consumption of an analogue-to-digital converter (ADC) (measured in Joules/sample) grows exponentially with its resolution (in bits/sample) [1,2]. When the available power is limited, for example, for mobile devices with limited battery capacity, or for wireless receivers that operate on limited energy harvested from ambient sources [3], the receiver circuitry may be constrained to operate with low resolution ADCs. The presence of a low-resolution ADC, in particular a one-bit ADC at the receiver, alters the channel characteristics significantly. Such a constraint not only limits the fundamental bounds on the achievable rate, but it also changes the nature of the communication and modulation schemes approaching these bounds. For example, in a real additive white Gaussian noise (AWGN) channel under an average power constraint on the input, if the receiver is equipped with a K-bin (i.e., $\log_2 K$-bit) ADC front end, it is shown in [4] that the capacity-achieving input distribution is discrete with at most $K + 1$ mass points. This is in contrast with the optimality of the Gaussian input distribution when the receiver has infinite resolution.

Especially with the adoption of massive multiple-input multiple-output (MIMO) receivers and the millimetre wave (mmWave) technology enabling communication over large bandwidths, communication systems with limited-resolution receiver front ends are becoming of practical importance. Accordingly, there has been a growing research interest in understanding both the fundamental information theoretic limits and the design of practical communication protocols for systems with finite-resolution ADC front ends. In [5], the authors showed that for a Rayleigh fading channel with a one-bit ADC and perfect channel state information at the receiver (CSIR), quadrature phase shift keying (QPSK) modulation is capacity-achieving. In case of no CSIR, [6] showed that QPSK modulation is optimal when the signal-to-noise ratio (SNR) is above a certain threshold, which depends on the coherence time of the channel, while for SNRs below this threshold, on-off QPSK achieves the capacity. For the point-to-point multiple-input multiple-output (MIMO) channel with a one-bit ADC front end at each receive antenna and perfect CSIR, [7] showed that QPSK is optimal at very low SNRs, while with perfect channel state information at the transmitter (CSIT), upper and lower bounds on the capacity are provided in [8].

To the best of our knowledge, the existing literature on communications with low-resolution ADCs focuses exclusively on point-to-point systems. Our goal in this paper is to understand the impact of low-resolution ADCs on the capacity region of a multiple access channel (MAC). In particular, we consider a two-transmitter Gaussian MAC with a one-bit quantizer at the receiver. The inputs to the channel are subject to average power constraints. We show that any point on the boundary of the capacity region is achieved by discrete input distributions. Based on the slope of the tangent line to the capacity region at a boundary point, we propose upper bounds on the cardinality of the support of these distributions. Finally, based on numerical analysis for the sum capacity, it is observed that we cannot obtain a sum rate higher than is achieved by time division with power control.

The paper is organized as follows. Section 2 introduces the system model. In Section 3, the capacity region of a general two-transmitter memoryless MAC under input average power constraints is investigated. The main result of the paper is presented in Section 3, and a detailed proof is given in Section 4. The proof has two parts: (1) it is shown that the support of the optimal distributions is bounded by contradiction; and (2) we make use of this boundedness to prove the finiteness of the optimal support by using Dubins' theorem [9]. Section 5 analyses the sum capacity, and finally, Section 6 concludes the paper.

Notations: Random variables are denoted by capital letters, while their realizations with lower case letters. $F_X(x)$ denotes the cumulative distribution function (CDF) of random variable X. The conditional probability mass function (pmf) $p_{Y|X_1,X_2}(y|x_1,x_2)$ will be written as $p(y|x_1,x_2)$. For integers $m \leq n$, we have $[m : n] = \{m, m+1, \ldots, n\}$. For $0 \leq t \leq 1$, $H_b(t) \triangleq -t \log_2 t - (1-t) \log_2(1-t)$ denotes the binary entropy function. The unit-step function is denoted by $s(\cdot)$.

2. System Model and Preliminaries

We consider a two-transmitter memoryless Gaussian MAC (as shown in Figure 1) with a one-bit quantizer Γ at the receiver front end. Transmitter $j = 1, 2$ encodes its message W_j into a codeword X_j^n and transmits it over the shared channel. The signal received by the decoder is given by:

$$Y = \Gamma(X_{1,i} + X_{2,i} + Z_i), \ i \in [1 : n],$$

where $\{Z_i\}_{i=1}^n$ is an independent and identically distributed (i.i.d.) Gaussian noise process, also independent of the channel inputs X_1^n and X_2^n with $Z_i \sim \mathcal{N}(0,1), i \in [1 : n]$. Γ represents the one-bit ADC operation given by:

$$\Gamma(x) = \begin{cases} 1 & x \geq 0 \\ 0 & x < 0 \end{cases}.$$

This channel can be modelled by the triplet $(\mathcal{X}_1 \times \mathcal{X}_2, p(y|x_1,x_2), \mathcal{Y})$, where $\mathcal{X}_1, \mathcal{X}_2 (= \mathbb{R})$ and $\mathcal{Y}$ $(= \{0,1\})$, respectively, are the alphabets of the inputs and the output. The conditional pmf of the channel output Y conditioned on the channel inputs X_1 and X_2 (i.e., $p(y|x_1,x_2)$) is characterized by:

$$p(0|x_1,x_2) = 1 - p(1|x_1,x_2) = Q(x_1 + x_2), \tag{1}$$

where $Q(x) \triangleq \frac{1}{\sqrt{2\pi}} \int_x^{+\infty} e^{-\frac{t^2}{2}} dt$.

We consider a two-transmitter stationary and memoryless MAC model $(\mathcal{X}_1 \times \mathcal{X}_2, p(y|x_1,x_2), \mathcal{Y})$, where $\mathcal{X}_1 = \mathcal{X}_2 = \mathbb{R}$, $\mathcal{Y} = \{0,1\}$, $p(y|x_1,x_2)$ is given in (1).

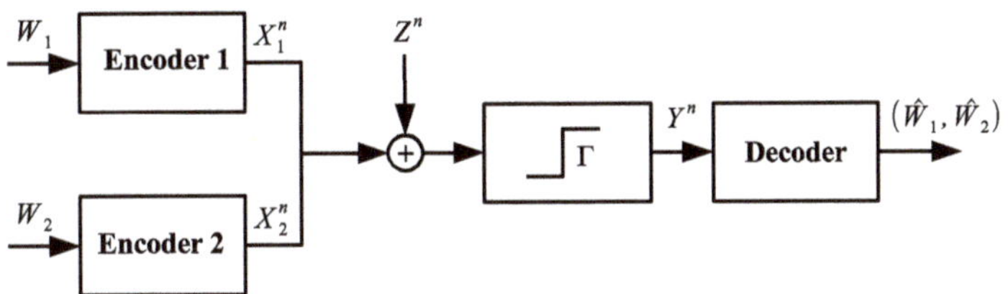

Figure 1. A two-transmitter Gaussian multiple access channel (MAC) with a one-bit analogue-to-digital converter (ADC) front end at the receiver.

A $(2^{nR_1}, 2^{nR_2}, n)$ code for this channel consists of (as in [10]):

- two message sets $[1 : 2^{nR_1}]$ and $[1 : 2^{nR_2}]$,
- two encoders, where encoder $j = 1, 2$ assigns a codeword $x_j^n(w_j)$ to each message $w_j \in [1 : 2^{nR_j}]$, and
- a decoder that assigns estimates $(\hat{w}_1, \hat{w}_2) \in [1 : 2^{nR_1}] \times [1 : 2^{nR_2}]$ or an error message to each received sequence y^n.

The stationary property means that the channel does not change over time, while the memoryless property indicates that $p(y_i|x_1^i, x_2^i, y^{i-1}, w_1, w_2) = p(y_i|x_{1,i}, x_{2,i})$ for any message pair (w_1, w_2).

We assume that the message pair (W_1, W_2) is uniformly distributed over $[1 : 2^{nR_1}] \times [1 : 2^{nR_2}]$. The average probability of error is defined as:

$$P_e^{(n)} \triangleq \Pr\left\{ (\hat{W}_1, \hat{W}_2) \neq (W_1, W_2) \right\}. \tag{2}$$

Average power constraints are imposed on the channel inputs as:

$$\frac{1}{n} \sum_{i=1}^{n} x_{j,i}^2(w_j) \leq P_j \, , \ \forall w_j \in [1 : 2^{nR_j}], \ j \in \{1, 2\}, \tag{3}$$

where $x_{j,i}(w_j)$ denotes the i-th element of the codeword $x_j^n(w_j)$.

A rate pair (R_1, R_2) is said to be achievable for this channel if there exists a sequence of $(2^{nR_1}, 2^{nR_2}, n)$ codes satisfying the average power constraints (3), such that $\lim_{n \to \infty} P_e^{(n)} = 0$. The capacity region $\mathscr{C}(P_1, P_2)$ of this channel is the closure of the set of achievable rate pairs (R_1, R_2).

3. Main Results

Proposition 1. *The capacity region $\mathscr{C}(P_1, P_2)$ of a two-transmitter stationary and memoryless MAC with average power constraints P_1 and P_2 is the set of non-negative rate pairs (R_1, R_2) that satisfy:*

$$R_1 \leq I(X_1; Y|X_2, U),$$
$$R_2 \leq I(X_2; Y|X_1, U),$$
$$R_1 + R_2 \leq I(X_1, X_2; Y|U), \tag{4}$$

for some $F_U(u)F_{X_1|U}(x_1|u)F_{X_2|U}(x_2|u)$, such that $\mathbb{E}[X_j^2] \leq P_j$, $j = 1, 2$. Furthermore, it is sufficient to consider $|\mathcal{U}| \leq 5$.

Proof of Proposition 1. The proof is provided in Appendix A. □

The main result of this paper is provided in the following theorem. It bounds the cardinality of the support set of the capacity-achieving distributions.

Theorem 1. *Let J be an arbitrary point on the boundary of the capacity region $\mathscr{C}(P_1, P_2)$ of the memoryless MAC with a one-bit ADC front end (as shown in Figure 1). J is achieved by a distribution in the form of*

$F_U^J(u)F_{X_1|U}^J(x_1|u)F_{X_2|U}^J(x_2|u)$. *Furthermore, let l_J be the slope of the line tangent to the capacity region at this point. For any $u \in \mathcal{U}$, the conditional input distributions $F_{X_1|U}^J(x_1|u)$ and $F_{X_2|U}^J(x_2|u)$ have at most n_1 and n_2 points of increase (a point Z is said to be a point of increase of a distribution if for any open set Ω containing Z, we have $\Pr\{\Omega\} > 0$), respectively, where:*

$$(n_1, n_2) = \begin{cases} (3,5) & l_J < -1 \\ (3,3) & l_J = -1 \\ (5,3) & l_J > -1 \end{cases}. \tag{5}$$

Furthermore, this result remains unchanged if the one-bit ADC has a non-zero threshold.

Proof of Theorem 1. The proof is provided in Section 4. $\square$

Proposition 1 and Theorem 1 establish upper bounds on the number of mass points of the distributions that achieve a boundary point. The significance of this result is that once it is known that the optimal inputs are discrete with at most a certain number of mass points, the capacity region along with the optimal distributions can be obtained via computer programs.

4. Proof of Theorem 1

In order to show that the boundary points of the capacity region are achieved, it is sufficient to show that the capacity region is a closed set, i.e., it includes all of its limit points.

Let $\mathcal{U}$ be a set with $|\mathcal{U}| \leq 5$ and Ω be defined as:

$$\Omega \triangleq \{F_{U,X_1,X_2} \mid U \in \mathcal{U}, \ X_1 - U - X_2, \ \mathbb{E}[X_j^2] \leq P_j, \ j = 1,2\}, \tag{6}$$

which is the set of all CDFs on the triplet (U, X_1, X_2), where U is drawn from $\mathcal{U}$, and the Markov chain $X_1 - U - X_2$ and the corresponding average power constraints hold.

In Appendix B, it is proven that Ω is a compact set. Since a continuous mapping preserves compactness, the capacity region is compact. Since the capacity region is a subset of $\mathbb{R}^2$, it is closed and bounded (note that a subset of $\mathbb{R}^k$ is compact if and only if it is closed and bounded [11]). Therefore, any point P on the boundary of the capacity region is achieved by a distribution denoted by $F_U^J(u)F_{X_1|U}^J(x_1|u)F_{X_2|U}^J(x_2|u)$.

Since the capacity region is a convex space, it can be characterized by its supporting hyperplanes. In other words, any point on the boundary of the capacity region, denoted by (R_1^b, R_2^b), can be written as:

$$(R_1^b, R_2^b) = \arg \max_{(R_1,R_2)\in\mathscr{C}(P_1,P_2)} R_1 + \lambda R_2,$$

for some $\lambda \in (0, \infty)$. Here, we have excluded the cases $\lambda = 0$ and $\lambda = \infty$, where the channel is not a two-transmitter MAC any longer, and boils down to a point-to-point channel, whose capacity is already known.

Any rate pair $(R_1, R_2) \in \mathscr{C}(P_1, P_2)$ must lie within a pentagon defined by (4) for some $F_U F_{X_1|U} F_{X_2|U}$ that satisfies the power constraints. Therefore, due to the structure of the pentagon, the problem of finding the boundary points is equivalent to the following maximization problem.

$$\max_{(R_1,R_2)\in\mathscr{C}(P_1,P_2)} R_1 + \lambda R_2 = \begin{cases} \max I(X_1;Y|X_2,U) + \lambda I(X_2;Y|U) & 0 < \lambda \leq 1 \\ \max I(X_2;Y|X_1,U) + \lambda I(X_1;Y|U) & \lambda > 1 \end{cases}, \tag{7}$$

where on the right-hand side (RHS) of (7), the maximizations are over all $F_U F_{X_1|U} F_{X_2|U}$ that satisfy the power constraints. It is obvious that when $\lambda = 1$, the two lines in (7) are the same, which results in the sum capacity.

For any product of distributions $F_{X_1} F_{X_2}$ and the channel in (1), let I_λ be defined as:

$$I_\lambda(F_{X_1}F_{X_2}) \triangleq \begin{cases} I(X_1; Y | X_2) + \lambda I(X_2; Y) & 0 < \lambda \le 1 \\ I(X_2; Y | X_1) + \lambda I(X_1; Y) & \lambda > 1 \end{cases}. \tag{8}$$

With this definition, (7) can be rewritten as:

$$\max_{(R_1,R_2)\in\mathscr{C}(P_1,P_2)} R_1 + \lambda R_2 = \max \sum_{i=1}^{5} p_U(u_i) I_\lambda(F_{X_1|U}(x_1|u_i) F_{X_2|U}(x_2|u_i)),$$

where the second maximization is over distributions of the form $p_U(u)F_{X_1|U}(x_1|u)F_{X_2|U}(x_2|u)$, such that:

$$\sum_{i=1}^{5} p_U(u_i)\mathbb{E}[X_j^2 | U = u_i] \le P_j, \ j = 1, 2.$$

Proposition 2. *For a given F_{X_1} and any $\lambda > 0$, $I_\lambda(F_{X_1}F_{X_2})$ is a concave, continuous and weakly differentiable function of F_{X_2}. In the statement of this proposition, F_{X_1} and F_{X_2} could be interchanged.*

Proof of Proposition 2. The proof is provided in Appendix C. □

Proposition 3. *Let P_1', P_2' be two arbitrary non-negative real numbers. For the following problem:*

$$\max_{\substack{F_{X_1} F_{X_2}: \\ \mathbb{E}[X_j^2]\le P_j', \, j=1,2}} I_\lambda(F_{X_1}F_{X_2}), \tag{9}$$

the optimal inputs $F_{X_1}^*$ and $F_{X_2}^*$, which are not unique in general, have the following properties,

(i) The support sets of $F_{X_1}^*$ and $F_{X_2}^*$ are bounded subsets of $\mathbb{R}$.
(ii) $F_{X_1}^*$ and $F_{X_2}^*$ are discrete distributions that have at most n_1 and n_2 points of increase, respectively, where:

$$(n_1, n_2) = \begin{cases} (5,3) & 0 < \lambda < 1 \\ (3,3) & \lambda = 1 \\ (3,5) & \lambda > 1 \end{cases}.$$

Proof of Proposition 3. We start with the proof of the first claim. Assume that $0 < \lambda \le 1$, and F_{X_2} is given. Consider the following optimization problem:

$$I_{F_{X_2}}^* \triangleq \sup_{\substack{F_{X_1}: \\ \mathbb{E}[X_1^2]\le P_1'}} I_\lambda(F_{X_1}F_{X_2}). \tag{10}$$

Note that $I_{F_{X_2}}^* < +\infty$, since for any $\lambda > 0$, from (8),

$$I_\lambda \le (\lambda + 1)H(Y) \le (1 + \lambda) < +\infty.$$

From Proposition 2, I_λ is a continuous, concave function of F_{X_1}. Furthermore, the set of all CDFs with bounded second moment (here, P_1') is convex and compact. The compactness follows from Appendix I in [12], where the only difference is in using Chebyshev's inequality instead of Markov's inequality. Therefore, the supremum in (10) is achieved by a distribution $F_{X_1}^*$. Since for any $F_{X_1}(x) = s(x - x_0)$ with $|x_0|^2 < P_1'$, we have $\mathbb{E}[X_1^2] < P_1'$, the Lagrangian theorem and the Karush–Kuhn–Tucker conditions state that there exists a $\theta_1 \ge 0$ such that:

$$I_{F_{X_2}}^* = \sup_{F_{X_1}} \left\{ I_\lambda(F_{X_1}F_{X_2}) - \theta_1 \left(\int x^2 dF_{X_1}(x) - P_1' \right) \right\}. \tag{11}$$

Furthermore, the supremum in (11) is achieved by $F_{X_1}^*$, and:

$$\theta_1 \left(\int x^2 dF_{X_1}^*(x) - P_1' \right) = 0. \tag{12}$$

Lemma 1. *The Lagrangian multiplier θ_1 is non-zero. From (12), this is equivalent to having $\mathbb{E}[X_1^2] = P_1'$, i.e., the first user transmits with its maximum allowable power (note that this is for $\lambda \leq 1$, as used in Appendix D).*

Proof of Lemma 1. In what follows, we prove that a zero Lagrangian multiplier is not possible. Having a zero Lagrangian multiplier means the power constraint is inactive. In other words, if $\theta_1 = 0$, (10) and (11) imply that:

$$\sup_{\substack{F_{X_1} \\ \mathbb{E}[X_1^2] \leq P_1'}} I_\lambda(F_{X_1} F_{X_2}) = \sup_{F_{X_1}} I_\lambda(F_{X_1} F_{X_2}). \tag{13}$$

We prove that (13) does not hold by showing that its left-hand side (LHS) is strictly less than one, while its RHS equals one. The details are provided in Appendix D. $\square$

$I_\lambda(F_{X_1} F_{X_2})$ $(0 < \lambda \leq 1)$ can be written as:

$$I_\lambda(F_{X_1} F_{X_2}) = \int_{-\infty}^{+\infty} \int_{-\infty}^{+\infty} \sum_{y=0}^{1} p(y|x_1, x_2) \log \frac{p(y|x_1, x_2)}{[p(y; F_{X_1} F_{X_2})]^\lambda [p(y; F_{X_1}|x_2)]^{1-\lambda}} dF_{X_1}(x_1) dF_{X_2}(x_2)$$

$$= \int_{-\infty}^{+\infty} \tilde{i}_\lambda(x_1; F_{X_1}|F_{X_2}) dF_{X_1}(x_1) \tag{14}$$

$$= \int_{-\infty}^{+\infty} i_\lambda(x_2; F_{X_2}|F_{X_1}) dF_{X_2}(x_2), \tag{15}$$

where we have defined:

$$\tilde{i}_\lambda(x_1; F_{X_1}|F_{X_2}) \triangleq \int_{-\infty}^{+\infty} \left(D\left(p(y|x_1, x_2)||p(y; F_{X_1} F_{X_2})\right) + (1-\lambda)\sum_{y=0}^{1} p(y|x_1, x_2) \log \frac{p(y; F_{X_1} F_{X_2})}{p(y; F_{X_1}|x_2)} \right) dF_{X_2}(x_2), \tag{16}$$

and:

$$i_\lambda(x_2; F_{X_2}|F_{X_1}) \triangleq \int_{-\infty}^{+\infty} D\left(p(y|x_1, x_2)||p(y; F_{X_1} F_{X_2})\right) dF_{X_1}(x_1) - (1-\lambda)D\left(p(y; F_{X_1}|x_2)||p(y; F_{X_1} F_{X_2})\right). \tag{17}$$

$p(y; F_{X_1} F_{X_2})$ is nothing but the pmf of Y with the emphasis that it has been induced by F_{X_1} and F_{X_2}. Likewise, $p(y; F_{X_1}|x_2)$ is the conditional pmf $p(y|x_2)$ when X_1 is drawn according to F_{X_1}. From (14), $\tilde{i}_\lambda(x_1; F_{X_1}|F_{X_2})$ can be considered as the density of I_λ over F_{X_1} when F_{X_2} is given. $i_\lambda(x_2; F_{X_2}|F_{X_1})$ can be interpreted in a similar way.

Note that (11) is an unconstrained optimization problem over the set of all CDFs. Since $\int x^2 dF_{X_1}(x)$ is linear and weakly differentiable in F_{X_1}, the objective function in (11) is concave and weakly differentiable. Hence, a necessary condition for the optimality of $F_{X_1}^*$ is:

$$\int \{\tilde{i}_\lambda(x_1; F_{X_1}^*|F_{X_2}) + \theta_1(P_1' - x_1^2)\} dF_{X_1}(x_1) \leq I_{F_{X_2}}^*, \quad \forall F_{X_1}. \tag{18}$$

Furthermore, (18) can be verified to be equivalent to:

$$\tilde{i}_\lambda(x_1; F_{X_1}^*|F_{X_2}) + \theta_1(P_1' - x_1^2) \leq I_{F_{X_2}}^*, \quad \forall x_1 \in \mathbb{R}, \tag{19}$$

$$\tilde{i}_\lambda(x_1; F_{X_1}^*|F_{X_2}) + \theta_1(P_1' - x_1^2) = I_{F_{X_2}}^*, \quad \text{if } x_1 \text{ is a point of increase of } F_{X_1}^*. \tag{20}$$

The justifications of (18)–(20) are provided in Appendix E.

In what follows, we prove that in order to satisfy (20), $F_{X_1}^*$ must have a bounded support by showing that the LHS of (20) goes to $-\infty$ with x_1. The following lemma is useful in the sequel for taking the limit processes inside the integrals.

Lemma 2. *Let X_1 and X_2 be two independent random variables satisfying $\mathbb{E}[X_1^2] \leq P_1'$ and $\mathbb{E}[X_2^2] \leq P_2'$, respectively ($P_1', P_2' \in [0, +\infty)$). Considering the conditional pmf in (1), the following inequalities hold.*

$$\left| D\left(p(y|x_1, x_2) \| p(y; F_{X_1} F_{X_2}) \right) \right| \leq 1 - 2 \log Q(\sqrt{P_1'} + \sqrt{P_2'}) \tag{21}$$

$$p(y; F_{X_1}|x_2) \geq Q\left(\sqrt{P_1'} + |x_2| \right) \tag{22}$$

$$\left| \sum_{y=0}^{1} p(y|x_1, x_2) \log \frac{p(y; F_{X_1} F_{X_2})}{p(y; F_{X_1}|x_2)} \right| \leq -2 \log Q\left(\sqrt{P_1'} + \sqrt{P_2'} \right) - 2 \log Q\left(\sqrt{P_1'} + |x_2| \right) \tag{23}$$

Proof of Lemma 2. The proof is provided in Appendix F. $\square$

Note that

$$\lim_{x_1 \to +\infty} \int_{-\infty}^{+\infty} D\left(p(y|x_1, x_2) \| p(y; F_{X_1}^* F_{X_2}) \right) dF_{X_2}(x_2) = \int_{-\infty}^{+\infty} \lim_{x_1 \to +\infty} D\left(p(y|x_1, x_2) \| p(y; F_{X_1}^* F_{X_2}) \right) dF_{X_2}(x_2) \tag{24}$$

$$= -\log p_Y(1; F_{X_1}^* F_{X_2}) \tag{25}$$

$$\leq -\log Q(\sqrt{P_1'} + \sqrt{P_2'}), \tag{26}$$

where (24) is due to the Lebesgue dominated convergence theorem [11] and (21), which permit the interchange of the limit and the integral; (25) is due to the following:

$$\lim_{x_1 \to +\infty} D\left(p(y|x_1, x_2) \| p(y; F_{X_1}^* F_{X_2}) \right) = \lim_{x_1 \to +\infty} \sum_{y=0}^{1} p(y|x_1, x_2) \log \frac{p(y|x_1, x_2)}{p(y; F_{X_1}^* F_{X_2})}$$

$$= -\log p_Y(1; F_{X_1}^* F_{X_2}),$$

since $p(0|x_1, x_2) = Q(x_1 + x_2)$ goes to zero when $x_1 \to +\infty$ and $p_Y(y; F_{X_1}^* F_{X_2})$ ($y = 0, 1$) is bounded away from zero by (A34) ; (26) is obtained from (A34) in Appendix F. Furthermore,

$$\lim_{x_1 \to +\infty} \int_{-\infty}^{+\infty} \sum_{y=0}^{1} p(y|x_1, x_2) \log \frac{p(y; F_{X_1}^* F_{X_2})}{p(y; F_{X_1}^*|x_2)} dF_{X_2}(x_2) = \int_{-\infty}^{+\infty} \lim_{x_1 \to +\infty} \sum_{y=0}^{1} p(y|x_1, x_2) \log \frac{p(y; F_{X_1}^* F_{X_2})}{p(y; F_{X_1}^*|x_2)} dF_{X_2}(x_2) \tag{27}$$

$$= \log p_Y(1; F_{X_1}^* F_{X_2}) - \int_{-\infty}^{+\infty} \log p(1; F_{X_1}^*|x_2) dF_{X_2}(x_2)$$

$$< -\log Q\left(\sqrt{P_1'} + \sqrt{P_2'} \right), \tag{28}$$

where (27) is due to the Lebesgue dominated convergence theorem along with (23) and (A39) in Appendix F; (28) is from (22) and the convexity of $\log Q(\alpha + \sqrt{t})$ in t when $\alpha \geq 0$ (see Appendix G).

Therefore, from (26) and (28),

$$\lim_{x_1 \to +\infty} \tilde{\imath}_\lambda(x_1; F_{X_1}^*|F_{X_2}) \leq -(2 - \lambda) \log Q(\sqrt{P_1'} + \sqrt{P_2'}) < +\infty. \tag{29}$$

Using a similar approach, we can also obtain:

$$\lim_{x_1 \to -\infty} \tilde{\imath}_\lambda(x_1; F_{X_1}^*|F_{X_2}) \leq -(2 - \lambda) \log Q(\sqrt{P_1'} + \sqrt{P_2'}) < +\infty. \tag{30}$$

From (29) and (30) and the fact that $\theta_1 > 0$ (see Lemma 1), the LHS of (19) goes to $-\infty$ when $|x_1| \to +\infty$. Since any point of increase of $F_{X_1}^*$ must satisfy (19) with equality and $I_{F_{X_2}}^* \geq 0$, it is proven that $F_{X_1}^*$ has a bounded support. Hence, from now on, we assume $X_1 \in [-A_1, A_2]$ for some $A_1, A_2 \in \mathbb{R}$ (note that A_1 and A_2 are determined by the choice of F_{X_2}).

Similarly, for a given F_{X_1}, the optimization problem:

$$I^*_{F_{X_1}} = \sup_{\substack{F_{X_2}: \\ \mathbb{E}[X_2^2] \leq P'_2}} I_\lambda(F_{X_1}, F_{X_2}),$$

boils down to the following necessary condition:

$$i_\lambda(x_2; F^*_{X_2}|F_{X_1}) + \theta_2(P'_2 - x_2^2) \leq I^*_{F_{X_1}}, \quad \forall x_2 \in \mathbb{R}, \tag{31}$$

$$i_\lambda(x_2; F^*_{X_2}|F_{X_1}) + \theta_2(P'_2 - x_2^2) = I^*_{F_{X_1}}, \quad \text{if } x_2 \text{ is a point of increase of } F^*_{X_2}, \tag{32}$$

for the optimality of $F^*_{X_2}$. However, there are two main differences between (32) and (20). First is the difference between i_λ and $\tilde{i}_\lambda$. Second is the fact that we do not claim θ_2 to be nonzero, since the approach used in Lemma 1 cannot be readily applied to θ_2. Nonetheless, the boundedness of the support of $F^*_{X_2}$ can be proven by inspecting the behaviour of the LHS of (32) when $|x_2| \to +\infty$.

In what follows, i.e., from (33)–(38), we prove that the support of $F^*_{X_2}$ is bounded by showing that (32) does not hold when $|x_2|$ is above a certain threshold. The first term on the LHS of (32) is $i_\lambda(x_2; F^*_{X_2}|F_{X_1})$. From (17) and (21), it can be easily verified that:

$$\lim_{x_2 \to +\infty} i_\lambda(x_2; F^*_{X_2}|F_{X_1}) = -\lambda \log p_Y(1; F_{X_1}, F^*_{X_2}) \leq -\lambda \log Q(\sqrt{P'_1} + \sqrt{P'_2}),$$

$$\lim_{x_2 \to -\infty} i_\lambda(x_2; F^*_{X_2}|F_{X_1}) = -\lambda \log p_Y(0; F_{X_1}, F^*_{X_2}) \leq -\lambda \log Q(\sqrt{P'_1} + \sqrt{P'_2}) \tag{33}$$

From (33), if $\theta_2 > 0$, the LHS of (32) goes to $-\infty$ with $|x_2|$, which proves that X^*_2 is bounded.

For the possible case of $\theta_2 = 0$, in order to show that (32) does not hold when $|x_2|$ is above a certain threshold, we rely on the boundedness of X_1, i.e., $X_1 \in [-A_1, A_2]$. Then, we prove that i_λ approaches its limit in (33) from below. In other words, there is a real number K such that $i_\lambda(x_2; F^*_{X_2}|F_{X_1}) < -\lambda \log p_Y(1; F_{X_1}, F^*_{X_2})$ when $x_2 > K$, and $i_\lambda(x_2; F^*_{X_2}|F_{X_1}) < -\lambda \log p_Y(0; F_{X_1}, F^*_{X_2})$ when $x_2 < -K$. This establishes the boundedness of X^*_2. In what follows, we only show the former, i.e., when $x_2 \to +\infty$. The latter, i.e., $x_2 \to -\infty$, follows similarly, and it is omitted for the sake of brevity.

By rewriting i_λ, we have:

$$i_\lambda(x_2; F^*_{X_2}|F_{X_1}) = -\lambda p(1; F_{X_1}|x_2) \log p_Y(1; F_{X_1}, F^*_{X_2})$$

$$- \int_{-A_1}^{A_2} H_b(Q(x_1 + x_2))dF_{X_1}(x_1) + (1 - \lambda) \underbrace{H(Y|X_2 = x_2)}_{H_b\left(\int Q(x_1+x_2)dF_{X_1}(x_1)\right)}$$

$$- \lambda \underbrace{p(0; F_{X_1}|x_2)}_{\int Q(x_1+x_2)dF_{X_1}(x_1)} \log p_Y(0; F_{X_1}, F^*_{X_2}). \tag{34}$$

It is obvious that the first term on the RHS of (34) approaches $-\lambda \log p_Y(1; F_{X_1}, F^*_{X_2})$ from below when $x_2 \to +\infty$, since $p(1; F_{X_1}|x_2) \leq 1$. It is also obvious that the remaining terms go to zero when $x_2 \to +\infty$. Hence, it is sufficient to show that they approach zero from below, which is proven by using the following lemma.

Lemma 3. *Let X_1 be distributed on $[-A_1, A_2]$ according to $F_{X_1}(x_1)$. We have:*

$$\lim_{x_2 \to +\infty} \frac{\int_{-A_1}^{A_2} H_b(Q(x_1 + x_2))dF_{X_1}(x_1)}{H_b\left(\int_{-A_1}^{A_2} Q(x_1 + x_2)dF_{X_1}(x_1)\right)} = 1. \tag{35}$$

Proof of Lemma 3. The proof is provided in Appendix H. □

From (35), we can write:

$$\int_{-A_1}^{A_2} H_b(Q(x_1+x_2))dF_{X_1}(x_1) = \gamma(x_2)H_b\left(\int_{-A_1}^{A_2} Q(x_1+x_2)dF_{X_1}(x_1)\right), \tag{36}$$

where $\gamma(x_2) \leq 1$ (due to the concavity of $H_b(\cdot)$), and $\gamma(x_2) \to 1$ when $x_2 \to +\infty$ (due to (35)). Furthermore, from the fact that $\lim_{x\to 0}\frac{H_b(x)}{cx} = +\infty$ $(c > 0)$, we have:

$$H_b\left(\int_{-A_1}^{A_2} Q(x_1+x_2)dF_{X_1}(x_1)\right) = -\eta(x_2)\log p_Y(0; F_{X_1}F_{X_2}^*)\int_{-A_1}^{A_2} Q(x_1+x_2)dF_{X_1}(x_1), \tag{37}$$

where $\eta(x_2) > 0$ and $\eta(x_2) \to +\infty$ when $x_2 \to +\infty$. From (36)–(37), the second and the third line of (34) become:

$$\left(1 - \gamma(x_2) + \frac{\lambda}{\eta(x_2)} - \lambda\right) \underbrace{\left(-\eta(x_2)\log p_Y(0; F_{X_1}F_{X_2}^*)\int_{-A_1}^{A_2} Q(x_1+x_2)dF_{X_1}(x_1)\right)}_{\geq 0}. \tag{38}$$

Since $\gamma(x_2) \to 1$ and $\eta(x_2) \to +\infty$ as $x_2 \to +\infty$, there exists a real number K such that $1 - \gamma(x_2) + \frac{\lambda}{\eta(x_2)} - \lambda < 0$ when $x_2 > K$. Therefore, the second and the third line of (34) approach zero from below, which proves that the support of X_2^* is bounded away from $+\infty$. As mentioned before, a similar argument holds when $x_2 \to -\infty$. This proves that X_2^* has a bounded support.

Remark 1. *We remark here that the order of showing the boundedness of the supports is important. First, for a given F_{X_2} (not necessarily bounded), it is proven that $F_{X_1}^*$ is bounded. Then, for a given bounded F_{X_1}, it is shown that $F_{X_2}^*$ is also bounded. Hence, the boundedness of the supports of the optimal input distributions is proven by contradiction. The order is reversed when $\lambda > 1$, and it follows the same steps as in the case of $\lambda \leq 1$. Therefore, it is omitted.*

We next prove the second claim in Proposition 3. We assume that $0 < \lambda < 1$, and a bounded F_{X_1} is given. We already know that for a given bounded F_{X_1}, $F_{X_2}^*$ has a bounded support denoted by $[-B_1, B_2]$. Therefore,

$$I_{F_{X_1}}^* = \sup_{\substack{F_{X_2}: \\ \mathbb{E}[X_2^2]\leq P_2'}} I_\lambda(F_{X_1}F_{X_2})$$

$$I_{F_{X_1}}^* = \sup_{\substack{F_{X_2}\in\mathscr{S}_2: \\ \mathbb{E}[X_2^2]\leq P_2'}} I_\lambda(F_{X_1}F_{X_2}), \tag{39}$$

where $\mathscr{S}_2$ denotes the set of all probability distributions on the Borel sets of $[-B_1, B_2]$. Let $p_0^* = p_Y(0; F_{X_1}F_{X_2}^*)$ denote the probability of the event $Y = 0$, induced by $F_{X_2}^*$ and the given F_{X_1}. Furthermore, let P_2^* denote the second moment of X_2 under $F_{X_2}^*$. The set:

$$\mathscr{F}_2 = \left\{F_{X_2} \in \mathscr{S}_2 \Big| \int_{-B_1}^{B_2} p(0|x_2)dF_{X_2}(x_2) = p_0^*, \int_{-B_1}^{B_2} x_2^2 dF_{X_2}(x_2) = P_2^*\right\} \tag{40}$$

is the intersection of $\mathscr{S}_2$ with two hyperplanes (note that $\mathscr{S}_2$ is convex and compact). We can write:

$$I_{F_{X_1}}^* = \sup_{F_{X_2}\in\mathscr{F}_2} I_\lambda(F_{X_1}F_{X_2}). \tag{41}$$

Note that having $F_{X_2} \in \mathscr{F}_2$, the objective function in (41) becomes:

$$\underbrace{\lambda H(Y)}_{\text{constant}} + \underbrace{(1-\lambda)H(Y|X_2) - H(Y|X_1,X_2)}_{\text{linear in } F_{X_2}}. \tag{42}$$

Since the linear part is continuous and $\mathscr{F}_2$ is compact (The continuity of the linear part follows similarly the continuity arguments in Appendix C. Note that this compactness is due to the closedness of the intersecting hyperplanes in $\mathscr{F}_2$, since a closed subset of a compact set is compact [11]. The hyperplanes are closed due to the continuity of x_2^2 and $p(0|x_2)$ (see (A16)).), the objective function in (41) attains its maximum at an extreme point of $\mathscr{F}_2$, which, by Dubins' theorem, is a convex combination of at most three extreme points of $\mathscr{S}_2$. Since the extreme points of $\mathscr{S}_2$ are the CDFs having only one point of increase in $[-B_1, B_2]$, we conclude that given any bounded F_{X_1}, $F_{X_2}^*$ has at most three mass points.

Now, assume that an arbitrary F_{X_2} is given with at most three mass points denoted by $\{x_{2,i}\}_{i=1}^3$. It is already known that the support of $F_{X_1}^*$ is bounded, which is denoted by $[-A_1, A_2]$. Let $\mathscr{S}_1$ denote the set of all probability distributions on the Borel sets of $[-A_1, A_2]$. The set:

$$\mathscr{F}_1 = \left\{ F_{X_1} \in \mathscr{S}_1 \,\Big|\, \int_{-A_1}^{A_2} p(0|x_1, x_{2,j}) dF_{X_1}(x_1) = p(0; \Gamma_{X_1}^* | x_{2,j}), \; j \in [1:3], \; \int_{-A_1}^{A_2} x_1^2 dF_{X_1}(x_1) = P_1' \right\}, \tag{43}$$

is the intersection of $\mathscr{S}_1$ with four hyperplanes. Note that here, since we know $\theta_1 \neq 0$, the optimal input attains its maximum power of P_1'. In a similar way,

$$I_{F_{X_2}}^* = \sup_{F_{X_1} \in \mathscr{F}_1} \left\{ I_\lambda(F_{X_1}, F_{X_2}) \right\}, \tag{44}$$

and having $F_{X_1} \in \mathscr{F}_1$, the objective function in (44) becomes:

$$I_\lambda = \underbrace{\lambda H(Y) + (1-\lambda) \sum_{i=1}^3 p_{X_2}(x_{2,i}) H(Y|X_2 = x_{2,i})}_{\text{constant}} - \underbrace{H(Y|X_1, X_2)}_{\text{linear in } F_{X_1}} \tag{45}$$

Therefore, given any F_{X_2} with at most three points of increase, $F_{X_1}^*$ has at most five mass points.

When $\lambda = 1$, the second term on the RHS of (45) disappears, which means that $\mathscr{F}_1$ could be replaced by:

$$\left\{ F_{X_1} \in \mathscr{S}_1 \Big| \int_{-A_1}^{A_2} p(0|x_1) dF_{X_1}(x_1) = \tilde{p}_0^*, \; \int_{-A_1}^{A_2} x_1^2 dF_{X_1}(x_1) = P_1' \right\},$$

where $\tilde{p}_0^* = p_Y(0; F_{X_1}^*, F_{X_2})$ is the probability of the event $Y = 0$, which is induced by $F_{X_1}^*$ and the given F_{X_2}. Since the number of intersecting hyperplanes has been reduced to two, it is concluded that $F_{X_1}^*$ has at most three points of increase. $\quad\square$

Remark 2. *Note that, the order of showing the discreteness of the support sets is also important. First, for a given bounded F_{X_1} (not necessarily discrete), it is proven that $F_{X_2}^*$ is discrete with at most three mass points. Then, for a given discrete F_{X_2} with at most three mass points, it is shown that $F_{X_1}^*$ is also discrete with at most five mass points when $\lambda < 1$ and at most three mass points when $\lambda = 1$. When $\lambda > 1$, the order is reversed, and it follows the same steps as in the case of $\lambda < 1$. Therefore, it is omitted.*

Remark 3. *If $\mathcal{X}_1, \mathcal{X}_2$ are assumed finite initially, similar results can be obtained by using the iterative optimization in the previous proof and the approach in Chapter 4, Corollary 3 of [13].*

5. Sum Rate Analysis

In this section, we propose a lower bound on the sum capacity of a MAC in the presence of a one-bit ADC front end at the receiver, which we conjecture to be tight. The sum capacity is given by:

$$C_{\text{sum}} = \sup\ I(X_1, X_2; Y|U),\tag{46}$$

where the supremum is over $F_U F_{X_1|U} F_{X_2|U}$ ($|\mathcal{U}| \leq 5$), such that $\mathbb{E}[X_j^2] \leq P_j$, $j = 1, 2$. We obtain a lower bound for the above by considering only those input distributions that are zero-mean per any realization of the auxiliary random variable U, i.e., $\mathbb{E}[X_j|U = u] = 0, \forall u \in \mathcal{U}, j = 1, 2$. Let P_1' and P_2' be two arbitrary non-negative real numbers. We have:

$$\sup_{\substack{F_{X_1}, F_{X_2}: \\ \mathbb{E}[X_j^2] \leq P_j' \\ \mathbb{E}[X_j] = 0,\, j=1,2}} I(X_1, X_2; Y) \leq \sup_{\substack{F_{\tilde{X}}: \\ \mathbb{E}[\tilde{X}^2] \leq P_1' + P_2'}} I(\tilde{X}; Y)\tag{47}$$

$$= 1 - H_b\left(Q\left(\sqrt{P_1' + P_2'}\right)\right)\tag{48}$$

where in (47), $\tilde{X} \triangleq X_1 + X_2$, $p_{Y|\tilde{X}}(0|\tilde{x}) = Q(\tilde{x})$; (48) follows from [4] for the point-to-point channel. Therefore, when $\mathbb{E}[X_j|U = u] = 0, \forall u \in \mathcal{U}, j = 1, 2$, we can write:

$$I(X_1, X_2; Y|U) = \sum_{i=1}^{5} p_U(u_i) I(X_1, X_2; Y|U = u_i)$$

$$\leq 1 - \sum_{i=1}^{5} p_U(u_i) H_b\left(Q\left(\sqrt{\mathbb{E}[X_1^2|U = u_i] + \mathbb{E}[X_2^2|U = u_i]}\right)\right)$$

$$\leq 1 - H_b\left(Q\left(\sqrt{\mathbb{E}[X_1^2] + \mathbb{E}[X_2^2]}\right)\right)\tag{49}$$

$$\leq 1 - H_b\left(Q\left(\sqrt{P_1 + P_2}\right)\right),\tag{50}$$

where (49) is due to the fact that $H_b\left(Q(\sqrt{x + y})\right)$ is a convex function of (x, y), and (50) follows from $\mathbb{E}[X_j^2] \leq P_j, j = 1, 2$.

The upper bound in (50) can be achieved by time division with power control as follows. Let $\mathcal{U} = \{0, 1\}$ and $p_U(0) = 1 - p_U(1) = \frac{P_1}{P_1 + P_2}$. Furthermore, let $F_{X_1|U}(x|1) = F_{X_2|U}(x|0) = s(x)$, where $s(\cdot)$ is the unit step function, and:

$$F_{X_1|U}(x|0) = F_{X_2|U}(x|1) = \frac{1}{2}s(x + \sqrt{P_1 + P_2}) + \frac{1}{2}s(x - \sqrt{P_1 + P_2}).$$

With this choice of $F_U F_{X_1|U} F_{X_2|U}$, the upper bound in (50) is achieved. Therefore,

$$C_{\text{sum}} \geq 1 - H_b\left(Q\left(\sqrt{P_1 + P_2}\right)\right).\tag{51}$$

A numerical evaluation of (46) is carried out as follows (the codes that are used for the numerical simulations are available at https://www.dropbox.com/sh/ndxkjt6h5a0yktu/AAAmfHkuPxe8rMNV1KzFVRgNa?dl=0). Although $\mathbb{E}[X_j^2]$ is upper bounded by P_j ($j = 1, 2$), the value of $\mathbb{E}[X_j^2|U = u]$ ($\forall u \in \mathcal{U}$) has no upper bound and could be any non-negative real number. However, in our numerical analysis, we further restrict our attention to the case $\mathbb{E}[X_j^2|U = u] \leq 20P_j, \forall u \in \mathcal{U}, j = 1, 2$. Obviously, as this upper bound tends to infinity, the approximation becomes more accurate (This further bounding of the conditional second moments is justified by the fact that the sum capacity is not greater than one, which is due to the one-bit quantization at the receiver. As a result, $I(X_1, X_2; Y | U = u)$ increases at most sublinearly with $\mathbb{E}[X_j^2 | U = u], j = 1, 2$, while $p_U(u)$ needs to decrease at least linearly to satisfy the average power constraints. Hence, the product $p_U(u) I(X_1, X_2; Y | U = u)$ decreases with $\mathbb{E}[X_j^2 | U = u]$ when $\mathbb{E}[X_j^2 | U = u]$ is above a threshold.). Each of the intervals $[0, 20P_1]$ and $[0, 20P_2]$ are divided into 201 points uniformly, which results in the discrete intervals $\frac{P_1}{10}[0 : 200]$ and $\frac{P_2}{10}[0 : 200]$, respectively. Afterwards, for any pair $(\alpha, \beta) \in \frac{P_1}{10}[0 : 200] \times \frac{P_2}{10}[0 : 200]$, the following is carried out for input distributions with at most three mass points.

$$\max_{\substack{F_{X_1} F_{X_2} : \\ \mathbb{E}[X_1^2] \leq \alpha, \mathbb{E}[X_2^2] \leq \beta}} I(X_1, X_2; Y) \tag{52}$$

The results are stored in a 201×201 matrix accordingly. In the above optimization, the MATLAB function fmincon is used with three different initial values, and the maximum of these three experiments is chosen. Then, the problem boils down to finding proper gains, i.e., the mass probabilities of U, that maximize $I(X_1, X_2; Y | U)$ and satisfy the average power constraints $\mathbb{E}[X_j^2] \leq P_j$. This is done via a linear program, which can be efficiently solved by the linprog function in MATLAB. Several cases were considered, such as $(P_1, P_2) = (1, 1), (P_1, P_2) = (1, 2), (P_1, P_2) = (3, 1)$, etc. In all these cases, the numerical evaluation of (46) leads to the same value as the lower bound in (51). Since the problem is not convex, it is not known whether the numerical results are the global optimum solutions; hence, we leave it as a conjecture that the sum capacity can be achieved by time division with power control.

6. Conclusions

We have studied the capacity region of a two-transmitter Gaussian MAC under average input power constraints and one-bit ADC front end at the receiver. We have derived an upper bound on the cardinality of this auxiliary variable, and proved that the distributions that achieve the boundary points of the capacity region are finite and discrete. Finally, a lower bound is proposed on the sum capacity of this MAC that is achieved by time division with power control. Through numerical analysis, this lower bound is shown to be tight.

Author Contributions: All three authors contributed to the paper. B.R. derived most of the claims with discussion with the M.V. and D.G.; The numerical results regarding the sum-rate analysis were obtained by the M.V.; Writing and editing of the paper has been done jointly, and all three authors have read and approved the final manuscript.

Funding: This research was supported in part by the European Research Council (ERC) through Starting Grant BEACON (Agreement No. 677854), by the U.K. Engineering and Physical Sciences Research Council (EPSRC) through the project COPES (EP/N021738/1) and by the British Council Institutional Link Program under Grant NO. 173605884.

Acknowledgments: The authors thank Professor Barbie and Professor Shirokov for their help in showing the preservation of the Markov chain in the weak convergence of the joint distributions in Appendix B.

Conflicts of Interest: The authors declare no conflict of interest.

Appendix A

The capacity region of the discrete memoryless (DM) MAC with input cost constraints has been addressed in Exercise 4.8 of [10]. If the input alphabets are not discrete, the capacity region is still the same because: (1) the converse remains the same if the inputs are from a continuous alphabet; (2) the region is achievable by coded time sharing and the discretization procedure (see Remark 3.8 in [10]). Therefore, it is sufficient to show the cardinality bound $|\mathcal{U}| \leq 5$.

Let $\mathscr{P}$ be the set of all product distributions (i.e., of the form $F_{X_1}(x_1)F_{X_2}(x_2)$) on $\mathbb{R}^2$. Let $\mathbf{g} : \mathscr{P} \to \mathbb{R}^5$ be a vector-valued mapping defined element-wise as:

$$
\begin{aligned}
g_1(F_{X_1|U}(\cdot|u)F_{X_2|U}(\cdot|u)) &= I(X_1; Y|X_2, U = u), \\
g_2(F_{X_1|U}(\cdot|u)F_{X_2|U}(\cdot|u)) &= I(X_2; Y|X_1, U = u), \\
g_3(F_{X_1|U}(\cdot|u)F_{X_2|U}(\cdot|u)) &= I(X_1, X_2; Y|U = u), \\
g_4(F_{X_1|U}(\cdot|u)F_{X_2|U}(\cdot|u)) &= \mathbb{E}[X_1^2|U = u], \\
g_5(F_{X_1|U}(\cdot|u)F_{X_2|U}(\cdot|u)) &= \mathbb{E}[X_2^2|U = u].
\end{aligned}
\tag{A1}
$$

Let $\mathscr{G} \subset \mathbb{R}^5$ be the image of $\mathscr{P}$ under the mapping $\mathbf{g}$ (i.e., $\mathscr{G} = \mathbf{g}(\mathscr{P})$). Given an arbitrary $(U, X_1, X_2) \sim F_U F_{X_1|U} F_{X_2|U}$, we obtain the vector $\mathbf{r}$ as:

$$
\begin{aligned}
r_1 &= I(X_1; Y|X_2, U) = \int_{\mathcal{U}} I(X_1; Y|X_2, U = u)dF_U(u), \\
r_2 &= I(X_2; Y|X_1, U) = \int_{\mathcal{U}} I(X_2; Y|X_1, U = u)dF_U(u), \\
r_3 &= I(X_1, X_2; Y|U) = \int_{\mathcal{U}} I(X_1, X_2; Y|U = u)dF_U(u), \\
r_4 &= \mathbb{E}[X_1^2] = \int_{\mathcal{U}} \mathbb{E}[X_1^2|U = u]dF_U(u), \\
r_5 &= \mathbb{E}[X_2^2] = \int_{\mathcal{U}} \mathbb{E}[X_2^2|U = u]dF_U(u).
\end{aligned}
$$

Therefore, $\mathbf{r}$ is in the convex hull of $\mathscr{G} \subset \mathbb{R}^5$. By Carathéodory's theorem [9], $\mathbf{r}$ can be written as a convex combination of six ($= 5 + 1$) or fewer points in $\mathscr{G}$, which states that it is sufficient to consider $|\mathcal{U}| \leq 6$. Since $\mathscr{P}$ is a connected set ($\mathscr{P}$ is the product of two connected sets; therefore, it is connected. Each of the sets in this product is connected because of being a convex vector space.) and the mapping $\mathbf{g}$ is continuous (this is a direct result of the continuity of the channel transition probability), $\mathscr{G}$ is a connected subset of $\mathbb{R}^5$. Therefore, the connectedness of $\mathscr{G}$ refines the cardinality of U to $|\mathcal{U}| \leq 5$.

It is also important to note that for the boundary points of $\mathscr{C}(P_1, P_2)$ that are not sum-rate optimal, it is sufficient to have $|\mathcal{U}| \leq 4$. The proof is as follows. Any point on the boundary of the capacity region that does not maximize $R_1 + R_2$ is either of the form $(I(X_1; Y|X_2, U), I(X_2; Y|U))$ or $(I(X_1; Y|U), I(X_2; Y|X_1, U))$ for some $F_U F_{X_1|U} F_{X_2|U}$ that satisfies $\mathbb{E}[X_j^2] \leq P_j, j = 1, 2$. In other words, it is one of the corner points of the corresponding pentagon in (4). As in the proof of Proposition 1, define the mapping $\mathbf{g} : \mathscr{P} \to \mathbb{R}^4$, where g_1 and g_2 are the coordinates of this boundary point conditioned on $U = u$, and g_3, g_4 are the same as g_4 and g_5 in (A1), respectively. The sufficiency of $|\mathcal{U}| \leq 4$ in this case follows similarly.

Appendix B

Since $|\mathcal{U}| \leq 5$, we assume $\mathcal{U} = \{0, 1, 2, 3, 4\}$ without loss of generality, since what matters in the evaluation of the capacity region is the mass probability of the auxiliary random variable U, not its actual values.

In order to show the compactness of Ω, we adopt a general form of the approach in [12].

First, we show that Ω is tight (a set of probability distributions Θ defined on $\mathbb{R}^k$, i.e., the set of CDFs $F_{X_1, X_2, \ldots, X_k}$, is said to be tight, if for every $\epsilon > 0$, there is a compact set $K_\epsilon \subset \mathbb{R}^k$ such that [14]:

$$\Pr\{(X_1, X_2, \ldots, X_k) \in \mathbb{R}^k \backslash K_\epsilon\} < \epsilon, \ \forall F_{X_1, X_2, \ldots, X_k} \in \Theta.$$

Choose $T_j, j = 1, 2$, such that $T_j > \sqrt{\frac{2P_j}{\epsilon}}$. Then, from Chebyshev's inequality,

$$\Pr\{|X_j| > T_j\} \leq \frac{P_j}{T_j^2} < \frac{\epsilon}{2}, \ j = 1, 2. \tag{A2}$$

Let $K_\epsilon = [0, 4] \times [-T_1, T_1] \times [-T_2, T_2] \subset \mathbb{R}^3$. It is obvious that K_ϵ is a closed and bounded subset of $\mathbb{R}^3$ and, therefore, compact. With this choice of K_ϵ, we have:

$$\Pr\{(U, X_1, X_2) \in \mathbb{R}^3 \backslash K_\epsilon\} \leq \Pr\{U \notin [0, 4]\} + \Pr\{X_1 \notin [-T_1, T_1]\} + \Pr\{X_2 \notin [-T_2, T_2]\}$$
$$< 0 + \frac{\epsilon}{2} + \frac{\epsilon}{2} = \epsilon, \tag{A3}$$

where (A3) is due to (A2). Hence, Ω is tight.

From Prokhorov's theorem [14] (p. 318), a set of probability distributions is tight if and only if it is relatively sequentially compact (a subset of topological space is relatively compact if its closure is compact). This means that for every sequence of CDFs $\{F_n\}$ in Ω, there exists a subsequence $\{F_{n_k}\}$ that is weakly convergent (the weak convergence of $\{F_n\}$ to F (also shown as $F_n(x) \xrightarrow{w} F(x)$) is equivalent to:

$$\lim_{n \to \infty} \int_{\mathbb{R}} \psi(x) dF_n(x) = \int_{\mathbb{R}} \psi(x) dF(x), \tag{A4}$$

for all continuous and bounded functions $\psi(\cdot)$ on $\mathbb{R}$. Note that $F_n(x) \xrightarrow{w} F(x)$ if and only if $d_L(F_n, F) \to 0$.) to a CDF F_0, which is not necessarily in Ω. If we can show that this F_0 is also an element of Ω, then the proof is complete, since we have shown that Ω is sequentially compact and, therefore, compact (Compactness and sequential compactness are equivalent in metric spaces. Note that Ω is a metric space with Lévy distance.).

Assume a sequence of distributions $\{F_n(\cdot, \cdot, \cdot)\}$ in Ω that converges weakly to $F_0(\cdot, \cdot, \cdot)$. In order to show that this limiting distribution is also in Ω, we need to show that both the average power constraints and the Markov chain $(X_1 - U - X_2)$ are preserved under F_0. The preservation of the second moment follows similarly to the argument in (Appendix I, [12]). In other words, since x^2 is continuous and bounded from below, from Theorem 4.4.4 in [15]:

$$\int x_j^2 d^3 F_0(u, x_1, x_2) \leq \liminf_{n \to \infty} \int x_j^2 d^3 F_n(u, x_1, x_2)$$
$$\leq P_j, \ j = 1, 2, \tag{A5}$$

Therefore, the second moments are preserved under the limiting distribution F_0.

For the preservation of the Markov chain $X_1 - U - X_2$, we need the following proposition.

Proposition A1. *Assume a sequence of distributions $\{F_n(\cdot, \cdot)\}$ over the pair of random variables (X, Y) that converges weakly to $F_0(\cdot, \cdot)$. Furthermore, assume that Y has a finite support, i.e., $\mathcal{Y} = \{1, 2, \ldots, |\mathcal{Y}|\}$. Then, the sequence of conditional distributions (conditioned on Y) converges weakly to the limiting conditional distribution (conditioned on Y), i.e.,*

$$F_n(\cdot|y) \overset{w}{\to} F_0(\cdot|y), \quad \forall y \in \mathcal{Y}, p_0(y) > 0. \tag{A6}$$

Proof of Proposition A1. The proof is by contradiction. If (A6) is not true, then there exists $y' \in \mathcal{Y}$, such that $p_0(y') > 0$ and $F_n(\cdot|y') \overset{w}{\nrightarrow} F_0(\cdot|y')$. This means, from the definition of weak convergence, that there exists a bounded continuous function of x, denoted by $g_{y'}(x)$, such that:

$$\int g_{y'}(x) dF_n(x|y') \nrightarrow \int g_{y'}(x) dF_0(x|y'). \tag{A7}$$

Let $f(x, y)$ be any bounded continuous function that satisfies:

$$f(x, y) = \begin{cases} 0 & y \in \mathcal{Y}, \ y \neq y' \\ g_{y'}(x) & y = y' \end{cases}. \tag{A8}$$

With this choice of $f(x, y)$, we have:

$$\int f(x, y) d^2 F_n(x, y) \nrightarrow \int f(x, y) d^2 F_0(x, y), \tag{A9}$$

which violates the assumption of the weak convergence of $F_n(\cdot, \cdot)$ to $F_0(\cdot, \cdot)$. Therefore, (A6) holds. $\square$

Since $\{F_n(\cdot, \cdot, \cdot)\}$ in Ω converges weakly to $F_0(\cdot, \cdot, \cdot)$ and $\mathcal{U}$ is finite, from Proposition A1, we have:

$$F_n(\cdot, \cdot|u) \overset{w}{\to} F_0(\cdot, \cdot|u), \quad \forall u \in \mathcal{U}, \tag{A10}$$

where it is obvious that the arguments are x_1 and x_2. Since $F_n \in \Omega$, we have $F_n(x_1, x_2|u) = F_n(x_1|u)F_n(x_2|u) \ \forall u \in \mathcal{U}$. Furthermore, since the convergence of the joint distribution implies the convergence of the marginals, we have (Theorem 2.7, [16,17]),

$$F_0(x_1, x_2|u) = F_0(x_1|u)F_0(x_2|u) \ \forall u \in \mathcal{U}, \tag{A11}$$

which states that under the limiting distribution F_0, the Markov chain $X_1 - U - X_2$ is preserved (Alternatively, this could be proven by the lower-semicontinuity of the mutual information as follow:

$$I_{F_0}(X_1; X_2|U = u) \leq \liminf_{n \to \infty} I_{F_n}(X_1; X_2|U = u) \tag{A12}$$

$$= 0, \ \forall u \in \mathcal{U}, \tag{A13}$$

where I_F denotes the mutual information under distribution F. The last equality is from the conditional independence of X_1 and X_2 given $U = u$ under F_n. Therefore, $I_{F_0}(X_1; X_2|U = u) = 0, \ \forall u \in \mathcal{U}$, which is equivalent to (A11).). This completes the proof of the compactness of Ω.

Appendix C

Appendix C.1. Concavity

When $0 < \lambda \leq 1$, we have:

$$I_\lambda(F_{X_1}, F_{X_2}) = \lambda H(Y) + (1 - \lambda)H(Y|X_2) - H(Y|X_1, X_2). \tag{A14}$$

For a given F_{X_1}, $H(Y)$ is a concave function of F_{X_2}, while $H(Y|X_2)$ and $H(Y|X_1, X_2)$ are linear in F_{X_2}. Therefore, I_λ is a concave function of F_{X_2}. For a given F_{X_2}, $H(Y)$ and $H(Y|X_2)$ are concave functions of F_{X_1}, while $H(Y|X_1, X_2)$ is linear in F_{X_1}. Since $(1 - \lambda) \geq 0$, I_λ is a concave function of F_{X_1}. The same reasoning applies to the case $\lambda > 1$.

Appendix C.2. Continuity

When $\lambda \leq 1$, the continuity of the three terms on the RHS of (A14) is investigated. Let $\{F_{X_2,n}\}$ be a sequence of distributions, which is weakly convergent to F_{X_2}. For a given F_{X_1}, we have:

$$\lim_{x_2 \to x_2^0} p(y; F_{X_1}|x_2) = \lim_{x_2 \to x_2^0} \int Q(x_1 + x_2) dF_{X_1}(x_1)$$

$$= \int \lim_{x_2 \to x_2^0} Q(x_1 + x_2) dF_{X_1}(x_1) \qquad (A15)$$

$$= p(y; F_{X_1}|x_2^0), \qquad (A16)$$

where (A15) is due to the fact that the Q function can be dominated by one, which is an absolutely integrable function over F_{X_1}. Therefore, $p(y; F_{X_1}|x_2)$ is continuous in x_2, and combined with the weak convergence of $\{F_{X_2,n}\}$, we can write:

$$\lim_{n \to \infty} p(y; F_{X_1} F_{X_2,n}) = \lim_{n \to \infty} \int p(y; F_{X_1}|x_2) dF_{X_2,n}(x_2)$$

$$= \int p(y; F_{X_1}|x_2) dF_{X_2}(x_2)$$

$$= p(y; F_{X_1} F_{X_2}).$$

This allows us to write:

$$\lim_{n \to \infty} - \sum_{y=0}^{1} p(y; F_{X_1} F_{X_2,n}) \log p(y; F_{X_1} F_{X_2,n}) = - \sum_{y=0}^{1} p(y; F_{X_1} F_{X_2}) \log p(y; F_{X_1} F_{X_2}),$$

which proves the continuity of $H(Y)$ in F_{X_2}. $H(Y|X_2 = x_2)$ is a bounded ($\in [0,1]$) continuous function of x_2, since it is a continuous function of $p(y; F_{X_1}|x_2)$, and the latter is continuous in x_2 (see (A16)). Therefore,

$$\lim_{n \to \infty} \int H(Y|X_2 = x_2) dF_{X_2,n}(x_2) = \int H(Y|X_2 = x_2) dF_{X_2}(x_2),$$

which proves the continuity of $H(Y|X_2)$ in F_{X_2}. In a similar way, it can be verified that $\int H(Y|X_1 = x_1, X_2 = x_2) dF_{X_1}(x_1)$ is a bounded and continuous function of x_2, which guarantees the continuity of $H(Y|X_1, X_2)$ in F_{X_2}, since:

$$H(Y|X_1, X_2) = \int \left(\int H(Y|X_1 = x_1, X_2 = x_2) dF_{X_1}(x_1) \right) dF_{X_2}(x_2) \qquad (A17)$$

Therefore, for a given F_{X_1}, I_λ is a continuous function of F_{X_2}. Exchanging the roles of F_{X_1} and F_{X_2}, also the case $\lambda > 1$ can be addressed similarly, so they are omitted for the sake of brevity.

Appendix C.3. Weak Differentiability

For a given F_{X_1}, the weak derivative of I_λ at $F_{X_2}^0$ is given by:

$$I_\lambda'(F_{X_1} F_{X_2})|_{F_{X_2}^0} = \lim_{\beta \to 0^+} \frac{I_\lambda(F_{X_1}((1 - \beta) F_{X_2}^0 + \beta F_{X_2})) - I_\lambda(F_{X_1} F_{X_2}^0)}{\beta}, \qquad (A18)$$

if the limit exists. It can be verified that:

$$I'_\lambda(F_{X_1}F_{X_2})\big|_{F^0_{X_2}}$$

$$= \lim_{\beta\to 0^+} \frac{\int i_\lambda(x_2;(1-\beta)F^0_{X_2}+\beta F_{X_2}|F_{X_1})d((1-\beta)F^0_{X_2}(x_2)+\beta F_{X_2}(x_2))-\int i_\lambda(x_2;F^0_{X_2}|F_{X_1})dF^0_{X_2}(x_2)}{\beta}$$

$$= \int i_\lambda(x_2;F^0_{X_2}|F_{X_1})dF_{X_2}(x_2) - \int i_\lambda(x_2;F^0_{X_2}|F_{X_1})dF^0_{X_2}(x_2)$$

$$= \int i_\lambda(x_2;F^0_{X_2}|F_{X_1})dF_{X_2}(x_2) - I_\lambda(F_{X_1}F^0_{X_2}),$$

where i_λ has been defined in (17). In a similar way, for a given F_{X_2}, the weak derivative of I_λ at $F^0_{X_1}$ is:

$$I'_\lambda(F_{X_1}F_{X_2})\big|_{F^0_{X_1}} = \int \tilde{i}_\lambda(x_1;F^0_{X_1}|F_{X_2})dF_{X_1}(x_1) - I_\lambda(F^0_{X_1}F_{X_2}), \tag{A19}$$

where $\tilde{i}_\lambda$ has been defined in (16). The case $\lambda > 1$ can be addressed similarly.

Appendix D

We have:

$$\sup_{\substack{F_{X_1}:\\ \mathbb{E}[X_1^2]\le P'_1}} I_\lambda(F_{X_1}F_{X_2}) \le \sup_{\substack{F_{X_1}F_{X_2}:\\ \mathbb{E}[X_j^2]\le P'_j,\, j=1,2}} I_\lambda(F_{X_1}F_{X_2})$$

$$\le \sup_{\substack{F_{X_1}F_{X_2}:\\ \mathbb{E}[X_j^2]\le P'_j,\, j=1,2}} I(X_1,X_2;Y) \tag{A20}$$

$$\le \sup_{\substack{F_{X_1}F_{X_2}:\\ \mathbb{E}[X_j^2]\le P'_j,\, j=1,2}} H(Y) - \inf_{\substack{F_{X_1}F_{X_2}:\\ \mathbb{E}[X_j^2]\le P'_j,\, j=1,2}} H(Y|X_1,X_2)$$

$$= 1 - \inf_{\substack{F_{X_1}F_{X_2}:\\ \mathbb{E}[X_j^2]\le P'_j,\, j=1,2}} \iint H_b\left(Q(x_1+x_2)\right)dF_{X_1}(x_1)dF_{X_2}(x_2)$$

$$= 1 - \inf_{\substack{F_{X_1}F_{X_2}:\\ \mathbb{E}[X_j^2]\le P'_j,\, j=1,2}} \iint H_b\left(Q\left(\sqrt{x_1^2}+\sqrt{x_2^2}\right)\right)dF_{X_1}(x_1)dF_{X_2}(x_2) \tag{A21}$$

$$\le 1 - \inf_{\substack{F_{X_1}F_{X_2}:\\ \mathbb{E}[X_j^2]\le P'_j,\, j=1,2}} \iint Q\left(\sqrt{x_1^2}+\sqrt{x_2^2}\right)dF_{X_1}(x_1)dF_{X_2}(x_2) \tag{A22}$$

$$= 1 - Q\left(\sqrt{P'_1}+\sqrt{P'_2}\right) \tag{A23}$$

$$< 1, \tag{A24}$$

where (A20) is from the non-negativity of mutual information and the assumption that $0 < \lambda \le 1$; (A21) is justified since the Q function is monotonically decreasing and the sign of the inputs does not affect the average power constraints; X_1 and X_2 can be assumed non-negative (or alternatively non-positive) without loss of optimality; in (A22), we use the fact that $Q\left(\sqrt{x_1^2}+\sqrt{x_2^2}\right)\le \frac{1}{2}$, and for $t \in [0,\frac{1}{2}]$, $H_b(t) \ge t$; (A23) is based on the convexity and monotonicity of the function $Q(\sqrt{u}+\sqrt{v})$ in (u,v), which is shown in Appendix G. Therefore, the LHS of (13) is strictly less than one.

Since X_2 has a finite second moment ($\mathbb{E}[X_2^2] \leq P_2'$), from Chebyshev's inequality, we have:

$$P(|X_2| \geq M) \leq \frac{P_2'}{M^2}, \quad \forall M > 0. \tag{A25}$$

Fix $M > 0$, and consider $X_1 \sim F_{X_1}(x_1) = \frac{1}{2}[s(x_1 + 2M) + s(x_1 - 2M)]$. By this choice of F_{X_1}, we get:

$$
\begin{aligned}
I_\lambda(F_{X_1} F_{X_2}) &= I(X_1; Y | X_2) + \lambda I(X_2; Y) \\
&\geq I(X_1; Y | X_2) \\
&= \int_{-\infty}^{+\infty} I(X_1; Y | X_2 = x_2) dF_{X_2}(x_2) \\
&\geq \int_{-M}^{+M} I(X_1; Y | X_2 = x_2) dF_{X_2}(x_2) \\
&\geq \inf_{F_{X_2}} \int_{-M}^{+M} H(Y | X_2 = x_2) dF_{X_2}(x_2) - \sup_{F_{X_2}} \int_{-M}^{+M} H(Y | X_1, X_2 = x_2) dF_{X_2}(x_2) \\
&\geq \left(1 - \frac{P_2'}{M^2}\right) H_b\left(\frac{1}{2} - \frac{1}{2}(Q(3M) + Q(M))\right) - H_b\left(Q(2M)\right),
\end{aligned}
\tag{A26, A27}
$$

where (A27) is due to (A25) and the fact that $H(Y | X_2 = x_2) = H_b(\frac{1}{2}Q(2M + x_2) + \frac{1}{2}Q(-2M + x_2))$ is minimized over $[-M, M]$ at $x_2 = M$ (or, alternatively at $x_2 = -M$) and $H(Y | X_1, X_2 = x_2) = \frac{1}{2}H_b(Q(2M + x_2)) + \frac{1}{2}H_b(Q(-2M + x_2))$ is maximized at $x_2 = 0$. (A27) shows that I_λ (≤ 1) can become arbitrarily close to one given that M is large enough. Hence, its supremum over all distributions F_{X_1} is one. This means that (13) cannot hold, and $\theta_1 \neq 0$.

Appendix E. Justification of (18), (19) and (20)

Let X be a vector space and Z be a real-valued function defined on a convex domain $D \subset X$. Suppose that x^* maximizes Z on D and that Z is Gateaux differentiable (weakly differentiable) at x^*. Then, from (Theorem 2, p. 178, [18]),

$$Z'(x)|_{x^*} \leq 0, \tag{A28}$$

where $Z'(x)|_{x^*}$ is the weak derivative of Z at x^*.

From (A19), we have the weak derivative of I_λ at $F_{X_1}^*$ as:

$$I_\lambda'(F_{X_1} F_{X_2})|_{F_{X_1}^*} = \int \tilde{i}_\lambda(x_1; F_{X_1}^* | F_{X_2}) dF_{X_1}(x_1) - I_\lambda(F_{X_1}^* F_{X_2}). \tag{A29}$$

Now, the derivation of (18) is immediate by inspecting that the weak derivative of the objective of (11) at $F_{X_1}^*$ is given by:

$$
\begin{aligned}
I_\lambda'(F_{X_1} F_{X_2})|_{F_{X_1}^*} - \theta_1 \left(\int x_1^2 dF_{X_1}(x_1) - \int x_1^2 dF_{X_1}^*(x_1)\right) &= \int \tilde{i}_\lambda(x_1; F_{X_1}^* | F_{X_2}) dF_{X_1}(x_1) - I_\lambda(F_{X_1}^* F_{X_2}) \\
&\quad - \theta_1 \left(\int x_1^2 dF_{X_1}(x_1) - \int x_1^2 dF_{X_1}^*(x_1)\right).
\end{aligned}
\tag{A30}
$$

Letting (A30) be lower than or equal to zero (as in (A28)) results in (18).

The equivalence of (18) to (19) and (20) follows similarly to the proof of Corollary 1 in (p. 210, [19]).

Appendix F

Equation (21) is obtained as follows.

$$\left| D\big(p(y|x_1, x_2) \| p(y; F_{X_1} F_{X_2})\big) \right| = \left| \sum_{y=0}^{1} p(y|x_1, x_2) \log \frac{p(y|x_1, x_2)}{p(y; F_{X_1} F_{X_2})} \right|$$

$$\leq \left| H(Y|X_1 = x_1, X_2 = x_2) \right| + \left| \sum_{y=0}^{1} p(y|x_1, x_2) \log p(y; F_{X_1} F_{X_2}) \right|$$

$$\leq 1 + \left| \sum_{y=0}^{1} \log p(y; F_{X_1} F_{X_2}) \right| \tag{A31}$$

$$= 1 - \sum_{y=0}^{1} \log p(y; F_{X_1} F_{X_2})$$

$$\leq 1 - 2 \min \left\{ \log p_Y(0; F_{X_1} F_{X_2}), \log p_Y(1; F_{X_1} F_{X_2}) \right\}$$

$$\leq 1 - 2 \log Q\big(\sqrt{P_1'} + \sqrt{P_2'}\big) \tag{A32}$$

$$< \infty,$$

where (A31) is due to the fact that the binary entropy function is upper bounded by one. (A32) is justified as follows.

$$\min \left\{ p_Y(0; F_{X_1} F_{X_2}), p_Y(1; F_{X_1} F_{X_2}) \right\} \geq \inf_{\substack{F_{X_1} F_{X_2}: \\ \mathbb{E}[X_j^2] \leq P_j'}} \min \left\{ p_Y(0; F_{X_1} F_{X_2}), p_Y(1; F_{X_1} F_{X_2}) \right\}$$

$$= \inf_{\substack{F_{X_1} F_{X_2}: \\ \mathbb{E}[X_j^2] \leq P_j'}} p_Y(0; F_{X_1} F_{X_2})$$

$$= \inf_{\substack{F_{X_1} F_{X_2}: \\ \mathbb{E}[X_j^2] \leq P_j'}} \iint Q(x_1 + x_2) dF_{X_1}(x_1) dF_{X_2}(x_2)$$

$$= \inf_{\substack{F_{X_1} F_{X_2}: \\ \mathbb{E}[X_j^2] \leq P_j'}} \iint Q\left(\sqrt{x_1^2} + \sqrt{x_2^2}\right) dF_{X_1}(x_1) dF_{X_2}(x_2) \tag{A33}$$

$$\geq Q\left(\sqrt{P_1'} + \sqrt{P_2'}\right), \tag{A34}$$

where (A34) is based on the convexity and monotonicity of the function $Q(\sqrt{u} + \sqrt{v})$, which is shown in Appendix G.

Equation (22) is obtained as follows.

$$p(y; F_{X_1}|x_2) \geq \min \left\{ p(0; F_{X_1}|x_2), p(1; F_{X_1}|x_2) \right\}$$

$$\geq \int Q\big(|x_1| + |x_2|\big) dF_{X_1}(x_1)$$

$$= \int Q\left(\sqrt{x_1^2} + |x_2|\right) dF_{X_1}(x_1)$$

$$\geq Q\left(\sqrt{P_1'} + |x_2|\right), \tag{A35}$$

where (A35) is due to the convexity of $Q(\alpha + \sqrt{x})$ in x for $\alpha \geq 0$.

Equation (23) is obtained as follows.

$$\left| \sum_{y=0}^{1} p(y|x_1, x_2) \log \frac{p(y; F_{X_1} F_{X_2})}{p(y; F_{X_1}|x_2)} \right| \leq -\sum_{y=0}^{1} p(y|x_1, x_2) \log p(y; F_{X_1}|x_2) - \sum_{y=0}^{1} p(y|x_1, x_2) \log p(y; F_{X_1} F_{X_2})$$

$$\leq -\sum_{y=0}^{1} \log p(y; F_{X_1}|x_2) - \sum_{y=0}^{1} \log p(y; F_{X_1} F_{X_2}) \tag{A36}$$

$$\leq -2\log Q\left(\sqrt{P_1'} + \sqrt{P_2'}\right) - 2\log Q\left(\sqrt{P_1'} + |x_2|\right), \tag{A37}$$

where (A36) is from $p(y|x_1, x_2) \leq 1$; and (A37) is from (A35) and (A34).

Note that (A37) is integrable with respect to F_{X_2} due to the concavity of $-\log Q(\alpha + \sqrt{x})$ in x for $\alpha \geq 0$ as shown in Appendix G. In other words,

$$\int_{-\infty}^{+\infty} \left(-2\log Q\left(\sqrt{P_1'} + \sqrt{P_2'}\right) - 2\log Q\left(\sqrt{P_1'} + |x_2|\right)\right) dF_{X_2}(x_2) < -4\log Q\left(\sqrt{P_1'} + \sqrt{P_2'}\right) \tag{A38}$$

$$< +\infty. \tag{A39}$$

Appendix G. Two Convex Functions

Let $f(x) = \log Q(a + \sqrt{x})$ for $x, a \geq 0$. We have,

$$f'(x) = -\frac{e^{-\frac{(a+\sqrt{x})^2}{2}}}{2\sqrt{2\pi x} Q(a + \sqrt{x})},$$

and:

$$f''(x) = \frac{e^{-\frac{(a+\sqrt{x})^2}{2}}}{4x\sqrt{2\pi} Q^2(a + \sqrt{x})} \left(\left(a + \sqrt{x} + \frac{1}{\sqrt{x}}\right) Q(a + \sqrt{x}) - \phi(a + \sqrt{x}) \right), \tag{A40}$$

where $\phi(x) = \frac{1}{\sqrt{2\pi}} e^{-\frac{x^2}{2}}$. Note that:

$$(1 + at + t^2) Q(a + t) + a\phi(a + t) > \left(1 + (a + t)^2\right) Q(a + t) \tag{A41}$$

$$> (a + t)\phi(a + t), \quad \forall a, t > 0, \tag{A42}$$

where (A41) and (A42) are, respectively, due to $\phi(x) > xQ(x)$ and $(1 + x^2)Q(x) > x\phi(x)$ $(x > 0)$. Therefore,

$$\left(a + \sqrt{x} + \frac{1}{\sqrt{x}}\right) Q(a + \sqrt{x}) > \phi(a + \sqrt{x}),$$

which makes the second derivative in (A40) positive and proves the (strict) convexity of $f(x)$.

Let $f(u, v) = Q(\sqrt{u} + \sqrt{v})$ for $u, v \geq 0$. By simple differentiation, the Hessian matrix of f is:

$$\mathbf{H} = \frac{e^{-\frac{(\sqrt{u}+\sqrt{v})^2}{2}}}{\sqrt{2\pi}} \begin{bmatrix} \frac{1}{2u\sqrt{u}} + \frac{\sqrt{u}+\sqrt{v}}{4u} & \frac{\sqrt{u}+\sqrt{v}}{4\sqrt{u}\sqrt{v}} \\ \frac{\sqrt{u}+\sqrt{v}}{4\sqrt{u}\sqrt{v}} & \frac{1}{2v\sqrt{v}} + \frac{\sqrt{u}+\sqrt{v}}{4v} \end{bmatrix}. \tag{A43}$$

It can be verified that $\det(\mathbf{H}) > 0$ and $\text{trace}(\mathbf{H}) > 0$. Therefore, both eigenvalues of $\mathbf{H}$ are positive, which makes the matrix positive definite. Hence, $Q(\sqrt{u} + \sqrt{v})$ is (strictly) convex in (u, v).

Appendix H

Let $A \triangleq \max\{A_1, A_2\}$.

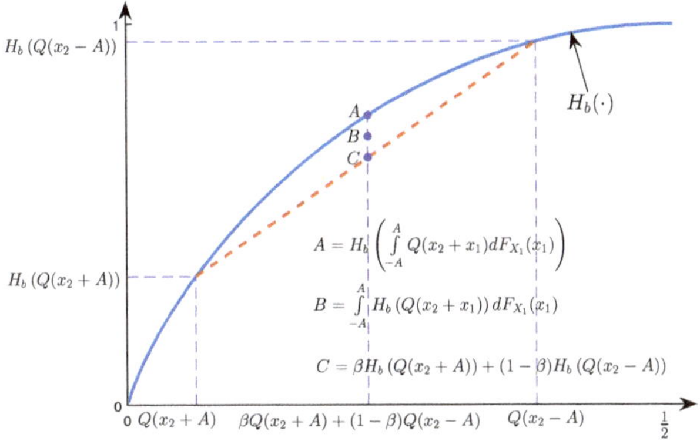

Figure A1. The figure depicting (A46) and (A48). Note that in the statement of Lemma 3, $x_2 \to +\infty$. Hence, we have assumed $x_2 > A$ in the figure.

It is obvious that:

$$Q(x_2 + A) \leq \int_{-A}^{A} Q(x_1 + x_2)dF_{X_1}(x_1) \leq Q(x_2 - A). \tag{A44}$$

Therefore, we can write:

$$\int_{-A}^{A} Q(x_1 + x_2)dF_{X_1}(x_1) = \beta Q(x_2 + A) + (1 - \beta)Q(x_2 - A), \tag{A45}$$

for some $\beta \in [0, 1]$. Note that β is a function of x_2. Furthermore, due to the concavity of $H_b(\cdot)$, we have:

$$H_b\left(\int_{-A}^{A} Q(x_1 + x_2)dF_{X_1}(x_1)\right) \geq \int_{-A}^{A} H_b(Q(x_1 + x_2))dF_{X_1}(x_1). \tag{A46}$$

From the fact that:

$$H_b(x) \geq \frac{H_b(p) - H_b(a)}{p - a}(x - a) + H_b(a), \ \forall x \in [a, \ p], \ \forall a, p \in [0, \ 1](a < p), \tag{A47}$$

we can also write:

$$\int_{-A}^{A} H_b(Q(x_1 + x_2))dF_{X_1}(x_1) \geq \frac{H_b(Q(x_2 - A)) - H_b(Q(x_2 + A))}{Q(x_2 - A) - Q(x_2 + A)}\left(\int_{-A}^{A} Q(x_1 + x_2)dF_{X_1}(x_1) - Q(x_2 + A)\right)$$

$$+ H_b(Q(x_2 + A))$$
$$= \beta H_b(Q(x_2 + A)) + (1 - \beta)H_b(Q(x_2 - A)), \tag{A48}$$

where (A45) and (A47) have been used in (A48). (A46) and (A48) are depicted in Figure A1. From (A45) and (A48), we have:

$$\frac{\beta H_b(Q(x_2 + A)) + (1 - \beta)H_b(Q(x_2 - A))}{H_b\left(\beta Q(x_2 + A) + (1 - \beta)Q(x_2 - A)\right)} \leq \frac{\int_{-A}^{A} H_b(Q(x_1 + x_2))dF_{X_1}(x_1)}{H_b\left(\int_{-A}^{A} Q(x_1 + x_2)dF_{X_1}(x_1)\right)} \leq 1. \tag{A49}$$

Let:

$$\beta^* \triangleq \arg\min_{\beta} \frac{\beta H_b(Q(x_2 + A)) + (1 - \beta) H_b(Q(x_2 - A))}{H_b\left(\beta Q(x_2 + A) + (1 - \beta) Q(x_2 - A)\right)}. \tag{A50}$$

This minimizer satisfies the following equality:

$$\frac{d}{d\beta}\left(\frac{\beta H_b(Q(x_2 + A)) + (1 - \beta) H_b(Q(x_2 - A))}{H_b\left(\beta Q(x_2 + A) + (1 - \beta) Q(x_2 - A)\right)}\right)\Bigg|_{\beta=\beta^*} = 0. \tag{A51}$$

Therefore, we can write:

$$\frac{\beta H_b(Q(x_2 + A)) + (1 - \beta) H_b(Q(x_2 - A))}{H_b\left(\beta Q(x_2 + A) + (1 - \beta) Q(x_2 - A)\right)} \geq \frac{\beta^* H_b(Q(x_2 + A)) + (1 - \beta^*) H_b(Q(x_2 - A))}{H_b\left(\beta^* Q(x_2 + A) + (1 - \beta^*) Q(x_2 - A)\right)} \tag{A52}$$

$$= \frac{\frac{H_b(Q(x_2-A))-H_b(Q(x_2+A))}{Q(x_2-A)-Q(x_2+A)}}{H_b'\left(\beta^* Q(x_2 + A) + (1 - \beta^*) Q(x_2 - A)\right)} \tag{A53}$$

$$\geq \frac{\frac{H_b(Q(x_2-A))-H_b(Q(x_2+A))}{Q(x_2-A)-Q(x_2+A)}}{H_b'(Q(x_2 + A))}, \tag{A54}$$

where (A52) is from the definition in (A50); (A53) is from the expansion of (A51), and $H_b'(t) = \log(\frac{1-t}{t})$ is the derivative of the binary entropy function; (A54) is due to the fact that $H_b'(t)$ is a decreasing function.

Applying L'Hôspital's rule multiple times, we obtain:

$$\begin{aligned}
\lim_{x_2 \to +\infty} \frac{\frac{H_b(Q(x_2-A))-H_b(Q(x_2+A))}{Q(x_2-A)-Q(x_2+A)}}{H_b'(Q(x_2+A))} &= \lim_{x_2 \to +\infty} \frac{H_b(Q(x_2-A))\left(1 - \frac{H_b(Q(x_2+A))}{H_b(Q(x_2-A))}\right)}{Q(x_2-A)\left(1 - \frac{Q(x_2+A)}{Q(x_2-A)}\right)\log(\frac{1-Q(x_2+A)}{Q(x_2+A)})} \\
&= \lim_{x_2 \to +\infty} -\frac{H_b(Q(x_2-A))}{Q(x_2-A)\log(Q(x_2+A))} \\
&= \lim_{x_2 \to +\infty} \frac{e^{-\frac{(x_2-A)^2}{2}}\log(Q(x_2-A))}{e^{-\frac{(x_2-A)^2}{2}}\log(Q(x_2+A))+\frac{Q(x_2-A)}{Q(x_2+A)}e^{-\frac{(x_2+A)^2}{2}}} \\
&= \lim_{x_2 \to +\infty} \frac{\log(Q(x_2-A))}{\log(Q(x_2+A))+1} \\
&= \lim_{x_2 \to +\infty} \frac{Q(x_2+A)e^{Ax_2}}{Q(x_2-A)e^{-Ax_2}} \\
&= 1
\end{aligned} \tag{A55}$$

From (A49), (A54) and (A55), (35) is proven. Note that the boundedness of X_1 is crucial in the proof. In other words, the fact that $Q(x_2 - A) \to 0$ as $x_2 \to +\infty$ is the very result of $A < +\infty$.

References

1. Walden, R.H. Analog-to-digital converter survey and analysis. *IEEE J. Sel. Areas Commun.* **1999**, *17*, 539–550. [CrossRef]
2. Murmann, B. ADC Performance Survey. *CoRR* 2014. Available online: http://web.stanford.edu/~murmann/adcsurvey.html (accessed on 20 May 2018).
3. Gunduz, D.; Stamatiou, K.; Michelusi, N.; Zorzi, M. Designing intelligent energy harvesting communication systems. *IEEE Comm. Mag.* **2014**, *52*, 210–216. [CrossRef]

4. Singh, J.; Dabeer, O.; Madhow, U. On the limits of communication with low-precision analog-to-digital Conversion at the Receiver. *IEEE Trans. Commun.* **2009**, *57*, 3629–3639. [CrossRef]

5. Krone, S.; Fettweis, G. Fading channels with 1-bit output quantization: Optimal modulation, ergodic capacity and outage probability. In Proceedings of the 2010 IEEE Information Theory Workshop, Dublin, Ireland, 30 August–3 September 2010; pp. 1–5.

6. Mezghani, A.; Nossek, J.A. Analysis of Rayleigh-fading channels with 1-bit quantized output. In Proceedings of the 2008 IEEE International Symposium on Information Theory, Toronto, ON, Canada, 6–11 July 2008.

7. Mezghani, A.; Nossek, J.A. On ultra-wideband MIMO systems with 1-bit quantized outputs: Performance analysis and input optimization. In Proceedings of the 2007 IEEE International Symposium on Information Theory, Nice, France, 24–29 June 2007; pp. 1286–1289.

8. Mo, J.; Heath, R. Capacity analysis of one-bit quantized MIMO systems with transmitter channel state information. *IEEE Trans. Signal Process.* **2015**, *63*, 5498–5512. [CrossRef]

9. Witsenhausen, H.S. Some aspects of convexity useful in information theory. *IEEE Trans. Inf. Theory* **1980**, *26*, 265–271. [CrossRef]

10. El Gamal, A.; Kim, Y.H. *Network Information Theory*; Cambridge University Press: Cambridge, UK, 2012.

11. Rudin, W. *Principles of Mathematical Analysis*; McGraw-Hill: New York, NY, USA, 1976.

12. Abou-Faycal, I.C.; Trott, M.D.; Shamai, S. The capacity of discrete-time memoryless Rayleigh-fading channels. *IEEE Trans. Inf. Theory* **2001**, *47*, 1290–1300. [CrossRef]

13. Gallager, R.G. *Information Theory and Reliable Communication*; Wiley: New York, NY, USA, 1968.

14. Shiryaev, A.N. *Probability*, 2nd ed.; Springer: Berlin, Germany, 1996.

15. Chung, K.L. *A Course in Probability Theory*, 2nd ed.; Academic Press: New York, NY, USA, 1974.

16. Billingsley, P. *Convergence of Probability Measures*, 2nd ed.; Wiley: New York, NY, USA, 1968.

17. Sagitov, S. Lecture Notes: Weak Convergence of Probability Measures. Available online: http://www.math.chalmers.se/~serik/C-space.pdf (accessed on 20 May 2018).

18. Luenberger, D. *Optimization by Vector Space Methods*; Wiley: New York, NY, USA, 1969.

19. Smith, J.G. The information capacity of amplitude and variance constrained scalar Gaussian channels. *Inf. Control* **1971**, *18*, 203–219. [CrossRef]

Article

Efficient Algorithms for Coded Multicasting in Heterogeneous Caching Networks

Giuseppe Vettigli [1] , Mingyue Ji [2,*], Karthikeyan Shanmugam [3], Jaime Llorca [4], Antonia M. Tulino [1,4] and Giuseppe Caire [5]

[1] Department of Electrical Engineering and Information Technology (DIETI), Universitá di Napoli Federico II, 80138 Napoli, Italy; g.vettigli86@gmail.com (G.V.); antoniamaria.tulino@unina.it (A.M.T.)
[2] Department of Electrical and Computer Engineering (ECE), University of Utah, Salt Lake City, UT 84112, USA
[3] IBM Research, New York, NY 10598, USA; karthikeyanshanmugam88@gmail.com
[4] Department of Math and Algorithms, Nokia Bell Labs, Murray Hill, NJ 07738, USA; jaime.llorca@nokia-bell-labs.com (J.L.); a.tulino@nokia-bell-labs.com (A.M.T.)
[5] Faculty of Electrical Engineering and Computer Science (EECS), Technical University of Berlin, 10587 Berlin, Germany; caire@tu-berlin.de
* Correspondence: mingyue.ji@utah.edu

Received: 14 January 2019; Accepted: 13 March 2019; Published: 25 March 2019

Abstract: Coded multicasting has been shown to be a promising approach to significantly improve the performance of content delivery networks with multiple caches downstream of a common multicast link. However, the schemes that have been shown to achieve order-optimal performance require content items to be partitioned into several packets that grows exponentially with the number of caches, leading to codes of exponential complexity that jeopardize their promising performance benefits. In this paper, we address this crucial performance-complexity tradeoff in a heterogeneous caching network setting, where edge caches with possibly different storage capacity collect multiple content requests that may follow distinct demand distributions. We extend the asymptotic (in the number of packets per file) analysis of shared link caching networks to heterogeneous network settings, and present novel coded multicast schemes, based on *local graph coloring*, that exhibit polynomial-time complexity in all the system parameters, while preserving the asymptotically proven multiplicative caching gain even for finite file packetization. We further demonstrate that the packetization order (the number of packets each file is split into) can be traded-off with the number of requests collected by each cache, while preserving the same multiplicative caching gain. Simulation results confirm the superiority of the proposed schemes and illustrate the interesting request aggregation vs. packetization order tradeoff within several practical settings. Our results provide a compelling step towards the practical achievability of the promising multiplicative caching gain in next generation access networks.

Keywords: caching networks; random fractional caching; coded caching; coded multicasting; index coding; finite-length analysis; graph coloring; approximation algorithms

---

1. Introduction

Recent information-theoretic studies [1–49] have characterized the fundamental limiting performance of several caching networks of practical relevance, in which throughput scales linearly with cache size, showing great promise to accommodate the exponential traffic growth experienced in today's communication networks [50]. In this context, a caching scheme is defined in terms of two phases: the *cache placement phase*, which operates at a large time-scale and determines the content to be placed at the network caches, and the *delivery phase*, during which user requests are served from the content caches and sources in the network. Some of the network topologies studied include shared link caching networks [1,2,8–14], device-to-device (D2D) caching networks [17–19,33,34], hierarchical caching networks [24], multi-server caching networks [29], and combination caching networks [36–42].

Consider a network with one source (e.g., base station) having access to m files, and n users (e.g., small-cell base stations or end user devices), each with a cache memory of M files. In [17], the authors showed that if the users can communicate between each other via D2D communications, a simple distributed random caching policy and TDMA-based unicast D2D delivery achieves the order-optimal throughput $\Theta\left(\max\{\frac{M}{m}, \frac{1}{m}, \frac{1}{n}\}\right)$ whose linear scaling with M when $Mn \geq m$ exhibits a remarkable multiplicative caching gain, in the sense that the per-user throughput grows proportionally to the cache size M for fixed library size m, and it is independent of the number of users n in the system. Moreover, in this scheme each user caches entire files without the need for partitioning files into packets, and missing files are delivered via unicast transmissions between neighbor nodes, making it efficiently implementable in practice. We recall that order-optimality refers to the fact that the multiplicative gap between information-theoretic converse and achievable performance can be bounded by a constant number when $m, n \to \infty$.

In the case that users cannot communicate between each other, but share a multicast link from the content source, the authors in [8,9] showed that the use of coded multicasting (also referred to as index coding [51]) allows achieving the same order-optimal worst-case throughput as in the D2D caching network. In this case, however, in order to create enough coding opportunities during the delivery phase, requested files are required to be partitioned into a number of packets that grows exponentially with the number of users, leading to coding schemes of exponential complexity [8,9,21].

In [10,12], the authors considered the same shared link caching network, but under random demands characterized by a probability distribution, and proposed a scheme consisting of random aggregate popularity (RAP) placement and chromatic number index coding (CIC) delivery, referred to as RAP-CIC, proved to be order-optimal in terms of average throughput. The authors further provided optimal average rate scaling laws under Zipf [52] demand distributions, whose analytical characterization required resorting to a polynomial-time approximation of CIC, referred to as greedy constrained coloring (GCC). Using RAP-GCC, the authors further established the regions of the system parameters, in which multiplicative caching gains are potentially achievable. While GCC exhibits polynomial complexity in the number of users and packets, the order-optimal performance guarantee still requires, in general, the packetization order (number of packets per file) to grow exponentially with the number of users, as showed in [21].

It is then key to understand if the promised multiplicative caching gain, shown to be asymptotically achievable by the above-referenced schemes, can be preserved in practical settings of finite packetization order. In this context, we shall differentiate between coded multicast schemes that assume a deterministic vs. a random cache placement phase. Deterministic placement policies determine where to store file packets according to a deterministic procedure that takes into account the ID of each packet. In contrast, random placement policies, after determining the number of packets to be cached of each file at each cache, choose the exact packet IDs uniformly at random. While the increased structure of deterministic placement policies can be exploited to design more efficient coded multicast algorithms, random placement policies are desirable in practice, as they provide increased robustness by requiring less cache configuration changes under system dynamics.

The seminal work of [21] showed that all previously proposed schemes (based on both deterministic and random cache placement) required exponential packetization, and that under random placement, no graph-coloring-based coded multicast algorithm can achieve multiplicative caching gains with sub-exponential packetization. Since the fundamental results of [21], several works have studied the now central problem in caching of finite file packetization. The authors in [53] connect the caching problem to resolvable combinatorial designs and derive a scheme that while improving exponentially over previous schemes [8,9,21], still requires exponential packetization. In [54], the authors introduce the combinatorial concept of Placement Delivery Array (PDA) and derive a caching scheme where the packetization scales super-polynomially with the number of users. The work in [22] establishes a connection with the construction of hypergraphs with extremal properties, and provides the first sub-exponential (but still intractable) scheme. Somewhat surprisingly, some of the authors of [21] introduced a new combinatorial design based on Ruzsa-Szemerédi graphs in [30] and showed that a linear scaling of the number of packets per file with n can be achieved for a throughput of $\Theta(n^{-\delta})$, where δ can be arbitrarily small. However, all the above studies focus on coded multicast algorithms that assume a deterministic cache placement phase. Under random cache placement, several coded multicast algorithms have been proposed in the context of homogenous shared link caching networks [55–60], including our previous work that serves as the basis for this paper.

In this work, we address the important problem of finite-length coded multicasting under random cache placement, focusing on a more general heterogeneous shared link caching network, in which caches with possibly different sizes collect possibly multiple requests according to possibly different demand distributions (see Figure 1). As shown in Figure 1, this scenario can be motivated by the presence of both end user caches and cache-enabled small-cell base stations or WLAN access points sharing a common multicast link. In this case, each small-cell base station can be modeled as a user cache placing multiple requests. In addition, multiple requests per user also arise in the presence of delay-tolerant content requests (e.g., file downloading). While there have been several information-theoretic studies of shared link caching networks with distinct cache sizes [61–63], and with multiple per-user requests [13,14,34,64,65], none of these works considered the finite-length regime nor addressed the joint effect of random demands, heterogenous cache sizes, and multiple per-user requests.

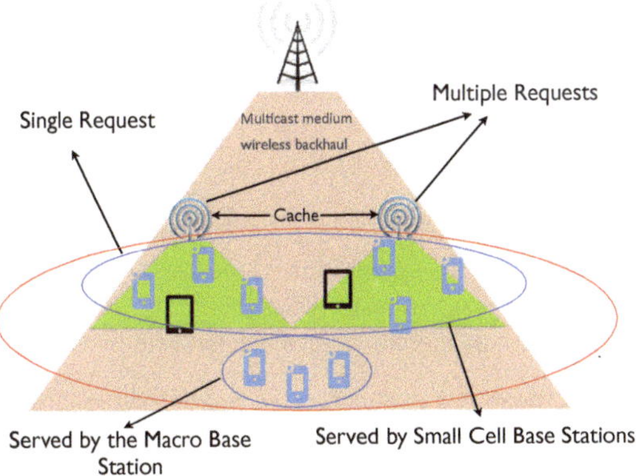

Figure 1. An example of the network model, which consists of a source node (base station in this figure) with access to the content library and connected to the users via a shared (multicast) link. Each user (end users and small-cell base stations) may have different cache size and request a different number of files according to their own demand distribution.

The contributions of this paper are as follows:

1. We provide a generalized model for heterogeneous shared link caching networks, in which users can have different cache sizes and make different number of requests according to different demand distributions.
2. We design two novel coded multicast algorithms based on *local graph coloring*, referred to as Greedy Local Coloring (GLC) and Hierarchical Greedy Local Coloring (HgLC) that exhibit polynomial-time complexity in both the number of caches and the packetization order. In combination with the Random Aggregate Popularity (RAP) placement policy of [10,12], we show that the overall schemes RAP-GLC and RAP-HgLC are order-optimal in the asymptotic file-length regime.
3. Focusing on the finite-length regime, in which content items can be partitioned into a finite number of packets, we show how the general advantage of local graph coloring is especially relevant when the number of per-user requests grow. We validate via simulations the superiority of RAP-GLC, especially with high number of per-user requests. We then show how RAP-HgLC, with a slight increase in the polynomial complexity order, further improves the caching gain of RAP-GLC, remarkably approaching the multiplicative gain that existing schemes can only guarantee in the asymptotic file-length regime.
4. We demonstrate that there is a tradeoff between the required packetization order and the number of requested files per user. In particular, for a given target gain, if the number of requests increases, then the number of packets per file can be reduced, while preserving the target gain. We further quantify the regime of per-user requests for which a caching scheme with unit packetization order (i.e., a scheme that treats only whole files) is order-optimal. Our analysis illustrates the key impact of content request aggregation in time and space on caching performance. That is, if edge caches can wait for collecting multiple requests over time and/or aggregate requests from multiple users, the same performance can be achieved with lower packetization order, and hence lower computational complexity.

The paper is organized as follows. Section 2 introduces the network model and problem formulation. Section 3 describes the construction of coded multicast algorithms using graph coloring, with special focus on the advantages of local graph coloring. Section 4 presents novel polynomial-time local-graph-coloring-based coded multicast schemes. Section 5 analyzes the effect of request aggregation on the performance–complexity tradeoff. Section 6 presents simulation results and related discussions. Finally, concluding remarks are given in Section 7.

2. Network Model and Problem Formulation

We consider a caching network formed by a source node with access to a content library, connected to several caching nodes/users via a single shared (multicast) link. Similar to previous works [8–10,12–14,21,22,30], we define a caching scheme in terms of two phases:

- *Placement phase,* which operates at a large time-scale and determines the content to be placed at the caching nodes,
- *Delivery phase,* during which users requests are served from the content caches and sources in the network.

However, differently from previous works, we generalize the model to a heterogeneous system in which each caching node has a possibly different cache size and requests a possibly different number of files. A practical example of our setting can be represented by a macro base station connected to several cache-enabled small-cell base stations, and a number of user devices served either by the macro base station or by the small-cell base stations. In this setting, each small cell acts as a super user requesting multiple files resulting from the requests of the users it serves.

Specifically, the heterogeneous caching network consists of a single source node storing a library of files $\mathcal{F} = \{1, \ldots, m\}$, each with entropy F bits, and n user nodes $\mathcal{U} = \{1, \ldots, n\}$, each with a cache of storage capacity $M_u F$ bits (i.e., each user caches up to M_u files). Each user u can requests L_u $(1 \leq L_u \leq m)$ different files according to its individual request probability distribution. We assume that the library files have finite length and consequently a finite packetization order. Our main objective is to design a caching scheme that minimizes the number of transmissions required to satisfy the demands of all users.

In a homogeneous network setting with infinite packetization order, recent works [8–10,12–14] have shown that it is possible to satisfy a scaling number of users with only a constant number of multicast transmissions. The achievable schemes configure user caches with complementary (side) information during the caching phase, such that the resulting coded multicasting opportunities that arise during the delivery phase can be used to minimize the transmission rate (or load) over the shared multicast link. Specifically, reference [12] showed that under Zipf file popularity, a properly optimized random fractional placement policy, referred to as Random Aggregate Popularity (RAP) caching, achieves order-optimality when combined with a graph-coloring-based coded multicast scheme. Unfortunately, even in the homogenous setting, it was shown in [21] that a central limitation of all previous works is that they require infinite packetization order: all existing caching schemes achieve at most a factor of two gain when the packetization order is finite.

In this work, inspired by the fundamental throughput-delay-memory tradeoff derived in [21], our goal is to design computationally efficient schemes that provide good performance in the finite packetization regime. For the caching phase, (1) we restrict our placement policies to the class of random fractional schemes described in [9,10,12–14], proved to be order-optimal in the homogeneous setting. For the delivery phase, (2) we focus on the class of graph-coloring-based index coding schemes, and design two novel polynomial-time algorithms that employ local graph coloring on the (index coding) conflict graph [51].

2.1. Random Fractional Cache Placement

The class of random fractional placement schemes is described as follows:

1. Packetization: Each file is partitioned into B packets of equal-size F/B bits, where the integer B is referred to as the packetization order. Each packet is represented by a symbol in finite field $\mathbb{F}_{2^{F/B}}$, where we assume that F/B is large enough.
2. Random Placement: Each user u caches $p_{f,u} M_u B$ packets independently at random from each file f, where $p_{f,u}$ is the probability that file f is cached at user u, and satisfies $0 \leq p_{f,u} \leq 1/M_u, \forall f \in \mathcal{F}$ such that $\sum_{f=1}^{m} p_{f,u} = 1, \forall u \in \mathcal{U}$.

We introduce a *caching distribution* matrix $\mathbf{P} = [p_{f,u}] \in \mathbb{R}_+^{m \times n}$, where $f \in \mathcal{F}$ and $u \in \mathcal{U}$. Please note that the number of packets of file f cached at user u, $p_{f,u} M_u B$, can be directly determined from the caching distribution matrix $\mathbf{P}$. As described in [10,12–14], the caching distribution must be properly optimized to balance the gains from local cache hits (where requested packets are served by the local cache) and coded multicast opportunities (where requested packets are served by coded transmissions that simultaneously satisfy distinct user requests). When this is the case, we refer to the cache placement scheme as Random Aggregate Popularity (RAP) caching (see e.g., [10,12–14]). Given the number of packets to be cached of a given file, the actual indices of the packets to be cached are chosen uniformly at random, and independently across users. We use $\mathbf{C}_{u,f}$ to denote the set of packets of file f cached at user u and $\mathbf{C} = \{\mathbf{C}_{u,f}\}$ with $u \in \mathcal{U}$ and $f \in \mathcal{F}$ to denote the *packet-level* cache placement realization.

The goal of the placement phase is to configure the user caches to create coding opportunities during the delivery phase that allow serving distinct user requests via common multicast transmissions.

Compared to deterministic placement [8], random placement schemes allow configuring user caches with lower complexity and increased robustness, i.e., changes in system parameters (e.g., number of users, number files, file popularity) require less changes in users' cache configurations [12].

Recall that the placement phase operates at a much larger time-scale than the delivery phase, and hence is unaware of the requests in the subsequent delivery rounds. Therefore, the placement phase can be designed according to the demand distribution, but must be independent of the requests realizations.

2.2. Random Multiple Requests

Each user $u \in \mathcal{U}$ requests L_u ($1 \leq L_u \leq m$) files independently from other users, following a probability distribution $q_{f,u}$ with $q_{f,u} \in [0,1]$ and $\sum_{f=1}^{m} q_{f,u} = 1$ (i.e., for each request of user u, file f is chosen with probability $q_{f,u}$). We introduce a *demand distribution* matrix $\mathbf{Q} = [q_{f,u}] \in \mathbb{R}_+^{m \times n}$, where $f \in \mathcal{F}$ and $u \in \mathcal{U}$. In the following, we use $\mathbf{W} = \{\mathbf{W}_{u,f}\}$, with $u \in \mathcal{U}$ and $f \in \mathcal{F}$, to denote the *packet-level* demand realization where $\mathbf{W}_{u,f}$ denotes the packets of file f requested by user u.

The multiple-request parameters $\{L_u\}$ have a key operational meaning, in that it captures the possibility of edge caches to collect requests across time and space. That is, L_u may represent the amount of requests collected over time (given the delay tolerance of some content requests) as well as the amount of requests collected across space from users served by the given edge cache (e.g., when edge caches are located at helper nodes or small-cell base stations serving multiple individual users).

2.3. Performance Metric

For given realizations of the random fractional cache placement and the random multiple requests, the goal is to design a delivery scheme that minimizes the rate over the shared multicast link required to satisfy all user requests. Since one placement phase is followed by an arbitrarily large number of delivery rounds (each characterized by a new independent request realization), the rate (or load) of the system refers only to the delivery phase (i.e., asymptotically the cache placement costs no rate). Furthermore, it makes sense to consider the average rate, where averaging with respect to the users request distribution takes on the meaning of a time-averaged rate, invoking an ergodicity argument.

At each request round, let $\mathbf{F} = \{\mathbf{f}_1, \mathbf{f}_2, \cdots, \mathbf{f}_n\}$ be the demand realization, where $\mathbf{f}_u = \{f_{1,u}, f_{2,u}, \cdots, f_{L_u,u}\}, u \in \mathcal{U}$. The source node computes a multicast codeword as a function of the library and the demand realization $\mathbf{F}$. We assume that the source node communicates to the user nodes through an error-free deterministic shared multicast link.

Given the demand realization $\mathbf{F}$, let the total number of bits transmitted by the source node be $J(\mathbf{F})$. We are interested in the average performance of the coded multicast scheme, and hence define the average rate (or load) as the number of transmitted bits normalized by the file size:

$$R = \frac{\mathbb{E}\left[J(\mathbf{F})\right]}{F}, \tag{1}$$

where the expectation is over the random demand distribution.

3. Graph-Coloring-Based Coded Multicast Delivery

It is important to note that for given cache placement and demand realizations, the delivery phase of a caching scheme reduces to an index coding problem *with a twist*. The only difference with the conventional index coding problem introduced in [51] is that the cache information may contain *part of* (as opposed to entire) requested files, and that users may request *multiple* (as opposed to single) files. Nevertheless, as in index coding, the problem can still be represented by a *conflict graph* [10,12–14], where vertices represent requested packets, and an edge between two vertices indicates a conflict, in the sense that the packet represented by one vertex is not present in the cache of the user requesting the packet represented by the other vertex. By construction, packets with no conflict in the graph can be simultaneously transmitted via an XOR operation. Performing graph coloring on the conflict graph and transmitting the packets via proper XOR operations, according to the graph coloring, results in an achievable linear index coding scheme, which we refer to as a coded multicast scheme.

In the following, we first illustrate how to construct the conflict graph, we then review classical linear index coding schemes, and then describe our proposed graph-coloring-based coded multicast schemes.

3.1. Conflict Graph Construction

Given cache placement realization $\mathbf{C}$ and demand realization $\mathbf{W}$, the directed conflict graph $\mathcal{H}_{\mathbf{C},\mathbf{W}}^{d} = (\mathcal{V}, \mathcal{E})$ can be constructed as follows:

- Vertices: For each packet request in $\mathbf{W}$, there is a vertex in $\mathcal{H}_{\mathbf{C},\mathbf{W}}^{d}$. Each vertex $v \in \mathcal{V}$ is uniquely identified or labeled by a *packet-user* pair $\{\rho(v), \mu(v)\}$, where $\rho(v)$ denotes the identity of the packet, and $\mu(v)$ the user requesting it. Hence, if a packet is requested by multiple users, such a packet is represented in as many vertices as the number of users requesting it. Such vertices have the same packet label $\rho(v)$, but different user label $\mu(v)$.
- Arcs: For any $v_1, v_2 \in \mathcal{V}$, there is an edge $(v_2, v_1) \in \mathcal{E}$ with direction from v_2 to v_1 if and only if $\rho(v_1) \neq \rho(v_2)$ and packet $\rho(v_1)$ is not in the cache of user $\mu(v_2)$.

To better understand the rationale behind the conflict graph and its construction, note that for any two vertices v_1 and v_2 that are labeled as $\{\rho(v_1), \mu(v_1)\}$ and $\{\rho(v_2), \mu(v_2)\}$, respectively, we have the following three possible cases:

- $\rho(v_1) \neq \rho(v_2)$ and $\mu(v_1) = \mu(v_2)$: This indicates that two different packets are requested by the same user. Then, v_1 and v_2 are mutually conflicting, in the sense that if sent within the same time-frequency resource they interfere with each other. Hence, in the conflict graph, they are connected with two directed edges, $(v_1, v_2) \in \mathcal{E}$ and $(v_2, v_1) \in \mathcal{E}$;
- $\rho(v_1) = \rho(v_2)$ and $\mu(v_1) \neq \mu(v_2)$: This indicates that the same packet is requested by two different users. Then, v_1 and v_2 are not conflicting, and hence not connected in the conflict graph; i.e., $(v_1, v_2) \notin \mathcal{E}$ and $(v_2, v_1) \notin \mathcal{E}$;
- $\rho(v_1) \neq \rho(v_2)$ and $\mu(v_1) \neq \mu(v_2)$: This indicates that two different packets are requested by two different users. In this case, if packet $\rho(v_1)$ is in the cache of user $\mu(v_2)$, then, even if $\rho(v_1)$ and $\rho(v_2)$ are sent within the same time-frequency resource, user $\mu(v_2)$ will not suffer from interference, since, using its cache information, it can cancel out the undesired packet $\rho(v_1)$ from the received signal. On the other hand, if packet $\rho(v_1)$ is not in the cache of user $\mu(v_2)$, then v_1 conflicts with v_2, and a directed edge is drawn from v_2 to v_1. Similarly, $(v_1, v_2) \in \mathcal{E}$ if and only if $\rho(v_2) \notin \mathbf{C}_{\mu(v_1)}$.

Based on the above construction, it follows that the number of interference dimensions faced by a given node is at most the number of its outgoing neighbors.

To illustrate the construction of the directed conflict graph $\mathcal{H}_{\mathbf{C},\mathbf{W}}^{d}$, we present the following example.

Example 1. *We consider a network with $n = 3$ users denoted as $\mathcal{U} = \{1,2,3\}$ and $m = 3$ files denoted as $\mathcal{F} = \{A, B, C\}$. We assume $M_u = 1, \forall u \in \mathcal{U}$ and partition each file into three packets. For example, $A = \{A_1, A_2, A_3\}$. Let $p_{A,u} = p_{B,u} = p_{C,u} = \frac{1}{3}$ for $u \in \mathcal{U}$, which means that one packet from each of A, B, C is stored in each user's cache. For the sake of notational convenience, we assume a symmetric caching realization, where the caching configuration $\mathbf{C}$ is given by $\mathbf{C}_{u,A} = \{A_u\}, \mathbf{C}_{u,B} = \{B_u\}, \mathbf{C}_{u,C} = \{C_u\}$). That is, the cache configuration of each user $u \in \mathcal{U}$ is $\mathbf{C}_u = \{A_u, B_u, C_u\}$. We let each user make two requests, i.e., $L_u = 2$ $(\forall u \in \mathcal{U})$. Specifically, we let user 1 request A, B, user 2 request B, C, and user 3 request C, A, i.e., $\mathbf{f}_1 = \{A, B\}, \mathbf{f}_2 =$*

$\{B,C\}, \mathbf{f}_3 = \{C,A\})$, *such that* $\mathbf{W}_1 = \{A_2, A_3, B_2, B_3\}, \mathbf{W}_2 = \{B_1, B_3, C_1, C_3\}, \mathbf{W}_3 = \{A_1, A_2, C_1, C_2\}$. *The associated directed conflict graph is shown in Figure* 2.

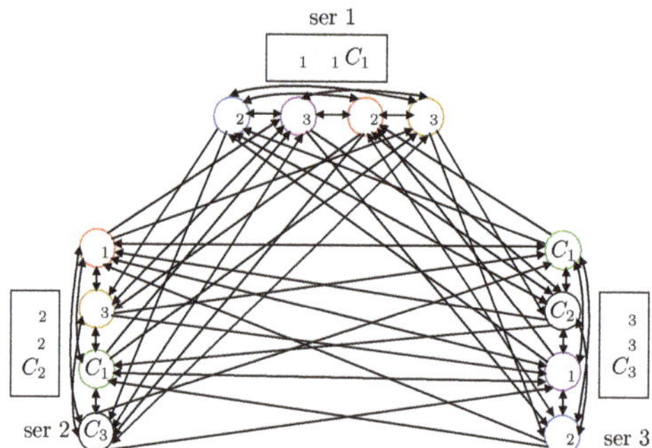

Figure 2. An example of the construction of the directed conflict graph (this figure needs to be viewed in color). The color of each circle in this figure represents the coloring of each vertex.

3.2. Code Construction

Let ω_v denote the content (or realization) of packet $\rho(v)$, $v \in \mathcal{V}$, represented by a symbol in $\mathbb{F}_q$. In general, in a linear index coding scheme of length ℓ, every vertex v is associated with a "coding" vector $\mathbf{g}_v \in \mathbb{F}_q^{\ell \times 1}$ where $v \in [1 : |\mathcal{V}|]$. Let $\mathbf{G} = [\mathbf{g}_1, \ldots \mathbf{g}_{|\mathcal{V}|}]$ and $\omega = [\omega_1, \ldots, \omega_{|\mathcal{V}|}]^{\mathsf{T}}$. Then, the transmitted codeword, $\mathbf{x} \in \mathbb{F}_q^{\ell \times 1}$, is built as follows:

$$\mathbf{x} = \sum_{v \in \mathcal{V}} \omega_v \mathbf{g}_v = \mathbf{G}\omega, \tag{2}$$

Let $\mathcal{N}(v) = \{w : (v,w) \in \mathcal{E}\}$ be the out-neighborhood of v. For any feasible scalar linear index coding scheme of the form (2), the following *interference alignment* condition is necessary: For every vertex v, the coding vector $\mathbf{g}_v$ should be linearly independent of all the coding vectors assigned to the out-neighborhood of v.

In the following, we describe how to construct coding vectors satisfying the interference alignment condition for every vertex. For ease of notation, we use $\mathcal{H}^d$ to denote the directed conflict graph, and $\mathcal{H}$ to represent its underlying undirected skeleton, where the direction of edges is ignored. Recall that an undirected skeleton of a directed graph $\mathcal{H}^d$ is an undirected graph where there is an undirected edge between v_1 and v_2 if, between v_1 and v_2, there is a directed edge in either or both directions in $\mathcal{H}^d$.

3.2.1. Graph Coloring and Chromatic Number

A well-known procedure to construct the coding vectors $\{\mathbf{g}_v, \ v \in \mathcal{N}(v)\}$ is the *coloring* of $\mathcal{H}^d$. In the following, when used without any qualification, a coloring of a directed graph is considered to be a proper (vertex) coloring of its underlying undirected skeleton $\mathcal{H}$, where a proper coloring is a labeling of the graph's vertices with colors, such that no two vertices sharing the same edge have the same color. Please note that by definition, any subset of nodes with the same color in a proper coloring form an *independent set* (i.e., a subset of nodes in a graph, no two of which share the same edge). A coloring using at most k colors is called a (proper) k-coloring. The smallest number of colors needed in a proper coloring of $\mathcal{H}^d$ is called its chromatic number, and is denoted by $\chi(\mathcal{H}^d)$. In the following, we explain why a coloring of $\mathcal{H}^d$ provides a way to design the coding vectors $\{\mathbf{g}_v, \ v \in \mathcal{N}(v)\}$. Let ξ be the total number of colors in a given coloring of $\mathcal{H}^d$. Let $\mathbf{e}_i$ be the i-th unit vector in the space $\mathbb{F}_q^{\ell \times 1}$, with $\ell = \xi$, i.e., $\mathbf{e}_i = [0, 0, \cdots, 1, \cdots, 0, 0]^T$, where the 1 is in the i-th position. Now, if vertex v is colored with color i, then, its coding vector is $\mathbf{g}_v = \mathbf{e}_i$. Making this choice for the coding vectors, the associated achievable rate is given by $\frac{\xi}{B}$. Since neighbors are assigned different colors, the interference alignment condition is satisfied for every vertex. Recalling the definition of $\chi(\mathcal{H}^d)$, it is immediate to see that the best achievable rate due to conflict graph coloring is given by $\frac{\chi(\mathcal{H}^d)}{B}$, and, according to the construction of the conflict graph, it is loosely bounded by:

$$\sum_{f \in \mathbf{f}_u} (1 - p_{f,u} M_u) \leq \frac{\chi(\mathcal{H}^d)}{B} \leq \sum_{u=1}^{n} \sum_{f \in \mathbf{f}_u} (1 - p_{f,u} M_u), \tag{3}$$

indicating that the achievable rate is a constant with regards to B. A much tighter bound will be given in Section 4.1.

3.2.2. Local Graph Coloring and Local Chromatic Number

More efficient sets of coding vectors can be constructed using the approach proposed in [66], which exploits the direction information in $\mathcal{H}_{\mathbf{C},\mathbf{W}}^d$, resulting in the following advanced coding scheme:

Definition 1 (Local Coloring Number). *Given a proper coloring* $\mathbf{c}$ *of* $\mathcal{H}^d$*, the associated local chromatic number is defined as:*

$$\xi_{\text{lc}}(\mathbf{c}) = \max_{v \in \mathcal{V}} |\mathbf{c}(\mathcal{N}^+(v))| \tag{4}$$

where $\mathcal{N}^+(v)$ *is the closed out-neighborhood of vertex* v *(i.e., vertex* v *and all its ongoing neighbors* $\mathcal{N}(v)$*) and* $|\mathbf{c}(\mathcal{N}^+(v))|$ *is the total number of colors in* $\mathcal{N}^+(v)$ *for a given proper color assignment* $\mathbf{c}$*.*

The minimum local coloring number over all proper colorings is referred to as the *local chromatic number* and is formally defined as follows:

Definition 2 (Local Chromatic Number). *The directed local chromatic number of a directed graph* $\mathcal{H}^d$ *is defined as:*

$$\chi_{\text{lc}}(\mathcal{H}^d) = \min_{\mathbf{c} \in \mathcal{C}} \xi_{\text{lc}}(\mathbf{c}) \tag{5}$$

where $\mathcal{C}$ *denotes the set of all proper coloring assignments of* $\mathcal{H}^d$*,* $\mathcal{N}^+(v)$ *is the closed out-neighborhood of vertex* v*, and* $|\mathbf{c}(\mathcal{N}^+(v))|$ *is the total number of colors in* $\mathcal{N}^+(v)$ *for a given proper color assignment* $\mathbf{c}$*.*

Encoding Scheme: For a given realization of the cache placement (**C**) and user requests (**W**), let us consider the conflict graph $\mathcal{H}_{\mathbf{C},\mathbf{W}}^d$ as in Section 3.1. Given a (proper) ξ-coloring (i.e., a proper coloring of graph $\mathcal{H}_{\mathbf{C},\mathbf{W}}^d$ with ξ colors), we compute the associated local coloring number ξ_{lc}. Set $\ell = \xi_{\text{lc}}$ and $p = \xi$. Then, consider the columns of the generator **H** of an $\ell \times p$ Maximum Distance Separable (MDS) [67] code over the field $\mathbb{F}_q : q > p$. If the color of a vertex v is i, then the coding vector $\mathbf{g}_v$ assigned to vertex v is given by i-th column $\mathbf{h}_i$ of **H**. Then, the transmitted multicast codeword, $\mathbf{x} \in \mathbb{F}_q^{\ell \times 1}$, is given by (2).

Decoding Scheme: In any closed out-neighborhood, there are at most ℓ different colors (from the definition of local coloring). Since every ℓ columns of **H** are linearly independent (from the defining property of MDS codes), the coding vectors in any closed out-neighborhood have full rank, satisfying the interference alignment condition. The message ω_v at vertex v is obtained at user v as follows: (1) Using side information at user v, cancel out message parts corresponding to all vertices outside $\mathcal{N}^+(v)$, i.e., $\mathbf{x}' = \mathbf{x} - \sum_{u \notin \mathcal{N}^+(v)} \omega_u \mathbf{g}_u$. This is possible because, by the definition of the conflict graph $\mathcal{H}^d$, the messages $\{\omega_u\}_{u \notin \mathcal{N}^+(v)}$ are available as side information at user v and the encoding mechanism is known to all the users. (2) Find a vector $\mathbf{z}$ in the dual space of $\{\mathbf{g}_u\}_{u \in \mathcal{N}^+(v) \setminus \{v\}}$ such that $\mathbf{z}^T \mathbf{x}' \neq 0$ (this is possible since $\mathbf{g}_v$ is linearly independent of $\{\mathbf{g}_u\}_{u \in \mathcal{N}^+(v) \setminus \{v\}}$ because of the local chromatic number-based construction). Now, $\mathbf{z}^T \mathbf{x}' = (\mathbf{z}^T \mathbf{g}_v)\omega_v$. Therefore, user v recovers its own message. It follows that all users can recover all the requested packets employing such linear scheme.

Achievable Rate: The coding scheme constructed as described above achieves a rate given by ξ_{lc}/B, where B is the number of packets per file.

Example 2. *We consider an example shown in Figure 3. First, we assign colors to each vertex such that the total number of colors $\xi = 5$, and count the local coloring number, which is $\xi_{lc} = 4$. Then, we construct the generator matrix $\mathbf{A}$ of a $(\xi = 5, \xi_{lc} = 4)$ MDS code, which is given by*

$$\mathbf{A} = \begin{pmatrix} 1 & 0 & 0 & 0 & 1 \\ 0 & 1 & 0 & 0 & 1 \\ 0 & 0 & 1 & 0 & 1 \\ 0 & 0 & 0 & 1 & 1 \end{pmatrix}. \tag{6}$$

After that, we assign the columns of $\mathbf{A}$ to $\mathbf{g}_v$, corresponding from the left to the right to the vertices with the packets $\{A_2, A_3, B_2, B_1, A_1\}$, as shown in Figure 3. Finally, the transmitted codewords can be generated which are the rows of the right-hand side of (2)

$$X(1) = A_1 \oplus A_2, X(2) = A_1 \oplus A_3 \tag{7}$$

$$X(3) = A_1 \oplus B_1, X(4) = A_1 \oplus B_2 \tag{8}$$

where the length of the code is $\xi_{lc}/B = 4/3$ file units. It can be easily verified that every user can decoded its desired packets with the cached ones.

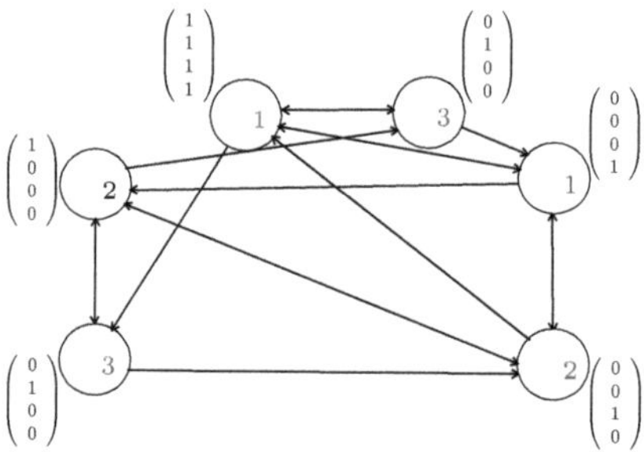

Figure 3. An illustration of coded multicast codewords construction based on local coloring (this figure needs to be viewed in color). The total number of colors is $\xi = 5$, and the local coloring number is $\xi_{lc} = 4$.

It is immediate to see that the best achievable rate due to local coloring is obtained by computing the local chromatic number of $\mathcal{H}^d_{C,W}$ and using its associated coloring to design the coding vectors, yielding a rate χ_{lc}/B. However, note that to compute χ_{lc}, we must optimize over all proper colorings to find the local chromatic number. As with the chromatic number, this can be cast as an *Integer Program* and it is hence an NP-hard problem. To overcome this limitation, in Section 4, we propose a greedy approach that (i) exhibits polynomial-time complexity in all the system parameters, (ii) achieves close to optimal performance for finite packetization order, and (iii) is asymptotically (i.e., for infinite packetization order) order-optimal.

3.3. Benefits of Local Coloring

Consider the following relation established for general directed graphs in [66]:

$$\chi_{lc}(\mathcal{H}^d) \leq \chi(\mathcal{H}^d) \leq \chi_{lc}(\mathcal{H}^d)O(\log n). \tag{9}$$

Focusing on the conflict graph of interest $\mathcal{H}^d_{\mathbf{C},\mathbf{W}}$, the number of vertices can be as large as $B\sum_{u\in\mathcal{U}} L_u$. It then follows from (9) that the gap between the local chromatic number and the chromatic number can be as large as $\log(B\sum_{u\in\mathcal{U}} L_u)$. Please note that this multiplicative factor grows with the number of packets per file B and the number of per-user requests L_u, supporting the extra benefit of local coloring in the multiple-request scenario. In addition, the higher the number of per-user requests, the higher the directionality of the conflict graph, which is the main factor exploited by local coloring to reduce the achievable rate (see Section 3.2.2), further supporting the suitability of local coloring in increasingly practical settings where there is some form of spatial or temporal request aggregation.

4. Proposed Algorithms and Performance Analysis

As stated earlier, computing the local chromatic number is NP-hard. To circumvent this challenge, in this section, we propose two greedy coded multicast schemes, which together with the cache placement described in Section 2.1, yield the following two caching schemes: Randomized Aggregate Popularity-Greedy Local Coloring (RAP-GLC) and Randomized Aggregate Popularity-Hierarchical greedy Local Coloring (RAP-HgLC). In both cases, the steps for obtaining the coded multicast scheme are as follow:

i. Given a realization of the cache placement (**C**) and of the user requests (**W**), build the conflict graph $\mathcal{H}^d_{\mathbf{C},\mathbf{W}}$ as in Section 3.1.
ii. Use any of the above algorithms (GLC or HgLC) to compute a proper coloring. Let ζ denote the number of colors used by either of the above algorithms to color $\mathcal{H}^d_{\mathbf{C},\mathbf{W}}$. Let ζ_{lc} be the associated local coloring number.
iii. Consider a (ζ, ζ_{lc}) MDS code and compute the corresponding coded multicast scheme as described in Section 3.2.2.

4.1. Randomized Aggregate Popularity-Greedy Local Coloring (RAP-GLC)

The RAP-GLC algorithm generalizes the RAP-GCC (Random Aggregate Popularity-Greedy Constrained Coloring) algorithm introduced in [12]. RAP-GCC is a caching scheme based on random fractional caching for the placement phase and a coded multicast scheme built on greedy-graph-coloring -based linear index coding [51,68] for the delivery phase. RAP-GLC is more general than RAP-GCC in two aspects: (1) conventional coloring is replaced by local coloring to leverage possible gains in the multiple-request scenario, as described in Sections 3.2.1 and 3.3, and (2) RAP-GLC adaptively (depending on the demand realization) chooses between naive or coded multicasting according to a threshold parameter, instead of sticking to one of them (as in RAP-GCC).

4.1.1. RAP-GLC Algorithm Description

The algorithm associates to each vertex v a label or tag, composed of two fields i.e., $\mathcal{K}_v \equiv (\mathcal{T}_D(v), \mathcal{T}_C(v))$ with $\mathcal{T}_C(v)$ denoting the subset of users caching the packet associated with vertex v, i.e.,

$$\mathcal{T}_C(v) \overset{\Delta}{=} \{u \in \mathcal{U} : \rho(v) \in \mathbf{C}_u\}, \tag{10}$$

and $\mathcal{T}_D(v)$ denoting the subset of users requesting the packet associated with vertex v, i.e.,

$$\mathcal{T}_D(v) \overset{\Delta}{=} \{u \in \mathcal{U} : \rho(v) \in \mathbf{W}_u\}, \tag{11}$$

which includes the user itself $\mu(v)$ who requests $\rho(v)$ and all the others requesting $\rho(v)$. Please note that the cardinality of $\mathcal{T}_D(v)$ indicates the popularity of packet $\rho(v)$. Furthermore, let

$$\mathcal{T}_v = \{\mu(v)\} \cup \mathcal{T}_C(v).$$

Given a vertex v, if the cardinality of $\mathcal{T}_D(v)$ is higher than a predetermined threshold parameter $t \in \{0 \cdots, n\}$ i.e., $|\mathcal{T}_D(v)| > t$, then all vertices v' such that $\rho(v) = \rho(v')$ are colored with the same color, leading to a naive multicast transmission scheme. If $|\mathcal{T}_D(v)| \leq t$, then RAP-GLC greedily looks for a maximal set of vertices with the same $\mathcal{T}_v$ (Algorithm 1, Line 14) and colors them with the same color if there is no conflict among the vertices (Algorithm 1, Line 15). The threshold parameter t is subject to optimization, as described in Section 4.1.2.

Doing this, RAP-GLC computes a valid coloring of the conflict graph $\mathcal{H}$. Finally, the algorithm computes its associated local coloring number (Algorithm 1, Line 24). The coding scheme employed is based on the MDS code described in Section 3.2.1 associated with the above local coloring.

Algorithm 1 RAP-GLC

1: Let $\mathcal{C} = \varnothing$;
2: Let $\mathbf{c} = \varnothing$;
3: **while** $\mathcal{V} \neq \varnothing$ **do**
4:     Pick an arbitrary vertex v in $\mathcal{V}$; Let $\mathcal{I} = \{v\}$;
5:     Let $\mathcal{V}' = \mathcal{V} \setminus \{v\}$;
6:     **if** $\{ |\mathcal{T}_D(v)| > t \}$ **then**
7:         **for all** $v' \in \mathcal{V}'$ with $\rho(v') = \rho(v)$ **do**
8:             $\mathcal{I} = \mathcal{I} \cup v'$;
9:         **end for**
10:        Color all the vertices in $\mathcal{I}$ by $c \notin \mathcal{C}$;
11:        Let $\mathbf{c}[\mathcal{I}'] = c$;
12:        $\mathcal{V} = \mathcal{V} \setminus \mathcal{I}'$.
13:     **else**
14:         **for all** $v' \in \mathcal{V}'$ with $\mathcal{T}_{v'} \equiv \mathcal{T}_v$ **do**
15:            **if** {There is no edge between v' and $\mathcal{I}$} **then**
16:                $\mathcal{I} = \mathcal{I} \cup v'$;
17:            **end if**
18:         **end for**
19:        Color all the vertices in $\mathcal{I}$ by $c \notin \mathcal{C}$;
20:        Let $\mathbf{c}[\mathcal{I}] = c$;
21:        $\mathcal{V} = \mathcal{V} \setminus \mathcal{I}$.
22:     **end if**
23: **end while**
24: **return** the local coloring number $\max_{v \in \mathcal{V}} |\mathbf{c}(\mathcal{N}^+(v))|$ and the corresponding color assignment $\mathbf{c}(\mathcal{N}^+(v))$ for
    each v;

Time Complexity: In Algorithm 1, both the outer while-loop starting at Line 3, and the inner for-loop starting at Line 6 iterate at most $|\mathcal{V}|$ times, and all other operations inside the loops take constant time. Therefore, the complexity of RAP-GLC is $O(|\mathcal{V}|^2)$ or, equivalently, $O(n^2 B^2)$, since $|\mathcal{V}| \leq nB$, which is polynomial in $|\mathcal{V}|$ (or n, B).

4.1.2. RAP-GLC Performance Analysis

In the following, we quantify the performance of RAP-GLC in the asymptotic regime when the number of users and files is kept constant while the packetization order is sent to infinity. Denoting by $\mathbb{E}[R^{\mathrm{RAP-GLC}}(\mathbf{P}, \mathbf{Q}, t)]$ the asymptotic average achievable rate of RAP-GLC for a fixed threshold t, the threshold parameter t is optimized to minimize $\mathbb{E}[R^{\mathrm{RAP-GLC}}(\mathbf{P}, \mathbf{Q}, t)]$. Hence, denoting by $\bar{R}^{\mathrm{RAP-GLC}}$ the average rate achieved by RAP-GLC with optimized t, i.e.,

$$\bar{R}^{\mathrm{RAP-GLC}} = \min_t \mathbb{E}[R^{\mathrm{RAP-GLC}}(\mathbf{P}, \mathbf{Q}, t)],$$

we have that

$$\bar{R}^{\mathrm{RAP-GLC}} \leq \min\{\mathbb{E}[R^{\mathrm{RAP-GLC}}(\mathbf{P}, \mathbf{Q}, n)], \mathbb{E}[R^{\mathrm{RAP-GLC}}(\mathbf{P}, \mathbf{Q}, 0)].$$

Since $\mathbb{E}[R^{\mathrm{RAP-GLC}}(\mathbf{P}, \mathbf{Q}, 0)]$ is just the rate achieved via naive multicasting, then, an upper bound on the average asymptotic performance of RAP-GLC can be obtained by upper bounding $\mathbb{E}[R^{\mathrm{RAP-GLC}}(\mathbf{P}, \mathbf{Q}, n)]$ which can be obtained by generalizing the asymptotic performance analysis of RAP-GCC derived in [10,12] using conventional graph coloring in the homogeneous shared link caching network to the case of using local coloring in the heterogeneous caching network. Specifically, we extend the order-optimality analysis under single per-user requests ($L = 1$) in the asymptotic regime of $B \to \infty$ [10,12], to that under multiple $L > 1$ per-use requests [13,14]. These theoretical results will serve as rate lower bounds for the finite-length performance of our proposed algorithms.

Let $L = \max_u L_u$ and order $L_u, u \in \mathcal{U}$ as a decreasing sequence $L_{[1]} \geq L_{[2]} \geq L_{[3]}, \ldots, L_{[n]}$, where $L_{[i]}$ is the i-th largest L_u and $[i] = u$ for some $u \in \mathcal{U}$. It can be seen that $L_{[1]} = \max_u L_u$ and $L_{[n]} = \min_u L_u$. Let $n_j = \sum_{[i]} \mathbb{1}\{L_{[i]} - j \geq 0\} > 0$, where $1 \leq j \leq L_{[1]}$ and $\mathbb{1}\{\cdot\}$ is the indicator function. Let $\mathcal{U}_{n_j} = \{[i] \in \mathcal{U} : \mathbb{1}\{L_{[i]} - j \geq 0\}\}$. In the next theorem, we provide a performance guarantee of the RAP-GLC algorithm.

Theorem 1. *For any given m, n, M_u, the random caching distribution $\mathbf{P}$ and the random request distribution $\mathbf{Q}$, the average achievable rate of the RAP-GLC algorithm, $\bar{R}^{\mathrm{RAP-GLC}}$ satisfies*

$$\bar{R}^{\mathrm{RAP-GLC}} \leq \min\{\psi(\mathbf{P}, \mathbf{Q}), \bar{m} - \bar{M}\}, \tag{12}$$

when $B \to \infty$, where,

$$\bar{m} = \sum_{f=1}^{m} \left(1 - \prod_{u=1}^{n} \left(1 - q_{f,u}\right)^{L_u}\right), \tag{13}$$

$$\bar{M} = \sum_{f=1}^{m} \left(1 - \prod_{u=1}^{n} \left(1 - q_{f,u}\right)^{L_u}\right) \min_u p_{f,u} M_u, \tag{14}$$

$$\psi(\mathbf{P}, \mathbf{Q}) = \sum_{j=1}^{L} \sum_{\ell=1}^{n} \sum_{\mathcal{U}^\ell \subset \mathcal{U}_{n_j}} \sum_{f=1}^{m} \sum_{u \in \mathcal{U}^\ell} p_{f,u,\mathcal{U}^\ell} \lambda(u, f, \mathcal{U}^\ell),$$

with $\mathcal{U}^\ell$ denoting a set of users with cardinality ℓ,

$$\lambda(u, f_u, \mathcal{U}^\ell) = (1 - p_{f_u,u} M_u) \times \prod_{k \in \mathcal{U}^\ell \setminus \{u\}} (p_{f_u,k} M_k) \prod_{k \in \mathcal{U} \setminus \mathcal{U}^\ell} (1 - p_{f_u,k} M_k) \tag{15}$$

and

$$\rho_{f,u,\mathcal{U}^\ell} \overset{\Delta}{=} \mathbb{P}(f = \arg\max_{f_u \in \mathbf{f}(\mathcal{U}^\ell)} \lambda(u, f_u, \mathcal{U}^\ell)),$$

denoting the probability that f is the file whose $p_{f,u}$ maximizes the term $\lambda(u, f_u, \mathcal{U}^\ell)$ among $\mathbf{f}(\mathcal{U}^\ell)$ (the set of files requested by $\mathcal{U}^\ell$).

Proof. See Appendix A. $\square$

Using the explicit expression for $\bar{R}^{\text{RAP-GLC}}$ in Theorem 1, we can optimize the caching distribution for a wide class of heterogeneous network models to minimize the number of transmissions. We use $\mathbf{P}^*$ to denote the caching distribution that minimizes $R^{\text{RAP-GLC}}$.

Remark 1. *For the sake of the numerical evaluation of $\psi(\mathbf{q}, \mathbf{p})$, it is worthwhile to note that the probabilities $\rho_{f,u,\mathcal{U}^\ell}$ can be easily computed as follows. Given the subset of users, $\mathcal{U}^\ell$ of cardinality ℓ, let $J_{u_1}, \ldots, J_{u_\ell}$ denote ℓ i.i.d. random variables each of them distributed over $\mathcal{F}$ with pmf $\mathbf{q}_{u_i}$, with $i = 1, \ldots, \ell$. Since $\lambda(u_1, J_{u_1}, \mathcal{U}^\ell), \cdots, \lambda(u_\ell, J_{u_\ell}, \mathcal{U}^\ell)$ are i.i.d., the CDF of $Y_\ell \overset{\Delta}{=} \max\{\lambda(u_1, J_{u_1}, \mathcal{U}^\ell), \cdots, \lambda(u_\ell, J_{u_\ell}, \mathcal{U}^\ell)\}$ is given by*

$$\mathbb{P}(Y_\ell \leq y) = \prod_{i=1}^{\ell} \mathbb{P}\left(\lambda(u_i, J_{u_i}, \mathcal{U}^\ell) \leq y\right)$$
$$= \prod_{i=1}^{\ell} \left(\sum_{j \in \mathcal{F}: \lambda(u_i, j, \mathcal{U}^\ell) \leq y} q_{u_i,j}\right). \tag{16}$$

Hence, it follows that

$$\rho_{f,u,\mathcal{U}^\ell} = \mathbb{P}(Y_\ell = \lambda(u, f, \mathcal{U}^\ell))$$
$$= \prod_{i=1}^{\ell} \left(\sum_{j \in \mathcal{F}: g_\ell(j) \leq \lambda(u, f, \mathcal{U}^\ell)} q_{u_i,j}\right) - \prod_{i=1}^{\ell} \left(\sum_{j \in \mathcal{F}: g_\ell(j) < \lambda(u, f, \mathcal{U}^\ell)} q_{u_i,j}\right), \tag{17}$$

which can be easily computed by sorting the values $\{\lambda(u_i, j, \mathcal{U}^\ell) : j \in \mathcal{F}, u_i \in \mathcal{U}^\ell\}$.

Nevertheless, as shown in [21], when B is finite or is not exponential in n, the performance of RAP-GLC can degrade significantly, compromising the promising multiplicative caching gain, although it is already an improved version of RAP-GCC in [12]. This brings us to the other main contribution where we propose a new algorithm that preserves the gain due to coded multicasting even when B is finite.

4.2. Randomized Aggregate Popularity-Hierarchical Greedy Local Coloring (RAP-HgLC) for Finite-Length Packetization

Similarly to RAP-GLC, RAP-HgLC has a predetermined parameter $t \in \{0, \cdots, n\}$ that is optimized to minimize its associated average achievable rate. However, in the RAP-HgLC algorithm, we arrange the vertices in a hierarchy and use this to design a more careful coloring algorithm. The key idea of

RAP-HgLC is to exploit the labeling of each vertex more efficiently. More specifically, as in RAP-GLC, RAP-HgLC associates to each vertex v a label or tag, composed by the two fields $\mathcal{K}_v \equiv (\mathcal{T}_D(v), \mathcal{T}_C(v))$, defined in (10) and (11).

4.2.1. RAP-HgLC Algorithm Description

Before jumping into the algorithm, we introduce the following useful notations and their definitions.

- $\mathcal{G}_i$: The i-th layer, $\mathcal{G}_i$ is initialized with the set of vertices $\{v : |\mathcal{T}_v| = i\}$ and at any point in the algorithm contains only vertices with $|\mathcal{K}_v| \geq i$. $\mathcal{G}_i$ is updated continuously in the algorithm. Therefore, higher numbered layers contain vertices with greater popularity.
- $\mathcal{W}_1 \subset \mathcal{G}_i$: a subset of $\mathcal{G}_i$ consists of all the vertices with $|\mathcal{K}_v| = i$ as well as a certain number of vertices with higher popularity (if available at any iteration), defined as

$$\mathcal{W}_1 = \left\{ v \in \mathcal{G}_i : \min_{v \in \mathcal{G}_i} |\mathcal{K}_v| \leq |\mathcal{K}_v| \leq \min_{v \in \mathcal{G}_i} |\mathcal{K}_v| + \left\lfloor a \left(\max_{v \in \mathcal{G}_i} |\mathcal{K}_v| - \min_{v \in \mathcal{G}_i} |\mathcal{K}_v| \right) \right\rfloor \right\}, \tag{18}$$

  where $a \in [0, 1]$ is a design parameter and $\mathcal{W}_1$ is updated with every iteration.
- $\mathcal{Q}_i$ (see Algorithm 2): another subset of $\mathcal{G}_i$ that is updated every iteration.
- $\mathcal{W}_2 \subset \mathcal{Q}_i$: a subset of vertices in $\mathcal{Q}_i$ defined as:

$$\mathcal{W}_2 = \left\{ v' \in \mathcal{Q}_i : \min_{v' \in \mathcal{Q}_i} |\mathcal{K}_{v'}| \leq |\mathcal{K}_{v'}| \leq \min_{v' \in \mathcal{Q}_i} |\mathcal{K}_{v'}| + \left\lfloor b \left(\max_{v' \in \mathcal{Q}_i} |\mathcal{K}_{v'}| - \min_{v' \in \mathcal{Q}_i} |\mathcal{K}_{v'}| \right) \right\rfloor \right\}, \tag{19}$$

  where $b \in [0, 1]$ is another design parameter.

Based on the above definitions, it follows that the total set vertices $\mathcal{V}$ forms an n-layer hierarchy with the i-th layer composed of the set of vertices $\mathcal{G}_i$.

Key Idea: Starting from layer n, at any layer $i \leq n$, the RAP-HgLC algorithm attempts to form an independent set of size at least i; when there are no more such independent sets, all remaining packets are dropped to layer $i - 1$, and transmission actions on those packets are *deferred* to later layers. This is the key difference between RAP-HgLC and RAP-GLC. That is, RAP-HgLC makes an extra effort to place nodes with large labels into large independent sets.

We will now describe how the above key idea is implemented in RAP-HgLC. The RAP-HgLC algorithm forms large independent sets in a "top-down" fashion, starting with the highest layer, and iteratively moving to lower layers until layer 1. The following two steps are performed at each layer:

1. **Step I**: The first step is similar to that in RAP-GLC algorithm. Given a vertex v, the algorithm first checks if the cardinality of $\mathcal{T}_D(v)$ is higher than t, i.e., $|\mathcal{T}_D(v)| > t$ then all the vertices v' such that $\rho(v) = \rho(v')$ are colored with the same color. If $|\mathcal{T}_D(v)| \leq t$ then the algorithm greedily finds independent sets of size i, where every vertex v in the independent set (Algorithm 2, Line 20) has the same $\mathcal{K}_v$ (Algorithm 2, Line 19). After removing these vertices, the rest of the vertices in $\mathcal{G}_i$ are left for the second step.
2. **Step II**: A candidate pool of vertices $\mathcal{W}_1 \subseteq \mathcal{G}_i$ is created. This set contains vertices v such that $|\mathcal{K}_v|$ being close to the smallest available $|\mathcal{K}_v|$'s. We randomly pick a vertex v from $\mathcal{W}_1$ (Algorithm 2, Line 31). The design parameter a determines how close is the picked $|\mathcal{K}_v|$ to the smallest available ones. We gradually form an independent set of size i with v included as follows: Form another set $\mathcal{W}_2$ (Algorithm 2, Line 34), excluding v, whose vertices have $|\mathcal{K}_{v'}|$ that is bigger but closer to that of v determined by b, sample repeatedly with replacement from it to grow the independent set. If an independent set of size at least i cannot be formed, we drop the vertex v to the lower layer $\mathcal{G}_{i-1}$, and take it into account in the next layer iteration. Otherwise, we assign a color to the independent set.

$\mathcal{W}_1$ is repeatedly formed and random sampling from $\mathcal{W}_1$ repeated till every vertex in $\mathcal{G}_i$ is dropped or colored.

Algorithm 2 HgLC

1: $\mathcal{C} = \varnothing$;
2: $\mathbf{c} = \varnothing$;
3: choose $a \in [0,1]$
4: choose $b \in [0,1]$
5: **for all** $i = n, n-1, \ldots, 2, 1$ **do**
6:     **for all** $v \in \mathcal{G}_i$ and $|\mathcal{K}_v| = i$ **do**
7:         $\mathcal{I} = \{v\}$;
8:         Let $\mathcal{V}' = \mathcal{V} \setminus \{v\}$;
9:         **if** $\{\,|\mathcal{T}_D(v)| > t\,\}$ **then**
10:             **for all** $v' \in \mathcal{V}'$ with $\rho(v') = \rho(v)$ **do**
11:                 $\mathcal{I} = \mathcal{I} \cup v'$;
12:             **end for**
13:             Color all the vertices in $\mathcal{I}$ by $c \notin \mathcal{C}$;
14:             Let $\mathbf{c}[\mathcal{I}] = c$;
15:             **for all** $i = n, n-1, \ldots, 2, 1$ **do**
16:                 $\mathcal{G}_i = \mathcal{G}_i \setminus \mathcal{I}$;
17:             **end for**
18:         **else**
19:             **for all** $v' \in \mathcal{G}_i \setminus \mathcal{I}$ with $\mathcal{K}_{v'} \equiv \mathcal{K}_v$ **do**
20:                 **if** {There is no edge between v' and $\mathcal{I}$} **then**
21:                     $\mathcal{I} = \mathcal{I} \cup v'$;
22:                 **end if**
23:             **end for**
24:             **if** $|\mathcal{I}| = i$ **then**
25:                 Color all the vertices in $\mathcal{I}$ by $c \notin \mathcal{C}$;
26:                 $\mathbf{c}[\mathcal{I}] = c, \mathcal{C} = \mathcal{C} \cup c$;
27:                 $\mathcal{G}_i = \mathcal{G}_i \setminus \mathcal{I}$;
28:             **end if**
29:         **end if**
30:     **end for**
31:     **for all** $v \in \mathcal{G}_i$ with v randomly picked from $\mathcal{W}_1 \subset \mathcal{G}_i$ **do**
32:         $\mathcal{I} = \{v\}$;
33:         $\mathcal{Q}_i = \mathcal{G}_i \setminus \mathcal{I}$;
34:         **for all** $v' \in \mathcal{Q}_i$ with v' randomly picked from $\mathcal{W}_2 \subset \mathcal{Q}_i$. **do**
35:             **if** $\{\mathcal{K}_{v'} \supset \mathcal{K}_v\} \cap \{\text{No edge between } v' \text{ and } \mathcal{I}\}$ **then**
36:                 $\mathcal{I} = \mathcal{I} \cup v'$;
37:                 $\mathcal{Q}_i = \mathcal{Q}_i \setminus \{v'\}$;
38:             **else**
39:                 $\mathcal{Q}_i = \mathcal{Q}_i \setminus \{v'\}$;
40:             **end if**
41:         **end for**
42:         **if** $|\mathcal{I}| \geq i$ **then**
43:             Color all the vertices in $\mathcal{I}$ by $c \notin \mathcal{C}$;
44:             $\mathbf{c}[\mathcal{I}] = c, \mathcal{C} = \mathcal{C} \cup c$;
45:             $\mathcal{G}_i = \mathcal{G}_i \setminus \mathcal{I}$;
46:         **else**
47:             $\mathcal{G}_i = \mathcal{G}_i \setminus \{v\}, \mathcal{G}_{i-1} = \mathcal{G}_{i-1} \cup \{v\}$;
48:         **end if**
49:     **end for**
50: **end for**
51: $\mathbf{c} = \text{LocalSearch}(\mathcal{H}_{\mathbf{C},\mathbf{W}}, \mathbf{c}, \mathcal{C})$;
52: **return** the local coloring number $\max_{v \in \mathcal{V}} |\mathbf{c}(\mathcal{N}^+(v))|$ and the corresponding color assignment $\mathbf{c}(\mathcal{N}^+(v))$ for each v;

Remark 2. *Please note that RAP-GLC goes through the same Step I as RAP-HgLC, and then simply assigns a different color to each remaining uncolored vertex. On the other hand, Step II in RAP-HgLC tries to find further independent sets among the remaining uncolored vertices. It is this extra step that guarantees the performance of RAP-HgLC to be no worse than that of RAP-GLC.*

The RAP-HgLC algorithm, when operating on the i-th layer, *always* colors at least i vertices with the same color. Please note that if there are remaining vertices when reaching layer 1, all such vertices will be colored, each with a different color.

To further reduce the required number of colors, we use a function called LocalSearch (Algorithm 2, Line 51), which is described in Algorithm 3. It works in an iterative fashion by replacing the current solution with a better one if there exists. It terminates when no better solutions can be found. In particular, the local search algorithm has the purpose of checking the redundancy of each color $c \in \mathcal{C}$, to eventually decrease the current objective function value $|\mathcal{C}|$. In more detail, the local search computes, iteratively for each color $c \in \mathcal{C}$, the set $\mathcal{J}_c$ of all vertices colored with color c, and performs the following steps:

1. For each vertex $i \in \mathcal{J}_c$, if there is a color $c' \in \mathcal{C}$, $c' \neq c$ that is not assigned to any adjacent vertex $j \in Adj(i)$, then assign vertex i with color c';
2. Color c is removed from the set $\mathcal{C}$ if and only if in the previous step it has been possible to replace c with some color $c' \neq c$ for all vertices in $\mathcal{J}_c$.

Finally, in Algorithm 2, Line 52, we compute the local coloring number.

Algorithm 3 LocalSearch($\mathcal{H}_{\mathbf{C,W}}, \mathbf{c}, \mathcal{C}$)

1: **for all** $c \in \mathcal{C}$ **do**

2:   Let $\mathcal{J}_c$ be the set of vertices whose color is c;
3:   Let $\mathcal{B} = \emptyset$;
4:   Let $\hat{\mathbf{c}} = \mathbf{c}$;
5:   **for all** $i \in \mathcal{J}_c$ **do**

6:     $\mathcal{A} = \emptyset$;
7:     **for all** $j \in \mathcal{N}(i)$ **do**

8:       $\mathcal{A} = \mathcal{A} \cup \mathbf{c}[j]$;
9:       **if** $\mathcal{C} \setminus \mathcal{A} \neq \emptyset$ **then**

10:         c' is chosen uniformly at random from $\mathcal{C} \setminus \mathcal{A}$;
11:         $\hat{\mathbf{c}}[i] = c'$;
12:         $\mathcal{B} = \mathcal{B} \cup \{i\}$;
13:       **end if**
14:     **end for**
15:     **if** $|\mathcal{B}| = |\mathcal{J}_c|$ **then**

16:       $\mathbf{c} = \hat{\mathbf{c}}$;
17:       $\mathcal{C} = \mathcal{C} \setminus c$;
18:     **end if**
19:   **end for**
20: **end for**
21: **return** $\mathbf{c}$;

To illustrate the RAP-HgLC algorithm, we present the following example.

Example 3. *Consider a shared link network with $n = 3$ users: $\mathcal{U} = \{1, 2, 3\}$, and $m = 3$ files: $\mathcal{F} = \{A, B, C\}$. Each file is partitioned into 4 packets. For example, $A = \{A_1, A_2, A_3, A_4\}$. For the caching part, let user 1 cache $\{A_1, B_1, B_2, B_3, B_4, C_2\}$, user 2 cache $\{A_1, A_2, A_3, A_4, B_1, C_2, C_3\}$, user 3 cache $\{A_1, A_2, A_3, C_1, B_2, B_3\}$. Then, let users $\{1, 2, 3\}$ request files $\{A, B, C\}$ respectively. Equivalently, user 1 requests A_2, A_3, A_4; user 2* *requests* B_2, B_3, B_4; *user 3 requests* C_2, C_3, C_4. *Then, we have* $\mathcal{K}_{A_2} = \{1, 2\ 3\}$; $\mathcal{K}_{A_3} = \{1, 2\ 3\}$; $\mathcal{K}_{A_4} = \{1, 2\}$; $\mathcal{K}_{B_2} = \{2, 1\ 3\}$; $\mathcal{K}_{B_2} = \{2, 1\ 3\}$; $\mathcal{K}_{B_4} = \{2, 1\}$; $\mathcal{K}_{C_2} = \{3, 1\ 2\}$; $\mathcal{K}_{C_3} = \{3, 2\}$; $\mathcal{K}_{C_4} = \{3\}$ *(here C_4 is requested by user 3 and not cached anywhere).*

The RAP-HgLC algorithm works as follows. For $i = n = 3$, $\mathcal{G}_3 = \{A_2, A_3, B_2, B_3, C_2\}$, let $v = A_2$, then it can be found that B_2 and C_2 would be in $\mathcal{I}$, hence $\mathcal{I} = \{A_2, B_2, C_2\}$. Now since $|\mathcal{I}| = n = 3$, we color A_2, B_2, C_2 by black (see Figure 4). Then $\mathcal{G}_i = \mathcal{G}_i \setminus \mathcal{I} = \{A_3, B_3\}$. In the following loop, since we cannot find a set $\mathcal{I}$ with $|\mathcal{I}| = n = 3$, we move to Line 19. Then since we cannot find a $\mathcal{I}$ with $|\mathcal{I}| \geq n = 3$, then we do $\mathcal{G}_2 = \mathcal{G}_2 \cup \{A_3\}$, and then $\mathcal{G}_2 = \mathcal{G}_2 \cup \{B_3\}$. Therefore, we obtain $\mathcal{G}_2 = \{A_3, A_4, B_3, B_4, C_3\}$. Now we go to Line 5 (start next loop). For $i = n - 1 = 2$, in this loop, we first pick $v = A_4$, then we can find $\mathcal{I} = \{A_4, B_4\}$. We color $\{A_4, B_4\}$ by blue (see Figure 4). Now $\mathcal{G}_2 = \mathcal{G}_2 \setminus \{A_4, B_4\} = \{A_3, B_3, C_3\}$. Then in Line 19, we find the vertex with smallest length of $\mathcal{K}_v$ (let $a = 0$), which is C_3 with $\mathcal{K}_{C_3} = \{3, 2\}$, then we have $\mathcal{I} = \{C_3\}$ and $\mathcal{Q}_2 = \{A_3, B_3\}$, then in the next loop, we can find $\mathcal{I} = \{C_3, B_3\}$. We color $\mathcal{I} = \{C_3, B_3\}$ by red (see Figure 4). Now $\mathcal{G}_2 = \mathcal{G}_2 \setminus \{C_3, B_3\} = \{A_3\}$. Since there is no $\mathcal{I}$ with $|\mathcal{I}| \geq 2$, then we do $\mathcal{G}_1 = \mathcal{G}_1 \cup \{A_3\} = \{C_4, A_3\}$. Then we go to next loop $i = n - 2 = 1$. Then we can see that $\mathcal{I} = \{C_4\}$, and we color $\{C_4\}$ by purple (see Figure 4). Then $\mathcal{G}_1 = \mathcal{G}_1 \setminus \{C_4\} = \{A_3\}$. Hence, we can find $\mathcal{I} = \{A_3\}$ and we color $\{A_3\}$ by brown.

According to Figure 4, the total number of required colors is 5, while the maximum number of colors required locally by each user is 4. For the naive multicasting, since it only allows the vertices represented the same packet to be colored by the same color, the total number of required colors is 9. The corresponding rate is given by $9/4$. Hence, the final rate achieved by RAP-HgLC with local coloring is no more than $\min\{4/4, 9/4\} = 1$. For the interested reader, it can be verified that if the GCC algorithm, designed for $B \to \infty$, as proposed in [10], is used, the corresponding number of required colors is 6.

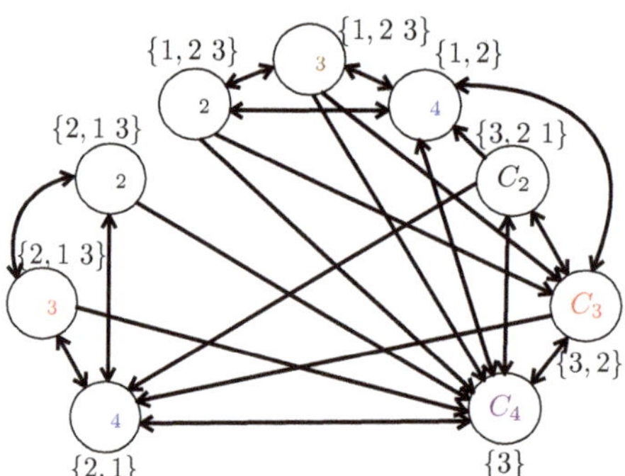

Figure 4. One example for the RAP-HgLC algorithm (this figure needs to be viewed in color).

The complexity of RAP-HgLC can be computed as follows. For the hierarchical coloring procedure (Line 5–50 in Algorithm 2), the complexity is $O(n|\mathcal{V}|^2)$, and the complexity of local search procedure is $O(|\mathcal{E}|)$. Therefore, the running time complexity of RAP-HgLC is given by $O(n|\mathcal{V}|^2 + |\mathcal{E}|) = O(n|\mathcal{V}|^2)$. Since $|\mathcal{V}| \leq nB$, the running time complexity of RAP-HgLC is $O\left(n^3 B^2\right)$.

4.2.2. RAP-HgLC Performance Analysis

For the general heterogeneous network setting, tight upper bounds on the asymptotic ($B \to \infty$) average achievable rate of RAP-HgLC are quite complex to derive, even though a simple (but not necessarily tight) upper bound on the asymptotic performance can be obtained considering the asymptotic average rate of RAP-GLC (see Remark 2).

Regarding the finite-length regime, in [21] we derived a tight upper bound on the performance of RAP-HgLC for the simpler case of homogenous networks under worst-case demands. Specifically, the bound in [21], requires B to be $\tilde{O}\left(\left(\frac{m}{M}\right)^{g+2}\right)$ (where $\tilde{O}$ hides some poly log terms) to achieve a worst-case rate of at most $\frac{n}{g}$. This approximately matches a lower bound of $\tilde{O}((\frac{m}{M})^g)$ derived in the same work for any coloring algorithm, showing, for the simpler homogenous network setting, the optimality of RAP-HgLC among all graph-coloring-based algorithms.

For the more complex setting where demands arise from popularity distributions and every user requests multiple files, the finite-length performance of RAP-HgLC is investigated in Section 6 via numerical analysis, where we show how the RAP-HgLC is able to recover most of the multiplicative caching gain even with very moderate packetization order.

5. Tradeoff between Number of Requests and Code Length

As mentioned earlier, in the simpler homogenous scenario, the authors in [21] showed that under worst-case demands, to achieve a gain g over conventional naive multicasting, it is necessary for B to grow exponentially with g. Intuitively, this is because a sufficiently large B is needed to create coded multicast transmissions that are useful for multiple users. However, when each user makes multiple requests, the number of requests $L_u = L$ can play a similar role to that of B, such that the requirement for B, and hence the resulting computational complexity can be reduced. For ease of analysis, in this section, we assume that all users place the same number of requests ($L_u = L$).

In the following, under either worst-case or uniform demands, we show the sufficient conditions on B and L that guarantee achieving a gain $g = \frac{Mn}{m}$. From this result, we can obtain the regime where B and L are interchangeable (L plays an equivalent role to B). Note that it can be shown that the number of file transmissions under both worst-case and uniform demands have the same order.

We consider two cases for the range of B: the case of $B = 1$, and the case of $B = \omega\left(\frac{m}{M}\right)$. The regime where $1 < B = O\left(\frac{m}{M}\right)$ is out of the scope of this paper.

When $B = 1$, the cache placement algorithm becomes scalar uniform cache placement (SUP), in which each user caches M entire files chosen uniformly at random. For simplicity, we let M be a positive integer. Then, as shown in [14], letting $L \to \infty$ as a function of n, m, M, we obtain the following theorem.

Theorem 2. *When $B = 1$ and $M = \omega(1)$, for the shared link caching network with n users, library size m, storage capacity M, and L distinct per-user requests ($nL \leq m$), if (i) $\frac{M}{m} \leq \frac{1}{2}$ and*

$$L = \omega\left(\max\left\{\frac{nM}{m}\left(\frac{m}{M}\right)^n \frac{1}{\left(1-\frac{M}{m}\right)}, \left(\frac{(nM)^{\frac{1}{2(1-\epsilon)}}}{\left(\frac{m}{M}-1\right)\left(1-\left(1-\frac{M}{m}\right)^n\right)}\right)\right\}\right), \tag{20}$$

or (ii) $\frac{M}{m} \geq \frac{\epsilon}{1+\epsilon}$ and

$$L = \omega\left(\max\left\{\left(\frac{m}{m-M}\right)^n, \left(\frac{(nM)^{\frac{1}{2(1-\epsilon)}}}{\left(\frac{m}{M}-1\right)\left(1-\left(1-\frac{M}{m}\right)^n\right)}\right)\right\}\right), \tag{21}$$

where ϵ is an arbitrarily small number,

then, the achievable rate of RAP-GLC is upper bounded by

$$\lim_{n,m\to\infty} \mathbb{P}\left(R^{\text{SUP-GLC}} \leq (1+o(1))\min\left\{L\left(\frac{m}{M}-1\right), Ln, m-M\right\}\right) = 1.$$

Proof. See Appendix B. $\square$

From Theorem 2, we can see that when L and M are large enough, instead of requiring a large B and packet-level coding, a simpler file-level coding scheme is sufficient to achieve the same order-optimal rate. We remark, however, that the range of the parameter regimes in which this result holds is limited due to the requirement of a large M and L. Next, we focus on another parameter regime, when $B = \omega\left(\frac{m}{M}\right)$, and find the achievable tradeoff between B and L.

Theorem 3. *When $B = \omega\left(\frac{m}{M}\right)$, for the shared link caching network with n users, library size m, storage capacity M, and L distinct per-user requests ($nL \leq m$), if (i) $\frac{M}{m} \leq \frac{1}{2}$, and*

$$B = \omega\left(\max\left\{\frac{nM}{Lm}\left(\frac{m}{M}\right)^n \frac{1}{\left(1-\frac{M}{m}\right)}, \frac{(nM)^{\frac{1}{2(1-\varepsilon)}}}{L\left(\frac{m}{M}-1\right)\left(1-\left(1-\frac{M}{m}\right)^n\right)}\right\}\right),$$ (22)

or (ii) $\frac{M}{m} \geq \frac{e}{1+e}$, and

$$B = \omega\left(\max\left\{\frac{1}{L}\left(\frac{m}{m-M}\right)^n, \frac{(nM)^{\frac{1}{2(1-\varepsilon)}}}{L\left(\frac{m}{M}-1\right)\left(1-\left(1-\frac{M}{m}\right)^n\right)}\right\}\right),$$ (23)

where ε is an arbitrarily small number,
Then, the achievable rate of RAP-GLC is upper bounded by

$$\lim_{n,m\to\infty} \mathbb{P}\left(R^{\mathrm{SUP-GLC}} \leq (1+o(1))\min\left\{L\left(\frac{m}{M}-1\right), Ln, m-M\right\}\right) = 1.$$ (24)

Proof. See Appendix C. $\square$

If we particularize Theorem 1 to the homogenous network setting under uniform demands, we see that the rate achieved by RAP-GLC is upper bounded by the same expression given in (24). Hence, from Theorem 2, we can see that when L is large enough, instead of requiring a very large B, an intermediate value of $B = \omega\left(\frac{m}{M}\right)$ is sufficient to achieve the same order-optimal rate. In practice, it is important to find the right balance and tradeoff between B and L given the remaining system parameters. In Section 6, we show via simulation that a similar tradeoff holds also for RAP-HgLC.

6. Simulations and Discussions

In this section, we numerically evaluate the performance of the two polynomial-time algorithms described in Section 4, RAP-GLC and RAP-HgLC, in the finite-length regime characterized by the number of packets per file B.

Recall that the caching distribution $\mathbf{P}^*$ is to be optimized to minimize the number of transmissions. Since the distribution $\mathbf{P}^*$ resulting from minimizing the right-hand side of (12) may not admit an analytically tractable expression in general, in the following numerical results, we restrict the caching distribution to take the form of a *truncated uniform distribution* $\tilde{\mathbf{p}}_u$, as described in [12]:

$$\begin{aligned} \tilde{p}_{f,u} &= \frac{1}{\tilde{m}_u}, \quad f \leq \tilde{m}_u \\ \tilde{p}_{f,u} &= 0, \quad f \geq \tilde{m}+1 \end{aligned}$$ (25)

where the cut-off index $\widetilde{m}_u \geq M$ is a function of the system parameters that is optimized to minimize the right-hand side of (12). The intuition behind the form of $\widetilde{\mathbf{p}}_u$ in (25) is that each user caches the same fraction of (randomly selected) packets from each of the most $\widetilde{m}_u$ popular files, and does not cache any packet from the remaining $m - \widetilde{m}_u$ least popular files. We point out that when $\widetilde{m}_u = M$, this cache placement coincides with the LFU (Least Frequently Used) caching policy. Thus, this cache placement is referred to as *Random LFU (RLFU)* [12], and the corresponding caching algorithms as RLFU-GLC and RLFU-HgLC. Recall that LFU discards the least frequently requested file upon the arrival of a new file to a full cache of size M_u files. In the long run, this is equivalent to caching the M_u most popular files [69].

In Figures 5 and 6, we plot the average achievable rate, i.e., the average number of transmissions (normalized by the file size) as a function of the cache size for RLFU-GLC and RLFU-HgLC. For comparison, we also simulate the following algorithms:

- LFU, which has been shown to be optimal in single cache networks;
- RLFU-GLC with infinite file packetization ($B \to \infty$), whose performance guarantee is given in Theorem 1, and it is shown to be order optimal.

Regarding the LFU algorithm, the average achievable rate is given by

$$\mathbb{E}[R^{\mathrm{LFU}}] = \sum_{f=\min_u\{M_u\}+1}^{m} \left(1 - \prod_{u \in \mathcal{U}_{\{M_u<f\}}} (1 - q_{f,u})^{L_u} \right), \tag{26}$$

where $\mathcal{U}_{\{M_u<f\}}$ denotes the set of users with $M_u < f$.

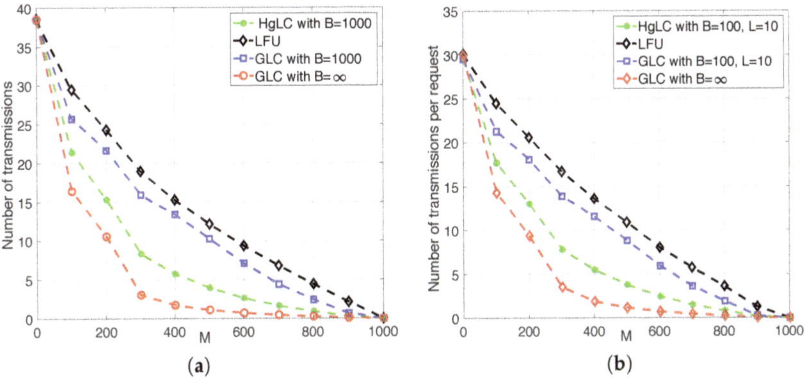

Figure 5. Average number of transmissions in a heterogeneous shared link caching network with $m = 1000$. (**a**) $n = 40$, $L = 1$, $\gamma = 0.5$; (**b**) $n = 40$, $L = 10$, $\gamma = 0.5$.

For simplicity, and to better illustrate the effectiveness of the proposed algorithms, especially under multiple per-user requests, we consider a scenario in which all users request files according to a Zipf demand distribution with parameter $\gamma \in \{0.2, 0,4, 0.5\}$, and all caches have size M files. Under Zipf demands, file f is requested with probability $\frac{f^{-\gamma}}{\sum_{i=1}^{m} i^{-\gamma}}$.

We consider two types of users. In Figures 5a and 6a, users represent end devices requesting only one file each ($L = 1$); while in Figures 5b and 6b, they represent helpers/small-cells, each serving 10 end user devices, and consequently collecting $L = 10$ requests.

In Figure 5a,b, we fix the total number of users n and the product between L and B ($L \times B = 1000$). Figure 5a plots the average rate for a network with $n = 40$ users, $\gamma = 0.5$, $L = 1$, and $B = 1000$. It is immediate to observe the impact of finite packetization on the multiplicative caching gain. In fact, as predicted by the theory (see [21]), the significant caching gain (with respect to LFU) quantified by the asymptotic performance of RAP-GLC (GLC with $B = \infty$) is completely lost when using RAP-GLC with finite packetization (GLC with $B = 1000$). On the other hand, RAP-HgLC remarkably preserves, at the expense of a slight increase in computational complexity, most of the multiplicative caching gain for the same value of file packetization. For example, in Figure 5a, if M doubles from $M = 200$ to $M = 400$, then the rate achieved by RAP-HgLC reduces from 15 to 5.7. Furthermore, RAP-HgLC can achieve a factor of 3.5 rate reduction from LFU for $M = 500$. For the same regime, it is straightforward to verify that neither RAP-GLC nor LFU exhibit this property. Note from Figure 5a that to guarantee a rate of 10, RAP-GLC requires a cache size of $M = 500$, while RAP-HgLC can reduce the cache size requirement to $M = 250$, a $2\times$ cache size reduction. Furthermore, while LFU can only provide an additive caching gain, additive and multiplicative gains may show indistinguishable when M is comparable to the library size m. Hence, one needs to pick a reasonably small M ($\frac{m}{n} < M \ll m$) to observe the multiplicative caching gain of RAP-HgLC.

Figure 5b shows the average rate for a network with $n = 40$ helpers/small-cells, each serving 10 users making requests according to a Zip distribution with $\gamma = 0.5$. Hence, the total number of distinct requests per helper is up to $L_u = 10, \forall u \in \{1, \ldots, 20\}$. In this case, we assume $B = 100$ (instead of $B = 1000$ in Figure 5a). In order to make easier the comparison with Figure 5a, we normalize the achievable rate (number of transmissions) by the file size and the number of requests.

Note from Figure 5a,b that as predicted by Theorem 3, when L_u increases (from $L_u = 1$ to $L_u = 10$), almost the same multiplicative caching gain can be achieved with a smaller B (from $B = 1000$ to $B = 100$). In fact, from Figure 5a,b, we see that under RAP-HgLC, the average rate per request for $B = 100$ and $L = 10$ is almost the same as the average rate per request for $B = 1000$ and $L = 1$. This confirms the interesting tradeoff between B and L established in Theorem 3.

We can observe a similar behavior in Figure 6a,b. Figure 6a plots the average rate for a network with $n = 80$ users, $\gamma = 0.4$, $L = 1$, and $B = 200$. RAP-HgLC is able to preserve most of the multiplicative caching gain for the same values of file packetization. For example, in Figure 6a, if M doubles from $M = 200$ to $M = 400$, then the rate achieved by RAP-HgLC essentially halves from 20 to 10. Furthermore, RAP-HgLC can achieve a factor of 5 rate reduction from LFU for $M = 500$. Note from Figure 6a that to guarantee a rate of 20, RAP-GLC requires a cache size of $M = 500$, while RAP-HgLC can reduce the cache size requirement to $M = 200$, a $2.5\times$ cache size reduction.

Figure 6b plots the average rate for a network with $n = 20$ helpers/small-cells, each serving 10 users making requests according to a Zip distribution with $\gamma = 0.2$. Hence, the total number of distinct requests per helper is up to $L_u = 10, \forall u \in \{1, \ldots, 20\}$. In this case, we assume $B = 100$. Differently from Figure 5b, here we plot the average rate without normalizing it by the number of requests.

Note from Figure 6a,b that, as predicted by Theorem 3, when L_u increases (from $L_u = 1$ to $L_u = 10$), almost the same multiplicative caching gain can be achieved with a smaller B (from $B = 200$ to $B = 100$). In fact, from Figure 6a,b, we see that under RAP-HgLC , the average rate per request for $B = 100$ and $L = 10$ is almost the same as the average rate per request for $B = 200$ and $L = 1$. For example, for $M = 200$, $B = 100$, and $L = 10$, the per request average rate achieved by RAP-HgLC is 0.3, while for $M = 200$ and $B = 200$, is 0.25. This again confirms the tradeoff between B and L stated in Theorem 3.

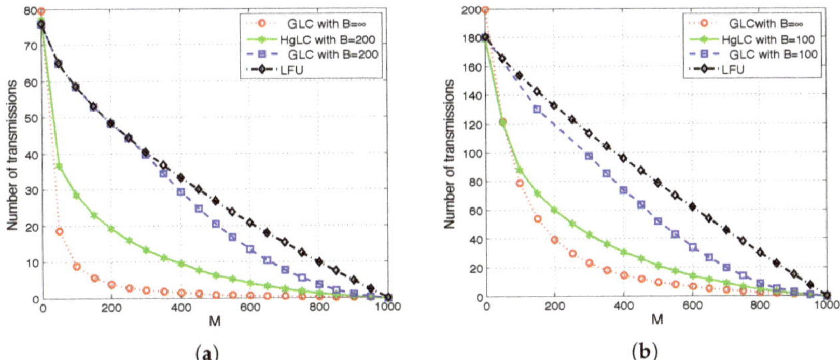

Figure 6. Average number of transmissions in a heterogeneous shared link caching network with $m = 1000$. (a) $n = 80$, $L = 1$, $\gamma = 0.4$; (b) $n = 20$, $L = 10$, $\gamma = 0.2$.

Furthermore, from Figures 5 and 6, we notice that increasing the Zip parameter reduces the gains with respect to LFU. This is explained by the fact that when aggregating multiple requests, there is a higher number of overlapping requests, which increases the opportunities for naive multicasting (as clearly characterized in [13]). Note, however, that RAP-HgLC can remarkably keep similar gains with respect to LFU in this multiple-request setting, and approach the asymptotic performance even with just $B = 100$ packets per file, confirming the effectiveness of the local graph coloring and extra processing procedures in RAP-HgLC.

7. Conclusions

Coded multicasting has been shown to be a promising approach to significantly reduce the traffic load in wireless caching networks. However, most existing schemes require the number of packets per file to grow exponentially with the number of users. To address this challenge, in this paper we focused on a heterogeneous shared link caching network model and designed novel coded multicast algorithms based on local graph coloring that exhibit polynomial-time complexity in all the system parameters, and preserve the asymptotically proven multiplicative caching gain for finite file packetization. We also demonstrated that the number of packets per file can be traded-off with the number of requests collected by each cache, such that the same multiplicative caching gain can be preserved. Simulation results confirm the superiority of the proposed schemes and illustrate the tradeoff between request aggregation and computational complexity (driven by the packetization order), shedding light into the practical achievability of the promising multiplicative caching gain in next generation wireless networks.

Author Contributions: All authors have contributed in equal part to the results of this paper.

Funding: This research was funded in part by NSF grants #1619129, #1817154, #1824558, and by the Alexander von Humboldt Professorship.

Conflicts of Interest: The authors declare no conflict of interest.

Appendix A. Proof of Theorem 1

To analytically characterize the performance of RAP-GLC, we consider two specific cases, where $t = n$ (i.e., the coded multicast only scheme) and $t = 0$ (i.e., the naive multicasting only scheme), and refer to these schemes as RAP-GLC$_1$ and RAP-GLC$_2$, respectively. In the following, we will compute the performance of these two cases respectively and take the minimum rate between these two cases. Obviously, this rate can serve as an upper bound of RAP-GLC.

Appendix A.1. Average Total Number of Colors for RAP-GLC$_1$

To compute the average total number of colors provided by RAP-GLC$_1$, we first see that for all $v \in \mathcal{I}$ obtained in this algorithm, $\mathcal{T}_v$ are identical. Based on Algorithm 1, by construction the independent sets $\mathcal{I} \subset V$ generated by RAP-GLC$_1$ have the same (unordered) label of users requesting or caching the packets $\{\rho(v) : v \in \mathcal{I}\}$. We shall refer to such unordered label of users as the user label of the independent set. Hence, we count the independent sets by enumerating all possible user labels, and upperbounding how many independent sets $\mathcal{I}$ Algorithm 1 generates for each user label. Consider a user label $\mathcal{U}^\ell \subset \mathcal{U}$ of size ℓ, and let $\mathcal{I}(\mathcal{U}^\ell, \mathbf{f}, i)$ the i-th independent set generated by Algorithm 1 with label $\mathcal{U}^\ell$ and while let $\mathcal{J}(\mathcal{U}^\ell, \mathbf{f}) = \{\mathcal{I}(\mathcal{U}^\ell, \mathbf{f}, i) : \forall i\}$.

Following Algorithm 1, for each $\mathcal{U}^\ell$, the number of used colors is $|\mathcal{J}(\mathcal{U}^\ell, \mathbf{f})|$. Given $\mathbf{f}$, we can see that $|\mathcal{J}(\mathcal{U}^\ell, \mathbf{f})|$ is a random variable which is a function of $\mathbf{C}$. Let the indicator $1\{\mathcal{T}_{v_{f_u}} = \mathcal{U}^\ell\}$ denote the event that vertex v_{f_u} from file f_u requested by user $u \in \mathcal{U}^\ell$ is available in all the users in $\mathcal{U}^\ell$ but u and the rest of the vertices $\mathcal{U} \setminus \mathcal{U}^\ell$, then $1\{\mathcal{T}_{v_{f_u}} = \mathcal{U}^\ell\}$ follows a Bernoulli distribution with parameter

$$\lambda(u, f_u) = (1 - p_{f_u,u} M_u) \prod_{k \in \mathcal{U}^\ell \setminus \{u\}} (p_{f_u,k} M_k) \prod_{k \in \mathcal{U} \setminus \mathcal{U}^\ell} (1 - p_{f_u,k} M_k) \tag{A1}$$

such that its expectation is $\lambda(u, f_u)$. Then, we can see that given f, $\sum_{\forall v_{f_u}} 1\{\mathcal{T}_{v_{f_u}} = \mathcal{U}^\ell\} = \lambda(u, f_u)B + o(B)$ with high probability [70]. Thus, as $B \to \infty$, we have that with high probability,

$$\begin{aligned}
|\mathcal{J}(\mathcal{U}^\ell, \mathbf{f})| &= \max_{f_u \in \mathbf{f}(\mathcal{U}^\ell)} \sum_{\forall v_{f_u}} 1\{\mathcal{T}_{v_{f_u}} = \mathcal{U}^\ell\} \\
&= \max_{f_u \in \mathbf{f}(\mathcal{U}^\ell)} \lambda(u, f_u)B + o(B),
\end{aligned} \tag{A2}$$

where $\mathbf{f}(\mathcal{U}^\ell)$ represent the set of files requested by $\mathcal{U}^\ell$.

Then, by averaging over the demand's distribution, we obtain that with high probability:

$$\begin{aligned}
\mathbb{E}[\chi(H_{M,W})] &\leq \mathbb{E}\left[\sum_{j=1}^{L} \sum_{\ell=1}^{n} \sum_{\mathcal{U}^\ell \subset \mathcal{U}_{n_j}} |\mathcal{J}(\mathcal{U}^\ell, \mathbf{f})|\right] \\
&= \sum_{j=1}^{L} \sum_{\ell=1}^{n} \sum_{\mathcal{U}^\ell \subset \mathcal{U}_{n_j}} \mathbb{E}\left[|\mathcal{J}(\mathcal{U}^\ell, \mathbf{f})|\right] \\
&\overset{(a)}{=} \sum_{j=1}^{L} \sum_{\ell=1}^{n} \sum_{\mathcal{U}^\ell \subset \mathcal{U}_{n_j}} \mathbb{E}\left[\max_{f_u \in \mathbf{f}(\mathcal{U}^\ell)} \lambda(u, f_u)B + o(B)\right] \\
&\overset{(b)}{=} \sum_{j=1}^{L} \sum_{\ell=1}^{n} \sum_{\mathcal{U}^\ell \subset \mathcal{U}_{n_j}} \sum_{f=1}^{m} \sum_{u \in \mathcal{U}^\ell} \rho_{f,u,\mathcal{U}^\ell} \lambda(u, f) + \delta_1(B),
\end{aligned} \tag{A3}$$

where (a) is by using (A2) and (b) is obtained by computing the probability that the requested file f_u in $\mathbf{f}(\mathcal{U}^\ell)$ maximizes $\lambda(u, f)$. $\delta_1(B)$ denotes a smaller order term of $\sum_{j=1}^{L} \sum_{\ell=1}^{n} \sum_{\mathcal{U}^\ell \subset \mathcal{U}_{n_j}} \sum_{f=1}^{m} \sum_{u \in \mathcal{U}^\ell} \rho_{f,u,\mathcal{U}^\ell} \lambda(u, f)$. For any $\mathcal{U}^\ell$, we obtain that $\sum_f \sum_{u \in \mathcal{U}^\ell} \rho_{f,u,\mathcal{U}^\ell} = 1$, and $\rho_{f,u,\mathcal{U}^\ell}$ denotes the probability that file f is the file with memory assignment $p_{f,u}$ such that $\rho_{f,u,\mathcal{U}^\ell} \triangleq \mathbb{P}(f = \arg \max_{f_u \in \mathbf{f}(\mathcal{U}^\ell)} \lambda(u, f_u))$, where $\mathbf{f}(\mathcal{U}^\ell)$ denotes the set of files requested by a subset users $\mathcal{U}^\ell$. Thus, we normalize (A3) by B and obtain that

$$
\begin{aligned}
R^{\text{RAP-GLC}_1} &= \frac{\mathbb{E}[\chi(H_{M,W})]}{B} \\
&\leq \sum_{j=1}^{L} \sum_{\ell=1}^{n} \sum_{\mathcal{U}^\ell \subset \mathcal{U}_{n_j}} \sum_{f=1}^{m} \sum_{u \in \mathcal{U}^\ell} \rho_{f,u,\mathcal{U}^\ell} \lambda(u,f), \\
&= \psi(\mathbf{P}, \mathbf{Q}),
\end{aligned}
\tag{A4}
$$

which is the first term inside the minimum in (12).

Appendix A.2. Average Total Number of Colors for RAP-GLC$_2$

As described in Section 4.1, RAP-GLC$_2$ computes the minimum coloring of $\mathcal{H}_{\mathbf{C},\mathbf{W}}$ subject to the constraint that only the vertices representing the same packet can have the same color. In this case, the total number of colors is equal to the number of distinct requested packets, and the coloring can be found in $O(|\mathcal{V}|^2)$. Starting from this valid coloring, GCLC$_2$ computes $\max_{v \in \mathcal{V}} |\mathbf{c}(\mathcal{N}^+(v))|$. To show that the performance of GCLC$_2$ are upper bounded by $\bar{m} - \bar{M}$ with $\bar{m}$ and $\bar{M}$ given as in (13) and (14) respectively, we note that:

$$
\begin{aligned}
\max_{v \in \mathcal{V}} |\mathbf{c}(\mathcal{N}^+(v))| &\overset{(a)}{\leq} \sum_{\mathbf{f} \in \mathcal{F}^n} \left(\prod_{u=1}^{n} q_{f,u} \right) \sum_{\hat{f}=1}^{m} \mathbb{1}\{\hat{f} \in \mathbf{f}\} B_f \\
&= \sum_{\hat{f}=1}^{m} \sum_{\mathbf{f} \in \mathcal{F}^n} \left(\prod_{u=1}^{n} q_{f,u} \right) \mathbb{1}\{\hat{f} \in \mathbf{f}\} B_{\hat{f},\mathbf{f}} \\
&\overset{(b)}{\leq} \sum_{f} \mathbb{P}(\text{file } f \text{ is requested})(B - \min_u p_{u,f} M_u B) \\
&= \sum_{f=1}^{m} \left(1 - \prod_{u=1}^{n} \left(1 - q_{f,u} \right)^{L_u} \right) (1 - \min_u p_{u,f} M_u) B,
\end{aligned}
\tag{A5}
$$

where $B_{\hat{f},\mathbf{f}}$ is number of chucks that are going to be transmitted of file $\hat{f}$ given that the demand vector is equal to $\mathbf{f}$, and (a) is due to the observation that given a file $\hat{f}$,

$$
\sum_{\mathbf{f} \in \mathcal{F}^n} \left(\prod_{u=1}^{n} q_{f,u} \right) \mathbb{1}\{\hat{f} \in \mathbf{f}\} = \mathbb{E}[\mathbb{1}\{\hat{f} \in \mathbf{f}\}]
$$
$$
= \mathbb{P}(\text{file } \hat{f} \text{ is requested}).
\tag{A6}
$$

Normalizing (A5) by B, we obtain that:

$$
\begin{aligned}
R^{\text{RAP-GLC}_2} &\leq \sum_{f=1}^{m} \left(1 - \prod_{u=1}^{n} \left(1 - q_{f,u} \right)^{L_u} \right) \\
&\quad \cdot (1 - \min_u p_{u,f} M_u) \\
&= \bar{m} - \bar{M},
\end{aligned}
\tag{A7}
$$

which is the second term inside the minimum in (12).

Appendix B. Proof of Theorem 2

In this section, we prove Theorem 2. Recall that for each $\mathcal{U}^\ell$, the number of used colors is given by $|\mathcal{J}(\mathcal{U}^\ell, \mathbf{f})|$, then we have we have $|\mathcal{J}(\mathcal{U}^\ell, \mathbf{f})| = \max_{u \in \mathcal{U}^\ell} \sum_{\forall v_{f_u}} 1\{\mathcal{T}_{v_{f_u}} = \mathcal{U}^\ell\}$, where $\mathbf{f}(\mathcal{U}^\ell)$ represent the set of files requested by $\mathcal{U}^\ell$. In this case, it is clear that

$$\mathbb{E}\left[\sum_{\forall v_{f_u}} 1\{\mathcal{T}_{v_{f_u}} = \mathcal{U}^\ell\}\right] = L\left(\frac{M}{m}\right)^{l-1}\left(1 - \frac{M}{m}\right)^{n-l+1}. \tag{A8}$$

Then let

$$Y_l \triangleq |\mathcal{J}(\mathcal{U}^\ell, \mathbf{f})| - \mathbb{E}\left[\sum_{\forall v_{f_u}} 1\{\mathcal{T}_{v_{f_u}} = \mathcal{U}^\ell\}\right]. \tag{A9}$$

The goal is to find a condition of L such that

$$|\mathbb{E}[Y_l]| = o\left(L\left(\frac{M}{m}\right)^{l-1}\left(1 - \frac{M}{m}\right)^{n-l+1}\right), \tag{A10}$$

which implies

$$\mathbb{E}\left[|\mathcal{J}(\mathcal{U}^\ell, \mathbf{f})|\right] \leq (1 + o(1))L\left(\frac{M}{m}\right)^{l-1}\left(1 - \frac{M}{m}\right)^{n-l+1}. \tag{A11}$$

Then following the similar step from Theorem 1 in [12], we can obtain the concentration result shown in Theorem 2, where we require $L = \omega\left(\dfrac{(nM)^{\frac{1}{2(1-\epsilon)}}}{\left(\frac{m}{M}-1\right)\left(1-\left(1-\frac{M}{m}\right)^n\right)}\right)$ and $\epsilon > 0$ is an arbitrarily small number.

To compute $|\mathbb{E}[Y_l]|$, we have

$$
\begin{aligned}
(\mathbb{E}[Y_l])^2 &\leq \mathbb{E}[(Y_l)^2] \\
&= \mathbb{E}\left[\left(|\mathcal{J}(\mathcal{U}^\ell, \mathbf{f})| - \mathbb{E}\left[\sum_{\forall v_{f_u}} 1\{\mathcal{T}_{v_{f_u}} = \mathcal{U}^\ell\}\right]\right)^2\right] \\
&= \mathbb{E}\left[\left(\max_{u \in \mathcal{U}^\ell}\left(\sum_{\forall v_{f_u}} 1\{\mathcal{T}_{v_{f_u}} = \mathcal{U}^\ell\} - \mathbb{E}\left[\sum_{\forall v_{f_u}} 1\{\mathcal{T}_{v_{f_u}} = \mathcal{U}^\ell\}\right]\right)\right)^2\right] \\
&\leq \sum_{u \in \mathcal{U}^\ell} \mathbb{E}\left[\left(\sum_{\forall v_{f_u}} 1\{\mathcal{T}_{v_{f_u}} = \mathcal{U}^\ell\} - \mathbb{E}\left[\sum_{\forall v_{f_u}} 1\{\mathcal{T}_{v_{f_u}} = \mathcal{U}^\ell\}\right]\right)^2\right] \\
&= l\mathbb{E}\left[\left(\sum_{\forall v_{f_u}} 1\{\mathcal{T}_{v_{f_u}} = \mathcal{U}^\ell\} - \mathbb{E}\left[\sum_{\forall v_{f_u}} 1\{\mathcal{T}_{v_{f_u}} = \mathcal{U}^\ell\}\right]\right)^2\right] \\
&\stackrel{(a)}{=} lL\left(\frac{M}{m}\right)^{l-1}\left(1 - \frac{M}{m}\right)^{n-l+1}\left(1 - \left(\frac{M}{m}\right)^{l-1}\left(1 - \frac{M}{m}\right)^{n-l+1}\right) + \delta_1,
\end{aligned}
\tag{A12}
$$

where (a) is because $M = \omega(1)$ such that δ_1 is a smaller order term compared to the first term in (A12) and use the variance for the Binomial distribution. Then, we let

$$\left(lL \left(\tfrac{M}{m}\right)^{l-1} \left(1 - \tfrac{M}{m}\right)^{n-l+1} \left(1 - \left(\tfrac{M}{m}\right)^{l-1} \left(1 - \tfrac{M}{m}\right)^{n-l+1}\right) \right)^{\frac{1}{2}} = o\left(L \left(\tfrac{M}{m}\right)^{l-1} \left(1 - \tfrac{M}{m}\right)^{n-l+1} \right), \tag{A13}$$

then, we can obtain

$$L = \omega \left(\frac{l \left(1 - \left(\tfrac{M}{m}\right)^{l-1} \left(1 - \tfrac{M}{m}\right)^{n-l+1}\right)}{\left(\tfrac{M}{m}\right)^{l-1} \left(1 - \tfrac{M}{m}\right)^{n-l+1}} \right) \tag{A14}$$

$$= \omega \left(\frac{\left(1 - \left(\tfrac{M}{m}\right)^{l-1} \left(1 - \tfrac{M}{m}\right)^{n-l+1}\right)}{\tfrac{1}{l} \left(\tfrac{M}{m}\right)^{l-1} \left(1 - \tfrac{M}{m}\right)^{n-l+1}} \right).$$

For sufficient condition, let

$$L = \omega \left(\frac{1}{\tfrac{1}{l} \left(\tfrac{M}{m}\right)^{l-1} \left(1 - \tfrac{M}{m}\right)^{n-l+1}} \right). \tag{A15}$$

Do the derivative of $s(l) = \tfrac{1}{l} \left(\tfrac{M}{m}\right)^{l-1} \left(1 - \tfrac{M}{m}\right)^{n-l+1}$ with respect to l, we obtain

$$\begin{aligned}
\frac{ds(l)}{dl} &= \frac{1}{l} \left(\left(\frac{M}{m}\right)^{l-1} \left(1 - \frac{M}{m}\right)^{n-l+1} \log \frac{M}{m} \right. \\
&\quad \left. - \log\left(1 - \frac{M}{m}\right) \left(\frac{M}{m}\right)^{l-1} \left(1 - \frac{M}{m}\right)^{n-l+1} \right) \\
&\quad - \frac{1}{l^2} \left(\frac{M}{m}\right)^{l-1} \left(1 - \frac{M}{m}\right)^{n-l+1} \\
&= \frac{1}{l} \log \left(\frac{\frac{M}{m}}{1 - \frac{M}{m}} \right) \left(\frac{M}{m}\right)^{l-1} \left(1 - \frac{M}{m}\right)^{n-l+1} \\
&\quad - \frac{1}{l^2} \left(\frac{M}{m}\right)^{l-1} \left(1 - \frac{M}{m}\right)^{n-l+1}.
\end{aligned} \tag{A16}$$

- When $\frac{M}{m} < \frac{1}{2}$, then we have $\frac{ds(l)}{l} < 0$. Hence, $s(l)$ is a decreasing function such that the minimum value of $s(l)$ take place when $l = n$. Thus, by using (A15), we obtain the sufficient condition is given by

$$L = \omega \left(\frac{n \left(\tfrac{m}{M}\right)^{n-1}}{1 - \tfrac{M}{m}} \right) = \omega \left(\frac{nM}{m} \left(\frac{m}{M}\right)^{n} \frac{1}{\left(1 - \tfrac{M}{m}\right)} \right). \tag{A17}$$

- When $\frac{M}{m} \geq \frac{e}{1+e}$, then we have $\frac{ds(l)}{l} > 0$. Hence, $s(l)$ is an increasing function such that the minimum value of $s(l)$ take place when $l = 1$. Thus, by using (A15) we obtain the sufficient condition is given by

$$L = \omega\left(\frac{1}{\left(1 - \frac{M}{m}\right)^n}\right) = \omega\left(\left(\frac{m}{m-M}\right)^n\right). \tag{A18}$$

Thus, we finished the proof of Theorem 2.

Appendix C. Proof of Theorem 3

In this proof, we follow the similar procedure in the proof of Theorem 2 in Appendix B, and obtain

$$\mathbb{E}\left[\sum_{\forall v_{f_u}} 1\{\mathcal{T}_{v_{f_u}} = \mathcal{U}^\ell\}\right] = LB\left(\frac{M}{m}\right)^{l-1}\left(1 - \frac{M}{m}\right)^{n-l+1}. \tag{A19}$$

Then let

$$Y_l \triangleq |\mathcal{J}(\mathcal{U}^\ell, \mathbf{f})| - \mathbb{E}\left[\sum_{\forall v_{f_u}} 1\{\mathcal{T}_{v_{f_u}} = \mathcal{U}^\ell\}\right]. \tag{A20}$$

The goal again is to find a condition of B and L such that

$$|\mathbb{E}[Y_l]| = o\left(LB\left(\frac{M}{m}\right)^{l-1}\left(1 - \frac{M}{m}\right)^{n-l+1}\right), \tag{A21}$$

which implies

$$\mathbb{E}\left[|\mathcal{J}(\mathcal{U}^\ell, \mathbf{f})|\right] \leq (1 + o(1))LB\left(\frac{M}{m}\right)^{l-1}\left(1 - \frac{M}{m}\right)^{n-l+1}. \tag{A22}$$

Then following the similar step from Theorem 1 in [12], we can obtain the concentration result shown in Theorem 2, where we require $LB = \omega\left(\frac{(nM)^{\frac{1}{2(1-\epsilon)}}}{\left(\frac{m}{M}-1\right)\left(1-\left(1-\frac{M}{m}\right)^n\right)}\right)$ and $\epsilon > 0$ is an arbitrarily small number.

Similar to (A12), to compute $|\mathbb{E}[Y_l]|$, we have

$$(\mathbb{E}[Y_l])^2 \leq \mathbb{E}[(Y_l)^2]$$
$$\stackrel{(a)}{=} lLB\left(\frac{M}{m}\right)^{l-1}\left(1 - \frac{M}{m}\right)^{n-l+1} \cdot \left(1 - \left(\frac{M}{m}\right)^{l-1}\left(1 - \frac{M}{m}\right)^{n-l+1}\right) + \delta_2, \tag{A23}$$

where (a) is because $B = \omega\left(\frac{m}{M}\right)$ such that δ_2 is a smaller order term compared to the first term in (A23) and use the variance for the Binomial distribution. Then, we let

$$\left(lLB\left(\frac{M}{m}\right)^{l-1}\left(1 - \frac{M}{m}\right)^{n-l+1}\left(1 - \left(\frac{M}{m}\right)^{l-1}\left(1 - \frac{M}{m}\right)^{n-l+1}\right)\right)^{\frac{1}{2}}$$
$$= o\left(LB\left(\frac{M}{m}\right)^{l-1}\left(1 - \frac{M}{m}\right)^{n-l+1}\right), \tag{A24}$$

then, we can obtain

$$
\begin{aligned}
LB &= \frac{\left(1 - \left(\frac{M}{m}\right)^{l-1}\left(1 - \frac{M}{m}\right)^{n-l+1}\right)}{\left(\frac{M}{m}\right)^{l-1}\left(1 - \frac{M}{m}\right)^{n-l+1}} \\
&= \frac{\left(1 - \left(\frac{M}{m}\right)^{l-1}\left(1 - \frac{M}{m}\right)^{n-l+1}\right)}{\frac{1}{l}\left(\frac{M}{m}\right)^{l-1}\left(1 - \frac{M}{m}\right)^{n-l+1}}.
\end{aligned}
\tag{A25}
$$

For sufficient condition, let

$$
LB = \omega\left(\frac{1}{\frac{1}{l}\left(\frac{M}{m}\right)^{l-1}\left(1 - \frac{M}{m}\right)^{n-l+1}}\right).
\tag{A26}
$$

Then following the similar steps as the proof of Theorem 2 in Appendix B, we fished the proof of Theorem 3.

References

1. Shanmugam, K.; Golrezaei, N.; Dimakis, A.; Molisch, A.; Caire, G. FemtoCaching: Wireless Video Content Delivery through Distributed Caching Helpers. *IEEE Trans. Inf. Theory* **2013**, *59*, 8402–8413. [CrossRef]
2. Llorca, J.; Tulino, A.; Guan, K.; Kilper, D. Network-Coded Caching-Aided Multicast for Efficient Content Delivery. In Proceedings of the 2013 IEEE International Conference on Communications (ICC), Budapest, Hungary, 9–13 June 2013.
3. Fadlallah, Y.; Tulino, A.M.; Barone, D.; Vettigli, G.; Llorca, J.; Gorce, J. Coding for Caching in 5G Networks. *IEEE Commun. Mag.* **2017**, *55*, 106–113. [CrossRef]
4. Liu, D.; Chen, B.; Yang, C.; Molisch, A.F. Caching at the wireless edge: Design aspects, challenges, and future directions. *IEEE Commun. Mag.* **2016**, *54*, 22–28. [CrossRef]
5. Tandon, R.; Simeone, O. Harnessing cloud and edge synergies: Toward an information theory of fog radio access networks. *IEEE Commun. Mag.* **2016**, *54*, 44–50. [CrossRef]
6. Maddah-Ali, M.A.; Niesen, U. Coding for caching: Fundamental limits and practical challenges. *IEEE Commun. Mag.* **2016**, *54*, 23–29. [CrossRef]
7. Paschos, G.; Bastug, E.; Land, I.; Caire, G.; Debbah, M. Wireless caching: Technical misconceptions and business barriers. *IEEE Commun. Mag.* **2016**, *54*, 16–22. [CrossRef]
8. Maddah-Ali, M.; Niesen, U. Fundamental Limits of Caching. *IEEE Trans. Inf. Theory* **2014**, *60*, 2856–2867. [CrossRef]
9. Maddah-Ali, M.; Niesen, U. Decentralized Coded Caching Attains Order-Optimal Memory-Rate Tradeoff. *IEEE/ACM Trans. Netw.* **2014**. [CrossRef]
10. Ji, M.; Tulino, A.; Llorca, J.; Caire, G. On the average performance of caching and coded multicasting with random demands. In Proceedings of the 2014 11th International Symposium on Wireless Communications Systems (ISWCS), Barcelona, Spain, 26–29 August 2014; pp. 922–926.
11. Niesen, U.; Maddah-Ali, M.A. Coded Caching With Nonuniform Demands. *IEEE Trans. Inf. Theory* **2017**, *63*, 1146–1158. [CrossRef]
12. Ji, M.; Tulino, A.M.; Llorca, J.; Caire, G. Order-Optimal Rate of Caching and Coded Multicasting with Random Demands. *IEEE Trans. Inf. Theory* **2017**, *63*, 3923–3949. [CrossRef]

13. Ji, M.; Tulino, A.; Llorca, J.; Caire, G. Caching and Coded Multicasting: Multiple Groupcast Index Coding. In Proceedings of the 2014 IEEE Global Conference on Signal and Information Processing (GlobalSIP), Atlanta, GA, USA, 3–5 December 2014; pp. 881–885.

14. Ji, M.; Tulino, A.M.; Llorca, J.; Caire, G. Caching-Aided Coded Multicasting with Multiple Random Requests. In Proceedings of the 2015 IEEE Information Theory Workshop (ITW), Jerusalem, Israel, 26 April–1 May 2015; pp. 1–5.

15. Wan, K.; Tuninetti, D.; Piantanida, P. On caching with more users than files. In Proceedings of the 2016 IEEE International Symposium on Information Theory (ISIT), Barcelona, Spain, 10–15 July 2016; pp. 135–139. [CrossRef]

16. Wan, K.; Tuninetti, D.; Piantanida, P. On the optimality of uncoded cache placement. In Proceedings of the 2016 IEEE Information Theory Workshop (ITW), Cambridge, UK, 11–14 September 2016; pp. 161–165. [CrossRef]

17. Ji, M.; Caire, G.; Molisch, A.F. The Throughput-Outage Tradeoff of Wireless One-Hop Caching Networks. *IEEE Trans. Inf. Theory* **2015**, *61*, 6833–6859. [CrossRef]

18. Ji, M.; Caire, G.; Molisch, A.F. Fundamental Limits of Caching in Wireless D2D Networks. *IEEE Trans. Inf. Theory* **2016**, *62*, 849–869. [CrossRef]

19. Ji, M.; Caire, G.; Molisch, A.F. Wireless Device-to-Device Caching Networks: Basic Principles and System Performance. *IEEE J. Sel. Areas Commun.* **2016**, *34*, 176–189. [CrossRef]

20. Cacciapuoti, A.S.; Caleffi, M.; Ji, M.; Llorca, J.; Tulino, A.M. Speeding Up Future Video Distribution via Channel-Aware Caching-Aided Coded Multicast. *IEEE J. Sel. Areas Commun.* **2016**, *34*, 2207–2218. [CrossRef]

21. Shanmugam, K.; Ji, M.; Tulino, A.M.; Llorca, J.; Dimakis, A.G. Finite-Length Analysis of Caching-Aided Coded Multicasting. *IEEE Trans. Inf. Theory* **2016**, *62*, 5524–5537. [CrossRef]

22. Shangguan, C.; Zhang, Y.; Ge, G. Centralized Coded Caching Schemes: A Hypergraph Theoretical Approach. *IEEE Trans. Inf. Theory* **2018**, *64*, 5755–5766. [CrossRef]

23. Chen, Z. Fundamental limits of caching: Improved bounds for users with small buffers. *IET Commun.* **2016**, *10*, 2315–2318. [CrossRef]

24. Karamchandani, N.; Niesen, U.; Maddah-Ali, M.A.; Diggavi, S.N. Hierarchical Coded Caching. *IEEE Trans. Inf. Theory* **2016**, *62*, 3212–3229. [CrossRef]

25. Pedarsani, R.; Maddah-Ali, M.A.; Niesen, U. Online Coded Caching. *IEEE/ACM Trans. Netw.* **2016**, *24*, 836–845. [CrossRef]

26. Sahraei, S.; Gastpar, M. K users caching two files: An improved achievable rate. In Proceedings of the 2016 Annual Conference on Information Science and Systems (CISS), Princeton, NJ, USA, 16–18 March 2016; pp. 620–624. [CrossRef]

27. Wang, C.; Lim, S.H.; Gastpar, M. Information-Theoretic Caching: Sequential Coding for Computing. *IEEE Trans. Inf. Theory* **2016**, *62*, 6393–6406. [CrossRef]

28. G., J. Fundamental limits of caching: Improved bounds with coded prefetching. *arXiv* **2016**, arXiv:1612.09071.

29. Shariatpanahi, S.P.; Motahari, S.A.; Khalaj, B.H. Multi-Server Coded Caching. *IEEE Trans. Inf. Theory* **2016**, *62*, 7253–7271. [CrossRef]

30. Shanmugam, K.; Tulino, A.M.; Dimakis, A.G. Coded caching with linear subpacketization is possible using Ruzsa-Szeméredi graphs. In Proceedings of the 2017 IEEE International Symposium on Information Theory (ISIT), Aachen, Germany, 25–30 June 2017; pp. 1237–1241. [CrossRef]

31. Ghasemi, H.; Ramamoorthy, A. Improved Lower Bounds for Coded Caching. *IEEE Trans. Inf. Theory* **2017**, *63*, 4388–4413. [CrossRef]

32. Lim, S.H.; Wang, C.; Gastpar, M. Information-Theoretic Caching: The Multi-User Case. *IEEE Trans. Inf. Theory* **2017**, *63*, 7018–7037. [CrossRef]

33. Jeon, S.; Hong, S.; Ji, M.; Caire, G.; Molisch, A.F. Wireless Multihop Device-to-Device Caching Networks. *IEEE Trans. Inf. Theory* **2017**, *63*, 1662–1676. [CrossRef]

34. Sengupta, A.; Tandon, R. Improved Approximation of Storage-Rate Tradeoff for Caching With Multiple Demands. *IEEE Trans. Commun.* **2017**, *65*, 1940–1955. [CrossRef]

35. Hachem, J.; Karamchandani, N.; Diggavi, S.N. Coded Caching for Multi-level Popularity and Access. *IEEE Trans. Inf. Theory* **2017**, *63*, 3108–3141. [CrossRef]

36. Ji, M.; Wong, M.F.; Tulino, A.M.; Llorca, J.; Caire, G.; Effros, M.; Langberg, M. On the fundamental limits of caching in combination networks. In Proceedings of the 2015 IEEE 16th International Workshop on Signal Processing Advances in Wireless Communications (SPAWC), Stockholm, Sweden, 28 June–1 July 2015; pp. 695–699. [CrossRef]

37. Ji, M.; Tulino, A.M.; Llorca, J.; Caire, G. Caching in combination networks. In Proceedings of the 2015 49th Asilomar Conference on Signals, Systems and Computers, Pacific Grove, CA, USA, 8 November 2015; pp. 1269–1273. [CrossRef]

38. Wan, K.; Ji, M.; Piantanida, P.; Tuninetti, D. Novel outer bounds for combination networks with end-user-caches. In Proceedings of the 2017 IEEE Information Theory Workshop (ITW), Kaohsiung, Taiwan, 6–10 November 2017; pp. 444–448. [CrossRef]

39. Wan, K.; Tuninetti, D.; Ji, M.; Piantanida, P. State-of-the-art in cache-aided combination networks. In Proceedings of the 2017 51st Asilomar Conference on Signals, Systems, and Computers, Pacific Grove, CA, USA, 29 October–1 November 2017; pp. 641–645. [CrossRef]

40. Wan, K.; Ji, M.; Piantanida, P.; Tuninetti, D. Caching in Combination Networks: Novel Multicast Message Generation and Delivery by Leveraging the Network Topology. In Proceedings of the 2018 IEEE International Conference on Communications (ICC), Kansas City, MO, USA, 20–24 May 2018; pp. 1–6. [CrossRef]

41. Wan, K.; Jit, M.; Piantanida, P.; Tuninetti, D. On the Benefits of Asymmetric Coded Cache Placement in Combination Networks with End-User Caches. In Proceedings of the 2018 IEEE International Symposium on Information Theory (ISIT), Vail, CO, USA, 17–22 June 2018; pp. 1550–1554. [CrossRef]

42. Wan, K.; Tuninetti, D.; Ji, M.; Piantanida, P. A Novel Asymmetric Coded Placement in Combination Networks with End-User Caches. In Proceedings of the 2018 Information Theory and Applications Workshop (ITA), San Diego, CA, USA, 11–16 February 2018; pp. 1–5. [CrossRef]

43. Tian, C.; Chen, J. Caching and Delivery via Interference Elimination. *IEEE Trans. Inf. Theory* **2018**, *64*, 1548–1560. [CrossRef]

44. Yu, Q.; Maddah-Ali, M.A.; Avestimehr, A.S. The Exact Rate-Memory Tradeoff for Caching With Uncoded Prefetching. *IEEE Trans. Inf. Theory* **2018**, *64*, 1281–1296. [CrossRef]

45. Zhang, K.; Tian, C. Fundamental Limits of Coded Caching: From Uncoded Prefetching to Coded Prefetching. *IEEE J. Sel. Areas Commun.* **2018**, *36*, 1153–1164. [CrossRef]

46. Wang, C.; Bidokhti, S.S.; Wigger, M. Improved Converses and Gap Results for Coded Caching. *IEEE Trans. Inf. Theory* **2018**, *64*, 7051–7062. [CrossRef]

47. Yu, Q.; Maddah-Ali, M.A.; Avestimehr, A.S. Characterizing the Rate-Memory Tradeoff in Cache Networks Within a Factor of 2. *IEEE Trans. Inf. Theory* **2019**, *65*, 647–663. [CrossRef]

48. Karat, N.S.; Thomas, A.; Rajan, B.S. Optimal Error Correcting Delivery Scheme for an Optimal Coded Caching Scheme with Small Buffers. In Proceedings of the 2018 IEEE International Symposium on Information Theory (ISIT), Vail, CO, USA, 17–22 June 2018; pp. 1710–1714. [CrossRef]

49. Tian, C. Symmetry, Outer Bounds, and Code Constructions: A Computer-Aided Investigation on the Fundamental Limits of Caching. *Entropy* **2018**, *20*, 603. [CrossRef]

50. Cisco. *The Zettabyte Era-Trends and Analysis*; Cisco White Paper; Cisco: San Jose, CA, USA, 2013.

51. Birk, Y.; Kol, T. Informed-source coding-on-demand (ISCOD) over broadcast channels. In Proceedings of the Conference on Computer Communications, Seventeenth Annual Joint Conference of the IEEE Computer and Communications Societies, Gateway to the 21st Century, San Francisco, CA, USA, 29 March–2 April 1998.

52. Breslau, L.; Cao, P.; Fan, L.; Phillips, G.; Shenker, S. Web caching and Zipf-like distributions: Evidence and implications. In Proceedings of the INFOCOM'99: Conference on Computer Communications, New York, NY, USA, 21–25 March 1999; Volume 1, pp. 126–134.

53. Tang, L.; Ramamoorthy, A. Coded Caching Schemes With Reduced Subpacketization From Linear Block Codes. *IEEE Trans. Inf. Theory* **2018**, *64*, 3099–3120. [CrossRef]

54. Yan, Q.; Cheng, M.; Tang, X.; Chen, Q. On the Placement Delivery Array Design for Centralized Coded Caching Scheme. *IEEE Trans. Inf. Theory* **2017**, *63*, 5821–5833. [CrossRef]

55. Vettigli, G.; Ji, M.; Tulino, A.M.; Llorca, J.; Festa, P. An efficient coded multicasting scheme preserving the multiplicative caching gain. In Proceedings of the 2015 IEEE Conference on Computer Communications Workshops (INFOCOM WKSHPS), Hong Kong, China, 26 April–1 May 2015; pp. 251–256. [CrossRef]

56. Ji, M.; Shanmugam, K.; Vettigli, G.; Llorca, J.; Tulino, A.M.; Caire, G. An efficient multiple-groupcast coded multicasting scheme for finite fractional caching. In Proceedings of the 2015 IEEE International Conference on Communications (ICC), London, UK, 8–12 June 2015; pp. 3801–3806. [CrossRef]

57. Ramakrishnan, A.; Westphal, C.; Markopoulou, A. An Efficient Delivery Scheme for Coded Caching. In Proceedings of the 2015 27th International Teletraffic Congress, Ghent, Belgium, 8–10 September 2015; pp. 46–54. [CrossRef]

58. Jin, S.; Cui, Y.; Liu, H.; Caire, G. Order-Optimal Decentralized Coded Caching Schemes with Good Performance in Finite File Size Regime. In Proceedings of the 2016 IEEE Global Communications Conference (GLOBECOM), Washington, DC, USA, 4–8 December 2016; pp. 1–7. [CrossRef]

59. Wan, K.; Tuninetti, D.; Piantanida, P. Novel delivery schemes for decentralized coded caching in the finite file size regime. In Proceedings of the 2017 IEEE International Conference on Communications Workshops (ICC Workshops), Paris, France, 21–25 May 2017; pp. 1183–1188. [CrossRef]

60. Asghari, S.M.; Ouyang, Y.; Nayyar, A.; Avestimehr, A.S. Optimal Coded Multicast in Cache Networks with Arbitrary Content Placement. In Proceedings of the 2018 IEEE International Conference on Communications (ICC), Kansas City, MO, USA, 20–24 May 2018; pp. 1–6. [CrossRef]

61. Amiri, M.M.; Yang, Q.; Gündüz, D. Decentralized coded caching with distinct cache capacities. In Proceedings of the 2016 50th Asilomar Conference on Signals, Systems and Computers, Pacific Grove, CA, USA, 6–9 November 2016; pp. 734–738. [CrossRef]

62. Amiri, M.M.; Yang, Q.; Gündüz, D. Decentralized Caching and Coded Delivery With Distinct Cache Capacities. *IEEE Trans. Commun.* **2017**, *65*, 4657–4669. [CrossRef]

63. Ibrahim, A.M.; Zewail, A.A.; Yener, A. Centralized Coded Caching with Heterogeneous Cache Sizes. In Proceedings of the 2017 IEEE Wireless Communications and Networking Conference (WCNC), San Francisco, CA, USA, 19–22 March 2017; pp. 1–6. [CrossRef]

64. Wei, Y.; Ulukus, S. Coded caching with multiple file requests. In Proceedings of the 2017 55th Annual Allerton Conference on Communication, Control, and Computing (Allerton), Monticello, IL, USA, 3–6 October 2017; pp. 437–442. [CrossRef]

65. Parrinello, E.; Unsal, A.; Elia, P. Fundamental Limits of Caching in Heterogeneous Networks with Uncoded Prefetching. *arXiv* **2018**, arXiv:1811.06247.

66. Shanmugam, K.; Dimakis, A.G.; Langberg, M. Local graph coloring and index coding. In Proceedings of the 2013 IEEE International Symposium on Information Theory, Istanbul, Turkey, 7–12 July 2013; pp. 1152–1156. [CrossRef]

67. Lin, S.; Costello, D.J. *Error Control Coding*; Prentice-hall Englewood Cliffs: Upper Saddle River, NJ, USA, 2004; Volume 123.

68. Bar-Yossef, Z.; Birk, Y.; Jayram, T.; Kol, T. Index coding with side information. *IEEE Trans. Inf. Theory* **2011**, *57*, 1479–1494. [CrossRef]

69. Lee, D.; Noh, S.; Min, S.; Choi, J.; Kim, J.; Cho, Y.; Kim, C. LRFU: A spectrum of policies that subsumes the least recently used and least frequently used policies. *IEEE Trans. Comput.* **2001**, *50*, 1352–1361.

70. Boucheron, S.; Lugosi, G.; Massart, P. *Concentration Inequalities: A Nonasymptotic Theory of Independence*; Oxford University Press: Oxford, UK, 2013.

Article

Cross-Entropy Method for Content Placement and User Association in Cache-Enabled Coordinated Ultra-Dense Networks

Jia Yu [1], Ye Wang [1,2,†], Shushi Gu [1,2,†], Qinyu Zhang [1,2,†], Siyun Chen [1] and Yalin Zhang [3,*]**

[1] Communication Engineering Research Centre, Harbin Institute of Technology (Shenzhen),
    HIT Campus of University Town of Shenzhen, Shenzhen 518055, China; yujia_hitsz@hotmail.com (J.Y.);
    wangye83@hit.edu.cn (Y.W.); gushushi@hit.edu.cn (S.G.); zqy@hit.edu.cn (Q.Z.);
    chensiyun@stu.hit.edu.cn (S.C.)
[2] Peng Cheng Laboratory, Shenzhen 518055, China
[3] School of Electronic and Communication Engineering, Shenzhen Polytechnic, Shenzhen 518055, China
*   Correspondence: zhangyalin@szpt.edu.cn; Tel.: +86-0755-2630-2145
†   These authors contributed equally to this work.

Received: 27 March 2019; Accepted: 1 June 2019; Published: 8 June 2019

Abstract: Due to the high splitting-gain of dense small cells, Ultra-Dense Network (UDN) is regarded as a promising networking technology to achieve high data rate and low latency in 5G mobile communications. In UDNs, each User Equipment (UE) may receive signals from multiple Base Stations (BSs), which impose severe interference in the networks and in turn motivates the possibility of using Coordinated Multi-Point (CoMP) transmissions to further enhance network capacity. In CoMP-based Ultra-Dense Networks, a great challenge is to tradeoff between the gain of network throughput and the worsening backhaul latency. Caching popular files on BSs has been identified as a promising method to reduce the backhaul traffic load. In this paper, we investigated content placement strategies and user association algorithms for the proactive caching ultra dense networks. The problem has been formulated to maximize network throughput of cell edge UEs under the consideration of backhaul load, which is a constrained non-convex combinatorial optimization problem. To decrease the complexity, the problem is decomposed into two suboptimal problems. We first solved the content placement algorithm based on the cross-entropy (CE) method to minimize the backhaul load of the network. Then, a user association algorithm based on the CE method was employed to pursue larger network throughput of cell edge UEs. Simulation were conducted to validate the performance of the proposed cross-entropy based schemes in terms of network throughput and backhaul load. The simulation results show that the proposed cross-entropy based content placement scheme significantly outperform the conventional random and Most Popular Content placement schemes, with with 50% and 20% backhaul load decrease respectively. Furthermore, the proposed cross-entropy based user association scheme can achieve 30% and 23% throughput gain, compared with the conventional N-best, No-CoMP, and Threshold based user association schemes.

Keywords: ultra dense network; cross-entropy; proactive caching; user association; CoMP

---

1. Introduction

Inspired by the development of intelligent terminal such as smart phones, the demand for data traffic in mobile communication systems is exponentially growing. To cater for this demand, a 1000-fold improvement of capacity per area in the next generation of mobile communication system (5G) compared to 4G is required. An Ultra-Dense Network (UDN) is capable of significantly improving the capacity per area under the limited spectrum resource due to the high splitting-gain of densely located small cells and is widely considered as one of the most promising techniques in the coming 5G. It also benefits load balance between Base Stations (BSs) since Small Base Stations (SBSs) can offload data traffic of Macro Base Stations (MBSs). Nevertheless, due to the short distance between BSs, the intercell interference in UDNs is severe, therefore making the user experience unsatisfactory. Coordinated Multi-Point (CoMP) transmissions technique is widely studied in academia and the industry, which can leverage the cooperation of multiple BSs to enhance the signal to interference and noise ratio (SINR), to counteract intercell interference and to enhance network capacity in UDNs.

Despite remarkable performance gain in network capacity, congestion on backhaul links caused by CoMP risks the mobile communication systems. In order to cooperatively serve users, BSs need to fetch more files from the Core Network (CN) via backhaul links in a CoMP-employed system, which brings heavy load to backhaul links between BSs and the CN and probably results in congestion. One way to alleviate backhaul load is to cache popular files on BSs. When BSs cache files requested by users, it does not need to fetch files from the CN, so the backhaul load can be dramatically reduced. In [1], an architecture based on distributed caching of content in SBSs was presented. Works on proactive caching in UDNs concentrate on two major issues: Content placement and content distribution.

Content placement focuses on how to distribute popular and hotspot files to the BSs' caching unit whose capacity is limited. In [2], an optimal content placement strategy is proposed to maximize the hit rate. In [3], the problem of content placement is studied to maximize energy efficiency. In [4], an approximation algorithm is proposed to jointly optimize routing and caching policy to maximize the fraction of requested files cached locally. In [5], a distributed algorithm is proposed to investigate content placement and user association jointly. In [6], a content placement strategy is investigated based on reinforcement learning. In [7], a content placement strategy is proposed under cooperation schemes of maximum ratio transmission (MRT) and zero-forcing beamforming (ZFBF). In [8], a caching space allocation scheme is proposed to improve the hit rate based on the categories of contents and UEs.

Content distribution is the study of how to associate UEs and BSs to improve the hit rate. In [9], the user association problem is modeled as an one-to-many game problem, based on which algorithm is proposed to maximize the average download rate under a given content placement strategy. In [10], an user association algorithm under a given content distribution in a CoMP enabled network is proposed to minimize the backhaul load under a guaranteed rate requirements of UEs. In [11], the content caching and user association schemes are proposed on two different scales: The caching algorithm operates in a long time scale and the user association algorithm operates frequently. In [12], user association is investigated to tradeoff between load balancing and backhaul savings in UDN.

To the best of our knowledge, related works on the content placement caching and user association have been investigated separately in small-scale networks. We are thus motivated to jointly investigate the tasks of caching and user association in more realistic large-scale UDNs. The main contributions of this paper are summarized as follows.

- The problem of content placement and user association is investigated jointly in large-scale cache-enabled coordinated ultra dense networks. We formulate the problem as a constrained non-convex combinatorial programming problem to maximize network throughput of cell edge UEs under the consideration of the backhaul load;
- A two-step heuristic algorithm based on the cross-entropy (CE) method is proposed to solve the problem: A content placement strategy is first proposed based on cross entropy under the assumption of the conventional N-Best scheme; given the proposed content placement strategy, a user association algorithm is then proposed based on the cross-entropy method. Extensive simulations are conducted to evaluate the performance of the proposed approach. Simulations are conducted to validate the performance of the proposed cross-entropy based schemes in terms of network throughput and backhaul load. Simulation results show that the proposed caching and user association algorithms can reduce backhaul load and improve network throughput of cell edge UEs simultaneously.

The rest of this paper is organized as follows. The system model is established in Section 2 with some basic assumptions. In Section 3, we formulate the problem and propose the algorithms. In Section 4, simulation results are presented and the system performance is evaluated. Section 5 concludes this paper.

2. System Model

2.1. Network

In this paper, we consider a heterogeneous Ultra-Dense Network consisting of N_{MBS} Macro BSs (MBS) and N_{SBS} Small Base Stations (SBS). The Macro BSs are uniformly distributed to provide coverage and to support capacity. The small BSs are randomly distributed within the covering area, following a Poisson Point Process (PPP) with a density of λ_{SBS}. Let $B^{\Omega} = \{b|b = 1, 2, \cdots, |B^{\Omega}|\}$ denote the set of BSs consisting of both Macro BSs and Small BSs, where $|B^{\Omega}| = N_{\mathrm{MBS}} + N_{\mathrm{SBS}}$ is the total number of BSs.

UEs are randomly distributed following a PPP with density of λ_{U}. The set of UEs is denoted by $M^{\Omega} = \{m|m = 1, 2, \cdots, |M^{\Omega}|\}$. UEs located at the edges of cells usually suffer from severe intercell interference and low capacity. To reduce the interference and enhance peak data rates of cell edge users, joint transmission Coordinated Multi-Point (termed as JT CoMP) is considered in the network architecture, which allows multiple BSs in the neighborhood to cooperatively serve a specific UE simultaneously.

The association relationship between UEs, m, and BS, b, is denoted by a bit number $x_{m,b}(m \in M^{\Omega}, b \in B^{\Omega})$ defined as:

$$x_{m,b} = \begin{cases} 1 & \text{UE } m \text{ is associated with BS } b \\ 0 & \text{otherwise} \end{cases}. \tag{1}$$

Thus, the entire association result between BSs and UEs in the considering network can be presented by:

$$\mathbf{X} = \begin{bmatrix} x_{1,1} & x_{1,2} & \cdots & x_{1,|B^{\Omega}|} \\ x_{2,1} & x_{2,2} & \cdots & x_{2,|B^{\Omega}|} \\ \vdots & \vdots & \ddots & \vdots \\ x_{|M^{\Omega}|,1} & x_{|M^{\Omega}|,2} & \cdots & x_{|M^{\Omega}|,|B^{\Omega}|} \end{bmatrix}. \tag{2}$$

The SINR of a specific UE in a downlink transmission can be presented in terms of $x_{m,b}$ as follows,

$$\Gamma_m = \frac{\sum\limits_{b \in B^{\Omega}} x_{m,b} P_b g_{m,b}}{\sum\limits_{b' \in B^{\Omega}} (1 - x_{m,b'}) P_{b'} g_{m,b} + \sigma^2}, \tag{3}$$

where P_b is the transmit power of BS b, $g_{m,b}$ is the channel gain between BS b and UE m, and σ^2 is the variance of additive white Gaussian noise (AWGN). Equation (3) suggests that the more BSs are associated with UE m, the better service it can obtain. However, the necessary overhead to accomplish a JT CoMP transmission involving too much BSs is unacceptable. Thus, it is better to narrow the associating BSs of a UE into N BSs in close proximity.

As in 4G LTE and 5G NR (New Radio) wireless standards, resource block (RB) is considered the unit of time and spectrum resource for allocation in this paper. Assume that the bandwidth of each RB is W and the total number of RBs is N_{RB}. Each UE in the network can occupy a part of resource for transmission. Let β_m denote the proportion that the resource assigned to UE m out of all. Then the data rate of a downlink transmission to UE m can be given by:

$$R_m = W \lfloor \beta_m N_{\mathrm{RB}} \rfloor \log_2 (1 + \Gamma_m), \tag{4}$$

where $\lfloor x \rfloor$ is the minimum integer smaller than or equal to x; Γ_m is defined by Equation (3).

Let $B_m^\Omega \subseteq B^\Omega$ denote the set consisting of BSs associated with UE m, i.e., $B_m^\Omega = \{b|b \in B^\Omega, \text{ and } x_{m,b} = 1\}$. Similarly, let $M_b^\Omega \subseteq M^\Omega$ denote the set consisting of UEs associated with BS b, i.e., $M_b^\Omega = \{m|m \in M^\Omega, \text{ and } x_{m,b} = 1\}$. The resource that BS b can assign to UE m should be no more than $\frac{1}{|M_b^\Omega|}$, where $|M_b^\Omega|$ represents the total number of UEs associated with BS b. In the case where JT CoMP is employed, the resource that UE m can be obtained is restricted by the most heavy-loading BS among those associated with UE m. As a result, the proportion β_m can be given by:

$$\beta_m = \min \left\{ \frac{1}{|M_b^\Omega|}, b \in B_m^\Omega \right\}. \tag{5}$$

2.2. Caching

UDNs can benefit from caching popular files on BSs in terms of throughput, delay, and traffic load on backhaul links. It is obvious that the more files cached on BSs, the better performance a network can achieve. The cache-enabled heterogeneous UDN we considered in this paper is illustrated as Figure 1. For simplicity, we assume that the files requested by all UEs in the networks are restricted into the set of $F^\Omega = \{f|f = 1, 2, \cdots, |F^\Omega|\}$, and each file is in the same size of $F_{\max}$ bits. We assume the popularity of files follows Zipf distribution [13]. Let p_f denote the probability mass function of popularity random variable F. The probability that file f is requested can be given by:

$$p_f = \frac{\left(\sum_{f=1}^{|F^\Omega|} f^{-\gamma} \right)^{-1}}{f^\gamma}, \tag{6}$$

where γ is the Shape Factor (SF) indicating the correlation between requests of UEs [13]. It is seen that the larger the shape factor γ is, the smaller the probability mass function p_f would be, the lower probability file f is requested out of the caching set F^Ω.

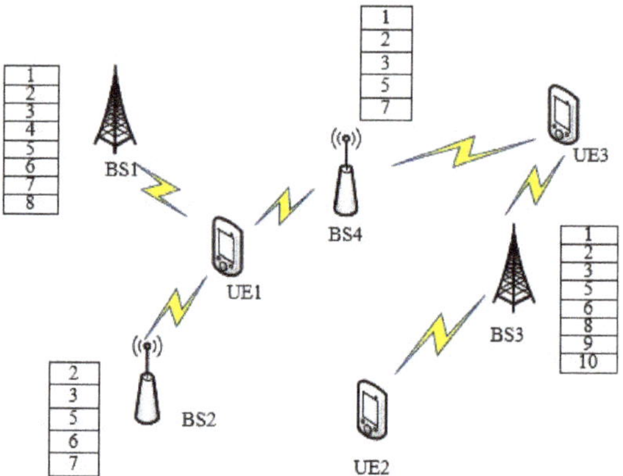

Figure 1. System model of caching-enabled Ultra-Dense Network (UDN) with joint transmission Coordinated Multi-Point (JT CoMP).

Define a caching index vector $y_b = [y_{b,1}, y_{b,2}, \cdots, y_{b,f}, \cdots y_{b,|F^\Omega|}]$, where $y_{b,f}$ indicates whether BS b caches file f in the caching set F^Ω or not. More specifically,

$$y_{b,f} = \begin{cases} 1 & \text{file } f \text{ is cached on BS } b \\ 0 & \text{otherwise} \end{cases}. \tag{7}$$

Then the File-BS caching matrix can be denoted by:

$$
\mathbf{Y} =
\begin{bmatrix}
y_{1,1} & y_{1,2} & \cdots & y_{|B^\Omega|,|F^\Omega|} \\
y_{2,1} & y_{2,2} & \cdots & y_{|B^\Omega|,|F^\Omega|} \\
\vdots & \vdots & \ddots & \vdots \\
y_{|B^\Omega|,1} & y_{|B^\Omega|,2} & \cdots & y_{|B^\Omega|,|F^\Omega|}
\end{bmatrix}.
\tag{8}
$$

As for UEs, we define a row vector $q_m = [q_{m,1}, q_{m,2}, \cdots, q_{m,f}, \cdots]$, where $q_{m,f}$ represents the request of UE m to file f, i.e.,

$$
q_{m,f} =
\begin{cases}
1 & \text{file } f \text{ is requested by UE } m \\
0 & \text{otherwise}
\end{cases}.
\tag{9}
$$

Suppose a UE can request one and only one file at each time, then we have $\sum_f q_{m,f} = 1, \forall m \in M^\Omega$. Then $q_m y_b^T$ represents whether the file requested by UE m is caching on BS b, where v^T represents the transpose of the vector v.

$$
q_m y_b^T =
\begin{cases}
1 & \text{file } f \text{ requested by UE } m \text{ is on BS } b \\
0 & \text{otherwise}
\end{cases}.
\tag{10}
$$

If UE m is associated with BS b (i.e., $x_{m,b} = 1$), we say that BS b misses a file if file f requested by UE m is not cached in it.

If BSs miss a file, they need to fetch the file from the Core Network (CN). We assume a centralized deployment of the considered UDN, where each BS is directly connected to the CN via backhaul links. The backhaul load is defined as the traffic carried by backhaul links between BSs and the CN [14]. In the case that BS b misses a file requested by UE m, BS b have to fetch the file from the CN through the backhaul link, which aggravates the backhaul load of BS b inevitably. In reality, the capacity of backhaul links is usually limited. Congestion occurs when the backhaul load of BS b exceeds the backhaul capacity $C_b^{\max}$.

Let U_{back} represent the increase of backhaul load due to fetching a file from the CN. The backhaul load caused by UE m can be given by:

$$
V_m = \sum_{b \in B_m^\Omega} (1 - q_m y_b^T) U_{\text{back}}.
\tag{11}
$$

It is obvious that the more files BSs cache (i.e., the less files BSs miss), the less heavier the backhaul load will be.

2.3. Delay

In addition to reducing backhaul load, caching also benefits from reducing the time delay of transmissions . In this paper, we consider the average time delay of UEs , which consists of two major parts: Wireless propagation delay and backhaul delay. Let d_1 denote the average wireless propagation delay of a UDN which is related to the size of a file $F_{\max}$ and the data rate of transmission.

$$
d_1 = \frac{1}{|M^\Omega|} \sum_{m \in M^\Omega} \frac{F_{\max}}{R_m}.
\tag{12}
$$

Backhaul delay, denoted by d_2, is related to whether the files are hit by the associating BSs of UEs. Due to joint transmission of CoMP, the backhaul delay for a specific UE m occurs when the requested file is not cached on all its associated BSs. That is, the minimum operation should be applied on backhaul delays due to file transmission between the associated BSs and the core network. More specifically, d_2 can be represented as follows,

$$d_2 = \frac{1}{|M^\Omega|} \sum_{m \in M^\Omega} \left(\min \left\{ \frac{F_{\max}}{U_{\text{back}}}, \sum_{b \in B_m^\Omega} (1 - \mathbf{q}_m \mathbf{y}_b^T) \frac{F_{\max}}{U_{\text{back}}} \right\} \right). \tag{13}$$

As a result, the total time delay of the network is:

$$D = d_1 + d_2. \tag{14}$$

3. Problem Formulation

3.1. Mathematical Formulation

In this paper, we aim to develop the optimal solution of content placement and user association that balances network throughput and the backhaul load simultaneously. Considering the fairness of UEs, the network throughput of cell edge UEs is used as the performance metric. Thus the objective function is a tradeoff between network throughput of cell edge UEs and backhaul load.

$$\max_{X,Y} \sum_{m \in M^\Omega} \log_{10}(R_m) - \lambda \frac{\sum_{m \in M^\Omega} V_m}{|M^\Omega|} \tag{15}$$

$$\text{s.t.} \quad \text{C1:} \quad 1 \leq |B_m^\Omega| \leq N, \forall m \in M^\Omega$$
$$\text{C2:} \quad \sum_m x_{m,b} \leq N_{\text{RB}}, \forall b \in B^\Omega, \forall m \in M^\Omega$$
$$\text{C3:} \quad \sum_m x_{m,b}(1 - \mathbf{q}_m \mathbf{y}_b^T) U_{\text{back}} \leq C_b^{\max}, \forall b \in B^\Omega$$

where the constraint C1 indicates that a specific UE m should be served by at least one BS, meanwhile the total number of BSs that cooperatively serve a specific UE should not be more than a given number N by considering the tradeoff between throughput gain and backhaul load due to the joint transmission of CoMP; C2 indicates that the total number of RBs allocated to UEs associating with a specific BS is limited to the maximum N_{RB}; C3 indicates that aggregate backhaul load of BS m should not be over the backhaul capacity $C_b^{\max}$.

$\lambda \geq 0$ in Equation (15) is a coefficient that influences the balance between network throughput and backhaul load. A larger λ suggests that we prefer improving network throughput than reducing backhaul load, and vice versa. Let $\sum_{m \in M^\Omega} \log_2(R_m^{(0)})$ and $\frac{\sum_{m \in M^\Omega} V_m^{(0)}}{|M^\Omega|}$ denote sum of logarithm of data rate and averaged load on backhaul links when $\lambda = 0$, respectively. We define λ with $\sum_{m \in M^\Omega} \log_2(R_m^{(0)})$ and $\frac{\sum_{m \in M^\Omega} V_m^{(0)}}{|M^\Omega|}$ as benchmarks, and then λ can be given by:

$$\lambda = \frac{\sum_{m \in M^\Omega} \log_2(R_m^{(0)}) |M^\Omega| \mu}{\sum_{m \in M^\Omega} V_m^{(0)}}. \tag{16}$$

where $\mu \in [0, 1]$ is a weight factor used for adjusting λ.

The problem in Equation (15) is a constrained non-convex combinatorial optimization problem, which requires extraordinary high complexity to trace its optimal solution. To obtain a practical solution, we decompose the problem into two steps based on the cross-entropy (CE) method. In this paper, the CE method is chosen because it is a simple, efficient, and general method for solving a great variety of estimation and optimization problems, especially NP-hard combinatorial deterministic and stochastic problems [15]. First, we minimize the backhaul load of the system under the assumption of the conventional N-Best user association strategy (By N-Best user association strategy, a CoMP UE will associate with $N_{\max}$ BSs which have N best SINRs [16]) and propose a content placement algorithm based on cross entropy, which is termed as the CPCE algorithm. Subsequently, under the given content placement strategy, we propose an user association algorithm based on cross entropy, which is referred to as UACE in the rest of this paper.

3.2. Cross-Entropy Method

The CE method was originally used in the context of rare event simulation [17] and has been extended as a Monte Carlo method for importance sampling and optimization [17,18].

The principle behind the CE method is to get as close as possible to the optimal importance sampling distribution by using the Kullback—Leibler (KL) distance as a measure of closeness. By repeatedly updating the Probability Density Function (PDF) of generated samples, the PDF of the samples can finally converge with the obtained optimal strategy solution [19]. Another method with a similar idea is logarithmic loss distortion measure [20,21], where logarithmic loss is also known as cross-entropy loss. The logarithmic loss distortion measure has been recently used in the study of Deep Neural Networks (DNNs) to approach an accurate classifier by minimizing the logarithmic loss. It also performs well on tradeoff between complexity and relevance in representation learning [22].

The main steps of the CE method can be depicted as follows:

STEP 1: (Encode Strategy Space). Consider a UDN constituted by B^{Ω} BSs, where each BS b is a decision-making entity. Suppose that each BS b can make a decision out of N_b possible strategies, then the strategy set at BS b can be expressed as $\mathbf{S}_b = [S_b^1, S_b^2, \cdots, S_b^{N_b}]$. For a specific decision-making entity BS b, S_b^i is one strategy that belongs to $\mathbf{S}_b$. The strategy set of B^{Ω} BSs entities in a UDN can be represented as $\mathcal{S}_{B^{\Omega}} = [\mathbf{S}_1, \mathbf{S}_2, \cdots, \mathbf{S}_{|B^{\Omega}|}]$, which is termed as the strategy space of the CE method. Then the samples of the strategy space of the CE method correspond to the strategies at all BSs in a UDN in current iteration.

Let $\mathbf{P_b} = [P_{b,1}, P_{b,2}, \cdots, P_{b,N_b}]$ denote the probability distribution of sample strategies at BS b. Let $P_{b,i}$ denote as the probability of strategy i at a specific BS b. First, $P_{b,i}$ is initialized to be equal as follows,

$$P_{b,i} = \frac{1}{N_b}, \text{ and } \sum_{i=1}^{N_b} P_{b,i} = 1, (\forall i = 1, \cdots, N_b). \tag{17}$$

STEP 2: (Generate Samples According to the Probability). In the second step of the CE method, sufficient strategy samples should be generated according to the given probability distribution. Denote the zth generated sample by $\mathbb{A}(z) = \left[\mathbf{A}_1^T(z), \mathbf{A}_2^T(z), \cdots, \mathbf{A}_b^T(z), \cdots, \mathbf{A}_{|B^{\Omega}|}^T(z)\right]$, where $\mathbf{A}_b(z)$ is the subsample set generated by decision-making entity element b. More specifically, the subsample $\mathbf{A}_b(z)$ can be represented as:

$$\mathbf{A}_b(z) = \left[\alpha_{b,1}(z), \alpha_{b,2}(z), \cdots, \alpha_{b,i}(z), \cdots, \alpha_{b,N_b}(z)\right], \tag{18}$$

where $\alpha_{b,i}(z)$ is a binary number. For each $\mathbf{A}_b(z)$, only one $\alpha_{b,i}(z)$ is "1" and others are "0" (i.e., $\sum_{i=1}^{N_b} \alpha_{b,i} = 1$), indicating that only one strategy out of all can be selected at each decision epoch. The probability of subsample $\mathbf{A}_b$ is $P_{b,i} \in \mathbf{P_b}$ with $\alpha_{b,i} = 1$ and $\alpha_{b,j} = 0 (j \neq i)$.

STEP 3: (Performance Evaluation). Fitness values of strategy samples can be calculated according to the result of the strategy in the current iteration. Let $F(z)$ denote the fitness value of strategy sample z, which can be expressed as:

$$F(z) = -\sum_{b}\sum_{m} x_{mb}(1 - \mathbf{q}_m \mathbf{y}_b^T) U_{back}. \tag{19}$$

Rearrange $F(z)$ in descending order as $F(1) \geq F(2) \cdots F(Z)$, where Z is the maximum number of strategy samples. Then calculate the ρ-quantile of the strategy samples in current iteration F_ρ and weed out the unexpected samples. The samples with fitness value $F(i) \geq F_\rho$ are selected for probability updating in the next iteration.

STEP 4: (Probability Updating). According to samples selected in the performance evaluation step, the probabilities of each strategy can be updated as follows,

$$P_{b,i}^{\text{update}} = \frac{\sum\limits_{z=1}^{Z} I_{F(z) \geq F_\rho} \alpha_{b,i}(z)}{\sum\limits_{z=1}^{Z} I_{F(z) \geq F_\rho}}, \tag{20}$$

where I is defined as,

$$I_{x \geq y} = \begin{cases} 1 & \text{if } x \geq y \\ 0 & \text{others} \end{cases}. \tag{21}$$

Go back to STEP 2, regenerate the samples based on the updated probability distribution and repeat STEP 2 to STEP 4.

STEP 5: (Convergence Conditions). The algorithm will come to an end when the fitness value reaches convergence or the algorithm reaches the maximum iteration number set in advance. The cross-entropy method is a global random search procedure, and asymptotical convergence can be achieved to find the optimal solution with probability arbitrarily close to 1 [23,24].

3.3. Content Placement Algorithm Based on the Cross-Entropy Method (CPCE)

Before joining in a network, a specific UE will measure Channel State Information (CSI) and choose the candidate BSs with the biggest reference signal received power (RSRP). To investigate the content placement strategies of BSs, we assume the conventional N-Best scheme for user association in the first place. BSs in the network are modeled as decision-making entities in the CE method, and feasible content placement candidates are strategies of each entity. The CPCE algorithm can be depicted as Algorithm 1.

The optimized content placement strategy $\mathbf{Y}$ can be given by Algorithm 1, under the assumption of the N-best user association scheme. User association results can be further optimized with the obtained $\mathbf{Y}$.

Algorithm 1 Content Placement based on CE method (CPCE)

1: User association under N-Best strategy.

2: Generate content placement request samples of UEs q_m based on Zipf distribution. Map BSs $\longleftrightarrow$ Decision-making entities in CE method.

3: Map the content placement strategy set $\longleftrightarrow$ Strategies in CE method.

4: Map the sum backhaul load $\longleftrightarrow$ Fitness value in CE method.

5: Execute **CPCE**.

6: Map the obtained solution into the best content placement strategies of BSs and output $\mathbf{Y}$.

3.4. User Association Algorithm Based on the Cross-Entropy Method (UACE)

By Algorithm 1, we obtain the optimal content placement strategy of each BS under the N-Best user association scheme. However, the N-Best user association scheme does not take into account load balancing and interference management. Under the obtained content placement result, we can further optimize the user association algorithm.

The user association problem is a constrained non-convex integer programming problem, which can also be solved by the CE method [15,19]. Considering that the maximum number of BSs associated with a UE is N, the amount of association strategies of a UE will be no more than $2^N - 1$. Similarly, the steps of the proposed UACE method is as follows.

STEP 1: (Encode Strategy Space). In UACE, the decision-making entities are UEs in the network. Suppose that each UE m can make a decision out of N_m possible strategies, then the strategy set at UE m can be expressed as $\mathbf{S}_m = [S_m^1, S_m^2, \cdots, S_m^{N_m}]$. Let $\mathbf{P_m} = [P_{b,1}, P_{b,2}, \cdots, P_{b,N_m}]$ denote the probability distribution of sample strategies at UE m, where $P_{m,i}$ denote the probability of strategy i at a specific UE m. $P_{m,i}$ can be initialized to be equal as follows,

$$P_{m,i} = \frac{1}{N_m}, \text{ and } \sum_{i=1}^{N_m} P_{m,i} = 1, (\forall i = 1, \cdots, N_m). \tag{22}$$

STEP 2: (Generate Samples According to the Probability). Samples are generated in this step in a similar way described in STEP 2 of Section 3.2.

STEP 3: (Encode Strategy Space). The fitness value of strategy samples in UACE is

$$\sum_{m \in M^\Omega} \log_{10}(R_m) - \lambda \frac{\sum_{m \in M^\Omega} V_m}{|M^\Omega|} \tag{23}$$

STEP 4 (Probability Updating) and **STEP 5 (Convergence Conditions)** of UACE are also similar to STEP 4 and STEP 5 in Section 3.2. Then the proposed UACE algorithm can be depicted as in Algorithm 2.

By applying Algorithms 1 and 2, we can obtain suboptimal solutions to problem (15). With the obtained results, optimized performance of the considered UDN in terms of both throughput and backhaul load is achieved.

Algorithm 2 User Association based on Cross-Entropy Algorithm (UACE)

1: Execute the proposed CPCE Algorithm 1 under popular contents' statistics.

2: Map UEs $\longleftrightarrow$ Decision-making entities in CE method.

3: Map association strategy set for a specific UE $\longleftrightarrow$ Strategies in CE method.

4: Map network throughput of all the cell edge UEs $\longleftrightarrow$ Fitness value in CE method.

5: Execute **UACE**.

6: Map the obtain solution into optimal user association solution and output **X**.

3.5. Complexity Analysis of the Cross-Entropy Method

From the description in the previous subsection, the computational complexity of the proposed CE algorithm is made up of 5 parts.

(1) Initialize the probability distribution of sample strategies. According to the size of encode strategy space and Equation (17), the computational complexity is $O(|B^\Omega|)$;

(2) Generate samples according to the probability. According to Equations (18) and (19), there are Z samples at most, and the size of each sample is $N_b \times |B^\Omega|$. Hence, the computational complexity is $O(Z \times N_b \times |B^\Omega|)$;

(3) Performance Evaluation. According to Equation (20), we should calculate the fitness value of each strategy sample according to Equation (20), and the computational complexity is $O(Z \times |M^\Omega| \times |B^\Omega|)$;

(4) Probability Updating. According to Equation (21) and the size of the probability distribution of the sample strategy, the computation complexity is $O(Z \times N_b \times |B^\Omega|)$;

(5) Iteration. The proposed algorithm will come to an end when the maximal iteration number V is reached. Hence, the computation complexity is V times the sum of the computation complexity from Equations (1)–(4), i.e., $O(V \times Z \times N_b \times |M^{\Omega}| \times |B^{\Omega}|)$.

According to the analysis above, the total computation complexity of the proposed algorithm based on the CE method is computable in polynomial time.

4. Simulation and Analysis

Extensive simulations are conducted to evaluate the performance of the proposed content placement and user association algorithm. In the simulation, 7 MBSs are uniformly distributed in the considered area, while SBSs and UEs randomly drops, the maximum number of BSs that a UE can be associated is 3 [25]. Major parameter settings are listed in Table 1.

Table 1. Parameters setting.

Parameters	Value
Plane of Topology	$1.5 \times 1.5 \, \text{km}^2$
Number of MBSs	7
Number of SBSs	40
Number of UEs	50–200
Channel Model	WINNER
Transmit Power of MBS	40 W
Transmit Power of SBS	2 W
Number of Available RB	100
Total Number of Files	20
Backhaul Capacity of MBS	1 Gbps
Backhaul Capacity of SBS	100 Mbps
Maximal Number of Caching Files on each BS	10
U_{back}	10 Mbps
N	3

4.1. System Performance under Different Content Placement Schemes

For content placement strategies, we compared the performance of the proposed CECP scheme to that of random scheme and Most Popular Content (MPC) scheme under different SFs (γ) of popularity of files. Network performance in terms of backhaul load and normalized time delay is shown in Figures 2 and 3 respectively.

When $\gamma = 0.2$, the backhaul load of the proposed CPCE scheme is two-times less than that of the MPC scheme and the random scheme. As the shape factor increases, the backhaul load and time delay decrease sharply for both the proposed CECP scheme and the MPC scheme. This is because as the shape factor increases, popular files tend to be prone to fewer files, thus more gain can be obtained by selecting proper caching strategies. When γ becomes larger and larger ($\gamma > 0.2$ in the simulation), the backhaul load and time delay of the CPCE scheme are comparable to that of the MPC, being about more than 13 times smaller than the Random scheme as shown in Figures 2 and 3.

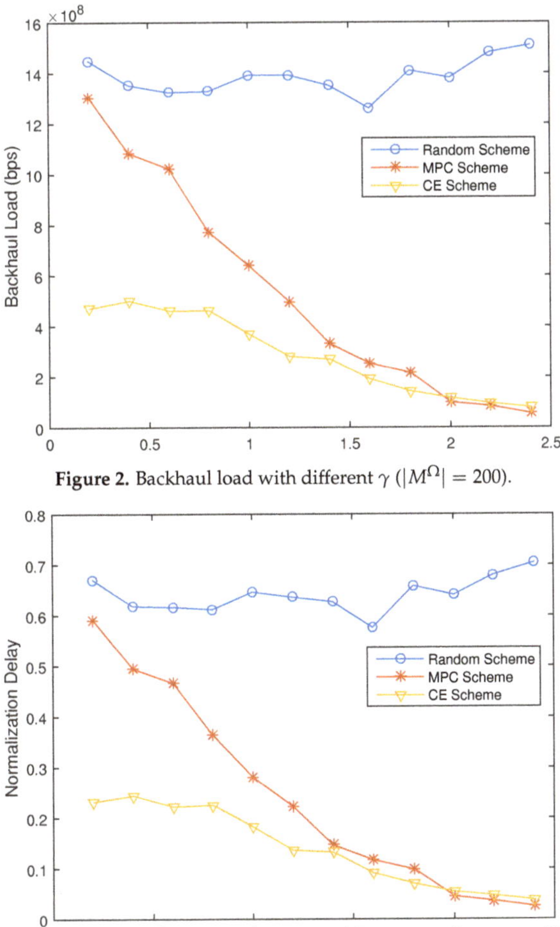

Figure 2. Backhaul load with different γ ($|M^{\Omega}| = 200$).

Figure 3. Time delay under different γ ($|M^{\Omega}| = 200$).

4.2. System Performance of CPCE with Different Numbers of UEs

A simulation was also conducted to evaluate the performance of the proposed CPCE scheme under different network scales with $\gamma = 1$. As expected, the proposed CPCE algorithm outperforms the MPC and random caching scheme in terms of the backhaul load under different scales of the network, as shown in Figure 4. When UEs are sparsely distributed in the network, the backhaul load of the CPCE scheme is almost ignorable. Even if the number of UEs increases up to 200, the backhaul load of CPCE is still a great deal lower than that of the random scheme and MPC. Figure 5 shows the normalized time delay of each content placement scheme under different network scales. It is observed that CPCE can achieve the lowest time delay compared to the MPC and random caching scheme.

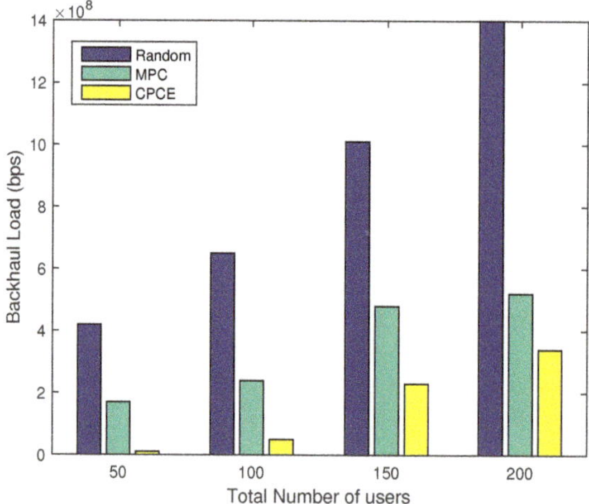

Figure 4. Backhaul load under different numbers of UEs ($\gamma = 1$).

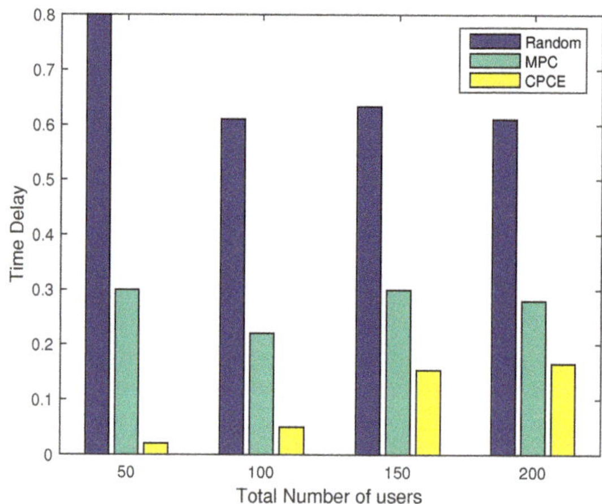

Figure 5. Normalized time delay under different numbers of UEs ($\gamma = 1$).

4.3. System Performance of CPCE under Different Storage Capacity of BSs

Figures 6 and 7 show the impact on network performance of the BSs' different storage capacity. It is clear that the more files that BSs can cache, the more possibility that BSs hit required files. We assume the total number of files is 20. When the storage capacity of BSs is half of total files, both backhaul load and time delay of the proposed CPCE is as a third as that of MPC. Even when storage capacity is very limited (for example, only 1 file can be cached), the backhaul load of CPCE is acceptable, as shown in Figure 6. Meanwhile, as shown in Figure 7, time delay decreases as the storage capacity increases, and the proposed CPCE outperforms the other two.

The performance of Random and MPC comparing with the proposed CPCE in terms of time delay and backhaul load is listed in Table 2.

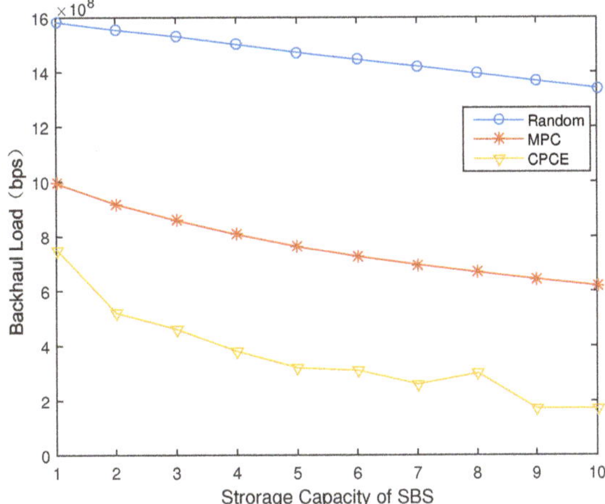

Figure 6. Backhaul load under different storage capacity of BSs ($\gamma = 1$).

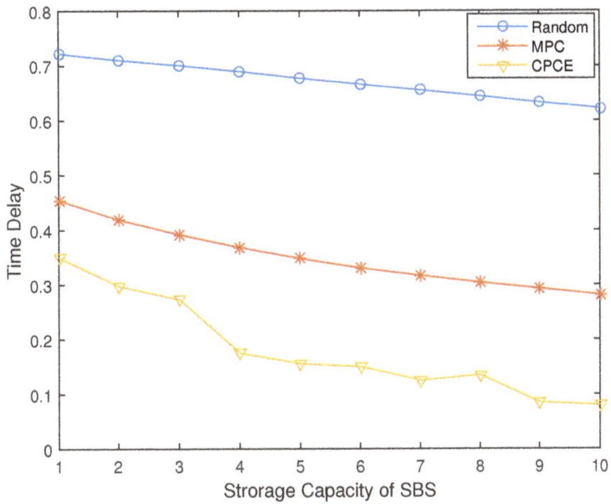

Figure 7. Normalized time delay under different storage capacity of BSs ($\gamma = 1$).

Table 2. Comparison of algorithms in terms of delay and backhaul load.

	Time Delay	**Backhaul Load**
Random	high	high
MPC	low to high	low to high
CPCE	low	low

4.4. System Performance of CPCE-UACE under Different Weight Factor

The performance of the entire CPCE-UACE algorithm is evaluated in the simulation and compared with N-best, No-CoMP, and Threshold user association schemes. No-CoMP scheme means each UE, no matter where it is located, can be associated only to the BS with the best RSRP. Threshold scheme allows a specific UE to be associated with multiple BSs whose RSRP is better than a given threshold [16]. We also assess backhaul load and network throughput of CPCE-UACE algorithm under a different weight factor μ (increases from 0 to 1 by step of 0.5). Generally speaking, the more BSs each UE in the network can be associated with, the better throughput can be achieved, while heavier the backhaul load will be. Fortunately, the proposed CPCE-UACE algorithm can balance backhaul load and network throughput by carefully selecting a weight factor μ. As shown in Figures 8 and 9, network throughput and backhaul load of CPCE-UACE decreases as the weight factor μ grows. When μ is very small (less than 0.05), throughput of CPCE-UACE is significantly better than the others, but the load of it is also outstandingly heavy. On the other hand, when μ is as large as 0.5, both throughput and the backhaul load of CPCE-UACE is lowest in the four schemes considered in the simulation. As a result, we can narrow the range of an optimal μ that perfectly balances network performance in terms of the two aspects into $[0.05, 0.5]$. We consider $\mu = 0.1$ as the almost optimal weight factor due to relatively high throughput, as well as the low backhaul load, of CPCE-UACE as shown in Figures 8 and 9.

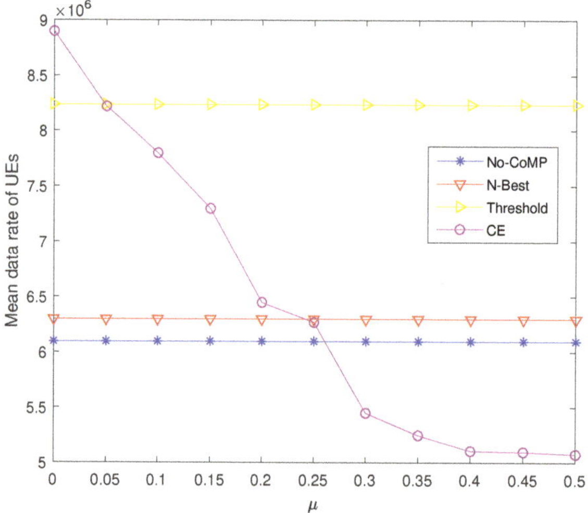

Figure 8. Network throughput under different μ.

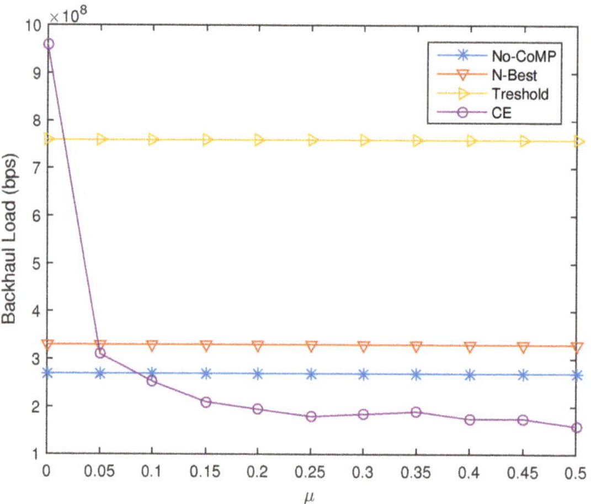

Figure 9. Backhaul load under different μ.

4.5. System Performance of CPCE-UACE under Different Numbers of UEs

In this subsection, we evaluated the proposed CPCE-UACE algorithm jointly in terms of throughput and backhaul load under different network scales with $\mu = 0.1$. The number of UEs in the network is set to be from 50 to 200 with an interval 50.

As shown in Figure 10, the average data rate of each UE decreases as the number of UEs in the network grows. It is obvious that the proposed CPCE-UACE algorithm can always achieve an outstanding performance compared to the No-CoMP and N-Best scheme, and 40% performance gain is obtained if $|M^{\Omega}| \leq 150$. Despite the threshold scheme being comparable to the CPCE-UACE algorithm in terms of the average data rate, the threshold scheme has the heaviest backhaul load as shown in Figure 11. It is observed that the proposed CPCE-UACE algorithm has the better performance of the backhaul load compared to the N-Best scheme and threshold scheme, as shown in Figure 11. Furthermore, the proposed CPCE-UACE algorithm is comparable with the No-CoMP scheme under a small number of UEs ($|M_{\Omega}| \leq 50$) and outperforms the other schemes in large scale networks ($|M_{\Omega}| \geq 100$).

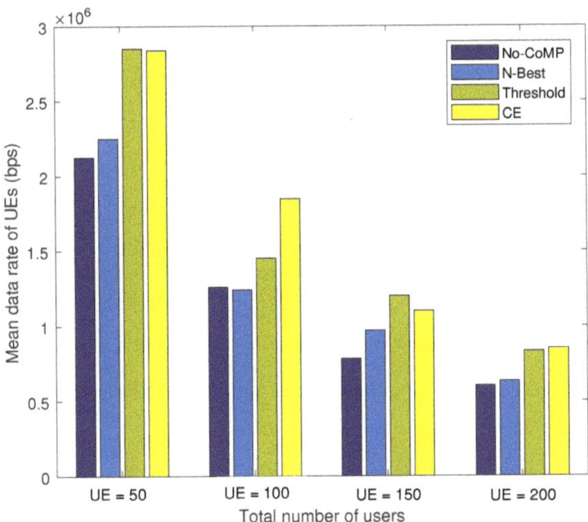

Figure 10. Network throughput under different numbers of UEs.

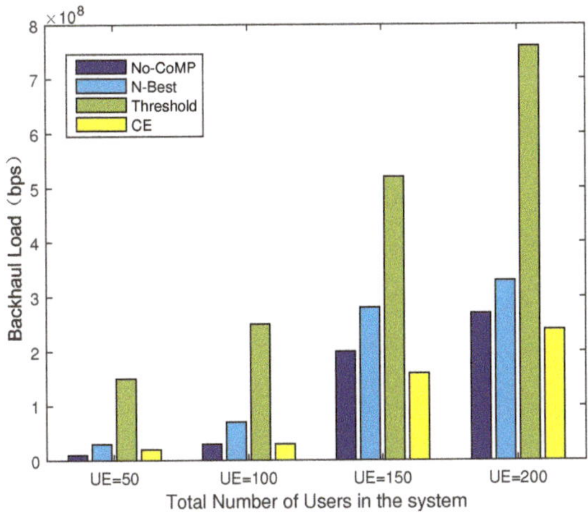

Figure 11. Backhaul load under different numbers of UEs.

The performance of No-CoMP, N-best, and Threshold compared with the proposed CPCE-UACE in terms of data rate and backhaul load is listed in Table 3.

Table 3. Comparison of algorithms in terms of data rate and backhaul load.

	Data Rate	Backhaul Load
No-CoMP	low	low to medium
N-best	low	low to medium
Threshold	medium to high	high
CPCE-UACE	high	low

5. Conclusions

This paper considered a problem involving content placement and user association in UDNs where proactive caching and CoMP are enabled. To alleviate the backhaul load and improve network performance, the CPCE-UACE algorithm was proposed to solve the problem. Simulation results demonstrated that the proposed algorithm was capable of decreasing the necessary backhaul traffic and improving network throughput simultaneously. Simulation results showed that the proposed cross-entropy based content placement scheme significantly outperformed the conventional random and MPC placement schemes, with a 50% and 20% backhaul load decrease respectively. Furthermore, the proposed cross-entropy based user association scheme could achieve 30% and 23% throughput gain, compared with the conventional N-best, No-CoMP, and Threshold based user association schemes.

Author Contributions: Conceptualization, Y.W.; Methodology, J.Y.; Project administration, S.G.; Supervision, Q.Z.; Writing—original draft, S.C.; Writing-review & editing, J.Y. and Y.Z.

Funding: This research received no external funding.

Acknowledgments: This work was supported in part by the National Natural Sciences Foundation of China (NSFC) under Grant 61701136; Shenzhen Basic Research Program under Grant JCYJ20170811154233370; and Shenzhen Science and Technology Projection under Grant JCYJ2016060815123996.

References

1. Golrezaei, N.; Molisch, A.F.; Dimakis, A.G.; Caire, G. Femtocaching and device-to-device collaboration: A new architecture for wireless video distribution. *IEEE Commun. Mag.* **2013**, *51*, 142–149. [CrossRef]
2. Golrezaei, N.; Shanmugam, K.; Dimakis, A.G.; Molisch, A.F.; Caire, G. FemtoCaching: Wireless video content delivery through distributed caching helpers. In Proceedings of the IEEE International Conference on Computer Communications (INFOCOM), Orlando, FL, USA, 25–30 March 2012; pp. 1107–1115.
3. Gabry, F.; Bioglio, V.; Land, I. On energy-efficient edge caching in heterogeneous networks. *IEEE J. Sel. Areas Commun.* **2016**, *34*, 3288–3298. [CrossRef]
4. Poularakis, K.; Iosifidis, G.; Tassiulas, L. Approximation Algorithms for Mobile Data Caching in Small Cell Networks. *IEEE Trans. Commun.* **2014**, *62*, 3665–3677. [CrossRef]
5. Wang, Y.; Tao, X.; Zhang, X.; Mao, G. Joint Caching Placement and User Association for Minimizing User Download Delay. *IEEE Access* **2016**, *4*, 8625–8633. [CrossRef]
6. ElBamby, M.S.; Bennis, M.; Saad, W.; Latva-aho, M. Content-aware user clustering and caching in wireless small cell networks. In Proceedings of the International Symposium on Wireless Communications Systems (ISWCS), Barcelona, Spain, 26–29 August 2014; pp. 945–949.
7. Ao, W.C.; Psounis, K. Fast Content Delivery via Distributed Caching and Small Cell Cooperation. *IEEE Trans. Mob. Comput.* **2018**, *17*, 1048–1061. [CrossRef]
8. Huo, R.; Xie, R.; Zhang, H.; Huang, T.; Liu, Y. What to cache: Differentiated caching resource allocation and management in information-centric networking. *China Commun.* **2016**, *13*, 261–276. [CrossRef]
9. Pantisano, F.; Bennis, M.; Saad, W.; Debbah, M. Cache-aware user association in backhaul-constrained small cell networks. In Proceedings of the International Symposium on Modeling and Optimization in Mobile, Ad Hoc, and Wireless Networks (WiOpt), Hammamet, Tunisia, 12–16 May 2014; pp. 37–42.
10. Yu, Y.; Tsai, W.; Pang, A. Backhaul Traffic Minimization under Cache-Enabled CoMP Transmissions over 5G Cellular Systems. In Proceedings of the IEEE Global Communications Conference (GLOBECOM), Washington, DC, USA, 4–8 December 2016; pp. 1–7.

11. Kwak, J.; Le, L.B.; Wang, X. Two Time-Scale Content Caching and User Association in 5G Heterogeneous Networks. In Proceedings of the IEEE Global Communications Conference (GLOBECOM), Singapore, 4–8 December 2017; pp. 1–6.

12. Dai, B.; Yu, W. Joint user association and content placement for Cache-enabled wireless access networks. In Proceedings of the IEEE International Conference on Acoustics, Speech and Signal Processing (ICASSP), Shanghai, China, 20–25 March 2016; pp. 3521–3525.

13. Breslau, L.; Cao, P.; Fan, L.; Phillips, G.; Shenker, S. Web caching and Zipf-like distributions: Evidence and implications. In Proceedings of the IEEE INFOCOM '99, New York, NY, USA, 21–25 March 1999; pp. 126–134.

14. Lakshmana, T.R.; Li, J.; Botella, C.; Papadogiannis, A.; Svensson, T. Scheduling for backhaul load reduction in CoMP. In Proceedings of the IEEE Wireless Communications and Networking Conference (WCNC), Shanghai, China, 7–10 April 2013.

15. Rubinstein, R.Y.; Kroese, D.P. *The Cross-Entropy Method: A Unified Approach to Combinatorial Optimization, Monte-Carlo Simulation, and Machine Learning*; Springer: New York, NY, USA, 2014; ISBN 978-0-387-21240-1.

16. Chen, S.; Zhao, T.; Chen, H.; Lu, Z.; Meng, W. Performance Analysis of Downlink Coordinated Multipoint Joint Transmission in Ultra-Dense Networks. *IEEE Netw.* **2017**, *31*, 106–114. [CrossRef]

17. Rubinstein, R.Y. Optimization of computer simulation models with rare events. *Eur. J. Oper. Res.* **1997**, *99*, 89–112. [CrossRef]

18. Rubinstein, R.Y. The cross-entropy method for combinatorial and continuous optimization. *Methodol. Comput. Appl. Probab.* **1999**, *1*, 127–190. [CrossRef]

19. De Boer, P.-T.; KroeseShie, D.P.; Rubinstein, M.R.Y. A Tutorial on the cross-entropy Method. *Ann. Oper. Res.* **2005**, *134*, 19–67. [CrossRef]

20. Ugur, Y.; Estella Aguerri, I.; Zaidi, A. Rate Distortion Region of the Vector CEO Problem under Logarithmic Loss. In Proceedings of the IEEE Information Theory Workshop (ITW 2018), Guangzhou, China, 25–29 November 2018.

21. Estella Aguerri, I.; Zaidi, A. Distributed Information Bottleneck Method for Discrete and Gaussian Sources. In Proceedings of the IEEE Int. Zurich Seminar on Information and Communications(IZS 2018), Zürich, Switzerland, 21–23 February 2018.

22. Estella Aguerri, I.; Zaidi, A. Distributed Variational Representation Learning. *arXiv* **2018**, arXiv:1807.04193.

23. Margolin, L. On the Convergence of the cross-entropy Method. *Ann. Oper. Res.* **2005**, *134*, 201–214. [CrossRef]

24. Costa, A.; Owen, J.; Kroese, D.P. Convergence Properties of the cross-entropy Method for Discrete Optimization. *Oper. Res. Lett.* **2007**, *35*, 573–580. [CrossRef]

25. Liu, L.; Garcia, V.; Tian, L.; Pan, Z.; Shi, J. Joint clustering and inter-cell resource allocation for CoMP in ultra dense cellular networks. In Proceedings of the IEEE International Conference on Communications (ICC), London, UK, 8–12 June 2015; pp. 2560–2564.

Article

Symmetry, Outer Bounds, and Code Constructions: A Computer-Aided Investigation on the Fundamental Limits of Caching

Chao Tian

Department of Electrical and Computer Engineering, Texas A&M University, College Station, TX 77843, USA; chao.tian@tamu.edu

Received: 26 June 2018; Accepted: 11 August 2018; Published: 13 August 2018

Abstract: We illustrate how computer-aided methods can be used to investigate the fundamental limits of the caching systems, which are significantly different from the conventional analytical approach usually seen in the information theory literature. The linear programming (LP) outer bound of the entropy space serves as the starting point of this approach; however, our effort goes significantly beyond using it to prove information inequalities. We first identify and formalize the symmetry structure in the problem, which enables us to show the existence of optimal symmetric solutions. A symmetry-reduced linear program is then used to identify the boundary of the memory-transmission-rate tradeoff for several small cases, for which we obtain a set of tight outer bounds. General hypotheses on the optimal tradeoff region are formed from these computed data, which are then analytically proven. This leads to a complete characterization of the optimal tradeoff for systems with only two users, and certain partial characterization for systems with only two files. Next, we show that by carefully analyzing the joint entropy structure of the outer bounds for certain cases, a novel code construction can be reverse-engineered, which eventually leads to a general class of codes. Finally, we show that outer bounds can be computed through strategically relaxing the LP in different ways, which can be used to explore the problem computationally. This allows us firstly to deduce generic characteristic of the converse proof, and secondly to compute outer bounds for larger problem cases, despite the seemingly impossible computation scale.

Keywords: computer-aided analysis; information theory

---

1. Introduction

We illustrate how computer-aided methods can be used to investigate the fundamental limits of the caching systems, which is in clear contrast to the conventional analytical approach usually seen in the information theory literature. The theoretical foundation of this approach can be traced back to the linear programming (LP) outer bound of the entropy space [1]. The computer-aided approach has been previously applied in [2–5] on distributed data storage systems to derive various outer bounds, which in many cases are tight. In this work, we first show that the same general methodology can be tailored to the caching problem effectively to produce outer bounds in several cases, but more importantly, we show that data obtained through computation can be used in several different manners to deduce meaningful structural understanding of the fundamental limits and optimal code constructions.

The computer-aided investigation and exploration methods we propose are quite general; however, we tackle the caching problem in this work. Caching systems have attracted much research attention recently. In a nutshell, caching is a data management technique that can alleviate the communication burden during peak traffic time or data demand time, by prefetching and prestoring certain useful content at the users' local caches. Maddah-Ali and Niesen [6] recently considered the problem in an information theoretical framework, where the fundamental question is the optimal tradeoff between local cache memory capacity and the content delivery transmission rate. It was shown in [6] that coding can be very beneficial in this setting, while uncoded solutions suffer a significant loss. Subsequent works extended it to decentralized caching placements [7], caching with nonuniform demands [8], online caching placements [9], hierarchical caching [10], caching with random demands [11], among other things. There have been significant research activities recently [12–21] in both refining the outer bounds and finding stronger codes for caching. Despite these efforts, the fundamental tradeoff had not been fully characterized except for the case with only two users and two files [6] before our work. This is partly due to the fact that the main focus of the initial investigations [6–9] was on systems operating in the regime where the number of files and the number of users are both large, for which the coded solutions can provide the largest gain over the uncoded counterpart. However, in many applications, the number of simultaneous data requests can be small, or the collection of users or files need to be divided into subgroups in order to account for various service and request inhomogeneities; see, e.g., [8]. More importantly, precise and conclusive results on such cases with small numbers of users or files can provide significant insights into more general cases, as we shall show in this work.

In order to utilize the computational tool in this setting, the symmetry structure in the problem needs be understood and used to reduce the problem to a manageable scale. The symmetry-reduced LP is then used to identify the boundary of the memory-transmission-rate tradeoff for several cases. General hypotheses on the optimal tradeoff region are formed from these data, which are then analytically proven. This leads to a complete characterization of the optimal tradeoff for systems with two users, and certain partial characterization for systems with two files. Next, we show that by carefully analyzing the joint entropy structure of the outer bounds, a novel code construction can be reverse-engineered, which eventually leads to a general class of codes. Moreover, data can also be used to show that a certain tradeoff pair is not achievable using linear codes. Finally, we show that outer bounds can be computed through strategically relaxing the LP in different ways, which can be used to explore the problem computationally. This allows us firstly to deduce generic characteristic of the converse proof, and secondly to compute outer bounds for larger problem cases, despite the seemingly impossible computation scale.

Although some of the tightest bounds and the most conclusive results on the optimal memory-transmission-rate tradeoff in caching systems are presented in this work, our main focus is in fact to present the generic computer-aided methods that can be used to facilitate information theoretic investigations in a practically-important research problem setting. For this purpose, we will provide the necessary details on the development and the rationale of the proposed techniques in a semi-tutorial (and thus less concise) manner. The most important contribution of this work is three methods for the investigation of fundamental limits of information systems: (1) computational and data-driven converse hypothesis, (2) reverse-engineering optimal codes, and (3) computer-aided exploration. We believe that these methods are sufficiently general, such that they can be applied to other coding and communication problems, particularly those related to data storage and management.

The rest of the paper is organized as follows. In Section 2, existing results on the caching problem and some background information on the entropy LP framework are reviewed. The symmetry structure of the caching problem is explored in detail in Section 3. In Section 4, we show how the data obtained through computation can be used to form hypotheses, and then analytically prove them. In Section 5, we show that the computed data can also be used to facilitate reverse-engineering new codes, and also to prove that a certain memory-transmission-rate pair is not achievable using linear codes. In Section 6, we provide a method to explore the structure of the outer bounds computationally, to obtain insights into the problem and derive outer bounds for large problem cases. A few concluding remarks are given in Section 7, and technical proofs and some computer-produced proof tables are relegated to the Appendices A–I .

2. Preliminaries

2.1. The Caching System Model

There are a total of N mutually independent files of equal size and K users in the system. The overall system operates in two phases: in the placement phase, each user stores in his/her local cache some content from these files; in the delivery phase, each user will request one file, and the central server transmits (multicasts) certain common content to all the users to accommodate their requests. Each user has a local cache memory of capacity M, and the contents stored in the placement phase are determined without knowing a priori the precise requests in the delivery phase. The system should minimize the amount of multicast information, which has rate R for all possible combinations of user requests, under the memory cache constraint M, both of which are measured as multiples of the file size F. The primary interest of this work is the optimal tradeoff between M and R. In the rest of the paper, we shall refer to a specific combination of the file requests of all users together as a demand, or a demand pattern, and reserve the word "request" as the particular file a user needs. Figure 1 provides an illustration of the overall system.

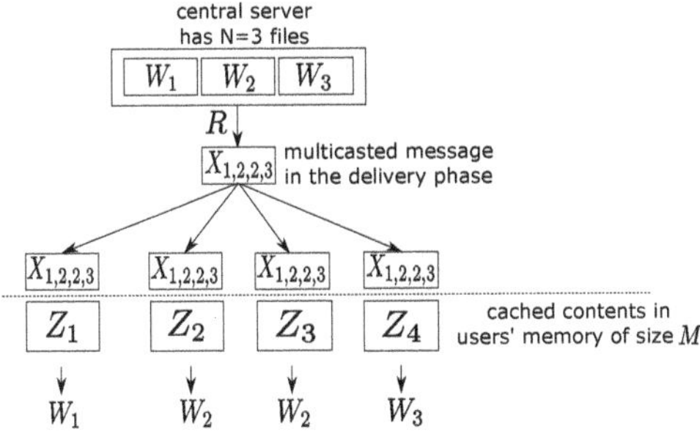

Figure 1. An example caching system, where there are $N = 3$ files and $K = 4$ users. In this case, the users request files $(1, 2, 2, 3)$, respectively, and the multicast common information is written as $X_{1,2,2,3}$.

Since we are investigating the fundamental limits of the caching systems in this work, the notation for the various quantities in the systems needs to be specified. The N files in the system are denoted as $\mathcal{W} \triangleq \{W_1, W_2, \ldots, W_N\}$; the cached contents at the K users are denoted as $\mathcal{Z} \triangleq \{Z_1, Z_2, \ldots, Z_K\}$; and the transmissions to satisfy a given demand are denoted as $X_{d_1, d_2, \ldots, d_K}$, i.e., the transmitted information $X_{d_1, d_2, \ldots, d_K}$ when user k requests file W_{d_k}, $k = 1, 2, \ldots, K$. For simplicity, we shall write $(W_1, W_2, \ldots, W_n)$ simply as $W_{[1:n]}$, and $(d_1, d_2, \ldots, d_K)$ as $d_{[1:K]}$; when there are only two users in the system, we write $(X_{i,1}, X_{i,2}, \ldots, X_{i,j})$ as $X_{i,[1:j]}$. There are other simplifications of the notation for certain special cases of the problem, which will be introduced as they become necessary.

The cache content at the k-th user is produced directly from the files through the encoding function f_k, and the transmission content from the files through the encoding function $g_{d_{[1:K]}}$, i.e.,

$$Z_k = f_k(W_{[1:N]}), \quad X_{d_{[1:K]}} = g_{d_{[1:K]}}(W_{[1:N]}),$$

the second of which depends on the particular demands $d_{[1:K]}$. Since the cached contents and transmitted information are both deterministic functions of the files, we have:

$$H(Z_k W_1, W_2, \ldots, W_N) = 0, \quad k = 1, 2, \ldots, K, \tag{1}$$

$$H(X_{d_1, d_2, \ldots, d_K} W_1, W_2, \ldots, W_N) = 0, \quad d_k \in \{1, 2, \ldots, N\}. \tag{2}$$

It is also clear that:

$$H(W_{d_k} Z_k, X_{d_1, d_2, \ldots, d_K}) = 0, \tag{3}$$

i.e., the file W_{d_k} is a function of the cached content Z_k at user k and the transmitted information when user k requests W_{d_k}. The memory satisfies the constraint:

$$M \geq H(Z_i), \quad i \in \{1, 2, \ldots, K\}, \tag{4}$$

and the transmission rate satisfies:

$$R \geq H(X_{d_1, d_2, \ldots, d_K}), \quad d_k \in \{1, 2, \ldots, N\}. \tag{5}$$

Any valid caching code must satisfy the specific set of conditions in (2)–(5). A slight variant of the problem definition allows vanishing probability of error, i.e., the probability of error asymptotically approaches zero as F goes to infinity; all the outer bounds derived in this work remain valid for this variant with appropriate applications of Fano's inequality [22].

2.2. Known Results on Caching Systems

The first achievability result on this problem was given in [6], which is directly quoted below.

Theorem 1 (Maddah-Ali and Niesen [6]). *For N files and K users each with a cache size* $M \in \{0, N/K, 2N/K, \ldots, N\}$,

$$R = K(1 - M/N) \cdot \min\left\{ \frac{1}{1 + KM/N}, \frac{N}{K} \right\} \tag{6}$$

is achievable. For general $0 \leq M \leq N$, *the lower convex envelope of these* (M, R) *points is achievable.*

The first term in the minimization is achieved by the scheme of uncoded placement together with coded transmission [6], while the latter term is by simple uncoded placement and uncoded transmission. More recently, Yu et al. [19] provided the optimal solution when the placement is restricted to be uncoded. Chen et al. [15] extended a special scheme for the case $N = K = 2$ discussed in [6] to the general case $N \leq K$, and showed that the tradeoff pair $\left(\frac{1}{K}, \frac{N(K-1)}{K} \right)$ is achievable. There were also several other notable efforts in attempting to find better binary codes [16–18,21]. Tian and Chen [20] proposed a class of codes for $N \leq K$, the origin of which will be discussed in more details in Section 5. Gómez-Vilardebó [21] also proposed a new code, which can provide further improvement in the small cache memory regime. Tradeoff points achieved by the codes in [20] can indeed be optimal in some cases. It is worth noting that while all the schemes [6,15–19,21] are binary codes, the codes in [20] use a more general finite field.

A cut-set outer bound was also given in [6], which is again directly quoted below.

Theorem 2 (Maddah-Ali and Niesen [6]). *For N files and K users each with a cache size* $0 \leq M \leq N$,

$$R \geq \max_{s \in \{1, 2, \ldots, \min\{N, K\}\}} \left(s - \frac{sM}{\lfloor N/s \rfloor} \right). \tag{7}$$

Several efforts to improve this outer bound have also been reported, which have led to more accurate approximation characterizations of the optimal tradeoff [12–14]. However, as mentioned earlier, even for the simplest cases beyond $(N, K) = (2, 2)$, complete characterizations was not available before our work (firstly reported in [23]). In this work, we specifically treat such small problem cases, and attempt to deduce more generic properties and outer bounds from these cases. Some of the most recent work [24,25] that were obtained after the publication of our results [23] provide even more accurate approximations, the best of which at this point of time is roughly a factor of 2 [24].

2.3. The Basic Linear Programming Framework

The basic linear programing bound on the entropy space was introduced by Yeung [1], which can be understood as follows. Consider a total of n discrete random variables $(X_1, X_2, \ldots, X_n)$ with a given joint distribution. There are a total of $2^n - 1$ joint entropies, each associated with a non-empty subset of these random variables. It is known that the entropy function is monotone and submodular, and thus, any valid $(2^n - 1)$ dimensional entropy vector must have the properties associated with such monotonicity and submodularity, which can be written as a set of inequalities. Yeung showed (see, e.g., [26]) that the minimal sufficient set of such inequalities is the so-called elemental inequalities:

$$H(X_i\{X_k, k \neq i\}) \geq 0, \quad i \in \{1, 2, \ldots, n\} \tag{8}$$

$$I(X_i; X_j\{X_k, k \in \Phi\}) \geq 0, \text{ where } \Phi \subseteq \{1, 2, \ldots, n\} \setminus \{i, j\}, i \neq j. \tag{9}$$

The $2^n - 1$ joint entropy terms can be viewed as the variables in a linear programming (LP) problem, and there is a total of $n + \binom{n}{2}2^{n-2}$ constraints in (8) and (9). In addition to this generic set of constraints, each specific coding problem will place additional constraints on the joint entropy values. These can be viewed as a constraint set of the given problem, although the problem might also induce constraints that are not in this form or even not possible to write in terms of joint entropies. For example, in the caching problem, the set of random variables are $\{W_i, i = 1, 2, \ldots, N\} \cup \{Z_i, i = 1, 2, \ldots, K\} \cup \{X_{d_1, d_2, \ldots, d_K} : d_k \in \{1, 2, \ldots, N\}\}$, and there is a total of $2^{N+K+N^K} - 1$ variables in this LP; the problem-specific constraints are those in (2)–(5), and there are $N + K + N^K + \binom{N+K+N^K}{2}2^{N+K+N^K-2}$ elemental entropy constraints, which is in fact doubly exponential in the number of users K.

2.4. A Computed-Aided Approach to Find Outer Bounds

In principle, with the aforedescribed constraint set, one can simply convert the outer bounding problem into an LP (with an objective function R for each fixed M in the caching problem, or more generally a linear combination of M and R), and use a generic LP solver to compute it. Unfortunately, despite the effectiveness of modern LP solvers, directly applying this approach on an engineering problem is usually not possible, since the scale of the LP is often very large even for simple settings. For example, for the caching problem, when $N = 2, K = 4$, there are overall 200 million elemental inequalities. The key observation used in [2] to make the problem tractable is that the LP can usually be significantly reduced, by taking into account the symmetry and the implication relations in the problem.

The details of the reductions can be found in [2], and here, we only provide two examples in the context of the caching problem to illustrate the basic idea behind these reductions:

- Assuming the optimal codes are symmetric, which will be defined more precisely later, the joint entropy $H(W_2, Z_3, X_{2,3,3})$ should be equal to the joint entropy $H(W_1, Z_2, X_{1,2,2})$. This implies that in the LP, we can represent both quantities using a single variable.
- Because of the relation (3), the joint entropy $H(W_2, Z_3, X_{2,3,3})$ should be equal to the joint entropy $H(W_2, W_3, Z_3, X_{2,3,3})$. This again implies that in the LP, we can represent both quantities using a single variable.

The reduced primal LP problem is usually significantly smaller, which allows us to find a lower bound for the tradeoff region for a specific instance with fixed file sizes. Moreover, after identifying the region of interest using these computed boundary points, a human-readable proof can also be produced computationally by invoking the dual of the LP given above. Note a feasible and bounded LP always has a rational optimal solution when all the coefficients are rational, and thus, the bound will have rational coefficients. More details can again be found in [2]; however, this procedure can be intuitively viewed as follows. Suppose a valid outer bound in the constraint set has the form of:

$$\sum_{\Phi \subseteq \{1,2,\ldots,n\}} \alpha_\Phi H(X_k, k \in \Phi) \geq 0, \tag{10}$$

then it must be a linear combination of the known inequalities, i.e., (8) and (9), and the problem-specific constraints, e.g., (2)–(5) for the caching problem. To find a human-readable proof is essentially to find a valid linear combination of these inequalities, and for the conciseness of the proof, the sparsest linear combination is preferred. By utilizing the LP dual with an additional linear objective, we can find within all valid combinations a sparse (but not necessarily the sparsest) one, which can yield a concise proof of the inequality (10).

It should be noted that in [2], the region of interest was obtained by first finding a set of fine-spaced points on the boundary of the outer bound using the reduced LP, and then manually identifying the effective bounding segments using these boundary points. This task can however be accomplished more efficiently using an approach proposed by Lassez and Lassez [27], as pointed out in [28]. This prompted the author to implement this part of the computer program using this more efficient approach. For completeness, the specialization of the Lassez algorithm to the caching problem, which is much simplified in this setting, is provided in Appendix A.

The proof found through this approach can be conveniently written in a matrix to list all the linear combination coefficients, and one can easily produce a chain of inequalities using such a table to obtain a more conventional human-readable proof. This approach of generating human-readable proofs has subsequently been adopted by other researchers [5,29]. Though we shall present several results thus obtained in this current work in the tabulation form, our main goal is to use these results to present the computer-aided approach, and show the effectiveness of our approach.

3. Symmetry in the Caching Problem

The computer-aided approach to derive outer bounds mentioned earlier relies critically on the reduction of the basic entropy LP using symmetry and other problem structures. In this section, we consider the symmetry in the caching problem. Intuitively, if we place the cached contents in a permuted manner at the users, it will lead to a new code that is equivalent to the original one. Similarly, if we reorder the files and apply the same encoding function, the transmissions can also be changed accordingly to accommodate the requests, which is again an equivalent code. The two types of symmetries can be combined, and they induce a permutation group on the joint entropies of the subsets of the random variables $\mathcal{W} \cup \mathcal{Z} \cup \mathcal{X}$.

For concreteness, we may specialize to the case $(N, K) = (3, 4)$ in the discussion, and for this case:

$$\mathcal{W} = \{W_1, W_2, W_3\}, \quad \mathcal{Z} = \{Z_1, Z_2, Z_3, Z_4\}, \quad \mathcal{X} = \{X_{d_1, d_2, d_3, d_4} : d_k \in \{1, 2, 3\}\}. \tag{11}$$

3.1. Symmetry in User Indexing

Let a permutation function be defined as $\tilde{\pi}(\cdot)$ on the user index set of $\{1, 2, \ldots, K\}$, which reflects a permuted placement of cached contents $\mathcal{Z}$. Let the inverse of $\tilde{\pi}(\cdot)$ be denoted as $\tilde{\pi}^{-1}(\cdot)$, and define the permutation on a collection of elements as the collection of the elements after permuting each element individually. The aforementioned permuted placement of cached contents can be rigorously defined through a set of new encoding functions and decoding functions. Given the original encoding functions f_k and $g_{d_{[1:K]}}$, the new functions $f_k^{\tilde{\pi}}$ and $g_{d_{[1:K]}}^{\tilde{\pi}}$ associated with a permutation $\tilde{\pi}$ can be defined as:

$$\begin{aligned}
\bar{Z}_k &\triangleq f_k^{\tilde{\pi}}(W_{[1:N]}) \triangleq f_{\tilde{\pi}(k)}(W_{[1:N]}) = Z_{\tilde{\pi}(k)}, \\
\bar{X}_{d_{[1:K]}} &\triangleq g_{d_{[1:K]}}^{\tilde{\pi}}(W_{[1:N]}) \triangleq g_{d_{\tilde{\pi}^{-1}([1:K])}}(W_{[1:N]}) = X_{d_{\tilde{\pi}^{-1}([1:K])}}.
\end{aligned} \tag{12}$$

To see that with these new functions, any demand $d_{([1:K])}$ can be correctly fulfilled as long as the original functions can fulfill the corresponding reconstruction task, consider the pair $(f_k^{\tilde{\pi}}(W_{[1:N]}), g_{d_{[1:K]}}^{\tilde{\pi}}(W_{[1:N]}))$, which should reconstruct W_{d_k}. This pair is equivalent to the pair $(f_{\tilde{\pi}(k)}(W_{[1:N]}), g_{d_{\tilde{\pi}^{-1}([1:K])}}(W_{[1:N]}))$, and in the demand vector $d_{\hat{\pi}^{-1}([1:K])}$, the $\hat{\pi}(k)$ position is in fact $d_{\hat{\pi}^{-1}(\hat{\pi}(k))} = d_k$, implying that the new coding functions are indeed valid.

We can alternatively view $\hat{\pi}(\cdot)$ as directly inducing a permutation on $\mathcal{Z}$ as $\hat{\pi}(Z_k) = Z_{\hat{\pi}(k)}$, and a permutation on $\mathcal{X}$ as:

$$\hat{\pi}(X_{d_1,d_2,\dots,d_K}) = X_{d_{\hat{\pi}^{-1}(1)},d_{\hat{\pi}^{-1}(2)},\dots,d_{\hat{\pi}^{-1}(K)}}. \tag{13}$$

For example, the permutation function $\hat{\pi}(1) = 2, \hat{\pi}(2) = 3, \hat{\pi}(3) = 1, \hat{\pi}(4) = 4$ will induce:

$$(d_1, d_2, d_3, d_4) \rightarrow (\bar{d}_1, \bar{d}_2, \bar{d}_3, \bar{d}_4) = (d_3, d_1, d_2, d_4). \tag{14}$$

Thus, it will map Z_1 to $\hat{\pi}(Z_1) = Z_2$, but map $X_{1,2,3,2}$ to $X_{3,1,2,2}$, $X_{3,2,1,3}$ to $X_{1,3,2,3}$, and $X_{1,1,2,2}$ to $X_{2,1,1,2}$.

With the new coding functions and the permuted random variables defined above, we have the following relation:

$$(\mathcal{W}^{\hat{\pi}}, \mathcal{Z}^{\hat{\pi}}, \mathcal{X}^{\hat{\pi}}) = (\mathcal{W}, \hat{\pi}(\mathcal{Z}), \hat{\pi}(\mathcal{X})), \tag{15}$$

where the superscript $\hat{\pi}$ indicates the random variables induced by the new encoding functions.

We call a caching code user-index-symmetric, if for any subsets $\mathcal{W}_o \subseteq \mathcal{W}, \mathcal{Z}_o \subseteq \mathcal{Z}, \mathcal{X}_o \subseteq \mathcal{X}$, and any permutation $\hat{\pi}$, the following relation holds:

$$H(\mathcal{W}_o, \mathcal{Z}_o, \mathcal{X}_o) = H(\mathcal{W}_o, \hat{\pi}(\mathcal{Z}_o), \hat{\pi}(\mathcal{X}_o)). \tag{16}$$

For example, for such a symmetric code, the entropy $H(W_2, Z_2, X_{1,2,3,2})$ under the aforementioned permutation is equal to $H(W_2, Z_3, X_{3,1,2,2})$; note that W_2 is a function of $(Z_2, X_{1,2,3,2})$, and after the mapping, it is a function of $(Z_3, X_{3,1,2,2})$.

3.2. Symmetry in File Indexing

Let a permutation function be defined as $\hat{\pi}(\cdot)$ on the file index set of $\{1, 2, \dots, N\}$, which reflects a renaming of the files $\mathcal{W}$. This file-renaming operation can be rigorously defined as a permutation of the input arguments to the functions f_k and $g_{d_{[1:K]}}$. Given the original encoding functions f_k and $g_{d_{[1:K]}}$, the new functions $f_k^{\hat{\pi}}$ and $g_{d_{[1:K]}}^{\hat{\pi}}$ associated with a permutation $\hat{\pi}$ can be defined as:

$$\begin{aligned}
\hat{Z}_k &\triangleq f_k^{\hat{\pi}}(W_{[1:N]}) \triangleq f_k(W_{\hat{\pi}^{-1}([1:N])}), \\
\hat{X}_{d_{[1:K]}} &\triangleq g_{d_{[1:K]}}^{\hat{\pi}}(W_{[1:N]}) \triangleq g_{\hat{\pi}(d_{[1:K]})}(W_{\hat{\pi}^{-1}([1:N])}).
\end{aligned} \tag{17}$$

We first show that the pair $(f_k^{\hat{\pi}}(W_{[1:N]}), g_{d_{[1:K]}}^{\hat{\pi}}(W_{[1:N]}))$ can provide reconstruction of W_{d_k}. This pair by definition is equivalent to $(f_k(W_{\hat{\pi}^{-1}([1:N])}), g_{\hat{\pi}(d_{[1:K]})}(W_{\hat{\pi}^{-1}([1:N])}))$, where the k-th position of the demand vector $\hat{\pi}(d_{[1:K]})$ is in fact $\hat{\pi}(d_k)$. However, because of the permutation in the input arguments, this implies that the $\hat{\pi}(d_k)$-th file in the sequence $W_{\hat{\pi}^{-1}([1:N])})$ can be reconstructed, which is indeed W_{d_k}.

Alternatively, we can view $\hat{\pi}(\cdot)$ as directly inducing a permutation on $\hat{\pi}(W_k) = W_{\hat{\pi}(k)}$, and it also induces a permutation on $\mathcal{X}$ as:

$$\hat{\pi}(X_{d_1,d_2,\dots,d_K}) = X_{\hat{\pi}(d_1),\hat{\pi}(d_2),\dots,\hat{\pi}(d_K)}. \tag{18}$$

For example, the permutation function $\hat{\pi}(1) = 2, \hat{\pi}(2) = 3, \hat{\pi}(3) = 1$ maps W_2 to $\hat{\pi}(W_2) = W_3$, but maps $X_{1,2,3,2}$ to $X_{2,3,1,3}$, $X_{3,2,1,3}$ to $X_{1,3,2,1}$, and $X_{1,1,2,2}$ to $X_{2,2,3,3}$.

With the new coding functions and the permuted random variables defined above, we have the following equivalence relation:

$$
\begin{aligned}
(W^{\hat{\pi}}, Z^{\hat{\pi}}, X^{\hat{\pi}}) &= \left(W_{([1:N])}, f_{[1:k]}(W_{\hat{\pi}^{-1}([1:N])}), \left\{ g_{\hat{\pi}(d_{[1:K]})}(W_{\hat{\pi}^{-1}([1:N])}) : d_{[1:K]} \in \mathcal{N}^K \right\} \right) \\
&\stackrel{d}{=} \left(W_{\hat{\pi}([1:N])}, f_{[1:k]}(W_{[1:N]}), \left\{ g_{\hat{\pi}(d_{[1:K]})}(W_{[1:N]}) : d_{[1:K]} \in \mathcal{N}^K \right\} \right) \\
&= (\hat{\pi}(W), Z, \hat{\pi}(X)),
\end{aligned}
\tag{19}
$$

where $\stackrel{d}{=}$ indicates equal in distribution, which is due to the the random variables in W being independently and identically distributed, thus exchangeable.

We call a caching code file-index-symmetric, if for any subsets $W_o \subseteq W, Z_o \subseteq Z, X_o \subseteq X$, and any permutation $\hat{\pi}$, the following relation holds:

$$
H(W_o, Z_o, X_o) = H(\hat{\pi}(W_o), Z_o, \hat{\pi}(X_o)).
\tag{20}
$$

For example, for such a symmetric code, $H(W_3, Z_3, X_{1,2,3,2})$ under the aforementioned permutation is equal to $H(W_1, Z_3, X_{2,3,1,3})$; note that W_3 is a function of $(Z_3, X_{1,2,3,2})$, and after the mapping, W_1 is a function of $(Z_3, X_{2,3,1,3})$.

3.3. Existence of Optimal Symmetric Codes

With the symmetry structure elucidated above, we can now state our first auxiliary result.

Proposition 1. *For any caching code, there is a code with the same or smaller caching memory and transmission rate, which is both user-index-symmetric and file-index-symmetric.*

We call a code that is both user-index-symmetric and file-index-symmetric a symmetric code. This proposition implies that there is no loss of generality to consider only symmetric codes. The proof of this proposition relies on a simple space-sharing argument, where a set of base encoding functions and base decoding function are used to construct a new code. In this new code, each file is partitioned into a total of $N!K!$ segments, each having the same size as suitable in the base coding functions. The coding functions obtained as in (12) and (17) from the base coding functions using permutations π and $\hat{\pi}$ are used on the i-th segments of all the files to produce random variables $W^{\hat{\pi} \cdot \hat{\pi}} \cup Z^{\hat{\pi} \cdot \hat{\pi}} \cup X^{\hat{\pi} \cdot \hat{\pi}}$. Consider a set of random variables $(W_o \cup Z_o \cup X_o)$ in the original code, and denote the same set of random variables in the new code as $(W_o' \cup Z_o' \cup X_o')$. We have:

$$
H(W_o' \cup Z_o' \cup X_o') = \sum_{\pi, \hat{\pi}} H(W_o^{\hat{\pi} \cdot \hat{\pi}} \cup Z_o^{\hat{\pi} \cdot \hat{\pi}} \cup X_o^{\hat{\pi} \cdot \hat{\pi}}) = \sum_{\pi, \hat{\pi}} H(\hat{\pi}(W_o) \cup \pi(Z_o) \cup \pi \cdot \hat{\pi}(X_o)),
\tag{21}
$$

because of (15) and (19). Similarly, for another pair of permutations $(\pi', \hat{\pi}')$, the random variables $\hat{\pi}'(W_o') \cup \pi'(Z_o') \cup \pi' \cdot \hat{\pi}'(X_o')$ in the new code will have exactly the same joint entropy value. It is now clear that the resultant code by space sharing is indeed symmetric, and it has (normalized) memory sizes and a transmission rate no worse than the original one. A similar argument was used in [2] to show, with a more detailed proof, the existence of optimal symmetric solution in regenerating codes. In a separate work [30], we investigated the properties of the induced permutation $\pi \cdot \hat{\pi}$, and particularly, showed that it is isomorphic to the power group [31]; readers are referred to [30] for more details.

3.4. Demand Types

Even for symmetric codes, the transmissions to satisfy different types of demands may use different rates. For example in the setting $N, K = (3, 4)$, $H(X_{1,2,2,2})$ may not be equal to $H(X_{1,1,2,2})$, and $H(X_{1,2,3,2})$ may not be equal to $H(X_{3,2,3,2})$. The transmission rate R is then chosen to be the maximum among all cases. This motivates the notion of demand types.

Definition 1. *In an (N, K) caching system, for a specific demand, let the number of users requesting file n be denoted as m_n, $n = 1, 2, \ldots, N$. We call the vector obtained by sorting the values $\{m_1, m_2, \ldots, m_N\}$ in a decreasing order as the demand type, denoted as $\mathcal{T}$.*

Proposition 1 implies that for optimal symmetric solutions, demands of the same type can always be satisfied with transmissions of the same rate; however, demands of different types may still require different rates. This observation is also important in setting up the linear program in the computer-aided approach outlined in the previous section. Because we are interested in the worst case transmission rate among all types of demands, in the symmetry-reduced LP, an additional variable needs to be introduced to constrain the transmission rates of all possible types.

For an (N, K) system, determining the number of demand types is closely related to the integer partition problem, which is the number of possible ways to write an integer K as the sum of positive integers. There is no explicit formula, but one can use a generator polynomial to compute it [32]. For several small (N, K) pairs, we list the demand types in Table 1.

Table 1. Demand types for small (N, K) pairs.

(N,K)	Demand Types
(2,3)	(3,0), (2,1)
(2,4)	(4,0), (3,1), (2,2)
(3,2)	(2,0,0), (1,1,0)
(3,3)	(3,0,0), (2,1,0), (1,1,1)
(3,4)	(4,0,0), (3,1,0), (2,2,0), (2,1,1)
(4,2)	(2,0,0,0), (1,1,0,0)
(4,3)	(3,0,0,0), (2,1,0,0), (1,1,1,0)

It can be seen that when $N \leq K$, increasing N induces more demand types, but this stops when $N > K$; however, increasing K always induces more demand types. This suggests it might be easier to find solutions for a collection of cases with a fixed K and arbitrary N values, but more difficult for that of a fixed N and arbitrary K values. This intuition is partially confirmed with our results presented next.

4. Computational and Data-Driven Converse Hypotheses

Extending the computational approach developed in [2] and the problem symmetry, in this section, we first establish complete characterizations for the optimal memory-transmission-rate tradeoff for $(N, K) = (3, 2)$ and $(N, K) = (4, 2)$. Based on these results and the known result for $(N, K) = (2, 2)$, we are able to form a hypothesis regarding the optimal tradeoff for the case of $K = 2$. An analytical proof is then provided, which gives the complete characterization of the optimal tradeoff for the case of $(N, 2)$ caching systems. We then present a characterization of the optimal tradeoff for $(N, K) = (2, 3)$ and an outer bound for $(N, K) = (2, 4)$. These results also motivate a hypothesis on the optimal tradeoff for $N = 2$, which is subsequently proven analytically to yield a partial characterization. Note that since both M and R must be nonnegative, we do not explicitly state their non-negativity from here on.

4.1. The Optimal Tradeoff for $K = 2$

The optimal tradeoff for $(N, K) = (2, 2)$ was found in [6], which we restate below.

Proposition 2 (Maddah Ali and Niesen [6]). *Any memory-transmission-rate tradeoff pair for the $(N, K) = (2, 2)$ caching problem must satisfy:*

$$2M + R \geq 2, \quad 2M + 2R \geq 3, \quad M + 2R \geq 2. \tag{22}$$

Conversely, there exist codes for any nonnegative (M, R) pair satisfying (22).

Our investigation thus starts with identifying the previously unknown optimal tradeoff for $(N, K) = (3, 2)$ and $(N, K) = (4, 2)$ using the computation approach outlined in Section 2, the results of which are first summarized below as two propositions.

Proposition 3. *Any memory-transmission-rate tradeoff pair for the $(N, K) = (3, 2)$ caching problem must satisfy:*

$$M + R \geq 2, \quad M + 3R \geq 3. \tag{23}$$

Conversely, there exist codes for any nonnegative (M, R) pair satisfying (23).

Proposition 4. *Any memory-transmission-rate tradeoff pair for the $(N, K) = (4, 2)$ caching problem must satisfy:*

$$3M + 4R \geq 8, \quad M + 4R \geq 4. \tag{24}$$

Conversely, there exist codes for any nonnegative (M, R) pair satisfying (24).

The proofs for Propositions 3 and 4 can be found in Appendix B, which are given in the tabulation format mentioned earlier. Strictly speaking, these two results are specialization of Theorem 3, and there is no need to provide the proofs separately; however, we provide them to illustrate the computer-aided approach.

The optimal tradeoff for these cases is given in Figure 2. A few immediate observations are as follows

- For $(N, K) = (3, 2)$ and $(N, K) = (4, 2)$, there is only one non-trivial corner point on the optimal tradeoff, but for $(N, K) = (2, 2)$, there are in fact two non-trivial corner points.
- The cut-set bound is tight at the high memory regime in all the cases.
- The single non-trivial corner point for $(N, K) = (3, 2)$ and $(N, K) = (4, 2)$ is achieved by the scheme proposed in [6]. For the $(N, K) = (2, 2)$ case, one of the corner point is achieved also by this scheme, but the other corner point requires a different code.

Given the above observations, a natural hypothesis is as follows.

Hypothesis 1. *There is only one non-trivial corner point on the optimal tradeoff for $(N, K) = (N, 2)$ caching systems when $N \geq 3$, and it is $(M, R) = (N/2, 1/2)$, or equivalently, the two facets of the optimal tradeoff should be:*

$$3M + NR \geq 2N, \quad M + NR \geq N. \tag{25}$$

We are indeed able to analytically confirm this hypothesis, as stated formally in the following theorem.

Theorem 3. *For any integer $N \geq 3$, any memory-transmission-rate tradeoff pair for the $(N, K) = (N, 2)$ caching problem must satisfy:*

$$3M + NR \geq 2N, \quad M + NR \geq N. \tag{26}$$

Conversely, for any integer $N \geq 3$, there exist codes for any nonnegative (M, R) pair satisfying (26). For $(N, K) = (2, 2)$, the memory-transmission-rate tradeoff must satisfy:

$$2M + R \geq 2, \quad 2M + 2R \geq 3, \quad M + 2R \geq 2. \tag{27}$$

Conversely, for $(N, K) = (2, 2)$, there exist codes for any nonnegative (M, R) pair satisfying (27).

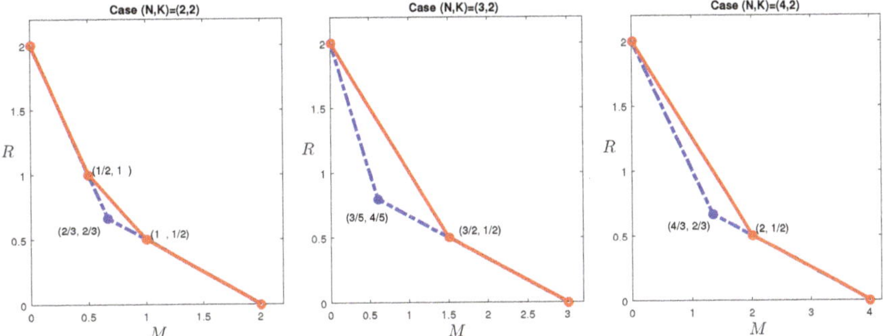

Figure 2. The optimal tradeoffs for $(N,K) = (2,2)$, $(N,K) = (3,2)$ and $(N,K) = (4,2)$ caching systems. The red solid lines give the optimal tradeoffs, while the blue dashed-dotted lines are the cut-set outer bounds, included here for reference.

Since the solution for the special case $(N,K) = (2,2)$ was established in [6], we only need to consider the cases for $N \geq 3$. Moreover, for the converse direction, only the bound $3M + NR \geq 2N$ needs to be proven, since the other one can be obtained using the cut-set bound in [6]. To prove the remaining inequality, the following auxiliary lemma is needed.

Lemma 1. *For any symmetric* $(N,2)$ *caching code where* $N \geq 3$, *and any integer* $n = \{1, 2, \ldots, N-2\}$,

$$(N-n)H(Z_1, W_{[1:n]}, X_{n,n+1}) \geq (N-n-2)H(Z_1, W_{[1:n]}) + (N+n). \tag{28}$$

Using Lemma 1, we can prove the converse part of Theorem 3 through an induction; the proofs of Theorem 3 and Lemma 1 can be found in Appendix C, both of which heavily rely on the symmetry specified in the previous section. Although some clues can be found in the proof tables for the cases $(N,K) = (3,2)$ and $(N,K) = (4,2)$, such as the effective joint entropy terms in the converse proof each having only a small number of X random variables, finding the proof of Theorem 3 still requires considerable human effort, and was not completed directly through a computer program. One key observation simplifying the proof in this case is that as the hypothesis states, the optimal corner point is achieved by the scheme given in [6], which is known only thanks to the computed bounds. In this specific case, the scheme reduces to splitting each file in half, and placing one half at the first user, and the other half at the second user; the corresponding delivery strategy is also extremely simple. We combined this special structure and the clues from the proof tables to find the outer bounding steps.

Remark 1. *The result in [12] can be used to establish the bound* $3M + NR \geq 2N$ *when* $K = 2$, *however only for the cases when* N *is an integer multiple of three. For* $N = 4$, *the bounds developed in [12–14] give* $M + 2R \geq 3$, *instead of* $3M + 4R \geq 8$, *and thus, they are loose in this case. After this bound was initially reported in [23], Yu et al. [24] discovered an alternative proof.*

4.2. A Partial Characterization for $N = 2$

We first summarize the characterizations of the optimal tradeoff for $(N,K) = (2,3)$, and the computed outer bound for $(N,K) = (2,4)$, in two propositions.

Proposition 5. *The memory-transmission-rate tradeoff for the* $(N,K) = (2,3)$ *caching problem must satisfy:*

$$2M + R \geq 2, \quad 3M + 3R \geq 5, \quad M + 2R \geq 2. \tag{29}$$

Conversely, there exist codes for any nonnegative (M,R) *pair satisfying (29).*

Proposition 6. *The memory-transmission-rate tradeoff for the* $(N, K) = (2, 4)$ *caching problem must satisfy:*

$$2M + R \geq 2, \; 14M + 11R \geq 20, \; 9M + 8R \geq 14, \; 3M + 3R \geq 5, \; 5M + 6R \geq 9, \; M + 2R \geq 2. \tag{30}$$

For Proposition 5, the only new bound $3M + 3R \geq 5$ is a special case of the more general result of Theorem 4, and we thus do not provide this proof separately. For Proposition 6, only the second and the third inequalities need to be proven, since the fourth coincides with a bound in the $(2, 3)$ case, the fifth is a special case of Theorem 4, and the others can be produced from the cut-set bounds. The proofs for these two inequalities are given in Appendix E. The optimal tradeoff for $(N, K) = (2, 2), (2, 3)$ and the outer bound for $(2, 4)$ are depicted in Figure 3. A few immediate observations and comments are as follows:

- There are two non-trivial corner points on the outer bounds for $(N, K) = (2, 2)$ and $(N, K) = (2, 3)$, and there are five non-trivial corner points for $(N, K) = (2, 4)$.
- The outer bounds coincide with known inner bounds for $(N, K) = (2, 2)$ and $(N, K) = (2, 3)$, but not $(N, K) = (2, 4)$. The corner points at $R = 1/K$ (and the corner point $(1, 2/3)$ for $(N, K) = (2, 4)$) are achieved by the scheme given in [6], while the corner points at $M = 1/K$ are achieved by the scheme given in [15]. For $(N, K) = (2, 4)$, two corner points at the intermediate memory regime cannot be achieved by either the scheme in [6] or that in [15].
- The cut-set outer bounds [6] are tight at the highest and lowest memory segments; a new bound for the second highest memory segment produced by the computer-based method is also tight.

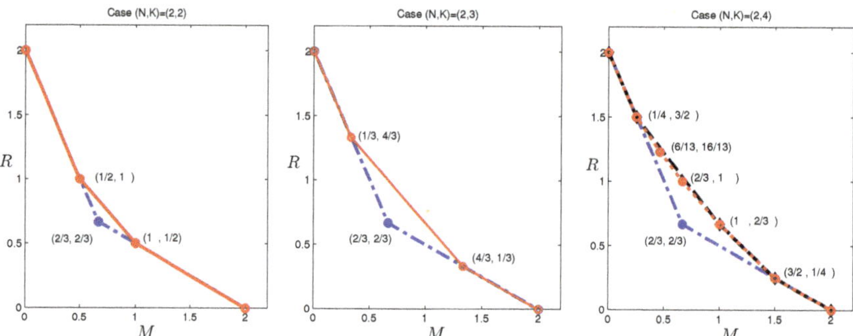

Figure 3. The optimal tradeoffs for $(N, K) = (2, 2)$, $(N, K) = (2, 3)$ and computed outer bound $(N, K) = (2, 4)$ caching systems. The red solid lines give the optimal tradeoffs for the first two cases, and the red dotted line gives the computed outer bound $(N, K) = (2, 4)$; the blue dashed-dot lines are the cut-set outer bounds, and the black dashed line is the inner bound using the scheme in [6,15].

Remark 2. *The bounds developed in [12–14] give* $2(M + R) \geq 3$ *for* $(N, K) = (2, 3)$ *and* $(N, K) = (2, 4)$, *instead of* $3M + 3R \geq 5$, *and thus, they are loose in this case. When specializing the bounds in [24], it matches Proposition 5 for* $(N, K) = (2, 3)$, *but it is weaker than Proposition 6 for* $(N, K) = (2, 4)$.

From the above observations, we can hypothesize that for $N = 2$, the number of corner points will continue to increase as K increases above four, and at the high memory regime, the scheme [6] is optimal. More precisely, we can establish the following theorem.

Theorem 4. *When* $K \geq 3$ *and* $N = 2$, *any* (M, R) *pair must satisfy:*

$$K(K + 1)M + 2(K - 1)KR \geq 2(K - 1)(K + 2). \tag{31}$$

As a consequence, the uncoded-placement-coded-transmission scheme in [6] (with space-sharing) is optimal when $M \geq \frac{2(K-2)}{K}$, for the cases with $K \geq 4$ and $N = 2$.

The first line segment at the high memory regime is $M + 2R \geq 2$, which is given by the cut-set bound; its intersection with (31) is indeed the first point in:

$$\left(\frac{2(K-1)}{K}, \frac{1}{K}\right) \quad \text{and} \quad \left(\frac{2(K-2)}{K}, \frac{2}{K-1}\right). \tag{32}$$

The proof of this theorem now boils down to the proof of the bound (31). This requires a sophisticated induction, the digest of which is summarized in the following lemma. The symmetry of the problem is again heavily utilized throughout the proof of this lemma. For notational simplicity, we use $X_{\to j}$ to denote $X_{1,1,\dots,1,2,1,\dots,1}$, i.e., when the j-th user requests the second file, and all the other users request the first file; we also write a collection of such variables $(X_{\to j}, X_{\to j+1}, \dots, X_{\to k})$ as $X_{\to [j:k]}$.

Lemma 2. *For $N = 2$ and $K \geq 3$, the following inequality holds for $k \in \{2, 3, \dots, K-1\}$:*

$$\begin{aligned}
&(K - k + 1)(K - k + 2)H(Z_1, W_1, X_{\to[2:k]}) \\
&\geq [(K - k)(K - k + 1) - 2]H(Z_1, W_1, X_{\to[2:k-1]}) + 2H(W_1, X_{\to[2:k-1]}) + 2(K - k + 1)H(W_1, W_2),
\end{aligned} \tag{33}$$

where we have taken the convention $H(Z_1, W_1, X_{\to[2:1]}) = H(Z_1, W_1)$

The proof of Lemma 2 is given in Appendix F. Theorem 4 can now be proven straightforwardly.

Proof of Theorem 4. We first write the following simple inequalities:

$$H(Z_1) + H(X_{\to 2}) \geq H(Z_1, X_{\to 2}) = H(Z_1, W_1, X_{\to 2}). \tag{34}$$

Now, applying Lemma 2 with $k = 2$ gives:

$$(K - 1)K[H(Z_1) + H(X_{\to 2})] \geq [K^2 - 3K]H(Z_1, W_1) + 2H(W_1) + 2(K - 1)H(W_1, W_2). \tag{35}$$

Observe that:

$$H(Z_1, W_1) = H(W_1 Z_1) + H(Z_1) \geq \frac{1}{2}H(W_1, W_2 Z_1) + H(Z_1) = \frac{1}{2}H(W_1, W_2) + \frac{1}{2}H(Z_1), \tag{36}$$

where in the first inequality the file index symmetry $H(W_1 Z_1) = H(W_2 Z_1)$ has been used. We can now continue to write:

$$(K - 1)K[H(Z_1) + H(X_{\to 2})] \geq \frac{K^2 - 3K}{2}[H(W_1, W_2) + H(Z_1)] + 2H(W_1) + 2(K - 1)H(W_1, W_2), \tag{37}$$

which has some a common term $H(Z_1)$ on both sizes with different coefficients. Removing the common term and multiplying both sides by two lead to:

$$\begin{aligned}
&K(K + 1)H(Z_1) + 2(K - 1)KH(X_{\to 2}) \\
&\geq [(K - 2)(K - 1) - 2 + 4(K - 1)]H(W_1, W_2) + 4H(W_1) \\
&= 2K^2 + 2K - 4,
\end{aligned} \tag{38}$$

where the equality relies on the assumption that W_1 and W_2 are independent files of unit size. Taking into consideration the memory and transmission rate constraints (4) and (5) now completes the proof. $\square$

Lemma 2 provides a way to reduce the number of X variables in $H(Z_1, X_{\to[2:k]})$, and thus is the core of the proof. Even with the hypothesis regarding the scheme in [6] being optimal, deriving the outer bound (particularly the coefficients in the lemma above) directly using this insight is far from being straightforward. Some of the guidance in finding our derivation was in fact obtained through a strategic computational exploration of the outer bounds. This information is helpful because the computer-generated proofs are not unique, and some of these solutions can appear quite arbitrary; however, to deduce general rules in the proof requires a more structured proof instead. In Section 6, we present in more detail this new exploration method, and discuss how insights can be actively identified in this particular case.

5. Reverse-Engineering Code Constructions

In the previous section, outer bounds of the optimal tradeoff were presented for the case $(N, K) = (2, 4)$, which is given in Figure 3. Observe that the corner points:

$$\left(\frac{2}{3}, 1\right) \quad \text{and} \quad \left(\frac{6}{13}, \frac{16}{13}\right), \tag{39}$$

cannot be achieved by existing codes in the literature. The former point can indeed be achieved with a new code construction. This construction was first presented in [20], where it was generalized more systematically to yield a new class of codes for any $N \leq K$, the proof and analysis of which are more involved. In this paper, we focus on how a specific code for this corner point was found through a reverse-engineering approach, which should help dispel the mystery on this seemingly arbitrary code construction.

5.1. The Code to Achieve $\left(\frac{2}{3}, 1\right)$ for $(N, K) = (2, 4)$

The two files are denoted as A and B, each of which is partitioned into six segments of equal size, denoted as A_i and B_i, respectively, $i = 1, 2, \ldots, 6$. Since we count the memory and transmission in multiples of the file size, the corner point $\left(\frac{2}{3}, 1\right)$ means the need for each user to store four symbols, and the transmission will use six symbols. The contents in the cache of each user are given in Table 2. By the symmetry of the cached contents, we only need to consider the demand (A, A, A, B), i.e., the first three users requesting A and User 4 requesting B, and the demand (A, A, B, B), i.e., the first two users requesting A and the other two requesting B.

Table 2. Caching content for $(N, K) = (2, 4)$.

User 1	$A_1 + B_1$	$A_2 + B_2$	$A_3 + B_3$	$A_1 + A_2 + A_3 + 2(B_1 + B_2 + B_3)$
User 2	$A_1 + B_1$	$A_4 + B_4$	$A_5 + B_5$	$A_1 + A_4 + A_5 + 2(B_1 + B_4 + B_5)$
User 3	$A_2 + B_2$	$A_4 + B_4$	$A_6 + B_6$	$A_2 + A_4 + A_6 + 2(B_2 + B_4 + B_6)$
User 4	$A_3 + B_3$	$A_5 + B_5$	$A_6 + B_6$	$A_3 + A_5 + A_6 + 2(B_3 + B_5 + B_6)$

Assume the file segments are in $\mathbb{F}_5$ for concreteness.

- For the demands (A, A, A, B), the transmission is as follows,

$$\text{Step 1: } B_1, B_2, B_4;$$

$$\text{Step 2: } A_3 + 2A_5 + 3A_6, A_3 + 3A_5 + 4A_6;$$

$$\text{Step 3: } A_1 + A_2 + A_4.$$

After Step 1, User 1 can recover (A_1, A_2); furthermore, he/she has $(A_3 + B_3, A_3 + 2B_3)$ by eliminating known symbols (A_1, A_2, B_1, B_2), from which A_3 can be recovered. After Step 2, he/she can obtain $(2A_5 + 3A_6, 3A_5 + 4A_6)$ to recover (A_5, A_6). Using the transmission in Step 3, he/she can obtain A_4 since he/she has (A_1, A_2). User 2 and User 3 can use a similar strategy to reconstruct all file segments in A. User 4 only needs B_3, B_5, B_6 after Step 1, which he/she already has in his/her cache; however, they are contaminated by file segments from A. Nevertheless, he/she knows $A_3 + A_5 + A_6$ by recognizing:

$$(A_3 + A_5 + A_6) = 2 \sum_{i=3,5,6} (A_i + B_i) - [A_3 + A_5 + A_6 + 2(B_3 + B_5 + B_6)]. \tag{40}$$

Together with the transmission in Step 2, User 4 has three linearly independent combinations of (A_3, A_5, A_6). After recovering them, he/she can remove these interferences from the cached content for (B_3, B_5, B_6).

- For the demand (A, A, B, B), we can send:

$$\text{Step 1: } B_1, A_6;$$
$$\text{Step 2: } A_2 + 2A_4, A_3 + 2A_5, B_2 + 2B_3, B_4 + 2B_5.$$

User 1 has A_1, B_1, A_6 after Step 1, and he/she can also form:

$$B_2 + B_3 = [A_2 + A_3 + 2(B_2 + B_3)] - (A_2 + B_2) - (A_3 + B_3),$$

and together with $B_2 + 2B_3$ in the transmission of Step 2, he/she can recover (B_2, B_3), and thus A_2, A_3. He/she still needs (A_4, A_5), which can be recovered straightforwardly from the transmission $(A_2 + 2A_4, A_3 + 2A_5)$ since he/she already has (A_2, A_3). Other users can use a similar strategy to decode their requested files.

5.2. Extracting Information for Reverse-Engineering

It is clear at this point that for this case of $(N, K) = (2, 4)$, the code to achieve this optimal corner point is not straightforward. Next, we discuss a general approach to deduce the code structure from the LP solution, which leads to the discovery of the code in our work. The approach is based on the following assumptions: the outer bound is achievable (i.e., tight); moreover, there is a (vector) linear code that can achieve this performance.

Either of the two assumptions above may not hold in general, and in such a case, our attempt will not be successful. Nevertheless, though linear codes are known to be insufficient for all network coding problems [33], existing results in the literature suggest that vector linear codes are surprisingly versatile and powerful. Similarly, though it is known that Shannon-type inequalities, which are the basis for the outer bounds computation, are not sufficient to characterize rate region for all coding problems [34,35], they are surprisingly powerful, particularly in coding problems with strong symmetry structures [36,37].

There are essentially two types of information that we can extract from the primal LP and dual LP:

- From the effective information inequalities: since we can produce a readable proof using the dual LP, if a code can achieve this corner point, then the information inequalities in the proof must hold with equality for the joint entropy values induced by this code, which reveals a set of conditional independence relations among random variables induced by this code;
- From the extremal joint entropy values at the corner points: although we are only interested in the tradeoff between the memory and transmission rate, the LP solution can provide the whole set of joint entropy values at an extreme point. These values can reveal a set of dependence relations among the random variables induced by any code that can achieve this point.

Though the first type of information is important, its translation to code constructions appears difficult. On the other hand, the second type of information appears to be more suitable for the purpose of code design, which we adopt next.

One issue that complicates our task is that the entropy values so extracted are not always unique, and sometimes have considerable slacks. For example, for different LP solutions at the same operating point of $(M, R) = \left(\frac{2}{3}, 1\right)$, the joint entropy $H(Z_1, Z_2)$ can vary between one and $4/3$. We can identify such a slack in any joint entropy in the corner point solutions by considering a regularized primal LP: for a fixed rate value R at the corner point in question as an upper bound, the objective function can be set as:

$$\text{minimize: } H(Z_1) + \gamma H(Z_1, Z_2) \tag{41}$$

instead of:

$$\text{minimize: } H(Z_1), \tag{42}$$

subject to the same original symmetric LP constraints at the target M. By choosing a small positive γ value, e.g., $\gamma = 0.0001$, we can find the minimum value for $H(Z_1, Z_2)$ at the same (M, R) point; similarly, by choosing a small negative γ value, we can find the maximum value for $H(Z_1, Z_2)$ at the same (M, R) point. Such slacks in the solution add uncertainty to the codes we seek to find and may indeed imply the existence of multiple code constructions. For the purpose of reverse-engineering the codes, we focus on the joint entropies that do not have any slacks, i.e., the "stable" joint entropies in the solution.

5.3. Reverse-Engineering the Code for $(N, K) = (2, 4)$

With the method outlined above, we identify the following stable joint entropy values in the $(N, K) = (2, 4)$ case for the operating point $\left(\frac{2}{3}, 1\right)$ listed in Table 3. The values are normalized by multiplying everything by six. For simplicity, let us assume that each file has six units of information, written as $W_1 = (A_1, A_2, \ldots, A_6) \triangleq A$ and $W_2 = (B_1, B_2, \ldots, B_6) \triangleq B$, respectively. This is a rich set of data, but a few immediate observations are given next.

- The quantities can be categorized into three groups: the first is without any transmission; the second is the quantities involving the transmission to fulfill the demand type $(3, 1)$; and the last for demand type $(2, 2)$.
- The three quantities $H(Z_1 W_1)$, $H(Z_1, Z_2 W_1)$ and $H(Z_1, Z_2, Z_3 W_1)$ provide the first important clue. The values indicate that for each of the two files, each user should have three units in his/her cache, and the combination of any two users should have five units in their cache, while the combination of any three users should have all six units in their cache. This strongly suggests placing each piece A_i (and B_i) at two users. Since each Z_i has four units, but it needs to hold three units from each of the two files, coded placement (cross files) is thus needed. At this point, we place the corresponding symbols in the caching, but keep the precise linear combination coefficients as undetermined.
- The next critical observation is that $H(X_{1,2,2,2} W_1) = H(X_{1,1,1,2} W_1) = H(X_{1,1,2,2} W_1) = 3$. This implies that the transmission has three units of information on each file alone. However, since the operating point dictates that $H(X_{1,2,2,2}) = H(X_{1,1,1,2}) = H(X_{1,1,2,2}) = 6$, it further implies that in each transmission, three units are for the linear combinations of W_2, and 3 units are for those of W_1; in other words, the linear combinations do not need to mix information from different files.
- Since each transmission only has three units of information from each file, and each user has only three units of information from each file, they must be linearly independent of each other.

The observation and deductions are only from the perspective of the joint entropies given in Table 3, without much consideration of the particular coding requirement. For example, in the last item discussed above, it is clear that when transmitting the three units of information regarding a file (say file W_2), they should be simultaneously useful to other users requesting this file, and to the users not requesting this file. This intuition then strongly suggests each transmitted linear combination of W_2 should be a subspace of the W_2 parts at some users not requesting it. Using these intuitions as guidance, finding the code becomes straightforward after trial-and-error. In [20], we were able to further generalize this special code to a class of codes for any case when $N \leq K$; readers are referred to [20] for more details on these codes.

Table 3. Stable joint entropy values at the corner point $\left(\frac{2}{3}, 1\right)$ for $(N, K) = (2, 4)$.

Joint Entropy	Computed Value
$H(Z_1 W_1)$	3
$H(Z_1, Z_2 W_1)$	5
$H(Z_1, Z_2, Z_3 W_1)$	6
$H(X_{1,2,2,2} W_1)$	3
$H(Z_1, X_{1,2,2,2} W_1)$	4
$H(X_{1,1,1,2} W_1)$	3
$H(Z_1, X_{1,1,1,2} W_1)$	4
$H(Z_1, Z_2, X_{1,1,1,2} W_1)$	5
$H(X_{1,1,2,2} W_1)$	3
$H(Z_1, X_{1,1,2,2} W_1)$	4
$H(Z_1, Z_2, X_{1,1,2,2} W_1)$	5

5.4. Disproving Linear-Coding Achievability

The reverse engineering approach may not always be successful, either because the structure revealed by the data is very difficult to construct explicitly, or because linear codes are not sufficient to achieve this operating point. In some other cases, the determination can be done explicitly. In the sequel, we present an example for $(N, K) = (3, 3)$, which belongs to the latter case. An outer bound for $(N, K) = (3, 3)$ is presented in the next section, and among the corner points, the pair $(M, R) = \left(\frac{2}{3}, \frac{4}{3}\right)$ is the only one that cannot be achieved by existing schemes. Since the outer bound appears quite strong, we may conjecture this pair to be also achievable and attempt to construct a code. Unfortunately, as we shall show next, there does not exist such a (vector) linear code. Before delving into the data provided by the LP, readers are encouraged to consider proving directly that this tradeoff point cannot be achieved by linear codes, which does not appear to be straightforward to the author.

We shall assume each file has $3m$ symbols in a certain finite field, where m is a positive integer. The LP produces the joint entropy values (in terms of the number of finite field symbols, not in multiples of file size as in the other sections of the paper) in Table 4 at this corner point, where only the conditional joint entropies relevant to our discussion next are listed. The main idea is to use these joint entropy values to deduce structures of the coding matrices, and then combining these structures with the coding requirements to reach a contradiction.

Table 4. Stable joint entropy values at the corner point $\left(\frac{2}{3}, \frac{4}{3}\right)$ for $(N, K) = (3, 3)$.

Joint Entropy	Computed Value
$H(Z_1 W_1)$	$2m$
$H(Z_1 W_1, W_2)$	m
$H(Z_1, Z_2 W_1, W_2)$	$2m$
$H(Z_1, Z_2, Z_3 W_1, W_2)$	$3m$
$H(X_{1,2,3})$	$4m$
$H(X_{1,2,3} W_1)$	$3m$
$H(X_{1,2,3} W_1, W_2)$	$2m$

The first critical observation is that $H(Z_1 W_1, W_2) = m$, and the user-index-symmetry implies that $H(Z_2 W_1, W_2) = H(Z_3 W_1, W_2) = m$. Moreover $H(Z_1, Z_2, Z_3 W_1, W_2) = 3m$, from which we can conclude that excluding file W_1 and W_2, each user stores m linearly independent combinations of the symbols of file W_3, which are also linearly independent among the three users. Similar conclusions hold for files W_1 and W_2. Thus, without loss of generality, we can view the linear combinations of W_i cached by the users, after excluding the symbols from the other two files, as the basis of file W_i. In other words, this implies that through a change of basis for each file, we can assume without loss of generality that user k stores $2m$ linear combinations in the following form:

$$V_k \cdot \begin{bmatrix} W_{1,[(k-1)m+1:km]} \\ W_{2,[(k-1)m+1:km]} \\ W_{3,[(k-1)m+1:km]} \end{bmatrix} \tag{43}$$

where $W_{n,j}$ is the j-th symbol of the n-th file and V_k is a matrix of dimension $2m \times 3m$; V_k can be partitioned into submatrices of dimension $m \times m$, which are denoted as $V_{k;i,j}$, $i = 1, 2$ and $j = 1, 2, 3$. Note that symbols at different users are orthogonal to each other without loss of generality.

Without loss of generality, assume the transmitted content $X_{1,2,3}$ is:

$$G \cdot \begin{bmatrix} W_{1,[1:3m]} \\ W_{2,[1:3m]} \\ W_{3,[1:3m]} \end{bmatrix} \tag{44}$$

where G is a matrix of dimension $4m \times 9m$; we can partition it into blocks of $m \times m$, and each block is referred to as $G_{i,j}$, $i = 1, 2, \ldots, 4$ and $j = 1, 2, \ldots, 9$. Let us first consider User 1, which has the following symbols:

$$\left[\begin{array}{ccccccccc} V_{k;1,1} & 0 & 0 & V_{k;1,2} & 0 & 0 & V_{k;1,3} & 0 & 0 \\ V_{k;2,1} & 0 & 0 & V_{k;2,2} & 0 & 0 & V_{k;2,3} & 0 & 0 \\ \hline G_{1,1} & G_{1,2} & & & \cdots & & & & G_{1,9} \\ \vdots & \vdots & & & \vdots & & & & \vdots \\ G_{4,1} & G_{4,2} & & & \cdots & & & & G_{4,9} \end{array} \right] \cdot \begin{bmatrix} W_{1,[1:3m]} \\ W_{2,[1:3m]} \\ W_{3,[1:3m]} \end{bmatrix} \tag{45}$$

The coding requirement states that $X_{1,2,3}$ and Z_1 together can be used to recover file W_1, and thus, one can recover all the symbols of W_1 knowing (45). Since W_1 can be recovered, its symbols can be eliminated in (45), i.e.,

$$\left[\begin{array}{cccccc} V_{k;1,2} & 0 & 0 & V_{k;1,3} & 0 & 0 \\ V_{k;2,2} & 0 & 0 & V_{k;2,3} & 0 & 0 \\ \hline G_{1,4} & G_{1,5} & & \cdots & & G_{1,9} \\ \vdots & \vdots & & \vdots & & \vdots \\ G_{4,4} & G_{4,4} & & \cdots & & G_{4,9} \end{array} \right] \cdot \begin{bmatrix} W_{2,[1:3m]} \\ W_{3,[1:3m]} \end{bmatrix} \tag{46}$$

in fact becomes known. Notice Table 4 specifies $H(Z_1 W_1) = 2m$, and thus, the matrix:

$$\begin{bmatrix} V_{k;1,2} & V_{k;1,3} \\ V_{k;2,2} & V_{k;2,3} \end{bmatrix} \tag{47}$$

is in fact full rank; thus, from the top part of (46), $W_{2,[1:m]}$ and $W_{3,[1:m]}$ can be recovered. In summary, through elemental row operations and column permutations, the matrix in (45) can be converted into the following form:

$$
\begin{bmatrix}
U_{1,1} & U_{1,2} & U_{1,3} & 0 & & \cdots & & & 0 \\
U_{2,1} & U_{2,2} & U_{2,3} & 0 & & \cdots & & & 0 \\
U_{3,1} & U_{3,2} & U_{3,3} & 0 & & \cdots & & & 0 \\
0 & 0 & 0 & U_{4,4} & U_{5,7} & 0 & 0 & 0 & 0 \\
0 & 0 & 0 & U_{4,4} & U_{5,7} & 0 & 0 & 0 & 0 \\
0 & 0 & 0 & 0 & 0 & U_{6,5} & U_{6,6} & U_{6,8} & U_{6,9}
\end{bmatrix}
\cdot
\begin{bmatrix}
W_{1,[1:3m]} \\
W_{2,[1:m]} \\
W_{3,[1:m]} \\
W_{2,[m+1:3m]} \\
W_{3,[m+1:3m]}
\end{bmatrix} ,
\tag{48}
$$

where diagonal block square matrices are of full rank $3m$ and $2m$, respectively, and $U_{i,j}$'s are the resultant block matrices after the row operations and column permutations. This further implies that the matrix $[U_{6,5}, U_{6,6}, U_{6,8}, U_{6,9}]$ has maximum rank m, and it follows that the matrix:

$$
\begin{bmatrix}
G_{1,5} & G_{1,6} & G_{1,8} & G_{1,9} \\
\vdots & \vdots & \vdots & \vdots \\
G_{4,5} & G_{4,6} & G_{4,8} & G_{4,9}
\end{bmatrix} ,
\tag{49}
$$

i.e., the submatrix of G by taking thick columns $(5,6,8,9)$ has only maximum rank m. However, due to the symmetry, we can also conclude that the submatrix of G taking only thick columns $(1,3,7,9)$ and that taking only thick columns $(1,2,4,5)$ both have only maximum rank m. As a consequence, the matrix G has rank no larger than $3m$, but this contradicts the condition that $H(X_{1,2,3}) = 4m$ in Table 4. We can now conclude that this memory-transmission-rate pair is not achievable with any linear codes

Strictly speaking, our argument above holds under the assumption that the joint entropy values produced by LP are precise rational values, and the machine precision issue has thus been ignored. However, if the solution is accurate only up to machine precision, one can introduce a small slack value δ into the quantities, e.g., replacing $3m$ with $(3 \pm \delta)m$, and using a similar argument show that the same conclusion holds. This extended argument however becomes notationally rather lengthy, and we thus omitted it here for simplicity.

6. Computational Exploration and Bounds for Larger Cases

In this section, we explore the fundamental limits of the caching systems in more detail using a computational approach. Due to the (doubly) exponential growth of the LP variables and constraints, directly applying the method outlined in Section 2 becomes infeasible for larger problem cases. This is the initial motivation for us to investigate single-demand-type systems where only a single demand type is allowed. Any outer bound on the tradeoff of such a system is an outer bound for the original one, and the intersection of these outer bounds is thus also an outer bound. This investigation further reveals several hidden phenomena. For example, outer bounds for different single-demand-type systems are stronger in different regimes, and moreover, the LP bound for the original system is not simply the intersection of all outer bounds for single-demand-type systems; however, in certain regimes, they do match.

Given the observations above, we take the investigation one step further by choosing only a small subset of demands instead of the complete set in a single demand type. This allows us to obtain results for cases which initially appear impossible to compute. For example, even for $(N, K) = (2, 5)$, there is a total of $2 + 5 + 2^5 = 39$ random variables, and the number of constraints in LP after symmetry reduction is more than 10^{11}, which is significantly beyond current LP solver capability (the problem can be further reduced using problem specific implication structures as outlined in Section 2, but our experience suggests that even with such additional reduction the problem may still too large for a start-of-the-art LP solver). However, by strategically considering only a small subset of the demand patterns, we are indeed able to find meaningful outer bounds, and moreover, use the clues obtained in such computational exploration to complete the proof of Theorem 4. We shall discuss the method we develop, and also present several example results for larger problem cases.

6.1. Single-Demand-Type Systems

As mentioned above, in a single-demand-type caching systems, the demand must belong to a particular demand type. We first present results on two cases $(N, K) = (2, 4)$ and $(N, K) = (3, 3)$, and then discuss our observations using these results.

Proposition 7. *Any memory-transmission-rate tradeoff pair for the* $(N, K) = (2, 4)$ *caching problem must satisfy the following conditions for single-demand-type* $(4, 0)$:

$$M + 2R \geq 2, \tag{50}$$

and conversely any non-negative (M, R) *pair satisfying (50) is achievable for single-demand-type* $(4, 0)$; *it must satisfy for single-demand-type* $(3, 1)$:

$$2M + R \geq 2, \quad 8M + 6R \geq 11, \quad 3M + 3R \geq 5, \quad 5M + 6R \geq 9, \quad M + 2R \geq 2, \tag{51}$$

and conversely any non-negative (M, R) *pair satisfying (51) is achievable for single-demand-type* $(3, 1)$; *it must satisfy for single-demand-type* $(2, 2)$

$$2M + R \geq 2, \quad 3M + 3R \geq 5, \quad M + 2R \geq 2, \tag{52}$$

and conversely any non-negative (M, R) *pair satisfying (52) is achievable for single-demand-type* $(2, 2)$.

The optimal (M, R) tradeoffs are illustrated in Figure 4 with the known inner bound, i.e., those in [6,15], and the one given in the last section, and the computed out bound of the original problem given in Section 4. Here, the demand type $(3, 1)$ in fact provides the tightest outer bound, which matches the known inner bound for $M \in [0, 1/4] \cup [2/3, 2]$. The converse proofs of (51) and (52) are obtained computationally, the details of which can be found in Appendix G. In fact, only the middle three inequalities in (51) and the second inequality in (52) need to be proven, since the others are due to the cut-set bound. Although the original caching problem requires codes that can handle all types of demands, the optimal codes for single-demand-type systems turn out to be quite interesting by their own right, and thus, we provide the forward proof of Theorem 7 in Appendix H.

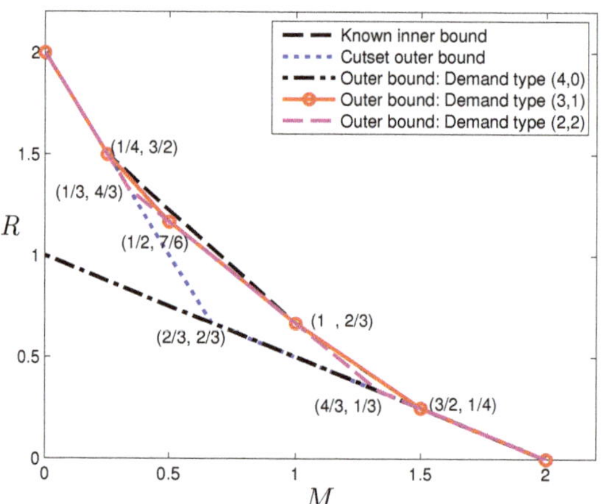

Figure 4. Tradeoff outer bounds for $(N, K) = (2, 4)$ caching systems.

The computed outer bounds for single-demand-type systems for $(N, K) = (3, 3)$ are summarized below; the proofs can be found in Appendix I.

Proposition 8. *Any memory-transmission-rate tradeoff pair for the $(N, K) = (3, 3)$ caching problem must satisfy the following conditions for single-demand-type $(3, 0, 0)$:*

$$M + 3R \geq 3, \tag{53}$$

and conversely any non-negative (M, R) pair satisfying (53) is achievable for single-demand-type $(3, 0, 0)$; it must satisfy for single-demand-type $(2, 1, 0)$:

$$M + R \geq 2, \quad 2M + 3R \geq 5, \quad M + 3R \geq 3, \tag{54}$$

and conversely any non-negative (M, R) pair satisfying (54) is achievable for single-demand-type $(2, 1, 0)$; it must satisfy for single-demand-type $(1, 1, 1)$:

$$3M + R \geq 3, \, 6M + 3R \geq 8, \, M + R \geq 2, \, 12M + 18R \geq 29, \, 3M + 6R \geq 8, \, M + 3R \geq 3. \tag{55}$$

These outer bounds are illustrated in Figure 5, together with the best known inner bound by combining [6,15], and the cut-set outer bound for reference. The bound is in fact tight for $M \in [0, 1/3] \cup [1, 3]$. Readers may notice that Proposition 8 provides complete characterizations for the first two demand types, but not the last demand type. As we have shown in Section 5, the point $(\frac{2}{3}, \frac{4}{3})$ in fact cannot be achieved using linear codes.

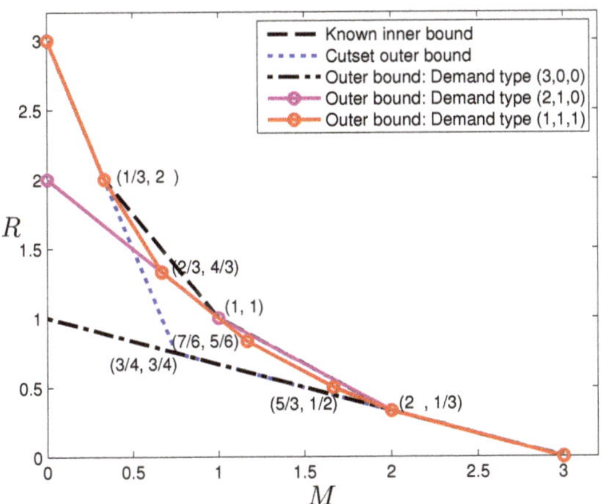

Figure 5. Tradeoff outer bounds for $(N, K) = (3, 3)$ caching.

Remark 3. *The bound developed in [13] gives $6M + 3R \geq 8$ and $2M + 4R \geq 5$, and that in [14] gives $(M + R) \geq 2$ in addition to the cut-set bound.*

We can make the following observations immediately:

- The single-demand-type systems for few files usually produce tighter bounds at high memory regimes, while those for more files usually produce tighter bounds at low memory regimes. For example, the first high-memory segment of the bounds can be obtained by considering only demands that request a single file, which coincidentally is also the cut-set bound; for

$(N, K) = (3, 3)$, the bound obtained from the demand type $(2, 1, 0)$ is stronger than that from $(1, 1, 1)$ in the range $M \in [1, 2]$.

- Simply intersecting the single-demand-type outer bounds does not produce the same bound as that obtained from a system with the complete set of demands. This can be seen from the case $(N, K) = (2, 4)$ in the range $M \in [1/4, 2/3]$.
- The outer bounds produced by single-demand-type systems in many cases match the bound when more comprehensive demands are considered. This is particularly evident in the case $(N, K) = (2, 4)$ in the range $M \in [0, 1/4] \cup [2/3, 2]$.

These observations provide further insights on the difficulty of the problem. For instance, for $(N, K) = (2, 4)$, the demand type $(3, 1)$ is the most demanding case, and code design for this demand type should be considered as the main challenge. More importantly, these observation suggests that it is possible to obtain very strong bounds by considering only a small subset of demands, instead of the complete set of demands. In the sequel, we further explore this direction.

6.2. Equivalent Bounds Using Subsets of Demands

Based on the observations in the previous subsection, we conjecture that in some cases, equivalent bounds can be obtained by using only a smaller number of requests, and moreover, these demands do not need to form a complete demand type class; next, we show that this is indeed the case. To be more precise, we are relaxing the LP, by including only elemental inequality constraints that involve joint entropies of random variables within a subset of the random variables $\mathcal{W} \cup \mathcal{Z} \cup \mathcal{X}$, and other constraints are simply removed. However, the symmetry structure specified in Section 3 is still maintained to reduce the problem. This approach is not equivalent to forming the LP on a caching system where only those files, users and demands are present, since in this alternative setting, symmetric solutions may induce loss of optimality.

There are many choices of subsets with which the outer bounds can be computed, and we only provide a few that are more relevant, which confirm our conjecture:

Fact 1. In terms of the computed outer bounds, the following facts were observed:

- For the $(N, K) = (2, 4)$ case, the outer bound in Proposition 6 can be obtained by restricting to the subset of random variables $\mathcal{W} \cup \mathcal{Z} \cup \{X_{1,1,1,2}, X_{1,1,2,2}\}$.
- For the $(N, K) = (2, 4)$ case, the outer bound in Proposition 7 in the range $M \in [1/3, 2]$ for single-demand-type $(3, 1)$ can be obtained by restricting to the subset of random variables $\mathcal{W} \cup \mathcal{Z} \cup \{X_{2,1,1,1}, X_{1,2,1,1}, X_{1,1,2,1}, X_{1,1,1,2}\}$.
- For the $(N, K) = (3, 3)$ case, the intersection of the outer bounds in Proposition 8 can be obtained by restricting to the subset of random variables $\mathcal{W} \cup \mathcal{Z} \cup \{X_{2,1,1}, X_{3,1,1}, X_{3,2,1}\}$.
- For the $(N, K) = (3, 3)$ case, the outer bound in Proposition 8 in the range $M \in [2/3, 3]$ for single-demand-type $(2, 1)$ can be obtained by restricting to the subset of random variables $\mathcal{W} \cup \mathcal{Z} \cup \{X_{2,1,1}, X_{3,1,1}\}$.

These observations reveal that the subset of demands can be chosen rather small to produce strong bounds. For example, for the $(N, K) = (2, 4)$ case, including only joint entropies involving eight random variables $\mathcal{W} \cup \mathcal{Z} \cup \{X_{1,1,1,2}, X_{1,1,2,2}\}$ will produce the strongest bound as including all 22 random variables. Moreover, for specific regimes, the same bound can be produced using an even smaller number of random variables (for the case $(N, K) = (3, 3)$), or with a more specific set of random variables (for the case $(N, K) = (2, 4)$, where in the range $[1/3, 2]$, including only some of the demand type $(3, 1)$ is sufficient). Equipped with these insights, we can attempt to tackle larger problem cases, for which it would have appeared impossible to produce computationally meaningful outer bounds. In the sequel, this approach is applied for two purposes: (1) to identify generic structures in converse proofs, and (2) to produce outer bounds for large problem cases.

6.3. Identifying Generic Structures in Converse Proofs

Recall our comment given after the proof of Theorem 4 that finding this proof is not straightforward. One critical clue was obtained when applying the exploration approach discussed above. When restricting the set of included random variables to a smaller set, the overall problem is relaxed; however, if the outer bound thus obtained remains the same, it implies that the sought-after outer bound proof only needs to rely on the joint entropies within this restricted set. For the specific case of $(N, K) = (2, 5)$, we have the following fact.

Fact 2. For $(N, K) = (2, 5)$, the bound $15M + 20R \geq 28$ in the range $M \in [6/5, 8/5]$ can be obtained by restricting to the subset of random variables $\mathcal{W} \cup \mathcal{Z} \cup \{X_{2,1,1,1,1}, X_{1,2,1,1,1}, X_{1,1,2,1,1}, X_{1,1,1,2,1}, X_{1,1,1,1,2}\}$.

Together with the second item in Fact 1, we can naturally conjecture that in order to prove the hypothesized outer bound, only the dependence structure within the set of random variables $\mathcal{W} \cup \mathcal{Z} \cup X_{\rightarrow[1:K]}$ needs to be considered, and all the proof steps can be written using mutual information or joint entropies of them alone. Although this is still not a trivial task, the possibility is significantly reduced, e.g., for the $(N, K) = (2, 5)$ case to only 12 random variables, with a much simpler structure than that of the original problem with 39 random variables. Perhaps more importantly, such a restriction makes it feasible to identify a common route of derivation in the converse proof and then generalize it, from which we obtain the proof of Theorem 4.

6.4. Computing Bounds for Larger Problem Cases

We now present a few outer bounds for larger problem cases, and make comparison with other known bounds in the literature. This is not intended to be a complete list of results we obtain, but these are perhaps the most informative.

In Figure 6, we provide results for $(N, K) = (4, 3)$, $(N, K) = (5, 3)$ and $(N, K) = (6, 3)$. Included are the computed outer bounds, the inner bound by the scheme in [6], the cut-set outer bounds, and for reference, the outer bounds given in [12]. We omit the bounds in [13,14] to avoid too much clutter in the plot; however, they do not provide better bounds than that in [12] for these cases. It can be seen that the computed bounds are in fact tight in the range $M \in [4/3, 4]$ for $(N, K) = (4, 3)$, $M \in [5/3, 5]$ for $(N, K) = (5, 3)$, and tight in general for $(N, K) = (6, 3)$; in these ranges, the scheme given in [6] is in fact optimal. Unlike our computed bounds, the outer bound in [12] does not provide additional tight results beyond those already determined using the cut-set bound, except the single point $(M, R) = (2, 1)$ for $(N, K) = (6, 3)$.

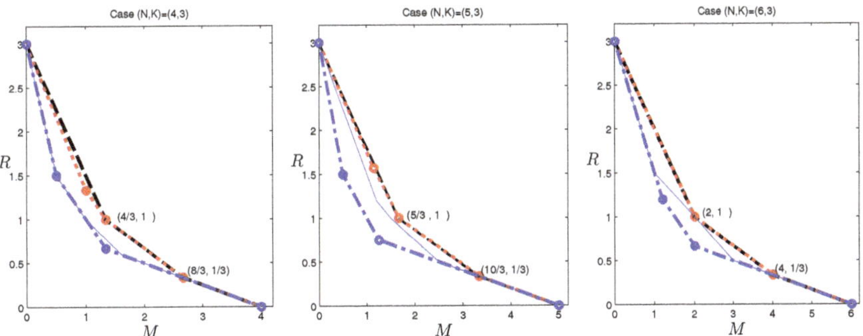

Figure 6. The computed outer bounds for $(N, K) = (4, 3)$, $(N, K) = (5, 3)$ and $(N, K) = (6, 3)$ caching systems. The red dotted lines give the computed outer bounds; the blue dashed-dot lines are the cut-set outer bounds; the black dashed lines are the inner bound using the scheme in [6]; and the thin blue lines are the outer bounds given in [12]. Only nontrivial outer bound corner points that match inner bounds are explicitly labeled.

In Figure 7, we provide results for $(N,K) = (3,4)$, $(N,K) = (3,5)$ and $(N,K) = (3,6)$. Included are the computed outer bounds, the inner bound by the code in [6] and that in [20], the cut-set outer bound, and for reference, the outer bounds in [12]. The bounds in [13,14] are again omitted. It can be seen that the computed bounds are in fact tight in the range $M \in [0,1/4] \cup [3/2,3]$ for $(N,K) = (3,4)$, $M \in [0,1/5] \cup [6/5,3]$ for $(N,K) = (3,5)$, and $M \in [0,1/6] \cup [3/2,3]$ for $(N,K) = (3,6)$. Generally, in the high memory regime, the scheme given in [6] is in fact optimal, and in the low memory regime, the schemes in [15,20] are optimal. It can be see that the outer bound in [12] does not provide additional tight results beyond those already determined using the cut-set bound. The bounds given above in fact provide grounds and directions for further investigation and hypotheses on the optimal tradeoff, which we are currently exploring.

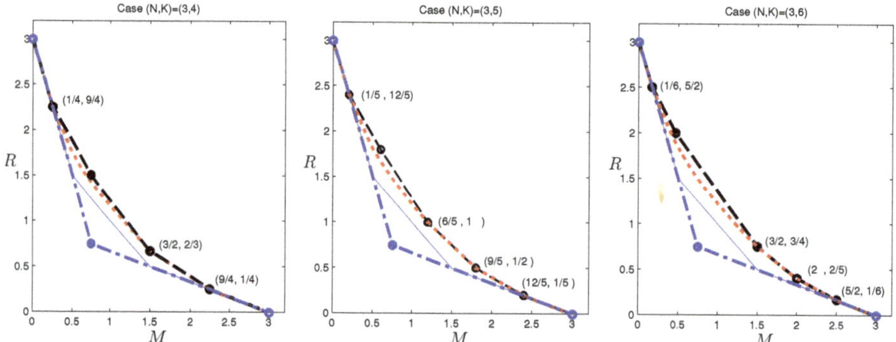

Figure 7. The computed outer bounds for $(N,K) = (3,4)$, $(N,K) = (3,5)$ and $(N,K) = (3,6)$ caching systems. The red dotted lines give the computed outer bounds; the blue dashed-dot lines are the cut-set outer bounds; the black dashed lines are the inner bound using the scheme in [6,20]; and the thin blue lines are the outer bounds given in [12]. Only nontrivial outer bound corner points that match inner bounds are explicitly labeled.

7. Conclusions

We presented a computer-aided investigation on the fundamental limit of the caching problem, including data-driven hypothesis forming, which leads to several complete or partial characterizations of the memory-transmission-rate tradeoff, a new code construction reverse-engineered through the computed outer bounding data and a computerized exploration approach that can reveal hidden structures in the problem and also enables us to find surprisingly strong outer bounds for larger problem cases.

It is our belief that this work provides strong evidence of the effectiveness of the computer-aided approach in the investigation of the fundamental limits of communication, data storage and data management systems. Although at first sight, the exponential growth of the LP problem would prevent any possibility of obtaining meaningful results on engineering problems of interest, our experience in [2,3] and the current work suggest otherwise. By incorporating the structure of the problem, we develop more domain-specific tools in such investigations and were able to obtain results that appear difficult for human experts to obtain directly.

Our effort can be viewed as both data-driven and computational, and thus, more advanced data analysis and machine learning technique may prove useful. Particularly, the computer-aided exploration approach is clearly a human-in-the-loop process, which can benefit from more automation based on reinforcement learning techniques. Moreover, the computed generated proofs may involve a large number of inequalities and joint entropies, and more efficient classification or clustering of these inequalities and joint entropies can reduce the human burden in the subsequent analysis. It is our hope that this work can serve as a starting point to introduce more machine intelligence and the corresponding computer-aided tools into information theory and communication research in the future.

Funding: This research was funded in part by the National Science Foundation under Grants CCF-15-26095 and CCF-18-32309.

Acknowledgments: The author wishes to thank Urs Niesen and Vaneet Aggarwal for early discussions, which partly motivated this work. He also wishes to thank Jun Chen for several discussions, as well as the insightful comments on an early draft. Additionally, the author wishes to thank the authors of [12] for making the source code to compute their proposed outer bounds available online, which was conveniently used to generate some of the comparisons.

Conflicts of Interest: The author declares no conflict of interest.

Appendix A. Finding Corner Points of the LP Outer Bounds

Since this is an LP problem, and also due to the problem setting, only the lower hull of the outer bound region between the two quantities M and R is of interest. The general algorithm in [27] is equivalent to the procedure given in Algorithm 1 in this specific setting. In this algorithm, the set $\mathcal{P}$ in the input is the initial extreme points of the tradeoff region, which are trivially known from the problem setting. The variables and constraints in the LP are given as outlined in Section 2 for a fixed (N, K) pair, which are populated and considered fixed. The output set $\mathcal{P}$ is the final computed extreme points of the outer bound. The algorithm can be intuitively explained as follows: starting with two known extreme points, if there are any other corner points, they must lie below the line segment connecting these two points, and thus, an LP that minimizes the bounding plane along the direction of this line segment must be able to find a lower value; if so, the new point is also an extreme point, and we can repeat this procedure again.

In the caching problem, the tradeoff is between two quantities M and R. We note here if there are more than two quantities that need to be considered in the tradeoff, the algorithm is more involved, and we refer the readers to [27,28] for more details on such settings.

Algorithm 1: An algorithm to identify the corner points of the LP outer bound.

   Input : $N, K, \mathcal{P} = \{(N,0), (0, \min(N,K))\}$
   Output: $\mathcal{P}$
1 $n = 2; i = 1;$
2 **while** $i < n$ **do**
3 | Compute the line segment connecting i-th and $(i+1)$-th (M, R) pairs in $\mathcal{P}$, as $M + \alpha R = \beta$;
4 | Set the objective of the LP as $M + \alpha R$, and solve LP for solution (M^*, R^*) and objective β^*;
5 | **if** $\beta^* < \beta$ **then**
6 | | Insert (M^*, R^*) in $\mathcal{P}$ between the i-th and $(i+1)$-th (M, R) pairs;
7 | | $n = n + 1;$
8 | **else**
9 | | $i = i + 1;$
10 | **end**
11 **end**

Appendix B. Proofs of Proposition 3 and Proposition 4

The proof of Proposition 3 is given in the Tables A1 and A2, and that of Proposition 4 is given in the Tables A3 and A4. Each row in Tables A2 and A4, except the last rows, are simple and known information inequalities, up to the symmetry defined in Section 3. The last rows in Tables A2 and A4 are the sum of all previous rows, which are the sought-after inequalities, and they are simply the consequences of the known inequalities summed together. When represented in this form, the correctness of the proof is immediate, since the columns representing quantities not present in the final bound cancel out each other when being summed together. The rows in Table A2 are labeled, and it has more details in order to illustrate the meaning and usage of the tabulation proof in the example we provide next.

As mentioned previously, each row in Table A2 is an information inequality, which involves multiple joint entropies, but can also be represented in a mutual information form. For example, Row (2) is read as:

$$2T_8 - T_4 - T_{10} \geq 0, \tag{A1}$$

and in the last, but one column of Table A2, an information inequality is given, which is an equivalent representation as a mutual information quantity:

$$I(Z_1; Z_2 W_1) \geq 0, \tag{A2}$$

which can be seen by simply expanding the mutual information as:

$$
\begin{aligned}
I(Z_1; Z_2 W_1) &= H(Z_1, W_1) + H(Z_2, W_1) - H(W_1) - H(Z_1, Z_2, W_1) \\
&= 2H(Z_1, W_1) - H(W_1) - H(Z_1, Z_2, W_1) \\
&= 2T_8 - T_4 - T_{10}.
\end{aligned}
\tag{A3}
$$

Directly summing up these information inequalities and canceling out redundant terms will directly result in the bound $2R + 2H(Z_1) - 4F \geq 0$, which clearly can be used to write $2R + 2M - 4F \geq 0$.

Using these proof tables, one can write down different versions of proofs, and one such example is provided next based on Tables A1 and A2 for Proposition 3 by invoking the inequalities in Table A2 one by one.

$$
\begin{aligned}
2M + 2R &\overset{(1)}{\geq} 2H(X_{1,2}) + 2H(Z_1) \overset{(3)}{\geq} 2H(Z_1, X_{1,2}) \\
&\overset{(5)}{\geq} 2H(Z_1, X_{1,2}) - 2I(X_{1,2}; Z_2 Z_1, W_1) \\
&= 2H(Z_1, X_{1,2}, W_1) - 2I(X_{1,2}; Z_2 Z_1, W_1) \\
&\overset{(c)}{=} 2H(Z_1, Z_2, W_1, X_{1,2}) + 2H(Z_1, W_1) - 2H(Z_1, Z_2, W_1) \\
&\overset{(2)}{\geq} 2H(Z_1, Z_2, W_1, X_{1,2}) + 2H(Z_1, W_1) - 2H(Z_1, Z_2, W_1) - I(Z_1; Z_2 W_1) \\
&= 2H(Z_1, Z_2, W_1, X_{1,2}) - H(Z_1, Z_2, W_1) + H(W_1) \\
&\overset{(4)}{\geq} 2H(Z_1, Z_2, W_1, X_{1,2}) - H(Z_1, Z_2, W_1) + H(W_1) - I(X_{1,2}; X_{1,3} Z_1, Z_2, W_1) \\
&= H(W_1) + H(W_1, W_2, W_3) \\
&\overset{(6,7)}{\geq} 4F,
\end{aligned}
\tag{A4}
$$

where the inequalities match precisely the rows in Table A2, and the equality labeled (c) indicates the decoding requirement is used. In this version of the proof, we applied the inequalities in the order of (1)-(3)-(5)-(2)-(4)-(6,7), but this is by no means critical, as any order will yield a valid proof. One can similarly produce many different versions of proofs for Proposition 4 based on Tables A3 and A4.

Table A1. Terms needed to prove Proposition 3.

T_1	F
T_2	R
T_3	$H(X_{1,2})$
T_4	$H(W_1)$
T_5	$H(W_1, W_2, W_3)$
T_6	$H(Z_1)$
T_7	$H(Z_1, X_{1,2})$
T_8	$H(Z_1, W_1)$
T_9	$H(Z_1, Z_2, X_{1,2})$
T_{10}	$H(Z_1, Z_2, W_1)$

Table A2. Proof by Tabulation of Proposition 3, with terms defined in Table A1.

T_1	T_2	T_3	T_4	T_5	T_6	T_7	T_8	T_9	T_{10}		
	2	−2								$2(R - H(X_{1,2})) \geq 0$	(1)
			−1				2		−1	$I(Z_1; Z_2 W_1) \geq 0$	(2)
		2			2	−2				$2I(X_{1,2}; Z_1) \geq 0$	(3)
				−1				2	−1	$I(X_{1,2}; X_{1,3} Z_1, Z_2, W_1) \geq 0$	(4)
						2	−2	−2	2	$2I(X_{1,2}; Z_2 Z_1, W_1) \geq 0$	(5)
−1			1							$H(W_1) - F \geq 0$	(6)
−3				1						$H(W_1, W_2, W_3) - 3F \geq 0$	(7)
−4	2					2				$2R + 2H(Z_1) - 4F \geq 0$	

Table A3. Terms needed to prove Proposition 4.

T_1	F
T_2	R
T_3	$H(X_{1,2})$
T_4	$H(W_1)$
T_5	$H(W_1, X_{1,2})$
T_6	$H(W_1, X_{1,3}, X_{2,1})$
T_7	$H(W_1, W_2)$
T_8	$H(W_1, W_3, X_{1,2})$
T_9	$H(W_1, W_2, W_3, W_4)$
T_{10}	$H(Z_1)$
T_{11}	$H(Z_1, X_{1,2})$
T_{12}	$H(Z_1, X_{1,3}, X_{2,1})$
T_{13}	$H(Z_1, X_{1,4}, X_{2,3})$
T_{14}	$H(Z_1, W_1)$
T_{15}	$H(Z_1, W_3, X_{1,2})$
T_{16}	$H(Z_1, W_2, X_{1,2})$
T_{17}	$H(Z_1, W_1, W_2)$
T_{18}	$H(Z_1, W_2, W_3, X_{1,2})$
T_{19}	$H(Z_1, Z_2, X_{1,2})$
T_{20}	$H(Z_1, Z_2, X_{1,3}, X_{2,1})$
T_{21}	$H(Z_1, Z_2, W_3, X_{1,2})$
T_{22}	$H(Z_1, Z_2, W_1, W_2)$

Table A4. Proof by Tabulation of Proposition 4, with terms defined in Table A3.

T_1	T_2	T_3	T_4	T_5	T_6	T_7	T_8	T_9	T_{10}	T_{11}	T_{12}	T_{13}	T_{14}	T_{15}	T_{16}	T_{17}	T_{18}	T_{19}	T_{20}	T_{21}	T_{22}
																		1			-1
	8	-8																			
		8																			
									8	-8											
									-2	4		-2									
								-1										-1	2		
					-2						4								-2		
			-2	2									2		-2						
						-1	1									1	-1				
										2	-2		-2		2						
														1		-1				-1	1
			-2	2						2	-2										
							-1							1			1			-1	
								-2				2		-2						2	
-2			2																		
-2							1														
-12									3												
-16	8								6												

Appendix C. Proofs of Lemma 1 and Theorem 3

Proofs of Lemma 1. We first write the following chain of inequalities:

$$(N-n)H(Z_1,W_{[1:n]},X_{n,n+1}) = (N-n)\left[H(Z_1,W_{[1:n]}) + H(X_{n,n+1}Z_1,W_{[1:n]})\right]$$

$$\overset{(a)}{=} (N-n)H(Z_1,W_{[1:n]}) + \sum_{i=n+1}^{N} H(X_{n,i}Z_1,W_{[1:n]})$$

$$\geq (N-n)H(Z_1,W_{[1:n]}) + H(X_{n,[n+1:N]}Z_1,W_{[1:n]}) \tag{A5}$$

$$= (N-n-1)H(Z_1,W_{[1:n]}) + H(X_{n,[n+1:N]},Z_1,W_{[1:n]}),$$

where (a) is because of the file-index-symmetry. Next, notice that by the user-index-symmetry:

$$H(Z_1,W_{[1:n]}) = H(Z_2,W_{[1:n]}), \tag{A6}$$

which implies that:

$$H(Z_1,W_{[1:n]}) + H(X_{n,[n+1:N]},Z_1,W_{[1:n]}) \geq H(Z_2,W_{[1:n]}) + H(X_{n,[n+1:N]},W_{[1:n]})$$

$$\overset{(b)}{\geq} H(W_{[1:n]}) + H(X_{n,[n+1:N]},Z_2,W_{[1:n]})$$

$$\overset{(c)}{=} H(W_{[1:n]}) + H(X_{n,[n+1:N]},Z_2,W_{[1:N]}) \tag{A7}$$

$$= H(W_{[1:n]}) + H(W_{[1:N]}) = N + n,$$

where (b) is by the sub-modularity of the entropy function, and (c) is because of (3). Now, substituting (A7) into (A5) gives (28), which completes the proof. $\square$

We are now ready to prove Theorem 3.

Proof of Theorem 3. For $N \geq 3$, it can be verified that the three corner points of the given tradeoff region are:

$$(0,2), \quad \left(\frac{N}{2},\frac{1}{2}\right), \quad (N,0), \tag{A8}$$

which are achievable using the codes given in [6]. The outer bound $M + NR \geq N$ can also be obtained as one of the cut-set outer bounds in [6], and it only remains to show that the inequality $3M + NR \geq 2N$ is true. For this purpose, we claim that for any integer $n \in \{1,2,\dots,N-2\}$:

$$3M + NR \geq 3 \sum_{j=1}^{n} \left[\frac{N+j}{N-j} \prod_{i=1}^{j-1} \frac{N-(i+2)}{N-i} \right] + 3 \prod_{j=1}^{n} \frac{N-(j+2)}{N-j} H(Z_1, W_{[1:n]})$$
$$+ \left[N - (n+2) \right] \prod_{j=1}^{n-1} \frac{N-(j+2)}{N-j} H(X_{1,2}), \tag{A9}$$

which we prove next by induction.

First, notice that:

$$3M + NR \geq 3H(Z_1) + NH(X_{1,2})$$
$$\geq 3H(Z_1, X_{1,2}) + (N-3)H(X_{1,2})$$
$$\overset{(3)}{=} 3H(Z_1, W_1, X_{1,2}) + (N-3)H(X_{1,2}) \tag{A10}$$
$$\overset{(d)}{\geq} \frac{3(N-3)}{N-1} H(Z_1, W_1) + \frac{3(N+1)}{N-1} + (N-3)H(X_{1,2}),$$

where we wrote (3) to mean by Equation (3), and (d) is by Lemma 1 with $n = 1$. This is precisely the claim when $n = 1$, when we take the convention $\prod_k^n(\cdot) = 1$ when $n < k$ in (A9).

Assume the claim is true for $n = n^*$, and we next prove it is true for $n = n^* + 1$. Notice that the second and third terms in (A9) have a common factor:

$$\frac{N-(n^*+2)}{N-n^*} \prod_{j=1}^{n^*-1} \frac{N-(j+2)}{N-j} = \prod_{j=1}^{n^*} \frac{N-(j+2)}{N-j}, \tag{A11}$$

and using this to normalize the last two terms gives:

$$3H(Z_1, W_{[1:n^*]}) + (N-n^*)H(X_{1,2})$$
$$\overset{(e)}{=} 3[H(Z_1, W_{[1:n^*]}) + H(X_{n^*+1,n^*+2})] + (N-n^*-3)H(X_{1,2})$$
$$\geq 3[H(Z_1, W_{[1:n^*]}, X_{n^*+1,n^*+2})] + (N-n^*-3)H(X_{1,2}) \tag{A12}$$
$$\overset{(3)}{=} 3[H(Z_1, W_{[1:n^*+1]}, X_{n^*+1,n^*+2})] + (N-n^*-3)H(X_{1,2})$$
$$\overset{(f)}{\geq} 3\frac{(N-n^*-3)}{N-n^*-1} H(Z_1, W_{[1:n^*+1]}) + 3\frac{N+n^*+1}{N-n^*-1} + (N-n^*-3)H(X_{1,2}),$$

where (e) is by the file-index-symmetry, and (f) is by Lemma 1. Substituting (A11) and (A12) into (A9) for the case $n = n^*$ gives exactly (A9) for the case $n = n^* + 1$, which completes the proof for (A9).

It remains to show that (A9) implies the bound $3M + NR \geq 2N$. For this purpose, notice that when $n = N - 2$, the last two terms in (A9) reduce to zero, and thus, we only need to show that:

$$Q(N) \triangleq 3 \sum_{j=1}^{N-2} \left[\frac{N+j}{N-j} \prod_{i=1}^{j-1} \frac{N-(i+2)}{N-i} \right] = 2N. \tag{A13}$$

For each summand, we have:

$$\frac{N+j}{N-j} \prod_{i=1}^{j-1} \frac{N-(i+2)}{N-i} = \frac{N+j}{N-j} \left[\frac{N-3}{N-1} \frac{N-4}{N-2} \frac{N-5}{N-3} \cdots \frac{N-j-1}{N-j+1} \right]$$
$$= \frac{(N-j-1)(N+j)}{(N-1)(N-2)}. \tag{A14}$$

Thus, we have:

$$Q(N) = \frac{3}{(N-1)(N-2)} \sum_{j=1}^{N-2} (N-j-1)(N+j) = 2N,$$

where we have used the well-known formula for the sum of integer squares. The proof is thus complete. $\square$

Appendix D. Proof of Proposition 5

We first consider the achievability, for which only the achievability of the following extremal points needs to be shown because of the polytope structure of the region:

$$(M, R) \in \left\{ (0,2), \left(\frac{1}{3}, \frac{4}{3} \right), \left(\frac{4}{3}, \frac{1}{3} \right), (2,0) \right\}. \tag{A15}$$

Achieving the rate pairs $(0,2)$ and $(2,0)$ is trivial. The scheme in [6] can achieve the rate pair $\left(\frac{4}{3}, \frac{1}{3} \right)$. The rate pair $\left(\frac{1}{3}, \frac{4}{3} \right)$ can be achieved by a scheme given in [15], which is a generalization of a special scheme given in [6]. To prove the converse, we note first that the cut-set-based approach can provide all bounds in (29) except

$$3M + 3R \geq 5, \tag{A16}$$

which is a new inequality. As mentioned earlier, this inequality is a special case of Theorem 4 and there is no need to prove it separately.

Appendix E. Proof of Proposition 6

The inequality $14M + 11R \geq 20$ is proven using Tables A5 and A6, and the inequality $9M + 8R \geq 14$ is proven using Tables A7 and A8.

Table A5. Terms needed to prove Proposition 6, inequality $14M + 11R \geq 20$.

T_1	F
T_2	R
T_3	$H(X_{1,1,1,2})$
T_4	$H(X_{1,1,2,2})$
T_5	$H(W_1)$
T_6	$H(W_1, X_{1,1,1,2})$
T_7	$H(W_1, X_{1,1,2,2})$
T_8	$H(W_1, W_2)$
T_9	$H(Z_1)$
T_{10}	$H(Z_1, X_{1,1,1,2})$
T_{11}	$H(Z_1, X_{1,1,2,2})$
T_{12}	$H(Z_4, X_{1,1,1,2})$
T_{13}	$H(Z_1, W_1)$
T_{14}	$H(Z_1, Z_2, X_{1,1,1,2})$
T_{15}	$H(Z_1, Z_2, X_{1,1,2,2})$
T_{16}	$H(Z_1, Z_2, W_1)$
T_{17}	$H(Z_1, Z_2, Z_3, X_{1,1,1,2})$
T_{18}	$H(Z_1, Z_2, Z_3, W_1)$

Table A6. Tabulation proof of Proposition 6 inequality $14M + 11R \geq 20$, with terms defined in Table A5.

T_1	T_2	T_3	T_4	T_5	T_6	T_7	T_8	T_9	T_{10}	T_{11}	T_{12}	T_{13}	T_{14}	T_{15}	T_{16}	T_{17}	T_{18}
																2	-2
	8	-8															
	3		-3														
		6						6	-6								
		4						4			-4						
			4					4		-4							
					-4				8				-4				
									-2				4			-2	
			-1			2	-1										
						-2				4				-2			
				-2	2		-2					2					
							-2				2	-2			2		
							-2							2	-2		2
		-2			2		-2				2						
-2				2													
-18							9										
-20	11						14										

Table A7. Terms needed to prove Proposition 6, inequality $9M + 8R \geq 14$.

T_1	F
T_2	R
T_3	$H(X_{1,1,1,2})$
T_4	$H(X_{1,1,2,2})$
T_5	$H(W_1)$
T_6	$H(W_1, X_{1,1,1,2})$
T_7	$H(W_1, X_{1,1,2,2})$
T_8	$H(W_1, X_{1,2,1,1}, X_{1,1,2,2})$
T_9	$H(W_1, W_2)$
T_{10}	$H(Z_1)$
T_{11}	$H(Z_1, X_{1,1,1,2})$
T_{12}	$H(Z_1, X_{1,1,2,2})$
T_{13}	$H(Z_1, X_{1,2,1,1}, X_{1,1,2,2})$
T_{14}	$H(Z_1, W_1)$
T_{15}	$H(Z_1, Z_2, X_{1,1,1,2})$
T_{16}	$H(Z_1, Z_2, X_{1,1,2,2})$
T_{17}	$H(Z_1, Z_2, W_1)$
T_{18}	$H(Z_1, Z_2, Z_3, X_{1,1,1,2})$
T_{19}	$H(Z_1, Z_2, Z_3, W_1)$

Table A8. Tabulation proof of Proposition 6 inequality $9M + 8R \geq 14$, with terms defined in Table A7.

T_1	T_2	T_3	T_4	T_5	T_6	T_7	T_8	T_9	T_{10}	T_{11}	T_{12}	T_{13}	T_{14}	T_{15}	T_{16}	T_{17}	T_{18}	T_{19}
																	1	-1
		4	-4															
		4		-4														
						-1						1						
			4						4	-4								
			5						5		-5							
					-2					4				-2				
										-1				2			-1	
		-1				2	-1						2			-1		
						-1					2				-1			
			-2	2			-2					2						
						-1				1	1	-1	-1			1		
											1	1	-1	-1				
						-1							1		1	-1		1
				-1	1	-1					1							
-2			2															
-12								6										
-14	8							9										

Appendix F. Proof of Lemma 2

Proof of Lemma 2. We prove this lemma by induction. First, consider the case when $k = K - 1$, for which we write:

$$
\begin{aligned}
2H(&Z_1, W_1, X_{\to[2:K-1]}) \\
&\overset{(a)}{=} H(Z_1, W_1, X_{\to[2:K-1]}) + H(Z_1, W_1, X_{\to[2:K-2]}, X_{\to K}) \\
&= H(X_{\to K-1}Z_1, W_1, X_{\to[2:K-2]}) + H(X_{\to K}Z_1, W_1, X_{\to[2:K-2]}) + 2H(Z_1, W_1, X_{\to[2:K-2]}) \\
&\geq H(Z_1, W_1, X_{\to[2:K]}) + H(Z_1, W_1, X_{\to[2:K-2]}),
\end{aligned}
\tag{A17}
$$

where (a) is by file-index symmetry. The first quantity can be lower bounded as:

$$
H(Z_1, W_1, X_{\to[2:K]}) \geq H(W_1, X_{\to[2:K]}),
\tag{A18}
$$

which leads to a bound on the following sum:

$$
\begin{aligned}
H(&Z_1, W_1, X_{\to[2:K]}) + H(Z_1, W_1, X_{\to[2:K-1]}) \\
&\geq H(W_1, X_{\to[2:K]}) + H(Z_1, W_1, X_{\to[2:K-1]}) \\
&\geq H(X_{\to K}W_1, X_{\to[2:K-1]}) + H(Z_1 W_1, X_{\to[2:K-1]}) + 2H(W_1, X_{\to[2:K-1]}) \\
&\overset{(b)}{=} H(X_{\to K}W_1, X_{\to[2:K-1]}) + H(Z_K W_1, X_{\to[2:K-1]}) + 2H(W_1, X_{\to[2:K-1]}) \\
&\geq H(Z_K, X_{\to K}W_1, X_{\to[2:K-1]}) + 2H(W_1, X_{\to[2:K-1]}) \\
&\overset{(c)}{=} H(Z_K, X_{\to K}, W_2 W_1, X_{\to[2:K-1]}) + 2H(W_1, X_{\to[2:K-1]}) \\
&\overset{(d)}{=} H(W_1, W_2) + H(W_1, X_{\to[2:K-1]}),
\end{aligned}
\tag{A19}
$$

where (b) is by the user index symmetry, (c) is because Z_K and $X_{1,1,\dots,2}$ can be used to produce W_2, and (d) is because all other variables are deterministic functions of (W_1, W_2). Adding $H(Z_1, W_1, X_{\to[2:K-1]})$ on both sides of (A17), and then applying (A19) lead to:

$$
\begin{aligned}
3H(Z_1, W_1, X_{\to[2:K-1]}) &\geq H(Z_1, W_1, X_{\to[2:K-2]}) + H(W_1, X_{\to[2:K-1]}) + H(W_1, W_2) \\
&\overset{(e)}{=} H(Z_{K-1}, W_1, X_{\to[2:K-2]}) + H(W_1, X_{\to[2:K-1]}) + H(W_1, W_2) \\
&\overset{(f)}{\geq} H(Z_{K-1}, W_1, X_{\to[2:K-1]}) + H(W_1, X_{\to[2:K-2]}) + H(W_1, W_2) \\
&= H(Z_{K-1}, W_1, X_{\to[2:K-1]}, W_2) + H(W_1, X_{\to[2:K-2]}) + H(W_1, W_2) \\
&= H(W_1, X_{\to[2:K-2]}) + 2H(W_1, W_2),
\end{aligned}
\tag{A20}
$$

which (e) follows from the user-index symmetry, and (f) by the sub-modularity of the entropy function. This is precisely (33) for $k = K - 1$.

Now, suppose (33) holds for $k = k^* + 1$; we next prove it is true for $k = k^*$ for $K \geq 4$, since when $K = 3$ there is nothing to prove beyond $k = K - 1 = 2$. Using a similar decomposition as in (A17), we can write:

$$
2H(Z_1, W_1, X_{\to[2:k^*]}) \geq H(Z_1, W_1, X_{\to[2:k^*+1]}) + H(Z_1, W_1, X_{\to[2:k^*-1]})
\tag{A21}
$$

Next, we apply the supposition for $k = k^* + 1$ on the first term of the right hand side, which gives:

$$2H(Z_1, W_1, X_{\to[2:k^*]}) \geq \frac{[(K - k^* - 1)(K - k^*) - 2]H(Z_1, W_1, X_{\to[2:k^*]})}{(K - k^*)(K - k^* + 1)} + \frac{2H(W_1, X_{\to[2:k^*]})}{(K - k^*)(K - k^* + 1)} \tag{A22}$$
$$+ \frac{2(K - k^*)H(W_1, W_2)}{(K - k^*)(K - k^* + 1)} + H(Z_1, W_1, X_{\to[2:k^*-1]})$$

Notice that the coefficient in front of $H(W_1, X_{\to[2:k^*]})$ is always less than one for $K \geq 4$ and $k^* \in \{2, 3, \ldots, K - 1\}$, and we can thus bound the following sum:

$$\frac{2H(W_1, X_{\to[2:k^*]})}{(K - k^*)(K - k^* + 1)} + H(Z_1, W_1, X_{\to[2:k^*-1]})$$
$$= \frac{2[H(W_1, X_{\to[2:k^*]}) + H(Z_1, W_1, X_{\to[2:k^*-1]})]}{(K - k^*)(K - k^* + 1)} + \frac{(K - k^*)(K - k^* + 1) - 2}{(K - k^*)(K - k^* + 1)}H(Z_1, W_1, X_{\to[2:k^*-1]}) \tag{A23}$$
$$\overset{(g)}{\geq} \frac{2[H(W_1, W_2) + H(W_1, X_{\to[2:k^*-1]})]}{(K - k^*)(K - k^* + 1)} + \frac{(K - k^*)(K - k^* + 1) - 2}{(K - k^*)(K - k^* + 1)}H(Z_1, W_1, X_{\to[2:k^*-1]}),$$

where (g) follows the same line of argument as in (A19). Substituting (A23) into (A22) and canceling out the common terms of $H(Z_1, W_1, X_{\to[2:k^*]})$ on both sides now give (33) for $k = k^*$. The proof is thus complete. $\square$

Appendix G. Proof for the Converse of Proposition 7

The inequalities $8M + 6R \geq 11$, $3M + 3R \geq 5$ and $5M + 6R \geq 9$ in (51) can be proven using Tables A9–A14, respectively. The inequality $3M + 3R \geq 5$ in (52) is proven using Tables A15 and A16. All other bounds in Proposition 7 follow from the cut-set bound.

Table A9. Terms needed to prove Proposition 7, inequality $8M + 6R \geq 11$.

T_1	F
T_2	R
T_3	$H(X_{1,1,1,2})$
T_4	$H(W_1)$
T_5	$H(W_1, X_{1,1,1,2})$
T_6	$H(W_1, W_2)$
T_7	$H(Z_1)$
T_8	$H(Z_1, X_{1,1,1,2})$
T_9	$H(Z_1, W_1)$
T_{10}	$H(Z_1, Z_2, X_{1,1,1,2})$
T_{11}	$H(Z_1, Z_2, W_1)$
T_{12}	$H(Z_1, Z_2, Z_3, X_{1,1,1,2})$
T_{13}	$H(Z_1, Z_2, Z_3, W_1)$

Table A10. Tabulation proof of Proposition 7 inequality $8M + 6R \geq 11$, with terms defined in Table A9.

T_1	T_2	T_3	T_4	T_5	T_6	T_7	T_8	T_9	T_{10}	T_{11}	T_{12}	T_{13}
											1	−1
	6	−6										
			2			2		−2				
		6				6	−6					
				−3			6		−3			
							−1		2		−1	
			−3	3	−3			3				
					−1		1	−1		1		
					−1					1	−1	1
−1			1									
−10					5							
−11	6					8						

Table A11. Terms needed to prove Proposition 7, inequality $3M + 3R \geq 5$ in (51).

T_1	F
T_2	R
T_3	$H(X_{1,1,1,2})$
T_4	$H(W_1)$
T_5	$H(W_1, X_{1,1,1,2})$
T_6	$H(W_1, X_{1,1,1,2}, X_{1,1,2,1})$
T_7	$H(W_1, W_2)$
T_8	$H(Z_1)$
T_9	$H(Z_1, X_{1,1,1,2})$
T_{10}	$H(Z_1, X_{1,1,1,2}, X_{1,1,2,1})$
T_{11}	$H(Z_1, W_1)$

Table A12. Tabulation proof of Proposition 7 inequality $3M + 3R \geq 5$ in (51), with terms defined in Table A11.

T_1	T_2	T_3	T_4	T_5	T_6	T_7	T_8	T_9	T_{10}	T_{11}
	3	−3								
					−1				1	
		3					3	−3		
								2	−1	−1
			−1	1		−1				1
				−1	1	−1		1		
−1			1							
−4						2				
−5	3					3				

Table A13. Terms needed to prove Proposition 7, inequality $5M + 6R \geq 9$.

T_1	F
T_2	R
T_3	$H(X_{1,1,1,2})$
T_4	$H(W_1)$
T_5	$H(W_1, X_{1,1,1,2})$
T_6	$H(W_1, X_{1,1,1,2}, X_{1,1,2,1})$
T_7	$H(W_1, X_{1,1,1,2}, X_{1,1,2,1}, X_{1,2,1,1})$
T_8	$H(W_1, W_2)$
T_9	$H(Z_1)$
T_{10}	$H(Z_1, X_{1,1,1,2})$
T_{11}	$H(Z_1, X_{1,1,1,2}, X_{1,1,2,1})$
T_{12}	$H(Z_1, X_{1,1,1,2}, X_{1,1,2,1}, X_{1,2,1,1})$
T_{13}	$H(Z_1, W_1)$

Table A14. Tabulation proof of Proposition 7 inequality $5M + 6R \geq 9$, with terms defined in Table A13.

T_1	T_2	T_3	T_4	T_5	T_6	T_7	T_8	T_9	T_{10}	T_{11}	T_{12}	T_{13}
	6	−6										
						−1					1	
							−1	−1				2
		6						6	−6			
									6	−3		−3
									−1	2	−1	
			−1	1			−1					1
				−1	1		−1		1			
					−1	1	−1			1		
−1			1									
−8							4					
−9	6						5					

Entropy **2018**, _20_, 603

Table A15. Terms needed to prove Proposition 7, inequality $3M + 3R \geq 5$ in (52).

T_1	F
T_2	R
T_3	$H(X_{1,1,2,2})$
T_4	$H(W_1)$
T_5	$H(W_1, X_{1,1,2,2})$
T_6	$H(W_1, X_{1,1,2,2}, X_{1,2,1,2})$
T_7	$H(W_1, W_2)$
T_8	$H(Z_1)$
T_9	$H(Z_1, X_{1,1,2,2})$
T_{10}	$H(Z_1, X_{1,1,2,2}, X_{1,2,1,2})$
T_{11}	$H(Z_1, W_1)$

Table A16. Tabulation proof of Proposition 7 inequality $3M + 3R \geq 5$ in (52), with terms defined in Table A15.

T_1	T_2	T_3	T_4	T_5	T_6	T_7	T_8	T_9	T_{10}	T_{11}
	3	−3								
					−1				1	
		3					3	−3		
								2	−1	−1
			−1	1		−1				1
				−1	1	−1		1		
−1			1							
−4						2				
−5	3						3			

Appendix H. Proof for the Forward of Proposition 7

Note that the optimal tradeoff for the single demand type $(3,1)$ system has the following corner points:

$$(M, R) = (0, 2), \left(\frac{1}{4}, \frac{3}{2}\right), \left(\frac{1}{2}, \frac{7}{6}\right), \left(1, \frac{2}{3}\right), \left(\frac{3}{2}, \frac{1}{4}\right), (2, 0).$$

The corner points $\left(1, \frac{2}{3}\right)$ and $\left(\frac{3}{2}, \frac{1}{4}\right)$ are achievable using the Maddah-Ali–Niesen scheme [6]. The point $\left(\frac{1}{4}, \frac{3}{2}\right)$ is achievable by the code given in [15] or [20]. The only remaining corner point of interest is thus $\left(\frac{1}{2}, \frac{7}{6}\right)$, in the binary field. This can be achieved by the following strategy in Table A17, where the first file has six symbols $(A_1, A_2, \ldots, A_6)$ and the second file $(B_1, B_2, \ldots, B_6)$.

Table A17. Code for the tradeoff point $\left(\frac{1}{2}, \frac{7}{6}\right)$ for demand type $(3,1)$ when $(N, K) = (2, 4)$.

User 1	$A_1 + B_1$	$A_2 + B_2$	$A_3 + B_3$
User 2	$A_1 + B_1$	$A_4 + B_4$	$A_5 + B_5$
User 3	$A_2 + B_2$	$A_4 + B_4$	$A_6 + B_6$
User 4	$A_3 + B_3$	$A_5 + B_5$	$A_6 + B_6$

By the symmetry, we only need to consider the demand when the first three users request A and the last user request B. The server can send the following symbols in this case:

$$A_3, A_5, A_6, B_1, B_2, B_4, A_1 + A_2 + A_4.$$

Let us consider now the single-demand-type $(2,2)$ system, for which the corner points on the optimal tradeoff are:

$$(M,R) = (0,2), \left(\frac{1}{3},\frac{4}{3}\right), \left(\frac{4}{3},\frac{1}{3}\right), (2,0).$$

Let us denote the first file as (A_1, A_2, A_3), and the second file as (B_1, B_2, B_3), which are in the binary field. To achieve the corner point $\left(\frac{1}{3},\frac{4}{3}\right)$, we use the caching code in Table A18.

Table A18. Code for the tradeoff point $\left(\frac{1}{3},\frac{4}{3}\right)$ for demand-type $(2,2)$ when $(N,K) = (2,4)$.

User 1	$A_1 + B_1$
User 2	$A_2 + B_2$
User 3	$A_3 + B_3$
User 4	$A_1 + A_2 + A_3 + B_1 + B_2 + B_3$

Again due to the symmetry, we only need to consider the case when the first two users request A, and the other two request B. For this case, the server can send:

$$B_1, B_2, A_3, A_1 + A_2 + A_3.$$

For the other corner point $\left(\frac{4}{3},\frac{1}{3}\right)$ the following placement in Table A19 can be used. Again for the case when the first two users request A, and the other two request B, the server can send:

$$A_1 - A_3 + B_2.$$

Table A19. Code for the tradeoff point $\left(\frac{4}{3},\frac{1}{3}\right)$ for demand-type $(2,2)$ when $(N,K) = (2,4)$.

User 1	A_1	A_2	B_1	B_2
User 2	A_2	A_3	B_2	B_3
User 3	A_1	A_3	B_1	B_3
User 4	$A_1 + A_2$	$A_2 + A_3$	$B_1 + B_2$	$B_2 + B_3$

Appendix I. Proof of Proposition 8

The inequalities $M + R \geq 2$ and $2M + 3R \geq 5$ in (54) are proven in Tables A20–A23, respectively. The inequalities $6M + 3R \geq 8$, $M + R \geq 2$, $12M + 18R \geq 29$ and $3M + 6R \geq 8$ in (55) are proven in Tables A24–A31, respectively. All other bounds in Proposition 8 can be deduced from the cut-set bound, and thus do not need a proof.

Table A20. Terms needed to prove Proposition 8, inequality $M + R \geq 2$ in (54).

T_1	F
T_2	R
T_3	$H(X_{1,1,2})$
T_4	$H(W_1)$
T_5	$H(W_2, X_{1,1,2})$
T_6	$H(W_1, W_2, W_3)$
T_7	$H(Z_1)$
T_8	$H(Z_3, X_{1,1,2})$
T_9	$H(Z_1, W_1)$
T_{10}	$H(Z_1, W_2, X_{1,1,2})$
T_{11}	$H(Z_1, Z_3, X_{1,1,2})$
T_{12}	$H(Z_1, Z_2, W_1)$

Table A21. Tabulation proof of Proposition 8 inequality $M + R \geq 2$ in (54), with terms defined in Table A20.

T_1	T_2	T_3	T_4	T_5	T_6	T_7	T_8	T_9	T_{10}	T_{11}	T_{12}
	2	-2									
		2				2	-2				
					-1					2	-1
			-1	1				1	-1		
							1	-1		-1	1
				-1			1		1	-1	
-1			1								
-3					1						
-4	2					2					

Table A22. Terms needed to prove Proposition 8, inequality $2M + 3R \geq 5$ in (54).

T_1	F
T_2	R
T_3	$H(X_{1,2,2})$
T_4	$H(X_{2,3,3}, X_{2,1,2})$
T_5	$H(W_1)$
T_6	$H(W_1, X_{1,2,2})$
T_7	$H(W_2, X_{1,2,2})$
T_8	$H(W_3, X_{1,3,3}, X_{2,3,3})$
T_9	$H(W_1, X_{1,3,3}, X_{2,1,1})$
T_{10}	$H(W_1, W_2)$
T_{11}	$H(W_2, W_3, X_{1,2,2})$
T_{12}	$H(W_1, W_2, W_3)$
T_{13}	$H(Z_1)$
T_{14}	$H(Z_2, X_{1,2,2})$
T_{15}	$H(Z_1, X_{1,2,2})$
T_{16}	$H(Z_2, X_{1,3,3}, X_{2,3,3})$
T_{17}	$H(Z_1, X_{1,3,3}, X_{2,1,1})$
T_{18}	$H(Z_2, X_{2,3,3}, X_{2,1,2})$
T_{19}	$H(Z_1, X_{2,3,3}, X_{2,1,2})$
T_{20}	$H(Z_1, X_{2,3,3}, X_{3,1,1}, X_{2,1,2})$
T_{21}	$H(Z_1, W_1)$
T_{22}	$H(Z_1, W_2, X_{1,2,2})$
T_{23}	$H(Z_1, W_3, X_{1,2,2})$
T_{24}	$H(Z_2, W_1, X_{1,2,2})$
T_{25}	$H(Z_2, W_3, X_{1,2,2})$
T_{26}	$H(Z_1, W_3, X_{2,3,3}, X_{2,1,2})$
T_{27}	$H(Z_1, W_1, X_{2,3,3}, X_{2,1,2})$
T_{28}	$H(Z_1, W_1, W_2)$
T_{29}	$H(Z_2, Z_3, X_{1,2,2})$
T_{30}	$H(Z_1, Z_2, X_{1,2,2})$
T_{31}	$H(Z_2, Z_3, X_{1,3,3}, X_{2,3,3})$
T_{32}	$H(Z_1, Z_3, X_{2,3,3}, X_{2,1,2})$
T_{33}	$H(Z_1, Z_2, W_1)$
T_{34}	$H(Z_2, Z_3, W_3, X_{1,2,2})$

Table A23. Tabulation proof of Proposition 8 inequality $2M + 3R \geq 5$ in (54), with terms defined in Table A22.

T_1	T_2	T_3	T_4	T_5	T_6	T_7	T_8	T_9	T_{10}	T_{11}	T_{12}	T_{13}	T_{14}	T_{15}	T_{16}	T_{17}	T_{18}	T_{19}	T_{20}	T_{21}	T_{22}	T_{23}	T_{24}	T_{25}	T_{26}	T_{27}	T_{28}	T_{29}	T_{30}	T_{31}	T_{32}	T_{33}	T_{34}
																												2				−2	
	27	−27																															
		8	−4																														
																														−3	3		
												−2							4										−2				
		8										8		−8																			
		15										15	−15																				
												12			−6				−6														
										−3																							−3
																									6					−3			
							−3																										
															6																		
				−2		2														2	−2												
				−1	1	1			−1																								
										−3	3	−3																					
												−3	3							3				−3									
													2							−2									−2		2		
												1	1					−1		−1													
																						1	1			−1	−1						
		−4	4																														
					−1			1					4			−4																	
						−3	3						1		−1						3												
												−3									−1				1				1				
																1	1		−1			−1											
												−1					1											1			−1		
												−1											−1			1			1				
												−3				3								−3									3
												−1						1							−1						1		
−3				3																													
−6									3																								
−36												12																					
−45	27											18																					

Table A24. Terms needed to prove Proposition 8, inequality $6M + 3R \geq 8$ in (55).

T_1	F
T_2	R
T_3	$H(X_{1,2,3,})$
T_4	$H(W_1, W_2)$
T_5	$H(W_1, W_2, X_{1,2,3})$
T_6	$H(W_1, W_2, W_3)$
T_7	$H(Z_1)$
T_8	$H(Z_1, X_{1,2,3,})$
T_9	$H(Z_1, W_2, X_{1,2,3})$
T_{10}	$H(Z_1, W_1, W_2)$
T_{11}	$H(Z_1, Z_2, X_{1,2,3})$
T_{12}	$H(Z_1, Z_2, W_1, W_2)$

Table A25. Tabulation proof of Proposition 8 inequality $6M + 3R \geq 8$ in (55), with terms defined in Table A24.

T_1	T_2	T_3	T_4	T_5	T_6	T_7	T_8	T_9	T_{10}	T_{11}	T_{12}
										1	−1
	3	−3									
				−1				1			
								−2		2	
		6				6	−6				
		−3					6			−3	
			−1	1	−1				1		
					−1			1	−1		1
−2			1								
−6					2						
−8	3					6					

Table A26. Terms needed to prove Proposition 8, inequality $M + R \geq 2$ in (55).

T_1	F
T_2	R
T_3	$H(X_{1,2,3})$
T_4	$H(W_1)$
T_5	$H(W_1, X_{1,2,3})$
T_6	$H(W_1, X_{1,2,3}, X_{1,3,2})$
T_7	$H(W_1, W_2, W_3)$
T_8	$H(Z_1)$
T_9	$H(Z_1, X_{1,2,3})$
T_{10}	$H(Z_1, X_{1,2,3}, X_{1,3,2})$
T_{11}	$H(Z_1, W_1)$
T_{12}	$H(Z_1, W_2, X_{1,2,3})$

Table A27. Tabulation proof of Proposition 8 inequality $M + R \geq 2$ in (55), with terms defined in Table A26.

T_1	T_2	T_3	T_4	T_5	T_6	T_7	T_8	T_9	T_{10}	T_{11}	T_{12}
	2	−2									
					−1				1		
		2					2	−2			
								2	−1	−1	
			−1	1						1	−1
				−1	1	−1					1
−1			1								
−3						1					
−8	3					6					

Table A28. Terms needed to prove Proposition 8, inequality $12M + 18R \geq 29$ in (55).

T_1	F
T_2	R
T_3	$H(X_{1,2,3})$
T_4	$H(X_{1,2,3}, X_{1,3,2})$
T_5	$H(X_{1,2,3}, X_{1,3,2}, X_{2,1,3})$
T_6	$H(W_1)$
T_7	$H(W_1, X_{1,2,3})$
T_8	$H(W_2, X_{1,2,3}, X_{1,3,2})$
T_9	$H(W_1, W_2)$
T_{10}	$H(W_1, W_2, X_{1,2,3})$
T_{11}	$H(W_1, W_2, X_{1,3,2}, X_{2,1,3})$
T_{12}	$H(W_2, W_3, X_{1,2,3}, X_{1,3,2})$
T_{13}	$H(W_1, W_2, X_{1,2,3}, X_{1,3,2}, X_{2,1,3})$
T_{14}	$H(W_1, W_2, W_3)$
T_{15}	$H(Z_1)$
T_{16}	$H(Z_1, X_{1,2,3})$
T_{17}	$H(Z_1, X_{1,3,2}, X_{2,1,3})$
T_{18}	$H(Z_2, X_{1,2,3}, X_{1,3,2})$
T_{19}	$H(Z_1, X_{1,2,3}, X_{1,3,2})$
T_{20}	$H(Z_1, X_{1,2,3}, X_{1,3,2}, X_{2,1,3})$
T_{21}	$H(Z_1, W_1)$
T_{22}	$H(Z_1, W_2, X_{1,2,3})$
T_{23}	$H(Z_1, W_2, X_{1,2,3}, X_{1,3,2})$
T_{24}	$H(Z_1, W_1, W_2)$

Table A29. Tabulation proof of Proposition 8 inequality $12M + 18R \geq 29$ in (55), with terms defined in Table A28.

T_1	T_2	T_3	T_4	T_5	T_6	T_7	T_8	T_9	T_{10}	T_{11}	T_{12}	T_{13}	T_{14}	T_{15}	T_{16}	T_{17}	T_{18}	T_{19}	T_{20}	T_{21}	T_{22}	T_{23}	T_{24}
		2	−1																				
	18	−18																					
										−1						1							
												−3							3				
		17												17	−17								
														−2	4		−2						
														−3	6	−3							
															2			−1		−1			
																					6	−3	−3
		−1	2	−1																			
						−3	3													3	−3		
								−1	1				−1										1
															2					−2	−2		2
						−3	3								3		−3						
								−1	1				−1								1		
																2	2		−2		−2		
				−1	1													1	−1				
							−3				3	−3										3	
											−3	3	−3			3							
−3				3																			
−2								1															
−24												8											
−29	**18**											**12**											

Entropy **2018**, _20_, 603

Table A30. Terms needed to prove Proposition 8, inequality $3M + 6R \geq 8$ in (55).

T_1	F
T_2	R
T_3	$H(X_{1,2,3})$
T_4	$H(W_1, W_2)$
T_5	$H(W_1, W_2, X_{1,2,3})$
T_6	$H(W_2, W_3, X_{1,2,3}, X_{1,3,2,})$
T_7	$H(W_1, W_2, X_{1,2,3}, X_{1,3,2}, X_{2,1,3})$
T_8	$H(W_1, W_2, W_3)$
T_9	$H(Z_1)$
T_{10}	$H(Z_1, X_{1,2,3})$
T_{11}	$H(Z_2, X_{1,2,3}, X_{1,3,2})$
T_{12}	$H(Z_1, X_{1,3,2}, X_{2,1,3})$
T_{13}	$H(Z_1, X_{1,2,3}, X_{1,3,2}, X_{2,1,3})$
T_{14}	$H(Z_1, W_2, X_{1,2,3})$
T_{15}	$H(Z_1, W_1, W_2)$

Table A31. Tabulation proof of Proposition 8 inequality $3M + 6R \geq 8$ in (55), with terms defined in Table A30.

T_1	T_2	T_3	T_4	T_5	T_6	T_7	T_8	T_9	T_{10}	T_{11}	T_{12}	T_{13}	T_{14}	T_{15}
												1	−1	
	6	−6												
				−1	1									
						−1						1		
		6						6	−6					
								−2	4	−2				
								−1	2		−1			
			−1	1			−1							1
										1	1	−1	−1	
					−1	1	−1			1				
−2			1											
−6							2							
−8	6						3							

References

1. Yeung, R.W. A framework for linear information inequalities. _IEEE Trans. Inf. Theory_ **1997**, _43_, 1924–1934. [CrossRef]
2. Tian, C. Characterizing the rate region of the (4, 3, 3) exact-repair regenerating codes. _IEEE J. Sel. Areas Commun._ **2014**, _32_, 967–975. [CrossRef]
3. Tian, C.; Liu, T. Multilevel diversity coding with regeneration. _IEEE Trans. Inf. Theory_ **2016**, _62_, 4833–4847. [CrossRef]
4. Tian, C. A note on the rate region of exact-repair regenerating codes. _arXiv_ **2015**, arXiv:1503.00011.
5. Li, C.; Weber, S.; Walsh, J.M. Multilevel diversity coding systems: Rate regions, codes, computation, & forbidden minors. _IEEE Trans. Inf. Theory_ **2017**, _63_, 230–251.
6. Maddah-Ali, M.A.; Niesen, U. Fundamental limits of caching. _IEEE Trans. Inf. Theory_ **2014**, _60_, 2856–2867. [CrossRef]
7. Maddah-Ali, M.A.; Niesen, U. Decentralized coded caching attains order-optimal memory-rate tradeoff. _IEEE/ACM Trans. Netw._ **2015**, _23_, 1029–1040. [CrossRef]
8. Niesen, U.; Maddah-Ali, M.A. Coded caching with nonuniform demands. _IEEE Trans. Inf. Theory_ **2017**, _63_, 1146–1158. [CrossRef]

9. Pedarsani, R.; Maddah-Ali, M.A.; Niesen, U. Online coded caching. *IEEE/ACM Trans. Netw.* **2016**, *24*, 836–845. [CrossRef]

10. Karamchandani, N.; Niesen, U.; Maddah-Ali, M.A.; Diggavi, S.N. Hierarchical coded caching. *IEEE Trans. Inf. Theory* **2016**, *62*, 3212–3229.

11. Ji, M.; Tulino, A.M.; Llorca, J.; Caire, G. Order-optimal rate of caching and coded multicasting with random demands. *IEEE Trans. Inf. Theory* **2017**, *63*, 3923–3949.

12. Ghasemi, H.; Ramamoorthy, A. Improved lower bounds for coded caching. *IEEE Trans. Inf. Theory* **2017**, *63*, 4388–4413. [CrossRef]

13. Sengupta, A.; Tandon, R. Improved approximation of storage-rate tradeoff for caching with multiple demands. *IEEE Trans. Commun.* **2017**, *65*, 1940–1955. [CrossRef]

14. Ajaykrishnan, N.; Prem, N.S.; Prabhakaran, V.M.; Vaze, R. Critical database size for effective caching. In Proceedings of the 2015 Twenty First National Conference on Communications (NCC), Mumbai, India, 27 February–1 March 2015; pp. 1–6.

15. Chen, Z.; Fan, P.; Letaief, K.B. Fundamental limits of caching: improved bounds for users with small buffers. *IET Commun.* **2016**, *10*, 2315–2318. [CrossRef]

16. Sahraei, S.; Gastpar, M. *K* users caching two files: An improved achievable rate. *arXiv* **2015**, arXiv:1512.06682.

17. Amiri, M.M.; Gunduz, D. Fundamental limits of caching: Improved delivery rate-cache capacity trade-off. *IEEE Trans. Commun.* **2017**, *65*, 806–815. [CrossRef]

18. Wan, K.; Tuninetti, D.; Piantanida, P. On caching with more users than files. *arXiv* **2016**, arXiv:1601.06383.

19. Yu, Q.; Maddah-Ali, M.A.; Avestimehr, A.S. The exact rate-memory tradeoff for caching with uncoded prefetching. *IEEE Trans. Inf. Theory* **2018**, *64*, 1281–1296. [CrossRef]

20. Tian, C.; Chen, J. Caching and delivery via interference elimination. *IEEE Trans. Inf. Theory* **2018**, *64*, 1548–1560. [CrossRef]

21. Gómez-Vilardebó, J. Fundamental limits of caching: Improved rate-memory trade-off with coded prefetching. *IEEE Trans. Commun.* **2018**. [CrossRef]

22. Cover, T.M.; Thomas, J.A. *Elements of Information Theory*, 1st ed.; Wiley: New York, NY, USA, 1991.

23. Tian, C. Symmetry, demand types and outer bounds in caching systems. In Proceedings of the 2016 IEEE International Symposium on Information Theory (ISIT), Barcelona, Spain, 10–15 July 2016; pp. 825–829.

24. Yu, Q.; Maddah-Ali, M.A.; Avestimehr, A.S. Characterizing the rate-memory tradeoff in cache networks within a factor of 2. *arXiv* **2017**, arXiv:1702.04563.

25. Wang, C.Y.; Bidokhti, S.S.; Wigger, M. Improved converses and gap-results for coded caching. *arXiv* **2017**, arXiv:1702.04834.

26. Yeung, R. *A First Course in Information Theory*; Kluwer Academic Publishers: New York, NY, USA, 2002.

27. Lassez, C.; Lassez, J.L. Quantifier Elimination for conjunctions of linear constraints via a convex hull algorithm. In *Symbolic and Numerical Computation for Artificial Intelligence*; Donald, B.R., Kapur, D., Mundy, J.L., Eds.; Academic Press: San Diego, CA, USA, 1992; Chapter 4, pp. 103–1199.

28. Apte, J.; Walsh, J.M. Exploiting symmetry in computing polyhedral bounds on network coding rate regions. In Proceedings of the International Symposium on Network Coding (NetCod), Sydney, Australia, 22–24 June 2015; pp. 76–80.

29. Ho, S.W.; Tan, C.W.; Yeung, R.W. Proving and disproving information inequalities. In Proceedings of the 2014 IEEE International Symposium on Information Theory (ISIT), Honolulu, HI, USA, 29 June–4 July 2014.

30. Zhang, K.; Tian, C. On the symmetry reduction of information inequalities. *IEEE Trans. Commun.* **2018**, *66*, 2396–2408. [CrossRef]

31. Harary, F.; Norman, R.Z. *Graph Theory as a Mathematical Model in Social Science*; University of Michigan Press: Ann Arbor, MI, USA, 1953.

32. Andrews, G.E. *The Theory of Partitions*; Cambridge University Press: Cambridge, UK, 1976.

33. Dougherty, R.; Freiling, C.; Zeger, K. Insufficiency of linear coding in network information flow. *IEEE Trans. Inf. Theory* **2005**, *51*, 2745–2759. [CrossRef]

34. Zhang, Z.; Yeung, R.W. A non-Shannon-type conditional inequality of information quantities. *IEEE Trans. Inf. Theory* **1997**, *43*, 1982–1986. [CrossRef]

35. Zhang, Z.; Yeung, R.W. On characterization of entropy function via information inequalities. *IEEE Trans. Inf. Theory* **1998**, *44*, 1440–1452. [CrossRef]

36. Yeung, R.W.; Zhang, Z. On symmetrical multilevel diversity coding. *IEEE Trans. Inf. Theory* **1999**, *45*, 609–621. [CrossRef]

37. Tian, C. Latent capacity region: A case study on symmetric broadcast with common messages. *IEEE Trans. Inf. Theory* **2011**, *57*, 3273–3285. [CrossRef]

MDPI
St. Alban-Anlage 66
4052 Basel
Switzerland
Tel. +41 61 683 77 34
Fax +41 61 302 89 18
www.mdpi.com

Entropy Editorial Office
E-mail: entropy@mdpi.com
www.mdpi.com/journal/entropy